T0192678

Diversity Resistance in Organizations

This new volume revisits diversity resistance 10 years later, examining the fluidity of diversity resistance in workplaces. Top-notch contributors provide insight about the motivations to resist diversity and inclusion as well as offer strategies for preventing and derailing diversity resistance and enhancing inclusion in organizations.

The current edition broadens the conversation about diversity resistance by demonstrating methods of counter-resistance and how diversity resistance manifests in everyday lives, as well as how it presents itself and limits the careers and lives of various stigmatized groups. Chapters also consider why, despite the often expressed value for diversity and inclusion, diversity resistance continues to persist. Contributors demonstrate the persistence of diversity resistance across time, context and for a variety of targets. For example, this volume addresses topics as well as marginalized groups not previously discussed in the first edition such as intersectionality, workers living with mental illness, gender identity, trans workers and the systemic resistance experienced by gay couples.

This volume will be of interest to scholars and practitioners as well as minoritized workers. It will function as a framework for understanding the continuum of exclusion, harassment and discrimination that occurs within organizational settings and the impact upon individual and organizational performance. Practitioners will find examples and cases for how diversity resistance manifests, but more importantly strategies and recommendations for derailing diversity resistance and enhancing inclusion.

Kecia M. Thomas is Professor of Industrial-Organizational Psychology at the University of Georgia U.S.A. She holds a joint appointment with the Institute of African-American Studies and is an affiliate of the Institute for Women's Studies. She is the Senior Associate Dean of the Franklin College of Arts and Sciences and is an elected Fellow of both the Society for Industrial and Organizational Psychology and the American Psychological Association.

Series in Applied Psychology

Bridging both academic and applied interests, the Applied Psychology Series offers publications that emphasize state-of-the-art research and its application to important issues of human behavior in a variety of societal settings. To date, more than 50 books in various fields of applied psychology have been published in this series.

Jeanette N. Cleveland, Colorado State University
Donald Truxillo, Portland State University
Edwin A. Fleishman, Founding Series Editor (1987-2010)
Kevin R. Murphy, Emeritus Series Editor (2010-2018)

Patterns of Life History
The Ecology of Human Individuality
Michael D. Mumford, Garnett Stokes and William A. Owens

Work Motivation
Uwe E. Kleinbeck, Hans-Henning Quast, Henk Thierry and Hartmut Häcker

Teamwork and the Bottom Line
Groups Make a Difference
Ned Rosen

Aging and Work in the 21st Century
Edited by Kenneth S. Schultz and Gary A. Adams

Employee Retention and Turnover
Peter W. Hom, David G. Allen and Rodger W. Griffeth

Diversity Resistance in Organizations, Second Edition
Edited by Kecia M. Thomas

Diversity Resistance in Organizations

Second Edition

Edited by Kecia M. Thomas

NEW YORK AND LONDON

Second edition published 2020
by Routledge
52 Vanderbilt Avenue, New York, NY 10017

and by Routledge
2 Park Square, Milton Park, Abingdon, Oxon, OX14 4RN

Routledge is an imprint of the Taylor & Francis Group, an informa business

© 2020 Taylor & Francis

The right of Kecia M. Thomas to be identified as the author of the editorial material, and of the authors for their individual chapters, has been asserted in accordance with sections 77 and 78 of the Copyright, Designs and Patents Act 1988.

Trademark notice: Product or corporate names may be trademarks or registered trademarks, and are used only for identification and explanation without intent to infringe.

First edition published by Routledge 2007

Library of Congress Cataloging-in-Publication Data
A catalog record for this title has been requested

ISBN: 978-0-367-34562-4 (hbk)
ISBN: 978-0-367-34560-0 (pbk)
ISBN: 978-1-003-02690-7 (ebk)

Typeset in Times New Roman
by Swales & Willis, Exeter, Devon, UK

Contents

Contributors

Danielle D. Dickens, Ph.D., is an Assistant Professor in the Department of Psychology at Spelman College. She earned her B.A. in psychology from Spelman College and her M.S. and Ph.D. from Colorado State University in Applied Social and Health Psychology. Her program of research focuses on Black womanhood and their implications for health behaviors, academic performance, experiences of discrimination, coping strategies and psychological well-being. Due to her productivity and innovations in teaching, research and service, she often serves as an expert scholar and invited speaker on issues affecting underserved and underrepresented individuals in education and in the workplace.

Bernardo M. Ferdman of Ferdman Consulting is passionate about helping to create an inclusive world where more of us can be fully ourselves and accomplish our goals effectively, productively and authentically. He is an accomplished leadership and organization development consultant and coach with over 35 years of experience working with diverse groups and organizations to increase individual and collective effectiveness and inclusion. Bernardo works with organizations to improve performance and leadership and to increase inclusion, focusing on assessment, coaching, dialogue, facilitation and training in the context of diversity.

Kayla Follmer, Ph.D., is an Assistant Professor of Management at the John Chambers College of Business and Economics at West Virginia University. Her research is focused on the work experiences of employees with concealable identities, with specific emphasis on employees with mental illness. She is particularly interested in the extent to which the work environment impacts individuals' strategies for managing their mental illness as well as their health-related outcomes including suicidal ideation and behaviors.

Plácida V. Gallegos is the founder of Solfire Consulting Group and has worked extensively with a wide range of corporate and non-profit clients across the country to build robust cultures that maximize the benefits of their differences. She provides inclusive leadership development, coaching and organizational consulting to companies in the areas of strategic culture change, supervisory and management skills, leadership styles, career development, conflict management and team building. Dr. Gallegos is also on the faculty at Fielding Graduate University in the Human and Organizational Development Program.

Janice Z. Gassam, Ph.D. is an Assistant Professor of Management in the Jack Welch College of Business and Technology at Sacred Heart University. Janice received a Ph.D. in Applied Organizational Psychology from Hofstra University in 2017. Her research areas of interest include diversity and inclusion, technology, social media, leadership and training and development. She is also the founder of BWG Business Solutions, a company created to help organizations and institutions design strategies to foster more diversity and inclusion. She is a Senior Contributor for Forbes, where her writing focuses on diversity, equity and inclusion in the workplace. Janice is also a TEDx speaker, having delivered a talk on Emotional Intelligence.

Emily Goldstein is a research analyst in the Culture, Diversity and Intergroup Relations Lab at the University of California, Berkeley. Previously, she was Research Coordinator & Data Analyst for the NIH-funded SF BUILD Grant at UCSF and SFSU. In this role, she managed research focused on psychosocial outcomes of teaching students about stereotype threat. She completed her B.A. in Industrial Design and M.A. in Linguistics, with an emphasis in Cognitive Science, at San Francisco State University. Her interests revolve around the application of statistical and machine-learning techniques to social science research, especially related to diversity and inclusion.

Brea M. Heidelberg is an arts management educator, consultant and researcher. She is an Assistant Professor & Program Director of the Entertainment & Arts Management program at Drexel University. Dr. Heidelberg earned her Ph.D. in Arts Administration, Education and Policy from The Ohio State University and a M.S. in Human Resource Development from Villanova University. She is the Founder of ISO Arts Consulting – a firm offering equity trainings, evaluation services and workforce solutions. Her research interests include professional development for arts administrators, human resource development in arts and cultural institutions and cultural equity.

Oscar Holmes IV is an Associate Professor of Management and Director of Access and Outreach in Business Education at Rutgers School of Business-Camden where he teaches executive education, graduate and undergraduate management courses. His research examines how leaders can maximize productivity and well-being through fostering more inclusive environments and has been published in several top-tier management journals and books. He earned a Ph.D. and M.A. in Management at The University of Alabama, a M.L.A. from the University of Richmond and a B.S. with honors from Virginia Commonwealth University.

Kyneshawau Hurd is a J.D./Ph.D. student in the Jurisprudence and Social Policy program and member of the Culture, Diversity and Intergroup Relations Lab at the University of California, Berkeley. Her research broadly examines the interactional role of social hierarchy and ideology in maintaining inequality and exclusion against marginalized groups in educational and organizational contexts (with an attention to intersectionality). Kyneshawau received her B.A. in Psychology from Baylor University and was a lab manager for the Social Interaction and Social Stigma Lab at the University of California, Los Angeles.

Zoe E. Johnston is a fourth year student at the University of Georgia majoring in psychology, she is interested in pursuing a career in I/O psychology with an emphasis on workplace diversity.

Kisha Jones, Ph.D., is an Assistant Professor of Industrial/Organizational (I/O) Psychology at Penn State University. She obtained her Ph.D. in I/O Psychology from the University of Illinois at Urbana-Champaign. Her research explores how race, gender, social class and mental illness impact worker experiences separately as well as how they intersect. In particular, she focuses on how these identities intersect to impact career interests, recruitment, employment discrimination and identity management. She teaches undergraduate and graduate courses on workplace diversity and personnel selection.

Justin A. Lavner, Ph.D., is an Associate Professor in the Department of Psychology at the University of Georgia. His research focuses on understanding factors that influence individual and family functioning among underserved, marginalized populations and developing interventions to promote well-being among these groups. He has been recognized as a "Rising Star" by the Association for Psychological Science and his work has been funded by the National Science Foundation and the National Institutes of Health. He received his B.A. from Williams College and his Ph.D. from the University of California, Los Angeles.

Christopher K. Marshburn is an Assistant Professor in the Department of Psychology at the University of Kentucky. He received his B.A. in Psychology from Randolph-Macon College (Ashland, VA) and his Ph.D. in Social Psychology from the University of California, Irvine. His research focuses on how Black Americans cope with racism. Specifically, he examines whether talking to same- and cross-race friends about racism has positive benefits for Black Americans. Additionally, he explores how White Americans think about, react to and perpetuate racism.

Candace P. Parrish, Ph.D., is the Director and Assistant Professor of the Strategic Communication and Public Relations Online Master's Program at Sacred Heart University. Candace received an interdisciplinary Ph.D. related to the areas of Public Relations, Health Communication and Visual Communication from Virginia Commonwealth University in 2016. She has been recognized nationally and internationally for her research centered around infographics used in public relations and health communication – although her interests span over a few areas (including sexual assault and diversity.) Additionally, Candace has published and presented research in top journals and at conferences and is a freelance writer who has contributed to PR Daily.

Victoria C. Plaut is the Claire Sanders Clements Professor of Law and Social Science at the University of California, Berkeley, School of Law, where she also serves as Director of the Culture, Diversity and Intergroup Relations Lab and is Affiliate Faculty in Psychology. A social psychologist, Dr. Plaut has conducted extensive empirical research on psychological processes relevant to diversity and inclusion in legal, educational and workplace contexts. Her research has been supported by the National Science Foundation, the National Institutes of Health, the Spencer Foundation and private organizations.

Celina A. Romano is a Ph.D. student and member of the Culture, Diversity and Intergroup Relations Lab at the University of California, Berkeley. Her research focuses on understanding the causes and consequences of discrimination and bias toward marginalized group members. Celina is devoted to researching ways to mitigate social inequality and cultivate more inclusive institutions and organizations. Celina earned her B. A. with honors from the University of Michigan, where she double majored in Psychology and Women's Studies and minored in Law, Justice and Social Change.

Enrica N. Ruggs is an Assistant Professor Management in the Fogelman College of Business and Economics at the University of Memphis. She received her Ph.D. from Rice University. Her research examines workplace diversity as well as the manifestation of workplace discrimination, its effects on employees with stigmatized identities and strategies to reduce discrimination.

Katina Sawyer, Ph.D., is an Assistant Professor of Management at The George Washington University in the School of Business. Her areas of expertise include diversity, work-life balance, leadership and negative workplace behaviors. Over the years, Dr. Sawyer has published numerous peer-reviewed articles and book chapters about diversity, leadership and work-family conflict. She received a dual-M.S. and dual-Ph.D. in Industrial/Organizational Psychology and Women's Studies from Penn State University.

Cambrilyn Scofield is a student at the University of Georgia. She is currently studying psychology and planning on obtaining a master's degree in consumer analytics. Cambrilyn is interested in pursuing a career in data science and consumer research. She aspires to make diversity an important platform in her career to ensure that marginalized voices are heard and respected.

Karoline M. Summerville is a doctoral student in the Organizational Science program at the University of North Carolina at Charlotte. She earned her B.A. in Communication from Queens University of Charlotte and her M.A. in Communication from Wake Forest University. Karoline studies minority leadership, diversity and inclusion in workgroups and big data methods in diversity research.

Kecia M. Thomas is a Professor of Industrial/Organizational (I/O) Psychology and African American Studies at the University of Georgia where she also serves as the Senior Associate Dean in the Franklin College of Arts and Sciences at the University of Georgia. Another major responsibility includes the leadership of the college's Office of Inclusion & Diversity Leadership.

Kecia is an expert in the psychology of workplace diversity.

Christian Thoroughgood, Ph.D., is an Assistant Professor of Psychology in Graduate Programs in Human Resource Development at Villanova University. His research interests primarily reside within the areas of leadership, counterproductive work behavior and diversity in organizations. Within these areas, his current work addresses the interactive effects of leaders, followers and environments on constructive and destructive leadership processes; the role of dispositional and situational factors in shaping supervisor aggression; as well as the intersections of race, gender and other demographic characteristics in the workplace. He received his M.S. and Ph.D. in Industrial/Organizational Psychology from Penn State University and his B.A. in Psychology and Economics from the University of Maryland, College Park.

Ilene C. Wasserman is Founder and Principal of ICW Consulting Group, a strategic organizational consulting firm that specializes in designing and facilitating culture change interventions in organizations that promote the engagement of diversity and organizational learning. She consults to a variety of industries in the corporate and non-profit arena. In her research and practice, Ilene focuses on relational and discursive processes that foster transformative meaning in the engagement of social, cultural and positional differences. Ilene is an active member of her community and serves on several community boards and professional organizations.

Veronica Y. Womack is an Inclusive Learning Specialist and Project Administrator at Northwestern University's Searle Center for Advancing Teaching and Learning. She has a Ph.D. in Social Psychology from Howard University and received postdoctoral training in cardiovascular epidemiology and prevention from Northwestern University. Her research interests pertain to coping strategy preferences for social identity-related stressors with a focus on microaggressions and "being the only one." She incorporates mindfulness, intersectionality, health behaviors and "safe space" development into her research and workplace-based initiatives. Dr. Womack's research and commentaries have been published in *Psychosomatic Medicine, Diabetes Care, Journal of Black Studies* and *TIME* magazine.

Foreword

The goal of the Applied Psychology Series is to create books that exemplify the use of scientific research, theory and findings to help solve real problems in organizations and society. In *Revisiting Diversity Resistance*, Dr. Kecia Thomas provides an excellent example and contribution to this Series.

Over 10 years ago, Thomas' *Diversity Resistance in Organizations* (2008) provided a critical resource that helped move diversity scholarship forward as both a legitimate academic science and a path towards organizational effectiveness. Today, not only is diversity research viewed as a core topic in organizational science, there is growing awareness that diversity provides a unique path for leadership excellence and improved organizational performance.

Despite changing attitudes toward diversity, diversity resistance continues to represent an important challenge requiring greater critical attention, especially in the face of significantly greater societal and political threats to diversity. In *Revisiting Diversity Resistance*, Dr. Thomas and a set of leading experts in the field address the pervasive, persistent, yet ever-evolving multi-level manifestations of diversity resistance.

This edited book includes 10 innovative chapters that provide emerging perspectives and language to understand the resurgence of overt as well as covert or subtle interpersonal and institutional dynamics of diversity intolerance and resistance, both within the U.S. and globally. In Chapter 1, Thomas, Lavner, Johnston and Scofield set the stage for the book, providing an overview of diversity resistance and its critical impact on health. In Chapter 2, Dickens and Womack examine the intersection of race and gender and potential pressure that Black women and other women of color may experience in terms of identity shifting at work. This chapter notes that Black women who choose not to alter their self-presentation to get ahead in their career and be their "authentic self" in a largely White workplace may be exhibiting both a form of wellness and an act of defiance. Romantic relationships among LGBTQ people and related terminology on the invisibility of such relationships is the topic of Chapter 3. In this chapter, Holmes introduces the topic of sexuality blindness and describes its relationships to diversity resistance and career experiences. In Chapter 4, Sawyer and Thoroughgood argue that organizations can break counterproductive methods of promoting gender inclusivity by taking a queer approach to management. Mental illness as a stigma and a distinct often invisible disability is addressed by Follmer and Jones in Chapter 5.

Members of dominant groups may believe that diversity programs give special treatment to some groups, while non-dominant group members may view such programs as not going far enough or inadvertently harming people that they are intended to help. This conflicting view is explored in Chapter 6, including the associated threats that some dominant groups perceive when they experience changing policies and practices to address historically underrepresented groups. Further, Plaut, Romano, Hurd and Goldstein identify how underrepresented groups may react when others have diluted the design and goals of diversity programs with the result of continued feelings of unsafety. Similarly, in Chapter 7, Ruggs, Summerville and Marshburn describe how organizational responses to social justice societal events depicting racial or sexual injustices can also lead to diversity resistance. The authors adeptly integrate organizational management response theory and white privilege/white identity management theory to understand and recommend how organizations can respond to such events in ways that support diversity goals. Identifying a set of personal- and organizational-level indicators of diversity resistance, Heidelberg (Chapter 8) provides suggestions for the identification and assessment of diversity resistance in organizations. Drawing from the field of nonprofit art management, Heidelberg describes how nonprofit performance management practice can be applied to evaluate equity work. In Chapter 9, Gallegos, Wasserman and Ferdman update their 2008 chapter and offer examples of how issues of social inequity and unequal power relations enhance the complexity, conflict and difficulty in maintaining equilibrium and fairness as organizational leaders. The public relations industry is showcased in Chapter 10 by Parrish and Gassam, who highlight the diversity needs in this field with particular focus upon African-American professionals.

We are extremely pleased to add this updated and timely new edition of Thomas' *Revisiting Diversity Resistance* to the Applied Psychology Series. This book accomplishes the goals that exemplify the challenge of balancing the science and practice goals of this Series, bringing together solid scholarship and new and emerging ideas to address problems that are becoming increasingly important in the workplace and in our global society.

Jeanette N. Cleveland
Donald Truxillo
Series Co-Editors

Preface

Revisiting Diversity Resistance

Kecia M. Thomas

The first volume of *Diversity Resistance in Organizations* was published in 2008 at a time when there was an increasing acceptance of diversity science (Plaut, 2010) as a legitimate area of academic study as well as a value to organizational effectiveness. Chief Diversity Officers (CDOs) became a growth industry (Williams & Wade-Golden, 2007). The introduction of diversity resistance in that initial volume was motivated by a desire to elevate diversity as an opportunity for organizational effectiveness while also shedding light on the variety of ways in which "differences" were marginalized and derailed, sometimes in ways that appeared relatively benign but often were ultimately damaging to organizations and to their bottom line nonetheless.

Contributors in that first volume elaborated upon how diversity resistance manifests for various demographic groups and highlighted how institutions and organizational practitioners might address it in their work. However, even during this period there was still a debate about the value of attending to diversity in organizations like workplaces. Specifically, the questions seemed to center on, was pursuing diversity a moral imperative and simply the "right thing to do," or could it actually contribute to organizational performance? Was diversity an opportunity to improve the quality of leadership's decision-making and expand organizational competitiveness (Phillips, Northcraft, & Neale, 2006)? Can an organization's diversity positively influence an institution's bottom line (Ng & Tung, 1998; Richard, 2000)?

For the most part, those questions have been answered affirmatively (Phillips, 2014) and diversity has remained one of the top workplace trends since 2015[1] and is among the most researched topics by social scientists like Industrial/Organizational Psychologists, yet resistance continues to persist in and outside of organizations. It could be argued that diversity resistance is no longer rooted in questions about diversity's value but rather that it is a reflection of increasing integration globally, the browning of the U.S. and the threat to the presumed dominance and leadership status of men and the inevitable shrinking status of Whites (Danbold & Huo, 2015). Even K-12 schools are re-segregating, discussions of race often remain taboo (Tatum, 2017) and attempts to maintain the cultural myth of colorblindness is seen as an attractive diversity-friendly ideology (Thomas, Plaut, & Tran, 2014).

Diversity resistance has never been just a U.S. phenomenon. Societies and organizations globally are struggling with how to incorporate the talents of new immigrants who differ in skin color, culture, language and religion against a backdrop of growing immigrant intolerance. Furthermore, the increasing voice of gay, lesbian, bi-sexual, transgender and queer (LGBTQ) workers reflects another dimension of diversity that organizations and many countries struggle to accept. For example, in the U.S. there is active resistance in the forms of attempts to repeal the modest human rights gains that have been made for LGBT workers and citizens.

The last 10 years has proven that resistance to diversity is fluid; it migrates and changes targets and expression in order to keep up with, silence and reject those who are "Othered." Perhaps more importantly, diversity resistance is a response to the demands for access to the privilege that dominant groups often unknowingly rely upon. Diversity resistance adapts and changes as the demands of a diverse society and workforce become more evident, expected and voiced. The first volume articulated that resistance to diversity occurs at both the individual and organizational levels as well as in subtle and overt ways. That volume mostly focused on gaining insight into the subtle and contemporary ways differences are resisted. It also focused on the ways in which this resistance is embedded in institutions through the maintenance of taboos and legitimizing myths and systems of silence like "don't ask, don't tell." Subtle ways in which resistance has been ingrained in ways that position diversity as a non-issue (often through colorblindness) has been replaced today by more overt and hostile movements to remove and outright reject LGBT soldiers, namely those who are transsexual. Outside of organizations the return to "in your face" resistance shows up as refusals of service and interpersonal-harassment, exclusion and outright rejection.

Technology and social media, especially Facebook and Twitter, have enabled formerly hidden communities and isolated groups to speak up and even organize globally. Perhaps even more importantly social media has created a platform for marginalized group to document their day-to-day experiences with intolerance, harassment, discrimination, and too frequently violence. Therefore, the strategy to downplay demands for justice, access and inclusion are no longer easily placed on the back burner or tabled for another day.

The U.S. campus protests of the Fall of 2015 highlighted that even higher education faces challenges with diversity resistance as evidenced by the national list of demands created by students seeking more inclusive campuses (see www.thedemands.org/). These institutions struggled with saving face, managing diversity and protecting their higher education brands. Following a hunger strike, marches and protests by student-athletes schools like the University of Missouri[2] saw tremendous drops in their subsequent application and enrollments that created a financial emergency that resulted in faculty furloughs and layoffs. Mizzou saw a 35% decrease in enrollment that led to closing seven dormitories and eliminating 400 positions.[3] Ultimately, many high-level leaders were forced to resign because they failed to listen to the voices of marginalized students and address the lack of inclusion and increasing diversity resistance that infiltrated their campuses.

The significant changes in the political climate globally has created an urgent need to return to discussions about diversity resistance. However, rather than focusing on primarily the covert and subtle nature of resistance like the first volume, in this volume the contributors and I are compelled to address the increasingly egregious, overt and sometimes hostile forms of resistance that show up in our daily newsfeeds like the re-emergence hangmen nooses in workplaces and on college campuses (e.g. the U.S. Mint; Stanford University), interpersonal acts of harassment and violence to sexual minority citizens (various retail stores) and Ku Klux Klan rallies like those held recently in the iconic college towns of Charlottesville, VA and Chapel Hill, NC.

People who identify as non-cisgender or non-binary must navigate individual and organizational forms of heterosexism as well as physical and mental abuse. Similarly, there is also growing attention to the benevolent sexism and more hostile forms of diversity resistance experienced by women and how it frequently shows up as sexual harassment and abuse. The institutionalization of patriarchy sometimes means that there is no demand for equity without fear of retaliation. Societies and organizations globally are struggling with how to incorporate the talents of new immigrants who differ in skin color, culture, language and religion against a backdrop of growing immigrant intolerance.

In 2019, over 900 hate groups exist throughout the U.S.[4] These groups are becoming more comfortable and visible given the national climate and leadership. How do organizations deal with employees who hold membership in groups that counter their organizational values of diversity and inclusion? How do organizations pursue their own values around diversity and inclusion against a backdrop in which the local environment may be increasingly tolerant of these groups? How do employees deal with organizations who resist supporting "Others" and communicating their support for diversity and inclusion (D&I), when they feel at risk of discrimination and harassment at work (Dover, Major, Kaiser, 2016)?

Despite the growing normalization of intolerance, there is some light in the growing visibility and leadership of women, particularly women of color whose intersectional minority statuses have provided them with unique opportunities to speak and lead. These leaders' intersectionality provides them with a legitimacy in building counter-resistance coalitions across various identity groups. Even though women of color continue to face both covert and overt resistance through both access and treatment discrimination in and out of work, increasingly they are the leaders of counter-resistance through initiating and organizing global and cross-generational movements like #MeToo and #BlackLivesMatter.

The Current Volume

Section one of the current volume explores the variety of ways in which diversity resistance is confronted by marginalized groups in this era of growing intolerance. This section highlights the experiences of a diversity of targets, responses to their resistance and venues for their inclusion. The first chapter (Thomas, Lavner, Johnston and Scofield, *Diversity perform-ance, social surveillance and rescinding human rights: Understanding the health outcomes of diversity resistance*) re-examines diversity resistance by spotlighting old and new forms that include diversity performance, social surveillance and the rescinding of modest human rights gains. Chapter 2, *Unapologetic authentic early career Black women: Challenging the domin-ant narrative* (Dickens and Womack) speaks directly to the professional and personal health risk faced by Black women who may feel pressured to engage in "identity-shifting" in order to maintain the status quo. Their authenticity is a form of counter-resistance to concrete and glass ceilings that resist their opportunities to ascend in their careers and lead.

In the same way that organizations' diversity resistance turns a blind eye to pressures confronted by Black women to be invisible, Chapter 3 (*Sexual blindness: A new frontier of diversity resistance*) speaks to the organiza-tional landscapes navigated by people who are LGBTQ to work and to exist as if their lives are void of romantic relationships. Oscar Holmes IV argues that this sexuality blindness has thwarted our full appreciation of the complexity of the experiences of LGBTQ people. This ubiquitous form of diversity resistance has therefore hindered our diversity science and oppor-tunity to build strategies to effectively serve this population. Likewise, Chapter 4, *Diversity resistance and gender identity: How far have we come and where do we still need to go?* speaks of the lack of inclusivity of gender identity minorities in organizations' diversity efforts as another example of diversity resistance. Sawyer and Thoroughgood argue for a queer approach to management and they make recommendations for organizations, allies and scientist-practitioners who want to enact truly inclusive work environments. Follmer and Jones in Chapter 5, *Mental ill-ness stigma*, turn to another under-investigated population, those with mental illness, to highlight the resistance and stigma confronted by this group and its impact upon opportunities for treatment. They too offer strat-egies for reducing resistance in organizations and creating more supportive and inclusive work environments.

Section two, dives more deeply into the question of why. Why is there a growing resistance to diversity and inclusion and an increasing acceptance of overt intolerance? Why does diversity resistance persist despite the growing diversity and our improved understanding of diversity science and the positive outcomes of inclusive climates? Chapter 6, *Diversity resistance redux: The nature and implications of dominant group threat for diversity and inclusion* (Plaut, Romano, Hurd and Goldstein), addresses the perceived threat experienced by dominant group members who might fear the loss of their own group's entitlements. Ruggs, Summerville and Marshburn's *The response to social justice issues in organizations as a form of diversity resistance* (Chapter 7) addresses the importance of if and how organizations respond to social and political events, as both sense-making devices for organizational members but also potentially as forms of resistance themselves. Heidelberg (Chapter 8, *Artful avoidance: Initial considerations for measuring diversity resistance in cultural organizations*) suggests that one form of organizational response in non-profit organizations, cultural equity statements, is an important vehicle for identifying and assessing resistance, specifically around the gap that can exist between equity intent and actual impact on marginalized groups.

Gallegos, Wasserman and Ferdman (Chapter 9, *The Dance of Inclusion: New ways of moving with resistance*) examine the work they published in the first volume and revisit the metaphor of "the dance" that leaders must perform in order to manage the tension of social inequity and unequal power relations with an even broader range of differences and social challenges. Finally, this volume concludes with a case by Parrish and Gassam (Chapter 10, *African-American professionals in public relations and the greater impacts*). Their case examines the diversity resistance dance in the industry of Public Relations (*The case of resistance in public relations*), which they regard as an opportunity to make more visible the lack of diversity and the chance to create a research and practice agenda that will serve underrepresented individuals and the industry.

We hope that this second volume will provide new insight, language and perspective to combat the growing culture of intolerance that seems to be spreading globally. We seek to build a community in which everyone experiences inclusion for both personal and communal good. However, without shining a light on all forms of resistance and resolving these challenges, the true positive potential of diversity will never be realized.

Notes

1 www.siop.org/Research-Publications/Items-of-Interest/ArtMID/19366/ArticleID/1639/It%E2%80%99s-the-Same-Only-Different.
2 Jaschik, S. (2015). What the protests mean. *Inside Higher Ed, 16.*
3 www.nytimes.com/2017/07/09/us/university-of-missouri-enrollment-protests-fall out.html.
4 www.splcenter.org/hate-map.

References

Danbold, F., & Huo, Y. J. (2015). No longer "all-American"? Whites' defensive reactions to their numerical decline. *Social Psychological and Personality Science, 6* (2), 210–218.

Dover, T. L., Major, B., & Kaiser, C. R. (2016). Members of high-status groups are threatened by pro-diversity organizational messages. *Journal of Experimental Social Psychology, 62*, 58–67.

Ng, E. S., & Tung, R. L. (1998). Ethno-cultural diversity and organizational effectiveness: A field study. *International Journal of Human Resource Management, 9*(6), 980–995.

Phillips, K. W. (2014). How diversity makes us smarter. *Scientific American, 311*(4), 43–47.

Phillips, K. W., Northcraft, G. B., & Neale, M. A. (2006). Surface-level diversity and decision-making in groups: When does deep-level similarity help? *Group Processes & Intergroup Relations, 9*(4), 467–482.

Plaut, V. C. (2010). Diversity science: Why and how difference makes a difference. *Psychological Inquiry, 21*(2), 77–99.

Richard, O. C. (2000). Racial diversity, business strategy and firm performance: A resource-based view. *Academy of Management Journal, 43*(2), 164–177.

Tatum, B. D. (2017). *Can we talk about race? And other conversations in an era of school resegregation.* Boston, MA: Beacon Press.

Thomas, K. M., Plaut, V. C., & Tran, N. M. (2014). Diversity ideologies in organizations: An introduction. In K. M. Thomas, V. C. Plaut, & N. M. Tran's (Eds.), *Diversity ideologies in organizations* (pp. 1–17). New York: Routledge.

Wasserman, I. C., Gallegos, P. V., & Ferdman, B. M. (2008). Dancing with resistance. *Diversity resistance in organizations.* In K. M. Thomas, V. C. Plaut, & N. M. Tran's (Eds.), *Diversity ideologies in organizations* (pp. 175–200). New York: Routledge.

Williams, D., & Wade-Golden, K. (2007). The chief diversity officer. *CUPA HR Journal, 58*(1), 38.

1 Diversity Performance, Social Surveillance and Rescinding Human Rights

Understanding the Health Outcomes of Diversity Resistance

Kecia M. Thomas, Justin A. Lavner,
Zoe E. Johnston and Cambrilyn Scofield

Since the first volume of *Diversity Resistance in Organizations* (2008b) was published, there has been a significant growth in the "mainstreaming" of diversity, seen in advertising, diversity functions in organizations and the increased hiring of diversity executives. In some ways, the growing racial diversity and more visible participation of women, linguistic minorities and sexual minorities in many areas of society signals some steps toward inclusion. Yet, even in those cases where diversity exists in terms of representation, resistance to both diversity and inclusion often persists. That is, as frequently as the topic appears in our schools, workplaces and communities, it often seems as though it is simply *performing diversity* rather than engaging in it. With an evolving political climate, there is also growing evidence of a return to old-fashioned and overt forms of discrimination, harassment and hate and efforts to remove human rights connected to areas such as voting and marriage. How does one navigate a culture that desires to promote diversity on the surface, yet challenges it at every turn? What are the health-related consequences of being in this context?

This chapter examines several examples of diversity resistance today. We build on Thomas and Plaut (2008), who defined diversity resistance as, " ... *a range of practices and behaviors within and by organizations that interfere, intentionally or unintentionally, with the use of diversity as an opportunity for learning and effectiveness* (p. 5)." These behaviors and practices were described as being perpetuated by both individuals and by organizations in covert and overt ways. Accordingly, we begin by examining diversity performance as evidenced by the growing presence of racial diversity in the advertising and staffing strategies used by many organizations. We then turn to the unexpected ways in which diversity resistance persists in often overt and hostile ways as perpetuated by individuals that at times seem validated and reinforced by broader systems of resistance. Given the mixed message conveyed through greater performance of diversity and persistent resistance to diversity and inclusion, we conclude the chapter by asking what consequences this climate of diversity resistance might present for the mental and physical health of members of minoritized groups.

Performing Diversity as Diversity Resistance

Evidence of diversity performance is rampant in organizational advertising. For example, consider the 2015 continuing education catalog for the University of North Georgia, titled *Success: Why Follow When You Can Lead.* The cover displayed four young professionals who appear to be at the end of a race on an outdoor track. Running are two White men in suits and ties (clearly winning the race), a White woman wearing a pantsuit and heels (coming in a distant third) and finally a young Black man without a jacket who is pulling up the rear. The university obviously thought about the importance of displaying a racially diverse scene of men and women in their efforts to attract more diverse students to their campus in the future. What the institution obviously did not consider was how the message of White men as winners and a (White) woman and a Black man as losers would be received. Essentially the ad replicated the current reality of power and leadership in many organizations: White men lead and others follow. In fact, rather than decreasing in representation, the proportion of White men seems to be increasing.[1]

Other well-meaning examples exemplify insensitivity to the reality and history of marginalized groups. A Chicago television affiliate, WGN-TV, sought to celebrate religious diversity by highlighting the beginning of the Jewish celebration Yom Kippur by including a mention of the holiday in its newscast. Unfortunately, the graphic that was shown behind the news anchor was actually the Star of David badge from a prisoner's uniform used during Nazi Germany to mark people of Jewish descent during the Holocaust. Again, the station attempted to demonstrate its commitment to religious diversity but did not invest the resources needed to accurately portray Jewish symbols or avoid minimizing the significance of the Holocaust.

Popular brands – even those known for racial diversity in advertising – have experienced similar missteps. For example, The Gap (a popular clothing store and brand) aired a commercial nationally in which three of four young women were shown as strong and active, with several of them engaging in gymnastic type exercises. The lone inactive girl was also the only Black actress in the ad. In fact, she was positioned more as a prop such that her head was used by the eldest and tallest White girl to rest her arm upon. Again, this commercial reinforced socialized stereotypes of Black women as servers and White women as being served (hooks, 2001).

The health and beauty organization Dove is known for taking great strives to celebrate size diversity by using racially diverse actresses with a variety of weights in their print and televised ads. However, Dove has likewise conveyed problematic messages around skin color and race. One print ad showed a Black woman literally removing her brown skin after using Dove soap, like she was removing a sweater in order to display her improved and now White skin. Another Dove ad for lotion included the text, " … for normal to dark skin," thus normalizing White skin and marking all other shades as somehow abnormal. Another health and beauty company, Nivea, had a print ad in which a heavily bearded Black man with a large afro was transformed by Nivea products into a smooth-shaven Black man with a low haircut. This ad was accompanied with the text, "Re-civilize yourself." This too reinforced Whiteness and standards of presentation as privileged and civilized, while the unique characteristics of Black men were presented as uncivilized.

Finally, a pool safety poster distributed by the American Red Cross highlighted a racially diverse group of swimmers by representing the participation of White, Black and Brown kids at what looks like a neighborhood pool. The title of the poster was, "Be Cool, Follow the Rules." Unfortunately, a closer look at the many examples of "cool" and "not cool" behaviors reflected in the poster indicates that all of the "cool" behaviors were demonstrated by White kids, while Black and Brown kids were responsible for *every* "not cool" behavior displayed. This unfortunate poster demonstrated the persistent stereotype of Black, and specifically Black kids, as unruly, difficult, and likely to be violent.

What is striking about these examples is that in each example provided, the organization went to great lengths to demonstrate its value for diversity – by casting racially diverse actors and actresses, acknowledging a Jewish holiday and creating racially integrated materials. Despite these good intentions, however, in each case history was ignored and negative messages about minorities were displayed, reinforced and highlighted. These cases may seem minor, but they reflect a type of diversity resistance in their marginalization of the important work of celebrating diversity and creating an inclusive society.

Another frequent way in which diversity is performed is through the recruitment and hiring of Chief Diversity Officers (CDO) in corporate offices and college campuses. Like the prior examples presented, the presence of CDOs and diversity functions in organizations signal the message that diversity and inclusion is important. However, the circumstances under which many CDOs ascend to these roles and the circumstances under which they work can also suggest a climate of diversity resistance.

There is not a simple nor linear career path nor set of competencies to define who would be an effective CDO. In fact, at the time when many of these positions were launched, there was not a common job description or set of expectations for this role (Williams & Wade-Golden, 2007; Wilson, 2013). Certainly, the role of many CDOs is to support and project an organization's image as one that is proactive and motivated to at least be diverse in its racial and gender representation (Wilson, 2013). However, a recent research report by consulting firm Russell Reynolds[2] (2019) suggests that most CDOs feel and are often set up to fail. Specifically, they report lacking the resources needed to be effective, including access to demographic data about their own organization, analytics staff, financial resources, decision-making authority and exposure to top decision makers, while also often working in climates that still question or negate the value of diversity. CDOs who are also members of minority groups themselves may not only find themselves ineffective in their work role but given their dual status may also find themselves overextended and burned out, in what is sometimes referred to as representation burnout.[3] Hiring CDOs who lack resources to be effective reflects how organizations simultaneously engage in diversity performance and diversity resistance.

Reemergence of Overt and Hostile Diversity Resistance

In addition to the subtle (and not so subtle) diversity resistance seen in diversity performance, recent years have also seen a reemergence of overt and hostile diversity resistance. The resistance to racial minorities, especially Blacks, is demonstrated in overt and hostile ways such as hangmen's nooses in the workplace (Thomas, 2008a). Although nooses are typically tied to lynchings, they reemerged in the 1990s in several large organizations where anti-Black racism was rampant. Black employees have been targeted by nooses at their desks or near their lockers or work-spaces in every state in the U.S. and across industries. Nooses have been found on a number of college campuses as well, most recently Stanford University[4] and the University of Maryland[5] among others. In fact, the Equal Employment Opportunity Commission began the Eradicating Racism and Colorism from Employment Initiative (E-RACE)[6] to address racial harassment in order to address the growing noose incidents and other forms of racial harassment. Thomas and Plaut (2008) suggested that noose cases are often vehicles for racial resistance when Black employee(s) have made an official complaint alleging discrimination and/or harassment; the noose is a form of retaliation and is a further threat for the target. Nooses are also used in cases where Blacks are promoted, often to positions of influence or leadership over their non-Black peers. Nooses are also used as threats outside of workplaces and universities. In one recent example, three (White) middle school teachers in New York who served almost exclusively Black and Latino children placed nooses on their bulletin boards as "back to school" necklaces.[7] When confronted by parents, they claimed it was a joke.

In many of these cases the organizational response was shock and then denial. The middle school teachers in fact claimed not to have known the historical significance of nooses. Likewise, in a very large case in which Georgia Power (owned by Southern Company) admitted to the presence of nooses for *decades*, the CEO admitted throughout his deposition that he had "no idea" a noose was a racist symbol and that he felt both Blacks and White would find the nooses humorous.[8]

Diversity resistance is also demonstrated in state-sanctioned discrimination connected to hair – a form of diversity resistance uniquely directed toward African-Americans. Hair has long been a civil rights issue for Blacks (Garrin, 2016). In many ways, natural hair for Blacks, especially Black women, is seen as signaling a lack of professionalism. Opie and Phillips (2015) conducted three experiments in which they demonstrated that Black women with Afrocentric hair were evaluated as more dominant and less professional than Black women who wore their hair in more Eurocentric (straightened) styles. As more and more Black women reject chemical hair relaxers in favor of styling their natural hair, there has been growing evidence of racial discrimination focused on hair. In 2010, Chastity Jones, a Black woman with dreadlocks, was offered a job as a customer service representative for which she would work over the phone. However, that job offer came with a stipulation that Ms. Jones would need to cut her dreads in order to be employed by the company. The courts sided with the employer. Currently, the judicial system only protects against hair discrimination in cases involving afros, since it is an immutable characteristic given that is its natural state. The law does not protect against forms of discrimination in cases where the hair has been altered, such as through braids or dreadlocks, even if these styles have a deep history for Black culture. The armed forces only recently allowed (Black) women to wear dreadlocks and braids (despite their neatness and convenience) without fear of being reprimanded. Here, too, we see state-sanctioned resistance to diversity, co-occurring alongside efforts to expand the diversity of the armed forces. There has, however, been some recent progress: in 2019, both the State of California and the City of New York outlawed employment discrimination based upon hair.

Surveillance as Diversity Resistance

We also see diversity resistance reflected in the heightened surveillance of minority group members. Increasingly there is evidence that racial minorities are policed by average White citizens. Cell phone video and social media posts have highlighted the many ways in which the mere presence of marginalized group members, especially Blacks, is policed. The most visible case of this was the untimely death of a Black teenager, Trayvon Martin, at the hands of a man who claimed to be on neighborhood watch. In February 2012, 28-year-old George Zimmerman was on community watch when he shot and killed Martin, a 17-year-old, unarmed Black boy. Zimmerman had dialed 9-1-1 to report his suspicions of Martin and the operator instructed him not to pursue him. Zimmerman disregarded the operator's advice, however, and instead followed Martin, and after a physical altercation with him, ended up shooting him in what he claimed was self-defense. The case went to court and Zimmerman was found not guilty, sparking protest and outcry among the African-American community and others. It seemed apparent that Zimmerman had profiled Martin and only thought he was suspicious because of the color of his skin. While Martin presented no clear threat, Zimmerman still followed him and used deadly force when Martin had no means of threatening Zimmerman's life ("Trayvon Martin," 2019).

This incident was the first highly publicized and arguably most tragic event in a phenomenon that Twitter users have dubbed #LivingWhileBlack. This hashtag describes the experience of Black people in America participating in everyday, mundane activities and having the police called on them, more often than not by White people. Black people are currently living in a time of hyper-surveillance, which stems back to the days of Jim Crow and "sundown towns," areas where black people had to leave the city by sundown (Loewen, 2018). Like sundown towns, #LivingWhileBlack exemplifies the control White people feel that they are able to enforce over people of color at their own will. Due to the stereotypes of criminality that bias perceptions of Black people, these perceived "threats" often result in police involvement (Welch, 2007). In 2018 alone, examples of when the police were called on Black people included when Blacks were "caught" doing things such as golfing too slowly, barbequing at a park and waiting for a friend at Starbucks (Griggs, 2018). Similar reactions to White people engaging in these activities are unlikely.

The overwhelming number of these responses have begun to cloud what really necessitates a 9-1-1 call. Many of these transgressions could have been resolved civilly between the two parties, but it is this perceived threat of criminality that makes people feel they need to involve the police. Some officers respond with similar levels of suspicion – for example, when someone called the police on a group of Black people leaving their Airbnb in Oakland, CA, the sergeant did not know what Airbnb was, did not believe them and proceeded to question them for 45 minutes (Victor, 2018). In other situations, however, the officers are on the side of the victims – officers told a woman who had called the police on a Black male real estate agent who was doing a home inspection that they would arrest her if she continued to harass the man (Lockhart, 2018). Nonetheless, the frequency of police involvement in these trivial incidents likely serves to increase anxiety and insecurity in daily life.

Among the recent #LivingWhileBlack episodes that have been particularly disturbing are those involving students and workers. Visible cases have involved a graduate student at Yale (Wootson, 2018) and a Black firefighter (Papenfuss, 2018). In May of 2018, Lolade Siyonbola, a 34-year-old Black graduate student at Yale University, fell asleep in her dorm's common room after a night of studying. She awoke to a White student turning the lights on and telling her that she had called the police because she was not supposed to be sleeping in there. The police arrived and after Siyonbola presented her key and unlocked her dorm room, the police asked for her identification to ensure that she belonged there. In a video of the incident taken by Siyonbola, she can be heard saying to the officers, "I deserve to be here. I pay tuition like everybody else. I'm not going to justify my existence here." Once the officers verified Siyonbola's identification, they determined it was not a police matter and left. Siyonbola shared the incident on Facebook, writing, "I know this incident is a drop in the bucket of trauma Black folk have endured since Day 1 America."

Also in 2018, a White citizen called the police on Kevin Moore, a Black firefighter who was in uniform doing routine safety inspections in a neighborhood in Oakland, CA. Another person filmed him and asked for identification. Moore told the neighbor that if he was still concerned, he could check the street for "a big red fire engine." Moore's colleague plans to partner with him on future inspections because she's worried about his safety, reporting that no one had ever called the police on her or the other White firefighters carrying out an inspection.

These episodes reflect indisputable racism, and their growing ubiquity has concerning implications for a national culture of trust and inclusion that is necessary for an engaged and effective society (see Downey, van der Werff, Thomas & Plaut, 2015). In direct response to these incidents, Oregon's only three Black state legislators introduced a bill that would allow victims of discrimination to sue over racially motivated 9-1-1 calls. It seeks to protect people of color's freedom and sense of security as well as limit anxiety-inducing interactions between them and the police. One of the co-sponsors of the proposal, state Sen. Lew Frederick, asserted that racially motivated 9-1-1 calls intensify the fear of police for people of color, saying, "It's not just an inconvenience when a police officer stops me. When a police officer stops me, I wonder whether I'm going to live for the rest of the day" (Arnold, 2019).

Rescinding Rights of LGBTQ Individuals

Racial minorities are not the only group who navigates a culture in which they confront diversity resistance. As the rights of Lesbian, Gay, Bisexual, Transgender and Queer (LGBTQ) individuals continue to advance, there is still some resistance to the diversity of their existence and experiences. As just one example, LGBTQ Americans face discrimination and lack comprehensive protections in housing, employment, healthcare and public places in 30 states (Movement Advancement Project, 2019). Throughout Obama's presidency, the U.S. saw a tremendous increase in support for the rights of LBGTQ people at the federal level. However, with the arrival of the Trump administration, as well as changes in state governments, that progress has been threatened.

Healthcare discrimination has been a longstanding problem for LGBTQ individuals. A 2017 study showed that roughly one in six (18%) LGBTQ respondents avoided seeking medical care because of concerns of discrimination in healthcare settings (NPR et. al., 2017). Even though these statistics still hold true today, some actions have been taken to remedy the issue. With the introduction of the Affordable Care Act (2010), provisions for insurance companies were set so that they were no longer able to discriminate against anyone due to a pre-existing condition. This meant that they could not turn anyone away based on their sexuality or gender identity. It also made it easier for people living with HIV and AIDS to obtain Medicaid, private health insurance and more qualified providers. These protections were overturned by the Department of Human and Health Services of the Trump administration in 2019 when they unveiled a proposal that removed all acknowledgment of federal protections for transgender people in regards to healthcare (U.S. DHHS, 2019). Within the same month, the administration published a final rule that emboldened healthcare professionals to deny care to patients based on religious or moral beliefs (U.S. DHHS, 2019), a change thought to affect the provision of care to LGBTQ individuals.

LGBTQ people also face discrimination in housing. To combat this trend, in 2012 (and again in 2015), Obama issued a final rule and subsequent guidance to ensure that the Department of Housing and Urban Development's (HUD) core housing programs and services were open to all persons regardless of sexual orientation or gender identity (U.S. HUD, 2012). This rule was meant to make federal housing more inclusive and welcoming to LGBTQ people, but this was soon challenged by the Trump administration. Along with an initiative to erase two policies that protect the homeless LGBTQ population, the HUD also announced new plans that would ultimately take away the rules prohibiting discrimination against transgender people in HUD-funded homeless shelters (U.S. HUD, 2019).

One of the most notable victories for LGBTQ rights in recent years was marriage equality. In 2011, the U.S. Department of Justice (DOJ) announced it would discontinue the legal defense of the Defense of Marriage Act's provision that defined marriage as only between a man and a woman (U.S. DOJ, 2011). This eventually led to one of the most important Supreme Court rulings of LGBTQ rights in the U.S., Obergefell v. Hodges. In 2015, this landmark decision required all states to issue marriage licenses to same-sex couples, as well as recognize marriages that were performed lawfully out-of-state, effectively legalizing same-sex marriage nationwide.

Since this decision, there have been numerous accounts of opposition coming from state and local governments across the country. Former Texas Attorney General Ken Paxton stated that county clerks have the right to turn away same-sex couples seeking marriage licenses if they have any religious objections. In Alabama, a bill was passed to abolish judge-signed marriage licenses in order to avoid judges facing legal backlash (S.B. 69, 2019). Going beyond county courts, in 2016, Tennessee lawmakers introduced the Natural Marriage Defense Act (H.B. 0892, 2017). The proposed bill – which did not move beyond the Tennessee House Civil Justice subcommittee – would have allowed the state to reject recognizing same-sex marriages as legal marriages, regardless of any court ruling. Legal battles over and legislation regarding issues related to the LGBTQ community continue to surface in areas ranging from religious freedom (e.g., Masterpiece Cakeshop v. Colorado Civil Rights Commission, 2017) to the so-called "bathroom bills" that mandated that people use the bathroom corresponding to the sex stated on their birth certificate. In each case, we see resistance to full societal inclusion of members of the LGBTQ community.

We see similar trends when considering LGBTQ workplace issues. In 2014, Obama signed an Executive Order that prohibited federal contractors from discriminating against employees and applicants on the basis of, "race, color, religion, sex, sexual orientation, gender identity, or natural origin." (E.O. 13672, 2014). This set a precedent for many government agencies, as they took various measures to ensure that sexual and gender minority Americans were treated fairly and were not discriminated against in the workplace, in official documents and in the healthcare system. However, these protections are still not available nationwide. Pending federal rule, the case of R.G. & G.R. Harris Funeral Homes, Inc. v. E.E.O.C. is waiting to be heard by the U.S. Supreme Court to determine if "Sex" covers sexual orientation and gender identity. Under the former Attorney General of the Trump administration, the DOJ argued that federal law does not ban sex discrimination in the workplace for transgender employees because sex discrimination does not include gender identity discrimination (Sessions J.B., 2017b). We see a move toward similar policy changes away from LGBTQ rights and inclusion in several other areas of federal policy, including whether transgender individuals can serve in the armed forces and the guidance provided to federal agency managers to help them understand how to support transgender federal workers.

Impacts Upon Health

Thus far we have discussed various forms of diversity resistance, from diversity performance to overt, hostile forms of resistance to surveillance to the rescinding of rights. In the remainder of this chapter, we discuss how the experience of discrimination resulting from diversity resistance affects individuals directly and indirectly. In short, the negative consequences of discrimination cannot be understated: Hundreds of studies now indicate that perceived discrimination is associated with worse mental and physical health (for meta-analyses, see Paradies et al., 2015; Pascoe & Richman, 2009; Schmitt, Branscombe, Postmes, & Garcia, 2014). These associations stem, at least in part, from increased psychological and physiological stress responses (e.g., decreased positive emotion, increased negative emotion, cardiovascular reactivity, cortisol response) and from discrimination's effects on health behaviors such as sleep and substance use (e.g., Clark,

Anderson, Clark, & Williams, 1999; Pascoe & Richman, 2009). In the sections that follow, we review these findings in detail to make clear how the experiences of discrimination resulting from diversity resistance – be they personal, vicarious, or anticipated – prove harmful to the health and well-being of minority group members.

Mental and Physical Health Correlates of Individual Experiences with Discrimination

Individuals' personal experiences with discrimination, ranging from subtle, everyday forms of discrimination such as being treated with less respect than others or receiving poor service (Williams, Yu, Jackson, & Anderson, 1997) to major forms of discrimination such as being unfairly fired from a job, denied a promotion, or being refused housing (Williams et al., 2008), show negative associations with mental health in a large number of studies (e.g., Pascoe & Richman, 2009; Schmitt et al., 2014; Williams et al., 2019). Perceived discrimination has been associated with a host of maladaptive psychological outcomes, including higher levels of negative outcomes such as psychological distress, depression, anxiety and negative moods, as well as lower levels of positive outcomes such as sense of control, self-esteem, life satisfaction, positive well-being and positive mood (Schmitt et al., 2014). Effects were found for racism, sexism, heterosexism, physical illness/disability, HIV status and weight, underscoring how discrimination proves pernicious across a wide range of interpersonal contexts. Other work indicates that everyday discrimination is associated with an increased likelihood of psychological disorders (Lewis, Cogburn, & Williams, 2015), including depression, post-traumatic stress disorder (PTSD), generalized anxiety disorder (GAD), social anxiety disorder, psychotic experiences and binge eating.

Research specifically on the psychological effects of experiencing discrimination in the workplace reveals similar patterns. In one study of LGBT employees (Galupo & Resnick, 2016), participants reported experiencing a range of microaggressions in their workplaces, including hearing derogatory language about the LGBT community from colleagues and supervisors and being excluded from social events. These experiences negatively impacted their mood and sense of well-being, impaired their relationships with colleagues, led them to question how they were viewed by others, decreased job satisfaction and productivity and even made them consider leaving their job entirely. Among racial minorities, experiences of racial microaggressions in the workplace led to feeling increased pressure to be perfect, less self-confidence and an internalization of these negative messages (Holder, Jackson, & Ponterotto, 2015). Immigrant professionals report experiencing microaggressions such as hearing stereotypes, skepticism about their qualifications and insensitive "jokes" about their nation/culture of origin (Shenoy-Packer, 2015); coping with these experiences led some workers to disengage through limiting their workplace contributions or creating a separate "work self." Daily diary research among Mexican immigrants living in the United States reveals that participants reported feeling more sad, angry and anxious on days when they experienced workplace discrimination (e.g., being stereotyped because they were Latino) relative to days when they did not have such experiences (Gassman-Pines, 2015). Consistent with these patterns, a recent meta-analysis (Dhanani, Beus, & Joseph, 2018) indicates that perceived workplace discrimination is associated with increased job stress, negative affectivity, turnover intentions and counterproductive work behavior and decreased job satisfaction, affective commitment and mental health.

Experiences of discrimination are also negatively associated with physical health, including self-reported physical health as well as objective indicators of health (for review and discussion, see Lick, Durso, & Johnson, 2013; Pascoe & Richman, 2009; Williams et al., 2019). Systematic and meta-analytic reviews indicate that perceived discrimination is negatively associated with indices of cardiovascular health (Lewis, Williams, Tamene, & Clark, 2014), including hypertension and nighttime ambulatory blood pressure (Dolezsar, McGrath, Herzig, & Miller, 2014) and biomarkers such as cortisol, C-reactive protein, interleukin 4 and blood glucose (Panza et al., 2019). Higher levels of self-reported discrimination are also associated with increased weight and body mass index (BMI) (de O. Bernardo, Bastos, González-Chica, Peres, & Paradies, 2017). In one example of how these linkages may emerge over time, recent longitudinal research among African-American adolescents indicates that perceived racial discrimination at ages 16–18 led to increased BMI from ages 19–21, which in turn led to insulin resistance (an indicator of cardiometabolic health) at ages 25 and 27 (Brody, Yu, Chen, Ehrlich, & Miller, 2018). These findings suggest that discrimination's negative effects on health unfold across development.

Perceived discrimination is also linked to health behaviors, which may partly contribute to these negative physical health outcomes. There is meta-analytic evidence linking experiences of discrimination with sleep problems (Slopen, Lewis, & Williams, 2016), including self-reported sleep difficulties, insomnia and fatigue, as well as objective indicators such as increased wakefulness after sleep onset (WASO). Discrimination is also associated with increased substance use. A recent meta-analysis among Black Americans (Desalu, Goodhines, & Park, 2019) shows that experiences of racial discrimination are associated with increased alcohol consumption, binge drinking, at-risk drinking and negative drinking consequences (e.g., fights, hangovers); among LGB youth, gay-related victimization such as homophobic teasing is also associated with increased substance use (Goldbach, Tanner-Smith, Bagwell, & Dunlap, 2014). Discrimination may also affect individuals' experience with and utilization of healthcare services. Perceived racial discrimination is associated with more negative healthcare experiences, including lower levels of trust in healthcare systems and professionals, poorer communication and relationships with these professionals and less satisfaction with services (Ben, Cormack, Harris, & Paradies, 2017). This meta-analysis also found some evidence that racial discrimination affects healthcare use, including delaying/not getting healthcare and not following recommended treatments. Experiences of identity-related stigma have also been associated with decreased healthcare utilization among LGBT populations (e.g., Anderson-Carpenter, Sauter, Luiggi-Hernández, & Haight, 2018; Whitehead, Shaver, & Stephenson, 2016). These patterns are concerning because they suggest that discrimination not only contributes directly to health problems, but it also undermines individuals' ability to access resources that could potentially lessen the impact of those problems.

Vicarious Discrimination and Health

Although the vast majority of research on discrimination and health focuses on how one's own discriminatory experiences are associated with health, a growing body of work is examining how *vicarious* experiences of discrimination can also serve to undermine health. Vicarious discrimination includes indirect exposure to the experiences of discrimination that happen to one's family, friends and/or strangers (e.g., Harrell, 2000). For example, given the close linkages between romantic partners' functioning and between that of parents and their children, we might expect that experiences of discrimination for one member of the family could affect the well-being of other family members. Several studies provide preliminary support for this idea. Among couples, partner discrimination – independent of one's own discrimination – has been associated with one's own self-reported health, depressive symptoms and relationship quality (e.g., Lavner, Barton, Bryant, & Beach, 2018; Wofford, Defever, & Chopik, 2019). Similar effects have been shown among parents and their children (for recent review, see Heard-Garris, Cale, Camaj, Hamati, & Dominguez, 2018). For example, in the daily diary study of Mexican immigrants described earlier (Gassman-Pines, 2015), children were rated as demonstrating more externalizing (e.g., temper outbursts) and internalizing (e.g., worrying) problems on days their parents reported more workplace discrimination. Among African-American parents and their children, parents' reports of discrimination are associated with children's distress, independent of children's own experiences of discrimination (Gibbons, Gerrard, Cleveland, Wills, & Brody, 2004) and with children's well-being (Ford, Hurd, Jagers, & Sellers, 2013). Mothers' experiences of racial discrimination are also negatively associated with indices of cardiometabolic health in their children, including higher levels of inflammation cross-sectionally (Condon et al., 2019) and over a two-year lag (Slopen et al., 2019). These patterns are concerning because they suggest that experiences of discrimination in one family member can "crossover" and affect the well-being among other family members.

Research assessing vicarious discrimination outside the family also reveals negative associations with health. One study examining experiences with unfair treatment by police found that participants who did not have this experience themselves but had a spouse, child, another relative, or close friend who had that experience (i.e., vicarious discrimination) had shorter telomere lengths (reflecting premature cellular aging) than those without these experiences (McFarland, Taylor, McFarland, & Friedman, 2018). A lab-based study found that Latino young adults exposed to indirect ethnic discrimination (i.e., overhearing confederates making ethnically-insensitive comments about a classmate) produced more cortisol during a standard stress task than those in a neutral condition without exposure to these discriminatory remarks (Huynh, Huynh, & Stein, 2017). These results are consistent with those from an earlier quasi-experimental study examining the influence of a naturally occurring race-based stressor on stress physiology (Richman & Jonassaint, 2008). Specifically, a study of African-American undergraduates at Duke University examined differences between students who participated before and after a widely-publicized incident in which an African-American woman accused several White members of the Duke lacrosse team of sexual assault. Results indicated that cortisol levels were elevated among those who participated in the study after the incident, suggesting that the racial stressor affected their physiological response. These same participants also identified less positively with their race, suggesting psychological consequences as well.

Lastly, recent research reveals negative associations between health and broad exposure to negative race-related events. Among Black Americans, for example, mental health was worse in the three months following police killings of unarmed Black Americans within the same state relative to participants assessed before the killings or more than three months after the killings (Bor, Venkataramani, Williams, & Tsai, 2018). Among Latino Americans, there is evidence that infants' birth outcomes are affected by the sociocultural context their mothers experience during pregnancy. Specifically, rates of preterm births increased among Latina women following the 2016 presidential election (Gemmill et al., 2019), likely reflecting the stress of anti-immigrant rhetoric during this period. Similar work focusing on women and infants living in a single community where an immigration raid took place found increased risk for infants' low birthweight after the raid compared to the same period one year earlier; effects were found for both U.S.A.-born and immigrant Latina mothers, with no corresponding increases for non-Latina White mothers (Novak, Geronimus, & Martinez-Cardoso, 2017). Taken together, this growing body of research on vicarious discrimination indicates that indirect experiences of discrimination, ranging from experiences of friends and family members to the broader cultural climate, can prove harmful for health and well-being.

Anticipated Discrimination and Health

Research also demonstrates that simply *anticipating* discrimination is associated with poor health outcomes. Higher levels of anticipated discrimination are associated with poorer self-reported psychological health (Lindström, 2008; Quinn et al., 2014), and research shows that racism-related vigilance (e.g., preparing for possible insults) is independently associated with sleep problems, above and beyond the effects of everyday discrimination and major experiences of discrimination (Hicken, Lee, Ailshire, Burgard, & Williams, 2013). A particularly powerful demonstration of the impact of anticipated discrimination comes from an experimental study by Sawyer and colleagues (Sawyer, Major, Casad, Townsend, & Mendes, 2012). In this study of Latina women preparing to give a speech in front of a partner, participants were initially led to believe that their partner had either negative racial attitudes or egalitarian attitudes, but then had identical actual interactions with their assigned partner (a White confederate). Nonetheless, participants who were led to believe that their partner had more negative racial attitudes reported more concern and threat emotions before the interaction, more stress after the interaction and greater cardiovascular responses than participants who were led to believe that their partner had egalitarian attitudes. Together, these findings suggest that even the anticipation of discrimination can lead to negative psychological and physiological responses. Given that some scholars have argued that vigilance to and the anticipation of discrimination may be as important in predicting negative health outcomes as the actual experience of discrimination (e.g., Williams & Mohammed, 2009), greater attention to the impact of anticipated discrimination on health is particularly important in the current sociopolitical climate of greater resistance to diversity.

Implications and Future Directions

It is clear from this discussion that resistance to diversity continues, despite efforts to promote diversity and inclusion. The continued expression of overt and covert resistance to diversity are especially concerning given robust evidence that experiences of discrimination – whether direct, vicarious, or even anticipated – are negatively associated with subjective and objective indices of psychological and physical health.

It is critical that we continue attending to the tangible and harmful consequences that diversity resistance causes for members of marginalized groups. We know much more about how overt and subtle discrimination are associated with health at a general level than we do about how discrimination affects individuals' emotional and behavioral functioning on a more micro, day-to-day level, or than we do about which individuals are particularly vulnerable to these harmful effects. We are also only just beginning to understand the effects of vicarious and anticipated discrimination – two domains that are likely heightened during periods of increased resistance to diversity. Future work addressing these issues is essential to provide a more complete picture of how discrimination affects the lives of marginalized individuals. For example, does working in an environment that seems to perform diversity but not enact inclusion lead to greater stereotype threat and in turn, worse performance (e.g., Schmader, Johns, & Forbes, 2008)? How might anticipating discrimination – a likely consequence of visible resistance to diversity – decrease minority group members' sense of authenticity and interfere with feelings of belonging and trust when working with members of majority groups? Are experiences of overt, hostile discrimination especially damaging for individuals who have a high level of internalized stigma? Addressing questions like these would help clarify how diversity resistance proves detrimental to well-being and better inform efforts to prevent these harmful effects from taking hold.

There is also a critical need to reverse the trend of growing diversity resistance, especially that which is increasingly overt, hostile and interpersonal. This return to "in your face" rejection and exclusion is too often supported by the broader community which gives the benefit of the doubt to the privileged and holds all others under suspicion. States like Oregon and California are taking proactive steps in allowing victims of social surveillance to sue those who call the police and to eliminate racial discrimination based upon attributes like hair, but much more needs to be done at a federal and state policy level to combat these trends. It is also essential for organizations to critically examine their own cultures and seek to eradicate lingering forms of diversity resistance therein. Only then will they be fully able to support the diverse employees and environments they seek to promote.

Notes

1 www.marketwatch.com/story/when-a-woman-or-person-of-color-becomes-ceo-white-men-have-a-strange-reaction-2018-02-23
2 www.russellreynolds.com/en/Insights/thought-leadership/Documents/Chief%20Diversity%20Officer_1218_FINAL.pdf
3 https://advice.shinetext.com/articles/we-need-to-talk-about-and-recognize-representation-burnout/?fbclid=IwAR0pQODXgdHaREGgcA0X4VnJ9CXKfs_23L7DRbm3UNLTZM6rSDaGr3PSX38
4 www.insidehighered.com/quicktakes/2019/07/26/noose-found-stanford-campus
5 www.washingtonpost.com/news/grade-point/wp/2017/05/04/beyond-the-realm-of-belief-noose-found-at-u-md-fraternity/?noredirect=on
6 www.eeoc.gov/eeoc/initiatives/e-race/upload/e-race-facts.pdf
7 www.usatoday.com/story/life/allthemoms/2019/02/19/new-york-teachers-suspended-on-leave-after-displaying-noose-images/2914173002/
8 www.cbsnews.com/news/racism-and-power-in-the-south/

References

Anderson-Carpenter, K. D., Sauter, H. M., Luiggi-Hernández, J. G., & Haight, P. E. (2018). Associations between perceived homophobia, community connectedness, and having a primary care provider among gay and bisexual men. *Sexuality Research and Social Policy*, 1–8.

Arnold, A. (2019, June 20). Oregon may allow victims of racist 911 calls to sue. Retrieved from www.huffpost.com/entry/oregon-racist-911-calls_n_5cf979d5e4b06af8b5059e13

Ben, J., Cormack, D., Harris, R., & Paradies, Y. (2017). Racism and health service utilisation: A systematic review and meta-analysis. *PLOS ONE, 12*, e0189900.

Bor, J., Venkataramani, A. S., Williams, D. R., & Tsai, A. C. (2018). Police killings and their spillover effects on the mental health of black Americans: A population-based, quasi-experimental study. *The Lancet, 392*, 302–310.

Brody, G. H., Yu, T., Chen, E., Ehrlich, K. B., & Miller, G. E. (2018). Racial discrimination, body mass index, and insulin resistance: A longitudinal analysis. *Health Psychology, 37*, 1107–1114.

Clark, R., Anderson, N. B., Clark, V. R., & Williams, D. R. (1999). Racism as a stressor for African Americans: A biopsychosocial model. *American Psychologist, 54*, 805–816.

Condon, E. M., Holland, M. L., Slade, A., Redeker, N. S., Mayes, L. C., & Sadler, L. S. (2019). Associations between maternal experiences of discrimination and biomarkers of toxic stress in school-aged children. *Maternal and Child Health Journal, 23*, 1–5.

de O. Bernardo, C., Bastos, J. L., González-Chica, D. A., Peres, M. A., & Paradies, Y. C. (2017). Interpersonal discrimination and markers of adiposity in longitudinal studies: A systematic review. *Obesity Reviews, 18*, 1040–1049.

Desalu, J. M., Goodhines, P. A., & Park, A. (2019). Racial discrimination and alcohol use and negative drinking consequences among Black Americans: A meta-analytical review. *Addiction, 114*, 957–967.

Dhanani, L. Y., Beus, J. M., & Joseph, D. L. (2018). Workplace discrimination: A meta-analytic extension, critique, and future research agenda. *Personnel Psychology, 71*, 147–179.

Dolezsar, C. M., McGrath, J. J., Herzig, A. J., & Miller, S. B. (2014). Perceived racial discrimination and hypertension: a comprehensive systematic review. *Health Psychology, 33*, 20–34.

Downey, S. L., van der Werff, L., Thomas, K. M., & Plaut, V. C. (2015). The roles of diversity practices and inclusion in the promotion trust and employee engagement. *Journal of Applied Social Psychology, 45*(1), 35–44.

Exec. Order No. 13672, 3 C.F.R. (2014).

Ford, K. R., Hurd, N. M., Jagers, R. J., & Sellers, R. M. (2013). Caregiver experiences of discrimination and African American adolescents' psychological health over time. *Child Development, 84*, 485–499.

Galupo, M. P., & Resnick, C. A. (2016). Experiences of LGBT microaggressions in the workplace: Implications for policy. In T. Köllen (Ed.), *Sexual orientation and transgender issues in organizations* (pp. 271–287). Cham, Switzerland: Springer.

Garrin, A. R. (2016). Hair and beauty choices of African American women during the Civil Rights Movement, 1960-1974. Doctoral Dissertation. Ames, IA: Iowa State University.

Gassman-Pines, A. (2015). Effects of Mexican immigrant parents' daily workplace discrimination on child behavior and family functioning. *Child Development, 86*, 1175–1190.

Gemmill, A., Catalano, R., Casey, J. A., Karasek, D., Alcalá, H. E., Elser, H., & Torres, J. M. (2019). Association of preterm births among U.S. Latina women with the 2016 presidential election. *JAMA Network Open, 2*, e197084–e197084.

Gibbons, F. X., Gerrard, M., Cleveland, M. J., Wills, T. A., & Brody, G. (2004). Perceived discrimination and substance use in African American parents and their children: A panel study. *Journal of Personality and Social Psychology, 86*, 517–529.

Goldbach, J. T., Tanner-Smith, E. E., Bagwell, M., & Dunlap, S. (2014). Minority stress and substance use in sexual minority adolescents: A meta-analysis. *Prevention Science, 15*, 350–363.

Griggs, B. (2018, December 28). Living while black. Retrieved from www.cnn.com/2018/12/20/us/living-while-black-police-calls-trnd/index.html

Harrell, S. P. (2000). A multidimensional conceptualization of racism-related stress: Implications for the well-being of people of color. *American Journal of Orthopsychiatry, 70*, 42–57.

Heard-Garris, N. J., Cale, M., Camaj, L., Hamati, M. C., & Dominguez, T. P. (2018). Transmitting trauma: A systematic review of vicarious racism and child health. *Social Science & Medicine, 199*, 230–240.

Hicken, M. T., Lee, H., Ailshire, J., Burgard, S. A., & Williams, D. R. (2013). "Every shut eye, ain't sleep": The role of racism-related vigilance in racial/ethnic disparities in sleep difficulty. *Race and Social Problems, 5*, 100–112.

Holder, A., Jackson, M. A., & Ponterotto, J. G. (2015). Racial microaggression experiences and coping strategies of black women in corporate leadership. *Qualitative Psychology, 2*, 164–180.

hooks, B. (2001). *Salvation: Black people and love*. New York, NY: William Morrow.

Huynh, V. W., Huynh, Q. L., & Stein, M. P. (2017). Not just sticks and stones: Indirect ethnic discrimination leads to greater physiological reactivity. *Cultural Diversity and Ethnic Minority Psychology, 23*, 425–434.

Lavner, J. A., Barton, A. W., Bryant, C. M., & Beach, S. R. (2018). Racial discrimination and relationship functioning among African American couples. *Journal of Family Psychology, 32*, 686–691.

Lewis, T. T., Cogburn, C. D., & Williams, D. R. (2015). Self-reported experiences of discrimination and health: Scientific advances, ongoing controversies, and emerging issues. *Annual Review of Clinical Psychology, 11*, 407–440.

Lewis, T. T., Williams, D. R., Tamene, M., & Clark, C. R. (2014). Self-reported experiences of discrimination and cardiovascular disease. *Current Cardiovascular Risk Reports, 8*, 365.

Lick, D. J., Durso, L. E., & Johnson, K. L. (2013). Minority stress and physical health among sexual minorities. *Perspectives on Psychological Science, 8*, 521–548.

Lindström, M. (2008). Social capital, anticipated ethnic discrimination and self-reported psychological health: A population-based study. *Social Science & Medicine, 66*, 1–13.

Lockhart, P. (2018, May 15). A white woman called the cops on a black real estate investor. Police defended him. Retrieved from www.vox.com/identities/2018/5/15/17358360/white-woman-police-black-real-estate-investor-racial-profiling-memphis-michael-hayes

Loewen, J. W. (2018). *Sundown towns: A hidden dimension of American racism*. New York City, NY: The New Press.

Masterpiece Cakeshop, Ltd. et. al. v. Colorado Civil Rights Commission. Et. al. 138 S. Ct. 1719. (2017).

McFarland, M. J., Taylor, J., McFarland, C. A., & Friedman, K. L. (2018). Perceived unfair treatment by police, race, and telomere length: A Nashville community-based sample of black and white men. *Journal of Health and Social Behavior, 59*, 585–600.

Movement Advancement Project. (2019). Employment non-discrimination [PDF file]. Retrieved from www.lgbtmap.org/equality-maps/non_discrimination_laws

Murphy, B. C. (2001). Anti-gay/Lesbian violence in the United States (28–38). In D. J. Christie, R. V. Wagner, & D. D. Winter (Eds.), *Peace, conflict, and violence: Peace psychology for the 21st century* (pp. 28–38). Saddle River, NJ: Prentice-Hall.

National Public Radio, Robert Wood Johnson Foundation, & Harvard T.H. Chan School of Public Health. (2017). Discrimination in America: Experiences and views of LGBTQ Americans [PDF file]. Retrieved from www.npr.org/documents/2017/nov/npr-discrimination-lgbtq-final.pdf

Natural Marriage Defense Act. H.B. 0892, 100th G.A. (2017).

Novak, N. L., Geronimus, A. T., & Martinez-Cardoso, A. M. (2017). Change in birth outcomes among infants born to Latina mothers after a major immigration raid. *International Journal of Epidemiology, 46*, 839–849.

Opie, T. R., & Phillips, K. W. (2015). Hair penalties: The negative influence of Afrocentric hair on ratings of Black women's dominance and professionalism. *Frontiers in Psychology, 6*, 1311.

Panza, G. A., Puhl, R. M., Taylor, B. A., Zaleski, A. L., Livingston, J., & Pescatello, L. S. (2019). Links between discrimination and cardiovascular health among socially stigmatized groups: A systematic review. *PLOS ONE, 14*, e0217623.

Papenfuss, M. (2018, June 26). People filmed and called the police on black Oakland firefighter. Retrieved from www.huffpost.com/entry/black-firefighter-oakland-police_n_5b316e53e4b0cb56051b7804

Paradies, Y., Ben, J., Denson, N., Elias, A., Priest, N., Pieterse, A., & Gee, G. (2015). Racism as a determinant of health: A systematic review and meta-analysis. *PLoS ONE, 10*, e0138511.

Pascoe, E. A., & Richman, L. S. (2009). Perceived discrimination and health: A meta-analytic review. *Psychological Bulletin, 135*, 531–554.

Patient Protection and Affordable Care Act Health-Related Portions of The Health Care and Education Reconciliation of 2010. H.R. 3590, 111th Cong. (2010).

Quinn, D. M., Williams, M. K., Quintana, F., Gaskins, J. L., Overstreet, N. M., & Chaudoir, S. R. (2014). Examining effects of anticipated stigma, centrality, salience, internalization, and outness on psychological distress for people with concealable stigmatized identities. *PLOS ONE, 9*, e96977.

Richman, L. S., & Jonassaint, C. (2008). The effects of race-related stress on cortisol reactivity in the laboratory: Implications of the Duke lacrosse scandal. *Annals of Behavioral Medicine, 35*, 105–110.

Sawyer, P. J., Major, B., Casad, B. J., Townsend, S. S., & Mendes, W. B. (2012). Discrimination and the stress response: Psychological and physiological consequences of anticipating prejudice in interethnic interactions. *American Journal of Public Health, 102*, 1020–1026.

S.B. 69, 2019 Reg. Sess. (Ala. 2019).

Schmader, T., Johns, M., & Forbes, C. (2008). An integrated process model of stereotype threat effects on performance. *Psychological Review, 115*, 336–356.

Schmitt, M. T., Branscombe, N. R., Postmes, T., & Garcia, A. (2014). The consequences of perceived discrimination for psychological well-being: A meta-analytic review. *Psychological Bulletin, 140*, 921–948.

Sessions, J. B. (2017b). *U.S. Department of Justice. Revised treatment of transgender employment discrimination claims under title VII of the civil rights act of 1964.* Washington, DC: The Attorney General.

Shenoy-Packer, S. (2015). Immigrant professionals, microaggressions, and critical sensemaking in the U.S. workplace. *Management Communication Quarterly, 29*, 257–275.

Slopen, N., Lewis, T. T., & Williams, D. R. (2016). Discrimination and sleep: A systematic review. *Sleep Medicine, 18*, 88–95.

Slopen, N., Strizich, G., Hua, S., Gallo, L. C., Chae, D. H., Priest, N., ... Daviglus, M. L. (2019). Maternal experiences of ethnic discrimination and child cardiometabolic outcomes in the Study of Latino Youth. *Annals of Epidemiology, 34*, 52–57.

Thomas, K. M. (2008a, August). Nooses (invited). *Communique* (Special Issue on Psychology and Racism). APA's Office of Ethnic Minority Affairs. X–XIII.

Thomas, K. M. (2008b). *Diversity resistance in organizations.* [Applied Psychology Series]. New York City, NY: LEA-Taylor Francis.

Thomas, K. M., & Plaut, V. C. (2008). The many faces of diversity resistance. In K. M. Thomas (Ed.), *Diversity resistance in organizations* (pp. 1–22). [Applied Psychology Series]. New York City, NY: LEA-Taylor Francis.

Trayvon Martin Shooting Fast Facts. (2019, February 28). Retrieved from www.cnn.com/2013/06/05/us/trayvon-martin-shooting-fast-facts/index.html

U.S. Department of Health and Human Services. (2019, May 24). HHS proposes to revise ACA section 1557 rule to enforce civil rights in healthcare, conform to law, and eliminate billions in unnecessary costs. Retrieved from www.hhs.gov/about/news/2019/05/24/hhs-proposes-to-revise-aca-section-1557-rule.html#

U.S. Department of Housing and Urban Development. (2012, February 3). Equal access to housing in HUD programs regardless of sexual orientation or gender identity. Retrieved from www.federalregister.gov/documents/2012/02/03/2012-2343/equal-access-to-housing-in-hud-programs-regardless-of-sexual-orientation-or-gender-identity

U.S. Department of Housing and Urban Development. (2019 Spring). Revised requirements under community planning and development housing programs (FR-6152). Retrieved from www.reginfo.gov/public/do/eAgendaViewRule?publ d=201904&RIN=2506-AC53

U.S. Department of Justice. (2011, February 23). Statement of the attorney general on litigation involving the Defense of Marriage Act. Retrieved from www.justice.gov/opa/pr/statement-attorney-general-litigation-involving-defense-marriage-act

Victor, D. (2018, May 08). A woman said she saw burglars. They were just black Airbnb guests. Retrieved from www.nytimes.com/2018/05/08/us/airbnb-black-women-police.html?action=click&module=inline&pgtype=Article

Welch, K. (2007). Black criminal stereotypes and racial profiling. *Journal of Contemporary Criminal Justice, 23*(3), 276–288.

Whitehead, J., Shaver, J., & Stephenson, R. (2016). Outness, stigma, and primary health care utilization among rural LGBT populations. *PLOS ONE, 11*, e0146139.

Williams, D., & Wade-Golden, K. (2007). The chief diversity officer. *CUPA HR Journal, 58*(1), 38–48.

Williams, D. R., González, H. M., Williams, S., Mohammed, S. A., Moomal, H., & Stein, D. J. (2008). Perceived discrimination, race and health in South Africa: Findings from the South Africa Stress and Health Study. *Social Science and Medicine*, *67*, 441–452.

Williams, D. R., & Mohammed, S. A. (2009). Discrimination and racial disparities in health: Evidence and needed research. *Journal of Behavioral Medicine*, *32*, 20–47.

Williams, D. R., Lawrence, J. A., & Davis, B. A. (2019). Racism and health: Evidence and needed research. *Annual Review of Public Health*, *40*, 105–125.

Williams, D. R., Yu, Y., Jackson, J. S., & Anderson, N. B. (1997). Racial differences in physical and mental health: Socioeconomic status, stress, and discrimination. *Journal of Health Psychology*, *2*, 335–351.

Wilson, J. L. (2013). Emerging trend: The chief diversity officer phenomenon within higher education. *The Journal of Negro Education*, *82*(4), 433–445.

Wofford, N., Defever, A. M., & Chopik, W. J. (2019). The vicarious effects of discrimination: How partner experiences of discrimination affect individual health. *Social Psychological and Personality Science*, *10*, 121–130.

Wootson, C. R., Jr. (2018, May 11). A black Yale student fell asleep in her dorm's common room. A white student called police. Retrieved from www.washingtonpost.com/news/grade-point/wp/2018/05/10/a-black-yale-student-fell-asleep-in-her-dorms-common-room-a-white-student-called-police/?noredirect=on&utm_term=.9200f559b629

2 Unapologetic Authentic Early Career Black Women

Challenging the Dominant Narrative

Danielle D. Dickens and Veronica Y. Womack

Introduction

The negative portrayals of Black women in mainstream media unconsciously and consciously dictate how Black women present themselves, particularly in the workplace. Studies have reported that Black women may struggle with the dual tasks of forming a positive sense of self, while counteracting negative images (e.g., Reynolds-Dobbs, Thomas, & Harrison, 2008). As an example, Kara, a 26-year-old Black female employee in higher education states that,

> If you can't be your authentic self at work, you are walking on eggshells for other people, which is okay at times if you want to spare people's feelings, but why can't I voice my opinion without repercussions?

Kara's story is unfortunately a common one among early career Black women who feel pressured to suppress their ideas and authentic self at work due to the fear of being misjudged or that it would impact their job performance review (Jones & Shorter-Gooden, 2003). This notion is consistent with previous research which suggests that women of color experience a "concrete ceiling," which is a thick and invisible barrier to advancement to senior level positions (Sanchez-Hucles & Davis, 2010). Three of the most cited barriers that influence the experiences of early career Black women in the workplace are unequal access to career opportunities (i.e. hiring, promotions, choice assignments), lack of mentorship or support and gendered racial stereotypical expectations (i.e. nurturing, aggressive, asexual, overly sexual, strong) (Hall, Everett, & Hamilton-Mason, 2012; Holder, Jackson, & Ponterotto, 2015; Lewis & Neville, 2015; Reynolds-Dobbs, Thomas, Harrison, 2008; Thomas & Hollenshead, 2001). Hence, it is possible that biases, such as the dominant narrative of Black women as the "mammy" or angry Black woman (Thomas, Witherspoon, & Speight, 2008), can impact their experiences in the workplace.

Dominant cultural narratives are defined as "overlearned stories communicated through mass media or other large social and cultural institutions and social networks" (Rappaport, 2000, p. 3). One dominant cultural narrative is the depiction of Black women as being strong, assertive and independent (Baker, Buchanan, Mingo, Roker, & Brown, 2014). Two studies in particular have shown that in cohorts of Black women, endorsement of the "Strong Black Woman" (SBW) stereotype predicts increased stress, depressive symptoms and psychological distress (Donovan & West, 2015; Watson-Singleton, 2017). These findings show that the pressure to uphold this one-dimensional image and not show vulnerability in the workplace can jeopardize one's mental health. Since a Black woman's workplace may be a setting where she feels the most pressure to "wear the mask" and adhere to societal expectations, Black early career women can resist by choosing to be their authentic self in the workplace by having honest communications at work.

Resistance by individuals from marginalized groups, refers to responses and behaviors that individuals and groups use to counteract oppression (Hasford, 2016). Moreover, Case and Hunter (2012) conceptualize resistance as adaptive responses to adversity, such as coping strategies and resilience (positive outcomes in spite of social adversity). Specifically, to offset the negative psychological outcomes associated with gendered racism at work, some Black women cope by engaging in identity shifting, the process of altering one's language and behavior in order to fit within the social norms of an environment, manage stereotypes and accommodate others (Dickens & Chavez, 2018; Jones & Shorter-Gooden, 2003). For example, early career Black women may decide to straighten their natural hair before an important meeting or alter their language while speaking with their non-Black colleagues. As supported in the literature, when individuals experience negative evaluations based on their identities, this may cause individuals to manage their identities to mitigate the negative discriminatory experiences (Ely & Roberts, 2008). In a research study that examined identity shifting among early career Black women, one woman spoke of changing the tone of her voice daily at work with non-Black colleagues so that she did not appear too aggressive (Dickens & Chavez, 2018). This further illustrates how Black women may be pressured to alter their appearance or behaviors to manage stereotypes associated with Black women, to not be perceived negatively by their colleagues in the workplace. Previous research has shown that consistently altering one's actions and speech can have negative outcomes, such as feelings of inauthenticity and a lack of sense of belonging (Dickens & Chavez, 2018; Whitfield-Harris, Lockhart, Zoucha, & Alexander, 2017). In challenging these stereotypical images and experiences of gendered racial discrimination, Black women may resist the internalization of these stereotypes by prioritizing one's self through the creation of a positive and authentic self-image at work. Research by Ménard and Brunet (2011) identified a positive association between being authentic at work and psychological well-being, showing further support for the benefits of authenticity.

Authenticity refers to being one's true self and acting in congruence with one's self and values (Ménard & Brunet, 2011). Specifically, when Black women show up to work as their authentic selves, they are "more comfortable in their skin" and are more likely to provide suggestions and recommendations from their unique vantage point, rather than focusing on how what they say might be perceived in a particular way from others (Smith, 2018). Goldman and Kernis (2002) add that authenticity is the unobstructed operation of one's true or core self in one's daily interactions. We view authenticity as both a cognitive (awareness) and behavioral (action) dimension (Ménard & Brunet, 2011) and as a form of resistance as it puts the power back into the hands of Black women, allowing them to speak their mind and to present themselves freely at work. In addition, research also suggests the importance of an authentic leadership style, a genuine interpersonal relationship between the leader and the follower, for women of color at work (Illies, Morgeson, & Nahrgang, 2005). As such, Giles (2017) interviewed senior career status Black women in the legal field and these women expressed the importance of being an authentic leader in relation to their career advancement. These women illustrated authentic leadership by leading with their values, such as being unapologetic about prioritizing their family over their career. In this chapter, we sought to emphasize the ways in which early career Black women may resist the dominant cultural narrative by highlighting resistance through coping strategies, the development of a positive identity, authenticity and authentic leadership in the workplace, and we provide recommendations for employers in building a culture of inclusion that promotes authenticity for early career Black women and other underrepresented groups in the workplace.

Resistant Coping Strategies among Early Career Black Women

In order to navigate the workplace, Black women proactively engage in strategies to manage their expectations, promote positive affect and move forward in their careers (Dickens & Chavez, 2018; Johnson et al., 2015; Shih, Young, & Bucher, 2013). For instance, qualitative studies show that Black women use a range of strategies to manage social stressors, including: altering outward appearance, behavior, or presentation also known as "shifting," relying on prayer and spirituality, avoiding contact with certain people or situations, standing up/using one's voice as power, sustaining a positive self-image/self-care and engaging in social support networks (Hall et al., 2011; Holder et al., 2015; Lewis, Medenhall, Harwood, Huntt, 2013; Shavers and Moore, 2014; Shorter-Gooden, 2004). In regards to the outcomes of the selected coping strategies among Black women, Thomas et al. (2008) reported that coping through avoidance of thinking about an issue or thinking about other things predicts psychological distress.

Alternatively, some coping strategies utilized by Black women are intentionally applied to resist disempowering gendered racial narratives or assumptions (Lewis et al., 2013). As such, Linnabery and colleagues (2014) revealed that both social support (including supervisor, co-worker, church and family) and self-help coping (i.e. strategies aimed at sustaining one's emotional well-being in stressful circumstances) predicted career satisfaction and life satisfaction in a sample of professional Black women. Hence, due to the high utilization of social media by Black Americans between the ages of 18 and 29 (Pew Research Center, 2014), social support and self-help coping strategies are not only more accessible, but can better position individuals to create counter-spaces in the midst of the dominant cultural narratives that they may be exposed to in the workplace (Solorzano, Ceja, & Yosso, 2000).

Contemporary scholarship has demonstrated how dominant cultural narratives in the media contribute to the marginalization of subordinate groups (McDonald, Keys, & Balcazar, 2007). Therefore, these narratives can increase feelings of isolation and create hostile and invalidating work environments for Black women (Hall, Everett, Hamilton-Mason, 2012; Holder et al., 2015; Terhune, 2008). We argue that the accessibility of affirming, culturally relevant social media content curated by other Black women may be a particularly useful form of resistance for early career Black women navigating gendered racial barriers in the workplace. For instance, *Black Career Women's Network* (https://bcwnetwork.com), a national organization dedicated to fostering the professional development of Black women, and *Beneath the Façade* (www.beneaththefacade.org), a psycho-educational resource for Black women coping with stereotypical expectations at work, are two examples of online resources that were designed by and for Black women to promote career development and help Black women combat discrimination in the workplace.

On various social media outlets, Black women also resist dominant narratives by contributing to national conversations by recalling experiences of workplace discrimination using the hashtag #BlackWomenAtWork (Moseley, 2017). In March 2017, politician Maxine Waters was mocked by Fox's Bill O'Reilly for her appearance on his show, comparing her hair to the singer James Brown's hair (Moseley, 2017). In turn, Black women all over the world started a virtual movement by sharing their stories of discrimination in the workplace via social media using the hashtag #BlackWomenAtWork. For instance, posts described the common scenario of Black women stereotyped as service workers, "#BlackWomenatWork A guy from another office told my mom to make him coffee. He didn't know she was the regional buyer aka HIS BOSS'S BOSS." Another comment stated, "#BlackWomenAtWork Constantly getting mistaken for his personal assistant/secretary when we are both qualified quantity surveyors." These examples of #BlackWomenatWork illustrate "using one's voice as power" (UVOP) as a form of resistance. Lewis and colleagues (2013) describe UOVP as a resistant coping strategy where the individual speaks up and directly addresses a microaggression, a subtle form of discrimination, to the perpetrator as a way to regain power in the situation. Although Black women may not directly address the perpetrator, publicly telling their stories about subtle and overt hostilities in the workplace and seeing similar stories from other Black women, can be cathartic, affirming and brings national attention to the plight of Black women in the workplace.

Resistance through the Integration of Positive Professional, Racial and Gender Identities

Another way in which early career Black women reject the dominant narrative is through the development of a positive sense of self. The Black feminist approach is a narrative inquiry that involves resisting stereotypical images of Black womanhood, allowing Black women to validate their power, thereby reclaiming their womanhood through positive self-definition and exploration (Hill Collins, 1990). One way in which Black women can develop a positive self-definition is through the development of their racial and gender identities. The Multidimensional Inventory of Black Identity (MIBI) characterizes Black racial identity according to four dimensions: salience, public regard, private regard, ideology and centrality (Sellers, Rowley, Chavous, Shelton, & Smith, 1997). In particular, studies have shown that private regard (the extent to which individuals feel positively or negatively towards African-Americans as well as being African-American) and public regard (the extent to which an individual feel that others view their racial group positively and/or negatively) predict psychological outcomes among Black women. For example, higher racial private regard, feeling positive about being Black, is related to lower perceived stress, lower depression (Caldwell, Zimmerman, Bernat, Sellers, & Notaro, 2002; Settles, Navarrete, Pagano, Abdou, & Sidanius, 2010), racial group belongingness and life satisfaction (Yap, Settles, Pratt-Hyatt, 2011). The MIBI has also been adapted to explore racial and gender identities among Black women. For example, Jones and Day (2018) used the MIBI model to examine racial and gender identity development among Black women and results showed that Black women perceive both their race and gender to be important components of their identity. Given what we know about the association between private regard and life satisfaction and the importance of race and gender identities, it is important for Black women to construct meaningful identities by embracing both their racial and gender identities.

In the development of a positive racial and gender identity, scholars have also explored how Black career women integrate their professional identity with their race and gender identities. As such, Shorter-Gooden (1996) conducted interviews with eight young women regarding their identity development and noted that career identity was important for two-thirds of the sample. All of the Black women in the study discussed having an identity that included domains of race, gender and career goals. Furthermore, research has shown that professional identity, one's professional self-concept based on one's attributes, beliefs, values, motives and experiences, is more adaptable early in one's career (Ibarra, 1999; Schein, 1978). When a professional identity is not yet fully developed, others' reactions or external evaluations can inform how an individual views themselves and their legitimacy in their given fields (Ibarra, 1999; Markus & Nurius, 1986). These findings highlight the key role that a positive professional identity can play for early career Black women in resisting the stereotypical narratives from the dominant culture.

Authenticity as a Form of Resistance: Benefits and Challenges for Early Career Black Women

In the development of a positive professional identity, authenticity is important to consider. Authenticity is conceptualized as owning one's personal experiences (values and thoughts) and acting in accordance with one's true self (Harter, 2002). Likewise, Gardner, Avolio, Luthans, May and Walumbwa (2005) define authenticity as optimal self-esteem characterized as genuine, true, stable and congruent high esteem. Previous research has shown that there are positive associations between authenticity, authentic leadership and psychological well-being (Illies et al., 2005). Furthermore, authentic leadership is described as achieving authenticity, and encompasses authentic relationships with followers and associates. These authentic relationships are depicted as transparent, open, trustworthy, as guidance towards worthy objectives and as an emphasis on follower development (Gardner et al., 2005). In general, these findings suggest that there may be psychological and career advancement benefits for having an authentic identity and engaging in authentic leadership at work for early career Black women.

Benefits of Being Authentic at Work

Being able to be authentic at work is not only essential for good leadership but promotes a sense of belonging which can lead to job satisfaction and job longevity. For people who hold marginalized identities, there is a pressure to assimilate to the dominant culture in the workplace. Thus, being authentic is a form of freedom and resistance to the cultural norms in a given workplace organization. Authenticity is an important component of the formation of social relations in the workplace to ensure that one's interests, characteristics and identities are recognized and valued by an organization (Archer, 2008). Trudy Bourgeois, CEO of the Center for Workforce Excellence, recommends that when Black women know their values and are able to communicate their values, it can open doors and creates opportunities that can level the playing field. Bourgeois also argues that when Black women know who they are, their values and how they can make an impact on the world, others will want to give them career opportunities (Smith, 2008).

In addition, when individuals engage in positive and authentic interactions more frequently, they are likely to thrive in their careers (Shih, Young, & Bucher, 2013). On the other hand, inauthenticity may pose psychological stress on employees, such as that minorities who feel that they should behave inauthentically (suppressing ideas and cultural values) may experience identity conflict (Bell & Nkomo, 2001; Roberts, Cha, Hewlin, & Settles, 2011). When a Black woman chooses to be her authentic self at work, she is resisting the counternarratives and taking pride in her cultural identities. In all, there is psychological empowerment in bringing one's authentic self to work, which can lead to genuine relationships at work, enhance work performance and lead to higher levels of job satisfaction (Giles, 2017). Therefore, this literature speaks to the benefits of the promotion of authenticity at work for organizations and its members, and the larger goal of increasing and broadening the participation and retention of individuals from underrepresented groups, such as early career Black women, into White male-dominated organizations.

Challenges of Being Authentic at Work

Despite the fact that there are physical, psychological and performance benefits associated with authenticity and authentic leadership at work, there can be some challenges for minorities, particularly early career Black women, with adopting an authentic professional identity at work. As an example, Black women in leadership positions are in a double bind because as leaders they may struggle to develop an appropriate and effective leadership style that integrates communal qualities (e.g., caring, sensitive and understanding), a preferred leadership style for women, with agentic qualities (e.g., dedicated, aggressive and competitive), a preferred leadership style for men. Consequently, it is difficult to maintain this modification while maintaining a sense of authenticity as a leader in an organization (Eagly & Carli, 2008). Hence, there is a no-win situation and there may be a penalty for Black women whose behaviors are consistent with the traditional male behaviors. Consider if a Black woman were to be aggressive to show authority, it is possible that she may be disliked by her non-woman and non-Black colleagues.

Furthermore, it is argued that early career professionals, due to their positions, are less powerful in the workplace, which brings up challenges with discourse around authenticity. For example, within academia it is argued that boundaries of authenticity are policed by people within powerful positions, such as administrators in leadership positions (e.g., department chair or dean of college) (Archer, 2008). Thus, there may be certain consequences for early career Black women who are being their authentic selves, such as being denied a promotion. Moreover, Phillips, Dumas, and Rothbard (2018) examined the obstacles to self-disclosure among young racial minority professionals, and the findings reflected a fear that revealing personal information highlighting their race might reinforce the stereotypes that can undermine performance reviews. As an example, Karen, a Black woman in higher education, was at a birthday party her colleagues threw for her and a White co-worker asked her what she did for her birthday. She did not respond to the question because she did not feel comfortable telling her colleagues that she went to see Kirk Franklin, a Black gospel artist, because she did not want to be misjudged (Phillips et al., 2018). Despite these challenges with presenting an authentic professional identity, the benefits of authenticity, such as job satisfaction and work enhancement, outweigh the costs.

The Role of Employers in Promoting Authenticity in Organizations

The question remains, what can be done to solve this issue of inequity and exclusivity that can take place within organizational cultures? It is important for

organizations to create inclusive environments that promote authenticity, especially for early career professionals, since authenticity at work is highly valued
by this generation, not only at work, but also in life (Hershatter & Epstein,
2010). Responding to this charge by hiring more members from underrepresented groups will not ensure that they will feel comfortable and have the necessary resources and opportunities for career advancement (Phillips, Dumas, &
Rothbard, 2018). Thus, companies must create an inclusive workplace culture
that embraces cultural differences and diverse perspectives.

In the recruitment and retention of Black women, it is critical to consider the role of mentors in their career advancement. Research shows
that having a mentor is linked to career advancement, salary increase and
greater career/life satisfaction (Marina, 2016). In turn, organizations also
benefit from mentorship in gaining future leaders that are better prepared
to enter and thrive in the workforce (Bova, 1998; Sanchez-Hucles &
Davis, 2010). As an example, Oliver, a millennial Black female entrepreneur, states that in order to bring one's authentic self to work, finding
a support system is key. "Find yourself in places where people are growing along with you and remove yourself from toxic places" (Gumbs,
2018). This speaks to the importance of formal and informal mentors.
Accordingly, companies should consider investing in formal mentorship
and sponsorship programs that target minorities to facilitate building relationships and community among employees of different cultural backgrounds, including but not limited to race, gender, sexual orientation, age,
ability, nationality and religion (Phillips et al., 2018). In the literature, it
shows that companies who are committed to attracting and retaining
ethnic minorities should consider a formal mentoring program (Heidrick
& Struggles, 2007). As an example, PayPal's Unity Mentorship program
was created to build an inclusive workplace culture for women. The
employee-led initiative matches mentors and mentees from the same or
different departments to create a space for women to share knowledge
and to find a support system in a male-dominated tech industry. We
would take it a step further and recommend that companies develop
a mentoring program that is geared towards women of color because this
population is disproportionately represented in leadership positions and
experience the double discrimination based on their race and gender (Sanchez-Hucles & Davis, 2010).

In addition to creating targeted diversity mentoring programs, organizations
should promote the use of microaffirmations. Much of the literature in psychology has focused on pro-racist behaviors known as microaggressions, which
are subtle forms of harsh and non-verbal exchanges that are "put downs" (Jones
& Rolón- Dow, 2018). An example of a microaggression may include asking
a Black woman, "Who did you know to get this job?" This microaggression
implies that she only got this position at her job because she knew someone.
Alternatively, a counterstory to microaggressions is the use of microaffirmations,
small acts of verbal and non-verbal acts of support and encouragement

(Pittinsky, 2016). An example of a microaffirmation includes, "I liked your opinion about … " or "I am glad that you are here." Microaffirmations have positive impacts on the lives of people from marginalized groups by promoting their success by affirming and validating their identities (Jones & Rolón- Dow, 2018). Taken together, these are just a few examples of ways in which organizations can create an inclusive workplace that promotes authenticity and authentic leadership for all of its employees. Though this literature review focuses on Black women, these suggestions apply to the professional experiences of other individuals who find themselves the token or minority at work, such as other women of color, Black men, persons who openly identify as LGBTQ and people with a disability, to name a few.

Conclusion

As the rates for college enrollment and graduation are increasing for Black women, the diversity of people entering into the workforce is changing and on the rise. The practical significance of this paper provides evidence of the unique challenges that early career Black women face in the workplace regarding the integration of a positive professional identity that is authentic. Presenting an authentic identity at work is a representation of "progressive, oppositional, and constructive political action taken on by women who challenge gender, racial, sexual, age, national, ethnic and other forms of oppression on multiple levels of scale" (Isoke, 2013, p. 27). The essence of wellness in the workplace promotes authenticity, vulnerability and strength and has positive implications for Black women's psychological well-being. Perhaps, a grasp of the mechanisms associated with identity formation and presentation of Black womanhood in the workplace can assist with the understanding and retention of Black women and other minorities in workplace organizations. More research on this topic can contribute to our understanding of how early career Black women face a difficult dilemma of simultaneously being successful Black women while confronting long-standing stereotypes and prejudices in the workplace. This work voices the politics of belonging by early career Black women in the workplace and their efforts to resist and disrupt hegemonic and patriarchal notions of Black womanhood and identity by choosing wellness and by being their true, authentic selves at work.

References

Archer, L. (2008). Younger academics' constructions of 'authenticity', 'success' and professional identity. *Studies in Higher Education*, *33*(4), 385–403. doi:10.1080/03075070802211729

Baker, T. A., Buchanan, N. T., Mingo, C. A., Roker, R., & Brown, C. S. (2014). Reconceptualizing successful aging among black women and the relevance of the strong black woman archetype. *The Gerontologist*, *55*(1), 51–57. doi:10.1093/geront/gnu105

Bell, E., & Nkomo, S. (2001). *Our separate ways: Black and white women and the struggles for professional identity*. Boston, MA: Harvard Business School Press.

Bova, B. M. (1998). Mentoring revisited: The African-American woman's experience. *Adult education research conference*. Retrieved from https://newprairiepress.org/aerc/1998/papers/1

Brack, J., & Kelly, K. (2012). Maximizing millennials in the workplace. *UNC Executive Development*, *22*(1), 2–14. www.gandyr.com/wp-content/uploads/2016/12/maximizing-millennials-in-the-workplace.pdf

Caldwell, C. H., Zimmerman, M. A., Bernat, D. H., Sellers, R. M., & Notaro, P. C. (2002). Racial identity, maternal support, and psychological distress among African American adolescents. *Child Development*, *73*(4), 1322–1336. doi:10.1111/1467-8624.00474

Case, A. D., & Hunter, C. D. (2012). Counterspaces: A unit of analysis for understanding the role of settings in marginalized individuals' adaptive responses to oppression. *American Journal of Community Psychology*, *50*(1-2), 257–270. doi:10.1007/s10464-012-9497-7

Cross, W. E., Jr. (1991). *Shades of black: Diversity in African-American identity*. Philadelphia, PA: Temple University Press.

Dickens, D. D., & Chavez, E. L. (2018). Navigating the workplace: The costs and benefits of shifting identities at work among early career us black women. *Sex Roles*, *78*(11-12), 760–774. doi:10.1007/s11199-017-0844-x

Donovan, R. A., & West, L. M. (2015). Stress and mental health: Moderating role of the strong black woman stereotype. *Journal of Black Psychology*, *41*(4), 384–396. doi:10.1177/0095798414543014

Eagly, A., & Carli, L. (2008). Women and the labyrinth of leadership. *Harvard Business Review*, 147–162. https://hbr.org/2007/09/women-and-the-labyrinth-of-leadership

Ely, R. J., & Roberts, L. M. (2008). Shifting frames in team-diversity research: From difference to relationships. In A. Brief (Ed.), *Diversity at work* (pp. 175–202). Cambridge, UK: Cambridge University Press.

Gardner, W. L., Avolio, B. J., Luthans, F., May, D. R., & Walumbwa, F. (2005). "Can you see the real me?" A self-based model of authentic leader and follower development. *The Leadership Quarterly*, *16*(3), 343–372. doi:10.1016/j.leaqua.2005.03.003

Giles, J. (2017). The intersection of race, gender, and leadership in the legal field. Retrieved from http://commons.emich.edu/honors/529

Goldman, B. M., & Kernis, M. H. (2002). The role of authenticity in healthy psychological functioning and subjective well-being. *Annals of the American Psychotherapy Association*, *5*(6), 18–20.

Gumbs, A. (2018, April 24). Millenial moves: How to stay authentic at work. Retrieved from www.blackenterprise.com/millennial-moves-authentic-work/

Hall, J. C., Everett, J. E., & Hamilton-Mason, J. (2012). Black women talk about workplace stress and how they cope. *Journal of Black Studies*, *43*(2), 207–226. doi:10.1177/0021934711413272

Harter, S. (2002). Authenticity. In C. R. Snyder, & S. Lopez (Eds.), *Handbook of positive psychology* (pp. 382–394). Oxford, UK: Oxford University Press.

Hasford, J. (2016). Dominant cultural narratives, racism, and resistance in the workplace: A study of the experiences of young Black Canadians. *American Journal of Community Psychology, 57*(1-2), 158–170. doi:10.1002/ajcp.12024

Heidrick and Struggles. (2007). Corporate governance in Europe: 2007 report, Paris.

Hershatter, A., & Epstein, M. (2010). Millennials and the world of work: An organization and management perspective. *Journal of Business and Psychology, 25*(2), 211–223. doi:10.1007/s10869-010-9160-y

Hill Collins, P. (1990). Black feminist thought in the matrix of domination. *Black feminist thought: Knowledge, consciousness, and the politics of empowerment*, 221–238.

Holder, A., Jackson, M. A., & Ponterotto, J. G. (2015). Racial microaggression experiences and coping strategies of Black women in corporate leadership. *Qualitative Psychology, 2*(2), 164–180. doi:10.1037/qup0000024

Ibarra, H. (1999). Provisional selves: Experimenting with image and identity in professional adaptation. *Administrative Science Quarterly, 44*(4), 764–791.

Illies, R., Morgeson, F. P., & Nahrgang, J. D. (2005). Authentic leadership and eudaemonic well being: Understanding leader–follower outcomes. *The Leadership Quarterly, 16*(3), 373–394. doi:10.1016/j.leaqua.2005.03.002

Isoke, Z. (2013). *Urban black women and the politics of resistance*. New York City, NY: Palgrave Macmillan.

Johnson, J. C., Gamst, G., Meyers, L. S., Arellano-Morales, L., & Shorter-Gooden, K. (2016). Development and validation of the African American Women's Shifting Scale (AAWSS). *Cultural Diversity and Ethnic Minority Psychology, 22*(1), 11–25. doi.org/10.1037/cdp0000039

Jones, C., & Shorter-Gooden, K. (2003). *Shifting: The double lives of African American women in America*. Darby, PA: Diane Publishing Company.

Johnson, J. C., Gamst, G., Meyers, L. S., Arellano-Morales, L., & Shorter-Gooden, K. (2016). Development and validation of the African American Women's Shifting Scale (AAWSS). *Cultural Diversity and Ethnic Minority Psychology, 22*(1), 11–25. doi.org/10.1037/cdp0000039

Jones, J., & Rolón- Dow, R. (2018). Multidimensional models of microaggressions and microaffirmations. In C. M. Capodilupo, K. L. Nadal, D. P. Rivera, D. W. Sue, & G. C. Torino *Microaggression theory: Influence and implications* (pp. 32–41). Hoboken, NJ: Wiley.

Jones, M. K., & Day, S. X. (2018). An exploration of Black women's gendered racial identity using a multidimensional and intersectional approach. *Sex Roles, 79*(1-2), 1–15. doi:10.1007/s11199-017-0854-8

Lewis, J. A., Mendenhall, R., Harwood, S. A., & Huntt, M. B. (2013). Coping with gendered racial microaggressions among Black women college students. *Journal of African American Studies, 17*(1), 51–73. doi:10.2307/2667055

Lewis, J. A., & Neville, H. A. (2015). Construction and initial validation of the gendered racial microaggressions scale for black women. *Journal of Counseling Psychology, 62* (2), 289–302. doi:10.1037/cou0000062

Linnabery, E., Stuhlmacher, A. F., & Towler, A. (2014). From whence cometh their strength: Social support, coping, and well-being of Black women professionals. *Cultural Diversity and Ethnic Minority Psychology, 20*(4), 541–549. doi:10.1037/a0037873

Marina, B. (2016). Mentoring away the glass ceiling in academia: A cultured critique edited by Brenda LH Marina. *InterActions: UCLA Journal of Education and Information Studies, 1*.

Markus, H., & Nurius, P. (1986). Possible selves. *American Psychologist*, *41*(9), 954.

McDonald, K. E., Keys, C. B., & Balcazar, F. E. (2007). Disability, race/ethnicity and gender: themes of cultural oppression, acts of individual resistance. *American Journal of Community Psychology*, *39*(1-2), 145–161. doi.org/10.1007/s10464-007-9094-3

Ménard, J., & Brunet, L. (2011). Authenticity and well-being in the workplace: A mediation model. *Journal of Managerial Psychology*, *26*(4), 331–346. doi:10.1108/02683941111124854

Moseley, M. (2017, March 29). #BlackWomenAtWork is trending on Twitter and proving that not much has changed. Retrieved from www.essence.com/news/black-women-at-work-maxine-waters-hashtag-twitter/

Pew Research Center. (2014). African Americans and technology use. Retrieved from www.pewinternet.org/2014/01/06/african-americans-and-technology-use/

Phillips, K. W., Dumas, T. L., & Rothbard, N. P. (2018). Minorities hesitate to share information about themselves at work. That's a problem for everyone. *Harvard Business Review*, *96*(2), 132–136.

Pittinsky, T. L. (2016). Backtalk: Why overlook microaffirmations? *Phi Delta Kappan*, *98*(2), 80–81. doi:10.1177/0031721716671918

Rappaport, J. (2000). Community narratives: Tales of terror and joy. *American Journal of Community Psychology*, *28*(1), 1–24. doi:0.1023/A:1005161528817

Reynolds-Dobbs, W., Thomas, K. M., & Harrison, M. S. (2008). From mammy to superwoman: Images that hinder black women's career development. *Journal of Career Development*, *35*(2), 129–150. doi:10.1177/0894845308325645

Roberts, L. M., Cha, S. E., Hewlin, P. F., & Settles, I. H. (2011). Bringing the inside out enhancing authenticity and positive identity in organizations. In J. Dutton, & L. M. Roberts (Eds.), *Exploring positive identities and organizations: Building a theoretical and research foundation* (pp. 149–169). New York, NY: Psychology Press.

Sanchez-Hucles, J. V., & Davis, D. D. (2010). Women and women of color in leadership: Complexity, identity, and intersectionality. *American Psychologist*, *65*(3), 171–181. doi:10.1037/a0017459

Schein, E. H. (1978). *Career dynamics: Matching individual and organizational needs*. Reading, MA: Addison Wesley Publishing Company.

Sellers, R. M., Rowley, S. A., Chavous, T. M., Shelton, J. N., & Smith, M. A. (1997). Multidimensional inventory of black identity: A preliminary investigation of reliability and construct validity. *Journal of Personality and Social Psychology*, *73*(4), 805–815. doi:10.1037/0022-3514.73.4.805

Settles, I. H., Navarrete, C. D., Pagano, S. J., Abdou, C. M., & Sidanius, J. (2010). Racial identity and depression among African American women. *Cultural Diversity and Ethnic Minority Psychology*, *16*(2), 248–255. doi:10.1037/a0016442

Shante, D. (2018, March 14). I'm exhausted from trying to be the 'right' kind of black girl at work. Retrieved May 29, 2018 from www.huffpost.com/entry/codes witching-while-black-at-work_n_5aa2b7dce4b07047bec60c5c

Shavers, M. C., & Moore, J. L., III. (2014). Black female voices: Self-presentation strategies in doctoral programs at predominately white institutions. *Journal of College Student Development*, *55*(4), 391–407. doi:10.1353/csd.2014.0040

Shih, M., Young, M. J., & Bucher, A. (2013). Working to reduce the effects of discrimination: Identity management strategies in organizations. *American Psychologist*, *68*, 145–157. doi:10.1037/a0032250

Shorter-Gooden, K. (2004). Multiple resistance strategies: How African American women cope with racism and sexism. *Journal of Black Psychology*, *30*(3), 406–425. doi:10.1177/0095798404266050

Shorter-Gooden, K., & Washington, N. C. (1996). Young, Black, and female: The challenge of weaving an identity. *Journal of adolescence*, *19*(5), 465–475. doi.org/10.1006/jado.1996.0044

Smith, M. (2018, April 25). Bring your most authentic self to work as a black woman- unpacked, part 2. Retrieved from www.blackenterprise.com/bringing-your-most-authentic-self-to-work-as-a-black-woman-unpacked-part-2/

Smith, Y. (2008). Womanist theology: Empowering black women through Christian education. *Black Theology*, *6*(2), 200–220. doi.org/10.1558/blth2008v6i2.200

Solorzano, D., Ceja, M., & Yosso, T. (2000). Critical race theory, racial microaggressions, and campus racial climate: The experiences of African American college students. *Journal of Negro Education*, 60–73. www.jstor.org/stable/2696265

Terhune, C. P. (2008). Coping in isolation: The experiences of black women in white communities. *Journal of Black Studies*, *38*(4), 547–564. doi:10.1177/0021934706288144

Thomas, A. J., Witherspoon, K. M., & Speight, S. L. (2008). Gendered racism, psychological distress, and coping styles of African American women. *Cultural Diversity and Ethnic Minority Psychology*, *14*(4), 307. doi:10.1037/1099-9809.14.4.307

Thomas, G. D., & Hollenshead, C. (2001). Resisting from the margins: The coping strategies of black women and other women of color faculty members at a research university. *Journal of Negro Education*, 166–175. doi:10.2307/3211208

Watson-Singleton, N. N. (2017). Strong Black woman schema and psychological distress: The mediating role of perceived emotional support. *Journal of Black Psychology*, *43*(8), 778–788.

Whitfield-Harris, L., Lockhart, J. S., Zoucha, R., & Alexander, R. (2017). The lived experience of black nurse faculty in predominantly white schools of nursing. *Journal of Transcultural Nursing*, *28*(6), 608–615. doi:10.1177/1043659617699064

Yap, S. C., Settles, I. H., & Pratt-Hyatt, J. S. (2011). Mediators of the relationship between racial identity and life satisfaction in a community sample of African American women and men. *Cultural Diversity and Ethnic Minority Psychology*, *17*(1), 89–97. doi:10.1037/a0022535

3 Sexuality Blindness

A New Frontier of Diversity Resistance

Oscar Holmes IV

Sheryl Sandberg, Chief Operating Officer of Facebook, is often credited for highlighting the important career implications spouses can have on each other when she stated at a 2011 *IGNITION* conference that "the most important career choice you'll make is who you marry" (Angelova, 2011). However, the idea that spouses (or significant others) play a critical role in one's career success dates back to the 1930s, if not earlier (Sostre, 2015). Indeed, the well-known (and sexist) transcultural adage, "Behind every great man, there's a great woman" has existed for at least seven decades (Martin, 2018). Considering this anecdotal support, scholars have begun to explicitly investigate the myriad effects that spouses or significant others have on each other as it relates to their careers. For example, research within the work-family conflict literature has found that experienced negativity in either domain can spillover to negatively affect the other domain (Carlson, Ferguson, Perrewé, & Whitten, 2011; Halbesleben, Harvey, & Bolino, 2009; Judge, Ilies, & Scott, 2006). Early research findings have found that unmarried employees tend to make less than their married counterparts (Korenman & Newumark, 1992). Identifying a reason for the marriage premium, Solomon and Jackson (2014) found that having a conscientious spouse improved both men's and women's job satisfaction, income and promotion opportunities because conscientious partners engage in a variety of positive behaviors that mitigate negative work-family conflict and create more opportunities for their spouses to invest in their careers without experiencing the negative drawbacks. Interestingly, Roussanov and Savor (2014) found that single CEOs are less risk averse than married CEOs, which led single CEOs to adopt more aggressive investment policies and be less reactive to changes in firm idiosyncratic risks than their married counterparts, lending further support to the idea that employees' relationship status can have a major impact on their own careers as well as the behaviors and decisions they enact within their workplaces. Though informative, this research omits lesbian, gay, bisexual, transgender and queer/

questioning (LGBTQ) individuals, so scholars can only speculate the impact that relationship status has on LGBTQ employees.

It is difficult to obtain an accurate estimate of the number of LGBTQ people in the workforce, but a 2018 Catalyst report based on surveys from around the world approximates that, between the ages of 18–59, 1.7% of Canadians (604,180) are gay or lesbian and 1.3% (462,020) are bisexual, 5.9% of the population in Japan (7,493,000) are LGBT, 2.0% of the population in the United Kingdom (1,320,400) are LGB and 4.5% of the population in the United States (14,722,200) are LGBT (Catalyst, 2018). Additionally, the 2018 Catalyst report estimates that in the United States, 1.4% of couple households (935,229) are same-sex couples, 0.5% of couple households in Germany (104,000) are same-sex couples, 0.9% of couple households in Canada (72,880) are same-sex couples and 0.9% of couple households in Australia (46,800) are same-sex couples. Due to the stigma and the discrimination risk of identifying as LGBTQ, these approximations likely significantly underestimate these populations. Notwithstanding the fallibility of these estimates, it is a fact that LGBTQ individuals have long participated in the workforce, yet few LGBTQ employees make it to the top leadership levels of mainstream organizations (Pichler & Holmes, 2017; Pichler & Ruggs, 2018).

Since the 1990s, organizational scholars have increasingly investigated the issues that LGBTQ employees face (Anteby & Anderson, 2014). The research in this area generally conclude that LGBTQ employees face increased discrimination due to their stigmatized sexual minority identity, current career theories do not adequately address the unique challenges these individuals face and that LGBTQ workers employ several identity management strategies to manage their experiences (King, Mohr, Peddie, Jones, & Kendra, 2017; Pichler & Holmes, 2017; Ragins, 2004). While this research has dramatically increased our understanding of LGBTQ employees' experiences, within the Industrial/Organizational (I/O) psychology and Organizational Behavior (OB) disciplines there are still significant gaps in the research when it comes to fully understanding LGBTQ employees' workplace experiences. In fact, in Volume One of this book, there was not a single chapter devoted to the diversity resistance issues that LGBTQ people might face. Furthermore, the I/O and OB research to date has overwhelmingly taken a "single" identity perspective with LGBTQ employees and has operated from the assumption that there are no important differences between the experiences of single and partnered LGBTQ people. I have named this phenomenon *sexuality blindness* and will argue in this chapter how and why it has significant implications for research, organizations and the careers and well-being of LGBTQ employees, thus representing a new frontier in diversity resistance.

Briefly defined, sexuality blindness is the invisibility, devaluation and/or denigration of the romantic lives of LGBTQ individuals. In this context, diversity resistance is defined as additional challenges and biases that LGBTQ individuals with significant others face in society and organizations that can affect the experiences they have and potentially thwart the opportunities afforded to them. In this chapter, I use a number of high- and low-profile examples to illustrate sexuality blindness in the real world, but my use of these examples does not suggest that sexuality blindness is limited to these specific examples, contexts, or cultures. For example, I analyze how Tim Cook, *came out* publicly as gay in his 2014 essay (Molina, 2014). Because Cook is a role model to many LGBTQ and heterosexual employees, I scrutinize the nuances of his role modeling and offer potential limitations his behavior might signal that can trigger diversity resistance. However, it is important to note that I am not disparaging *how* Cook decided to come out and that this chapter is not about Cook. Simply, I use Cook as a convenient example throughout the chapter as I think readers would more quickly understand the phenomenon and significance of sexuality blindness and how it relates to diversity resistance due to Cook's top executive position and high societal status.

As mentioned earlier, far less is known about the experiences of LGBTQ people in romantic relationships in organizational research. The assumption in the research is that the experiences, opportunities and challenges of single LGBTQ individuals are exactly the same as partnered LGBTQ individuals. But examining Sandberg's comment on career success and the popular transcultural adage on how people become great, having (or not having) a romantic partner should have some significant effects on one's career experiences. Importantly, I am neither suggesting that there are no shared experiences, opportunities, challenges among single and partnered LGBTQ individuals nor that single LGBTQ individuals no longer face any challenges or discrimination. This chapter simply suggests that we have significant gaps in our knowledge of LGBTQ people because our research has largely investigated LGBTQ people as single individuals. To make this case, I introduce sexuality blindness as a new construct, review relevant literature on how significant others impact one's career experiences in three fields and highlight specific issues LGBTQ couples might face in organizations. I contend that the devaluation and/or denigration of the romantic lives of LGBTQ people inherent in sexuality blindness ideology presents a new challenge of diversity resistance for LGBTQ people. Taken together, this chapter contributes to the diversity literature by identifying and outlining an understudied, yet important, area of bias that affects the lives of LGBTQ people.

Sexuality Blindness, Homophobia and Heterosexism: What's the Difference?

Cumulatively, sexuality blindness entails: 1) devaluing same-sex relationships in relation to cis-gender heterosexual relationships; 2) rendering invisible the unique identity, and subsequently, challenges and opportunities of partnered LGBTQ couples with respect to their single counterparts and cis-gender heterosexual partnered counterparts; 3) preferring LGBTQ couples mute or keep their relationships private; and 4) preferring LGBTQ couples not show public displays of affection.[1] Similar to color blindness (for color blindness reviews see Plaut, Thomas, & Goren, 2009; St. Antoine, 1961), sexuality blindness is an ideology that can be adopted and perpetuated at various degrees for various reasons through the process of socialization and maintenance of the status quo by higher status promulgators (i.e., heterosexual people) and the targets (i.e., LGBTQ people) of the ideology (Jost & Hunyady, 2002). Rendering an identity *invisible* can be interpreted to mean that the identity is overlooked such that it does not negatively affect a person's outcomes. However, in the context of sexuality blindness (similar to color blindness), rendering an identity invisible is not a positive action, but instead, serves to invalidate and disregard a person's identity-related experiences.

It is important to differentiate any new construct from among similar constructs within its nomological network (DeVillis, 2003). Unsurprisingly, sexuality blindness, homophobia and heterosexism are related, but distinct constructs. Over the years, several definitions of homophobia and heterosexism have been introduced into the literature. Despite their significant overlap, some scholars have identified more pronounced differences in the terms than others. For example, the most contrasting definitions of homophobia highlights the belief that LGBTQ people are "ill" or "pathological" and the prejudiced individual holds an irrational fear or belief toward LGBTQ people (Fassinger, 1991; Pharr, 1988; Simoni & Walters, 2001), whereas heterosexism entails blatant or subtle discrimination and bias towards LGBTQ people and situates heterosexuality as the normative sexual orientation (Herek, 1992; Morrison, Morrison, & Franklin, 2009).

At the root of the distinction between sexuality blindness and homophobia and heterosexism is the fact that sexuality blindness recognizes that an important identity change (Holmes, Whitman, Campbell, & Johnson, 2016; Petriglieri, 2011) takes place when LGBTQ individuals enter romantic relationships,[2] particularly when these relationships are known to others. As a result of this identity change, assumptions and attributions made about LGBTQ individuals may not necessarily apply to partnered LGBTQ individuals, which have important implications for the study of both groups. Similar to other forms of modern prejudice (Avery, Volpone, & Holmes, 2018), people who endorse sexuality blindness can also espouse support for LGBTQ policies and see themselves as *non-heterosexist* or *non-homophobic*. In this regard, sexuality blindness endorsement maintains the status quo and active or passive invisibility of partnered LGBTQ people, and ensures discrimination persists against LGBTQ people while proponents can simultaneously express support for LGBTQ people. In fact, sexuality blindness ideology recognizes that some *out and proud* LGBTQ people also believe their own romantic relationships and those of other LGBTQ couples should be less valued than heterosexual romantic relationships, that LGBTQ people should downplay (i.e., make invisible) their romantic relationships and that LGBTQ couples should not engage in public displays of affection. Therefore, traditional survey measures validated to detect explicit or indirect heterosexism or homophobia will likely prove inadequate to fully detect sexuality blindness endorsement as the endorsement can coexist with or can be independent of heterosexist or homophobic attitudes. Therein, the differentiation of sexuality blindness from homophobia and heterosexism is fundamentally rooted in intersectionality (Crenshaw, 1991).

In an effort to integrate the various definitions and to provide guidance to scholars to differentiate between the constructs, I propose scholars use homophobia to reference a prejudiced individual's fear of or belief that homosexuality is an illness, heterosexism to reference explicit or subtle prejudice towards individuals with non-heterosexual sexual orientations and sexuality blindness to reference the invisibility and explicit or subtle bias against the romantic lives of partnered LGBTQ individuals. To further elucidate important nuances between sexuality blindness and heterosexism/homophobia, I reference Cook's (2014) public coming out essay he wrote for *Bloomsberg Businessweek*. While making a point that people have important identities other than their sexual orientation, race, or gender, Cook (2014) wrote of himself, "I'm an engineer, an uncle, a nature lover, a fitness nut, a son of the South, a sports fanatic and many other things." Cook's diminution of important social identities like sexual orientation, race and gender to highlight his other important individualized identities is a commonly used rhetorical tactic to mitigate potential identity threats (Holmes et al., 2016). But if we stop our analysis here, we would miss important nuances in the language he employs in his essay, nuances that I argue align with the ideology of sexuality blindness. Of all the identifiers Cook used to describe himself in his public coming out story (e.g., uncle, nature lover, son of the South, etc.), curiously, he never speaks about his romantic life. Indeed, it is possible that Cook was single at the time, which might render speaking about his romantic life unnecessary or a moot point. But the language he used in his essay leads me to speculate that even if he were in a relationship, he would have avoided speaking about his romantic life (e.g., a loving partner/boyfriend, etc.). I base this speculation on the individualized identifiers he used to describe himself in his essay and his incessant evocation of privacy concerns.

Throughout his essay (and in other interviews), Cook referenced privacy concerns several times stating, "Throughout my professional life, I've tried to maintain a basic level of privacy," " … I've come to realize that my desire for personal privacy has been holding me back from doing something more important," " … it's worth the trade-off with my own privacy" and "Privacy remains important to me, and I'd like to hold on to a small amount of it." Certainly, privacy concerns are real and understandable for LGBTQ people who have not publicly come out and it is understandable that people do not want every detail of their romantic lives in the public sphere. But most journalists do not want, and certainly do not publish, every detail of someone's romantic life and Cook was coming out as gay to the world so his incessant evocation of privacy and wanting to maintain it could not have been solely about keeping others from knowing that he was gay. Which begs the question, what else about being gay does he insist on maintaining as private? Considering the many accounts of other

LGBTQ people evoking the privacy refrain (BBC, 2015; Browne, 2014; Woods & Lucas, 1994), I conclude it is his romantic life. As defined earlier, the endorsement of sexuality blindness ideology renders invisible the romantic lives of LGBTQ people and specifically advocates for them to mute or keep private their relationships and avoid public displays of affection. Representing an additional form of diversity resistance, we can see that even highly influential and high status *out* LGBTQ people confront strong normative and descriptive norms that present challenges to their full expression of sexuality. In many LGBTQ people's stories, they state that people often express support of their (single) LGBTQ identity, but admonish that they should not "flaunt" their sexuality in others' faces (BBC, 2015; Browne, 2014; Woods & Lucas, 1994). So, while heterosexual engagements and weddings are frequently announced and celebrated in media outlets and in personal interactions with others without being characterized as "flaunting" sexuality, LGBTQ relationships, engagements and marriages published and celebrated in media outlets and highlighted in personal interactions with others often face such criticism and receive backlash (Grinberg, 2012). An important point for intersectionality is that it is often the LGBTQ relationship identity that garners this backlash rather than the LGBTQ identity.

Providing additional examples to understand the utility of the nuances between sexuality blindness and heterosexism/homophobia, Cook (2014) admits in his essay that "for years, I've been open with many people about my sexual orientation. Plenty of colleagues at Apple know I'm gay, and it doesn't seem to make a difference in the way they treat me." He goes on to write, "While I have never denied my sexuality, I haven't publicly acknowledged it either, until now. So let me be clear: I'm proud to be gay, and I consider being gay among the greatest gifts God has ever given me." Cook's language obviously indicates that he was out to several colleagues in his workplace before he penned this essay, that he finds pride in being gay and his language clearly does not suggest that he is heterosexist or homophobic. But his continued insistence on privacy after publicly coming out and his reticence to publicly speak of his romantic life, suggests that he has endorsed, at least to some degree, sexuality blindness ideology. As a role model, what potential message could this send to LGBTQ and heterosexual people? One plausible message is that LGBTQ people can make it to the top as long as they are single or keep their romantic lives muted or private. Interestingly, this message contradicts our conventional wisdom and the advice of top executives like Sheryl Sandberg and Warren Buffet, as it is suggested that having supportive significant others is not only personally pleasurable and expected, but is also a key reason why many people become great or make it to the top at all. For LGBTQ employees in romantic relationships and those who would like to be in them and not mute or keep their relationships private, this type of role modeling potentially does more harm than good and can create diversity resistance.

Why do some people render same-sex romantic relationships invisible despite their purported support of LGBTQ individuals? In their book examining the professional lives of gay men, Woods and Lucas (1994, p. 37) highlighted a popular syllogism that I contend that both heterosexual and LGBTQ individuals who endorse sexuality blindness believe: "Sexuality is private; offices are public; therefore sexuality doesn't belong in the office." Woods and Lucas stated that many partnered gay men evoked this syllogism when asked why they did not discuss their significant other with coworkers and other professional colleagues. Woods and Lucas recounted, "Martin invoked the same binary logic: 'Sex belongs in the bedroom, not the boardroom' (p. 37)." Evoking the same syllogism, John Browne (2014, p. 21), the former CEO of BP who was forced to resign when a story broke that he was in a relationship with a male escort stated, "For the past forty-one years of my career at BP, I have kept my private life separate from my business life. I have always regarded my sexuality as a personal matter, to be kept private." Even after the Prime Minister of Luxembourg, Xavier Bettel, married his husband in 2015, a friend of the couple stated, "Mr. Bettel was keen to keep his private life out of the public spotlight" (BBC, 2015). Unlike the normal media coverage of heterosexual political spouses, the same article noted that the media and public were respecting the Prime Minister's wishes to keep his relationship out of the spotlight. Prime Minister Bettel was quoted in the article as stating, "It is something quite ingrained in society – we are quite reserved and do not like things to be too public" (BBC, 2015). The significance of this syllogism is that it is meant to only govern the romantic relationships of LGBTQ individuals and not those of heterosexual individuals. Considering these norms, unsurprisingly, Tim Cook's essay, John Browne's statement, Prime Minister Bettel's quote and a number of other international LGBTQ narratives suggest that sexuality blindness ideology is a universal phenomenon.

Romantic Relationships and Careers in Politics, Ministry and Higher Education

Although workplace romances can occur between LGBTQ couples, Pierce and Aguinis (2009) highlighted that the workplace romance literature has exclusively focused on heterosexual couples. With approximately 10 million couples who are in workplace romances, work colleagues often affirm and celebrate heterosexual romantic relationships save for the concern when these relationships are hierarchical (e.g., possibility of sexual harassment concerns, conflicts of interests, etc.) and abusive (Pierce & Aguinis, 2009; Powell, 2001). For example, research from Byron and Laurence (2015) has found that people often personalize their work spaces with objects, such as photos of their spouses or significant others, not only to communicate what or who they value to others, but these objects and symbols also serve to communicate important values to themselves.

Furthermore, it is common that as employees ascend the ranks in their careers, they are likely to become more visible within and outside of their organizations (Offermann, 1986). What is less acknowledged, yet also common, is that with this ascension, employees' significant others also become more visible and, at times, are expected to take on explicit and/or tacit roles to support the job roles and careers of their significant other (Anderson, 2002). In the following sections, I will highlight some examples of how significant others can impact one's career experiences in three fields (politics, ministry and higher education) and how via sexuality blindness this can present diversity resistance issues for LGBTQ individuals in romantic relationships.

Sexuality Blindness and Politics

In politics, a candidate's significant other's behavior, reputation and actions can cause candidates to gain or lose votes (Pergram, 2016). For example, on *NPR Day to Day* Karen Grigsby Bates (2008) hosted a broadcast that was specifically dedicated to answering the question of whether Michelle Obama was an asset or a liability to Barack Obama's 2008 presidential campaign. She noted, "Michelle Obama has been lauded as 'Barack's Rock,' but she's also been deemed a potential liability to his campaign. She speaks her mind, even when her opinions may be controversial." In many countries, it is often expected that politicians' significant others avail themselves to media interviews, campaign on their partner's behalf, raise funds, entertain guests, build or repair their political partner's reputation, establish and lead special initiatives and, in some cases, assist in policy adoption and implementation (Chai, 2012; Smith & Higgins, 2013). In speaking on the role of political spouses, University of Manitoba Political Studies Professor Royce Koop explained, " ... a presidential wife plays prominent roles that play into a party's platform" (Chai, 2012). Political Consultant and Professor of Political Communications Carrie Pergram noted that "the job description of political spouse isn't an attractive one. It involves long hours on the campaign trail ... " (Pergram, 2016).

In their study on the role of the U.S. presidential and vice presidential spouses in the 2004 election, MacManus and Quecan (2008) found that the Bush-Cheney and Kerry-Edwards campaigns strategically used the wives as surrogates for their husbands' campaign. In fact, they found that as Teresa Heinz Kerry's reputation declined, Elizabeth Edwards, spouse of Vice-Presidential Democratic candidate, John Edwards, ended up making more appearances (112 to Heinz Kerry's 76), indicating to scholars "the need to follow all four candidates' wives, not just the spouses of the presidential nominees" in order to get an accurate understanding of how campaigns strategically use spouses (MacManus & Quecan, 2008, p. 339). Finding further evidence of the strategic use of political spouses, MacManus and Quecan found that all wives most frequently visited battleground states while the 16 states that none of them visited were due to the states' presumed "safe" red or blue state status (e.g., Alabama, Connecticut, etc.; MacManus & Quecan, 2008).

By 2009, according to the Gay and Lesbian Victory Fund, there were more than 450 openly LGBTQ elected officials in the United States (Haider-Markel, 2010). In 2016, 191 openly LGBTQ candidates ran for office in the U.S. and less than half of them (87) won their races (Walker, 2017). Despite this progress, diversity resistance issues due to sexuality blindness can present challenges for partnered LGBTQ politicians. For example, one tactic that LGBTQ politicians have used to win their elections was to wait to come out publicly until after they have already been elected (Haider-Markel, 2010). In his book *Out and Running: Gay and Lesbian Candidates, Elections, and Policy Representation*, Donald Haider-Markel explained that many openly LGBTQ candidates and legislators prefer to play down their sexual orientation to boost their chances of career success and want "to avoid being seen as one-issue legislators" (p. 1443). As an indication of the challenges that LGBTQ politicians face in obtaining the highest political offices, to date, there have only been five openly gay and lesbian world leaders (all prime ministers): Johanna Sigurðardóttir of Iceland (2009–2013); Elio Di Rupo of Belgium (2011–2014); Xavier Bettel of Luxembourg (2013-Present); Leo Varadkar of Ireland (2017-Present); and Ana Brnabić of Serbia (2017-Present) (Trimble, 2017).

Although Sigurðardóttir was married to her wife years before becoming Prime Minister, Di Rupo's and Brnabić's relationship status remains elusive, whereas Varadkar is in a long-term (unmarried) relationship with his partner. Bettel married his partner two years after he became prime minister and a year after same-sex marriage was legal in Luxembourg (Trimble, 2017). In fact, reminiscent of Tim Cook's sexuality blindness language, the press stated, "few details have emerged about the wedding, which Bettel had aimed to keep private. Press photographers have been banned" (Gayle, 2015). Despite publicly coming out as gay in 2008, Bettel has remarked, "What happens at home remains private" and one of Bettel's friends who attended his wedding told the press, "He does not want to put his private life in the public spotlight and he has turned down requests from the celebrity magazines to cover the event" (Gayle, 2015). Although Varadkar has spoken positively about his partner to the press stating, "Matt [Barrett] makes me a better man," the press has noted that Barrett's choice to continue his medical education in the United States, thousands of miles away from Ireland,

> might perhaps in a way suit the very private political leader that is Leo Varadkar. It mightn't suit his heart, but it will suit his head, especially if there is a general election at some point in the next 12 months.
>
> (Egan, 2017)

Again, even though Varadkar is openly gay, the press suggested that his political career could be benefitted if his partner stays away during the election season and they mute their relationship. After Sigurðardóttir left public office in 2014, her wife, author Jónína Leósdóttir, published a book about their relationship. Despite being out and married, Leósdóttir stated, "We have always tried to keep our personal life private but now we believe it's time to make this unusual story, which spans almost three decades, public" (Reykjavik, 2014). Interestingly, Brnabić stated in an interview, "I've been openly gay throughout my life and I've never had a problem in Serbia. I would like to think that Serbia is not that conservative or homophobic, or xenophobic for that matter" (Duffy, 2017). Perhaps most in line with sexuality blindness ideology and despite her being openly gay and her beliefs about Serbia, Brnabić refused to support a proposal to make same-sex partnerships legal in Serbia (Duffy, 2017), and similar to other openly LGBTQ politicians, she's been on record stating that she did not want to be "branded as Serbia's gay PM" (Wintour, 2017).

Although strategies to downplay one's sexuality can be used by all LGBTQ politicians, they may be more easily employed by single LGBTQ politicians. First, it would be much harder for partnered LGBTQ politicians to hide their sexual orientation. Second, the presence of an LGBTQ politician's significant other might constantly remind voters of the candidate's *otherness*, whereas single LGBTQ politicians might more easily be able to quickly acknowledge their sexual orientation (e.g., like Tim Cook), but quickly pivot back to "focus on the issues" (and remain silent on their romantic life). This sentiment seems to be supported by Ireland's press when it was suggested that Barrett's physical distance away from Varadkar might be politically beneficial for the openly gay prime minister. Third, since the lives of politicians' significant others are also heavily scrutinized, partnered LGBTQ politicians are potentially open to more reputational risk attacks. Finally, scripts are yet to be normalized as to how much they should reference and how much public affection LGBTQ politicians should display toward their significant others or the roles that LGBTQ significant others should play in their partners' campaigns and political careers. Therefore, partnered LGBTQ politicians may expend considerable resources attempting to navigate these potential political and personal land mines. In fact, in Haider-Markel's entire book and Walker's (2017) article on how LGBTQ politicians can win elections, neither make any mention of the role, challenges and/or opportunities of the significant others of LGBTQ politicians – in other words, they are invisible.

Despite being aware of several openly gay politicians, 2018 was the first year that I have ever seen a political ad of an openly gay politician mentioning and/or showing affection toward his/her significant other, behaviors that are quite common and expected in ads of heterosexual politicians. In her ad, Democrat Ashley Lunkenheimer, who ran for the Pennsylvania 5th District congressional seat, mentioned that she and her wife were raising three children (Lunkenheimer, 2018). In a strategic attempt to "take a jab at Trump," Maryland Democrat State Senator Richard Madaleno, Jr. kissed his husband in his gubernatorial ad (Dresser, 2018). Former Houston Mayor Annise Parker, who was the first openly gay mayor of the city, commented on the ad, "I kissed my spouse on stage at my inauguration as mayor. But I don't think I would have put it in a campaign eight years ago." Of course it is impossible to identify any single cause, but it is important to note that both Lunkenheimer and Madaleno lost their contests. Jared Polis was the only openly LGBTQ candidate running in a gubernatorial election in 2018. In Tim Cook-like style, in his biography on his website, he listed several identities (e.g., Coloradan, entrepreneur, child education advocate, U.S. Representative) accompanied with a paragraph of text further expounding upon the identities, yet, he devoted only two sentences in speaking about his partner, particularly only referencing him in relation to their children and his identity of being a dad (Polis, 2018). Curiously, Polis only shared his partner's first name (Marlon) on his website and there was neither a picture of him or them together on his website nor was Marlon Reis featured in the campaign ad prominently displayed on his website (Polis, 2018). Polis did win his gubernatorial election, becoming the first openly gay person to be elected governor in the United States. To his credit, he did thank his partner and children for their support in his inaugural speech and Reis participated in Polis' swearing in ceremony. Although there are too few cases to definitively state a pattern, in each of the aforementioned cases, the openly gay politicians also referenced or showed their children when they referenced their significant others. If this type of "partner-children" coupling persists, it might point to a strategy that openly LGBTQ partnered individuals employ to mitigate any real or perceived bias they might experience due to sexuality blindness.

Although imperfect, public opinion survey data give us some indication of how sexuality blindness might impact people's voting intentions of LGBTQ presidential candidates. In a 1999 Gallup poll, 59% of participants responded "yes" to the question, "If your party nominated a generally well-qualified person for president who happened to be a homosexual, would you vote for that person?" (Bowman & Foster, 2008). Interestingly, in a 2000 Fox News/Opinion Dynamics poll, 53% of participants responded "would still vote for" to the question, " If you were considering a candidate whom you would otherwise support, and you discovered that they had had a homosexual relationship, would you still vote for them, probably vote against them, or definitely vote against them?"(Bowman & Foster, 2008). The seven point decrease of a hypothetical homosexual presidential candidate in a relationship versus a hypothetical homosexual presidential candidate can be significant when it comes to actual votes and winning elections. Of course, since the survey sources and questions were not identical, one cannot draw perfect inferences regarding the *true* difference between these intersectional LGBTQ identities. However, these percentage differences are informative in highlighting the nuances that sexuality blindness might account for and diversity resistance that partnered LGBTQ candidates might experience versus single LGBTQ political candidates. Next, I turn to the role that romantic relationships play in ministry.

Sexuality Blindness and Ministry

Ministry is the field in which diversity resistance against LGBTQ individuals and couples is perhaps most prevalent (Creed, Dejordy, & Lok, 2010). Despite the passing of anti-LGBTQ legislation, in many countries, religious organizations may still receive exemptions from following these laws (Eckes, 2007). Coincidentally, ministry is also the field where nuances between discriminatory experiences and outcomes due to sexuality blindness and heterosexism/homophobia might be most lucid. In fact, despite their subscription to anti-LGBTQ religious doctrine and policies, religious organizations often hire openly LGBTQ people to serve in various aspects of ministry (Creed et al., 2010; Moore, 2008, 2012).

Although people typically think of their participation in the music ministry, sizable numbers of openly LGBTQ ministers are on the preaching and teaching staff, with some even serving in senior pastoral positions in religious institutions (Enroth, 1974; van Loggerenberg, 2015; Ward, 2005). However, when openly LGBTQ ministers are fired/dismissed, it is most often due to their entering romantic relationships or making their relationships public, not because of their LGBTQ identity (van Loggerenberg, 2015; Ward, 2005). Such cases are practical examples where sexuality blindness better explains these outcomes than heterosexism or homophobia. As another example, the 2007 Dutch Reformed Church Synod Resolution accepted gay ministers and gay candidate ministers, but had a requirement that gay (but not heterosexual) ministers must remain celibate (van Loggerenberg, 2015). There are even out LGBTQ Christians in ministry who call themselves "B siders" who agree that being a gay minister is acceptable but that they should remain single and celibate (Urquhart, 2014). Furthermore, a cursory *Google* search will uncover numerous headlines where LGBTQ teachers (most often of parochial schools) were fired after they married their same-sex partner (Caron, 2018; Morris-Young, 2014). In one such case, in the official lawsuit documents for Mark Zmuda v. Eastside Catholic School (ECS), Morris-Young reported that

> the school president told Zmuda about a week later that he would be "terminated unless he filed for a divorce," but also that "if he were to divorce his husband, ECS would pay the costs of holding a 'commitment ceremony' in place of a wedding."

This case spectacularly highlights sexuality blindness (as opposed to simply homophobia/heterosexism), as in the official lawsuit documents ECS admits that they were not firing Zmuda because he was gay and actually would not fire him if he got a divorce. Instead, they were firing him because he legally married his same-sex partner, which is an important and legally-recognizable identity change.

Taken together, these examples briefly highlight how relationships can affect one's career in ministry. Specifically, they indicate that while hetero-sexism/homophobia does not prevent some religious organizations from hiring openly LGBTQ ministers and teachers, sexuality blindness can cause those same institutions to discriminate against openly LGBTQ individuals by firing/dismissing them if they enter or do not mute their romantic relation-ships. Ironically, because of sexuality blindness being single may assist LGBTQ ministers to make it to the top in ministry, whereas being single may adversely affect the careers of heterosexual ministers who want to make it to the top (e.g., excluding religious denominations that require their leaders to be single and celibate, most congregations prefer their heterosexual religious leader to be married) (DeWitt, 2014). Next, I turn to examine the role that romantic relationships might play in higher education.

Sexuality Blindness and Higher Education

Despite it being illegal to use the information when making hiring decisions, it is still quite common for academics to explicitly or implicitly ask about one's marital/partner status during interviews (Lundsteen, 2015). Many academics report that they inquire about academic's relationship status due to quell any mobility concerns as it is common for some higher education institutions to help significant others also find employment when they are recruiting academ-ics and administrators. Unfortunately, relationship status questions in higher education recruitment are also asked with malevolent intentions as well as it could also raise mobility concerns and confirm prejudicial attitudes among hiring committee members (Lundsteen, 2015; Rivera, 2017). In her study of junior faculty searches at a Research-1 university, Rivera found that due to committees' reliance on gender scripts that value men's careers over women's, relationship status often evoked mobility concerns, thus negatively impacted (heterosexual) female academics, particularly when their husbands were also academics or held other high status jobs, but a partnered relationship status did not harm (heterosexual) male academics.

Notwithstanding the relationship status concerns raised during the recruit-ment process, much of the research on the impact of relationship status in higher education is focused on how relationship status affects the research productivity of faculty members. Astin and Davis (1985) found that marital status positively impacted faculty productivity for (heterosexual) male *and* female academics, a finding that countered previous assumptions that only male academics benefitted from the marriage premium (Sowell, 1975). Inter-estingly, Astin and Davis (1985) attributed the married female productivity boost to the fact that women are much more likely than men to be married to other academics, suggesting that "[married] women had a relative advantage over academic women without male partners because their partners put them in closer contact with male colleagues and collegial networks to which they might not otherwise have had access" (Astin & Milem, 1997, p. 129). In fact, they found that "men with academic spouses/partners tend to be slightly less productive than men with nonacademic spouses/partners ... . There seems to be a clear advantage, however, for women with academic spouses/partners" (Astin & Milem, 1997, p. 133). Indeed, access to social networks that could provide information, resources and sponsorship has been found to have a positive effect on one's career outcomes (Seibert, Kraimer, & Liden, 2001). Inconsistent with some previous research (Bellas, 1992; Toutkoushian, 1998), a more recent study has found that like men, married/cohabiting women academics also earned more than their single counterparts (Toutkoushian, Bellas, & Moore, 2007). The aforementioned research suggests that relationship status can have a positive or negative impact depending on one's gender and context in higher education, but what about partnered LGBTQ academics in higher education? How does sexuality blindness present diversity resistance issues for them?

Considering that same-sex marriage has become legal in the past two decades in fewer than 30 countries around the world, it is understandable that there is a dearth of research on married same-sex academics. However, the dearth of research on partnered LGBTQ academics is a curious omission, as LGBTQ individuals have long committed themselves to romantic and cohabiting relationships and a study has found that lesbian and gay people are over-represented in academia (Bochner & Hesketh, 1994; Cox, 1964; Tilcsik, Anteby, & Knight, 2015). Consistent with sexuality blindness ideology, a study by Toutkoushian and colleagues (2007) included cohabiting (non-married) heterosexual academic couples but did not include any LGBTQ couples.

Certainly, there are several entry points and roles in which sexuality blindness can impact partnered LGBTQ people's careers in academia. Nonetheless, I will focus on the position of president/chancellor due to the high status, responsibility and salience of the role, although my arguments are applicable to other senior academic leadership roles such as that of deans, provosts and vice presidents. In addition to the internal importance of these roles to a college and university, these roles are also externally important as individuals in these positions, including their significant other, often interface with various stakeholders within the broader academic and non-academic community. As such, strong descriptive and injunctive norms held by various stakeholders might present several unique challenges to partnered LGBTQ academics in these roles. In support of this viewpoint, Grinnell College's openly gay and married president, Dr. Raynard Kington, remarked in an article interview, "There's no culture [for LGBTQ presidents], and colleges are whole communities that have been ones of tradition and informal rules and they're complex social organizations" (Abdul-Alim, 2017).

Considering the overrepresentation of gay, bisexual and lesbians in academia (Tilcsik et al., 2015), one would think more openly gay, bisexual and lesbian academics would be in top administrative positions in academia. In 2010, Dr. Raymond Crossman, President of Adler University, cofounded LGBTQ Presidents in Higher Education to combat the dearth of LGBTQ academics in top leadership positions in higher education (Abdul-Alim, 2017). After starting with approximately 12 members, today the group has a little over 80 members (Abdul-Alim, 2017). Within Abdul-Alim's article on the progress that the group has made, he presented anecdotes that perfectly highlight cases of sexuality blindness. Kington informed that it had been Grinnell College's tradition since 1862 that the president's wife chair the college's oldest surviving organization, the Ladies Education Society. Kington shared that he and his husband had been at the college for seven years before his husband, Dr. Peter Daniolos, was asked to join and chair the organization, to which Daniolos agreed (Abdul-Alim, 2017). Kington stated that he did not think hatred was behind the delay in the invitation, but rather it "was likely driven by sensitivity or uncertainty over how to approach the matter" (Abdul-Alim, 2017). While the true intentions of the group's delayed invitation may never be known, what is certain is that being openly gay and married did not prevent Kington from being hired as the college president, but his relationship status (i.e., sexuality blindness) did present other obstacles and experiences above and beyond his openly gay identity. In speaking about his married status and the situation, Kington replied, " … it's another layer of complexity that's compounded by the fact that so few LGBTQ folks have been presidents" (Abdul-Alim, 2017). In recognizing the pertinent need to address relationship issues, the LGBTQ Presidents in Higher Education annual conference includes specific workshops for LGBTQ academics on how to navigate leader-spouse relationships (Abdul-Alim, 2017).

Similar to faculty searches, questions of one's relationship status frequently come up in academic administrative searches. Although some on the search committee knew that Roosevelt University's former president, Dr. Charles Middleton, was openly gay, it is clear that not everyone did and Middleton believed that he would not be hired for the job because of his sexual orientation (Blickensderfer, 2014). During an interview with a board trustee, Middleton recalled that the trustee asked him, "I don't know anything about you personally. Can you tell me about your wife?" (Wilson, 2011). Middleton knew it was a risk, but he responded, "I'm not married. I'm in a long-term relationship with another man" (Wilson, 2011). Fortunately, the trustee did not respond unfavorably and moved on to ask for Middleton's advice on how he could show support for gay employees at his bank (Wilson, 2011). Importantly, Middleton was comfortable revealing his sexual orientation *and* relationship status, whereas not all LGBTQ candidates might want to reveal this information. When this is the case, it presents a more challenging situation for partnered LGBTQ candidates than single LGBTQ candidates.

For example, single LGBTQ candidates in the same situation who do not want to reveal their sexual orientation could state, "I'm not married or I'm not in a relationship" and steer the conversation to a more comfortable topic. However, for partnered LGBTQ candidates who do not want to reveal their sexual orientation, their options are more complicated. First, they could respond with, "I'm not married" (if they truly are not) and try to steer the conversation to a more comfortable topic. This tactic could potentially work, but I posit that these candidates might immediately experience guilt and feelings of inauthenticity which could impair their confidence answering questions going forward (Creary, Caza, & Roberts, 2015). This guilt and feelings of inauthenticity may come into play more significantly for partnered LGBTQ candidates than for single LGBTQ candidates because in the latter situation, the single candidates are answering the full question honestly and are not currently denying the existence of a significant other, particularly since the existence of a significant other would come up at some point in the hiring process and/or as one goes over the negotiated terms of the job (Rivard, 2014). Second, they could respond with, "I'm not married, but I am in a relationship." Although it might buy them some time, this response almost certainly evokes follow-up queries to which their relationship status will reveal their sexual orientation, thus this would not be an effective option for candidates who do not want to reveal their sexual orientation. Finally, they could refuse to answer the question, appealing to their discomfort, questioning the relevance, or suggesting the illegality of the question, but these types of responses typically backfire on the candidates by creating suspicion, distrust and lack of fit – concerns which could decrease their chances of being hired (Lundsteen, 2015). Therefore, it is important to recognize that experiences and options available to single LGBTQ applicants will not always be the same for partnered LGBTQ applicants. Again, the point of this exercise was not to suggest that single LGBTQ individuals do not face any discrimination or any challenges, it is simply to recognize that sexuality blindness presents important nuances for partnered LGBTQ individuals that lends itself to unique or more complicated diversity resistance issues.

Despite Roosevelt University being known for its staunch social justice mission, a dean on the search committee revealed that a board member asked her during the search process, "Does it matter that he's gay, with fundraising, and that he doesn't have a wife to entertain?" While the first part of the question is focused specifically on how his sexual orientation might impact a critically important task of a college president – fundraising; the second part of the question focuses on the role the significant other of a college president is often expected to enact, thus it is specifically relevant to his relationship status and sexuality blindness. In spite of these hesitancies, Roosevelt University did hire Middleton as president and he served in the role for 13 years until he retired. Interestingly, in describing Middleton's office, Wilson reported that

> The windowsill of the president's eighth-floor office – which looks out over Lake Michigan – bears paperweights and photos of him posing

with Democratic politicians, including President Obama and Hillary Rodham Clinton. Tucked into the corner near his desk is a photo of his partner, Mr. Geary, from before the two first met.

It is curious that although Middleton prominently displayed photos of him- self *with* politicians, he chose not to prominently display a picture of him and his partner together on his desk and the photo of his partner that he did choose to display was "tucked into the corner" and was of one before they ever met, thus his partner's identity and relationship to him likely goes unknown to many of his office visitors until Middleton chooses to reveal that information.

Unlike in Middleton's case, unfortunately, search committees do not always get past their hesitancies and biases. In his study examining how openly gay White men have successfully navigated the college presidency search process, Leipold (2014) presented quotes from his informants that reveal sexuality blindness. In one example, one of the openly gay presidents stated, "I don't think a large [university] system in the South is going to hire a gay man and his boyfriend to be president" (Leipold, 2014, p. 51). In another example, a single openly gay president stated, "I've talked to people that have partners that say, 'Yeah, I had a lot of issues because I have a partner'" (Leipold, 2014, p. 54). In a final example, an openly gay president stated,

> The searches that went well were searches where they engaged him [my partner], when they invited him to campus, they made him very much part of the interview, they were very committed to his happiness in the process and very validating as a couple.
>
> (Leipold, 2014, p. 53)

The final example shows that some search committees got past sexuality blindness issues as the committees engaged his partner throughout the pro- cess, but he implicitly reveals that sexuality blindness presented issues for some of his presidential searches and those searches did not "go well."

When top academic leaders are hired, their significant others are often named in news releases. In contrast to this norm, the University of Wash- ington official news release fails to mention Dr. Ana Mari Cauce's spouse, even though within the same article she is applauded for being the univer- sity's first woman and Latina appointed to the role, despite also being the first openly gay person in the role (Balta, 2015). In a *BBC News* article that was written to specifically highlight her being the first lesbian in the role, the only mention of Cauce's wife, Dr. Susan Joslyn, is a quote in which Cauce shared in speaking about her mother's gradual acceptance of her, stating, "She grew to love my partner, now spouse" (Coughlan, 2016) and an *Associated Press* article only references that Cauce is a lesbian, but does not name or mention her spouse at all (Warren, 2015). Again, even for openly LGBTQ academics, their partnered identity is often rendered invisible even when other aspects of their identities – even their LGBTQ identities – are being celebrated.

Reinforcing the particular challenges that LGBTQ academics face in higher education, respondents in Leipold's (2014, p. 55) study agreed that "the current sitting presidents are only at small or boutique-like institutions; gay presidents have access to leadership positions, but they have only limited access to high profile institutions in higher education." In 2015, Cauce became the first and is still the only openly gay person to lead a major Research-1 university. Interestingly, Kington's, Middleton's and Cauce's significant others are also academics, which raises important questions as to whether their partners' career similarity gave them systematic advantages that might not be available to LGBTQ academics whose partners are not fellow academics (or who may not have other high status jobs). As the previous examples highlight, sexuality blindness presents several challenges for partnered LGBTQ academics, particularly as they move into administrative leadership positions.

Conclusion

In this chapter, I presented a new construct called sexuality blindness which I posit better explains the invisibility and devaluation of the romantic relationships of LGBTQ individuals. To date, much of the research in organizational studies investigates issues related to LGBTQ people using a "single" lens perspective and subsequently employs equivalent assumptions across the groups. I argue that while there are similarities between single and partnered LGBTQ individuals, researchers should cease adopting a single lens perspective and making equivalent assumptions across the groups as this perspective severely limits our complete knowledge of LGBTQ people and obscures important nuances between the groups. For example, when researchers investigate LGBTQ identity-management strategies (King et al., 2017), they typically do not address the reality that some commonly used identity management strategies are no longer available to partnered LGBTQ people, their efficacy is greatly reduced and/or they present additional complexities to them that could lead to other negative outcomes. Additionally, when organizational researchers investigate romantic relationships, they hardly ever include LGBTQ couples. In addition to the LGBTQ literature, the workplace romance, work-family conflict, sexual harassment and domestic partner abuse, discrimination and bias, recruitment and socialization, expatriate assignment and career management literatures would all benefit from including high quality investigations of partnered LGBTQ individuals. In identifying a reason for the dearth of this type of research in organizational studies, I argued that sexuality blindness presents a new frontier of diversity resistance. By shining a light on this phenomenon, I hope that this chapter encourages researchers to further investigate sexuality blindness as a construct within LGBTQ research and to begin investigating the significance of LGBTQ romantic partners and significant others in organizational research.

Notes

1 I understand that when bisexual and transgender individuals are in romantic relationships with an opposite gender partner, it is often viewed as a heterosexual relationship. I include the entire acronym LGBTQ to be purposefully inclusive as well as to acknowledge that bisexual and transgender individuals also experience unique bias surrounding their romantic relationships.
2 In this manuscript, I use relationships to mean the romantic relations between consenting adults and do not take an evaluative judgment of the length or nature of the relationship.

References

Abdul-Alim, J. (2017). LGBTQ college presidents: Work is not over for those who are "out." Retrieved August 1, 2018 from http://diverseeducation.com/article/97138/

Anderson, K. V. (2002). From spouses to candidates: Hillary Rodham Clinton, Elizabeth Dole, and the gendered office of U.S. president. *Rhetoric & Public Affairs, 5*(1), 105–132. doi:10.1353/rap.2002.0001

Angelova, K. (2011). Facebook COO Sheryl Sandberg and her husband have done the impossible. Retrieved February 22, 2017 from www.businessinsider.com/sheryl-sandberg-husband-2011-12

Anteby, M., & Anderson, C. (2014). The shifting landscape of LGBT organizational research. *Research in Organizational Behavior, 34*, 3–25. doi:10.1016/j.riob.2014.08.001

Astin, H. S., & Davis, D. E. (1985). Research productivity across the life and career cycles: Facilitators and barriers for women. In M. F. Fox (Ed.), *Scholarly writing and publishing: Issues, problems, and solutions* (pp. 147–160). Boulder, CO: Westview Press.

Astin, H. S., & Milem, J. (1997). The status of academic couples in U.S. institutions. In M. Ferber, & J. Loeb, (Eds.), *Academic couples: Problems and promises* (pp. 128–155). Urbana, IL: University of Illinois Press.

Avery, D. R., Volpone, S. D., & Holmes, O., IV (2018). Racial discrimination in organizations. In A. J. Colella, & E. B. King (Eds.), *The Oxford handbook of workplace discrimination* (pp. 89–109). New York: Oxford University Press. doi:10.1093/oxfordhb/9780199363643.013.8

Balta, V. (2015). UW regents name Ana Mari Cauce president. Retrieved February 27, 2017 from www.washington.edu/news/2015/10/13/uw-regents-name-ana-mari-cauce-president/

Bates, K. G. (2008). Is Michelle Obama an asset or liability? Retrieved June 21, 2018 from www.npr.org/templates/story/story.php?storyId=87943583?storyId=87943583

BBC. (2015). Luxembourg PM first EU leader to marry same-sex partner. Retrieved February 27, 2017 from www.bbc.com/news/world-europe-32753014

Bellas, M. L. (1992). The effects of marital status and wives' employment on the salaries of faculty men. *Gender & Society, 6*(4), 609–622.

Blickensderfer, G. R. (2014). Openly gay Roosevelt Univ. president to retire. Retrieved August 2, 2018 from www.windycitymediagroup.com/lgbt/Openly-gay-Roosevelt-Univ-president-to-retire/49031.html

Bochner, S., & Hesketh, B. (1994). Power distance, individualism/collectivism, and job-related attitudes in a culturally diverse work group. *Journal of Cross Cultural Psychology, 25*(2), 233–257.

Bowman, K., & Foster, A. (2008). Attitudes about homosexuality and gay marriage. Retrieved August 2, 2018 from www.aei.org/wp-content/uploads/2011/11/ 20050520_HOMOSEXUALITY0520.pdf

Browne, J. (2014). *The glass closet: Why coming out is good business*. London: Harper Collins.

Byron, K., & Laurence, G. A. (2015). Diplomas, photos, and tchotchkes as symbolic self-representations: Understanding employees' individual use of symbols. *Academy of Management Journal, 58*(1), 298–323. doi:10.5465/amj.2012.0932

Carlson, D. S., Ferguson, M., Perrewé, P. L., & Whitten, D. (2011). The fallout from abusive supervision: An examination of subordinates and their partners. *Personnel Psychology, 64*(4), 937–961. doi:10.1111/j.1744-6570.2011.01232.x

Caron, C. (2018). Teacher marries her girlfriend, and then Catholic school fires her. Retrieved June 22, 2018 from www.nytimes.com/2018/02/17/us/gay-teacher-fired. html

Catalyst. (2018). Lesbian, Gay, Bisexual, and Transgender workplace issues. Retrieved January 29, 2019 from www.catalyst.org/knowledge/lesbian-gay-bisex ual-transgender-workplace-issues

Chai, C. (2012). How U.S. presidential candidates' wives help and hurt a campaign. Retrieved June 22, 2018 from https://globalnews.ca/news/281395/how-us-presiden tial-candidates-wives-help-and-hurt-a-campaign/

Cook, T. (2014). Tim cook speaks up. Retrieved June 19, 2018 from www.bloomberg. com/news/articles/2014-10-30/tim-cook-speaks-up

Coughlan, S. (2016). The gay Cuban-American breaking barriers in Washington. Retrieved August 2, 2018 from www.bbc.com/news/business-36053068

Cox, J. A. (1964). Application of a method of evaluating training. *Journal of Applied Psychology, 48*(2), 84–87. doi:10.1037/h0046001

Creary, S., Caza, B. B., & Roberts, L. M. (2015). Out of the box? How managing a subordinate's multiple identities affects the quality of a manager-subordinate relationship. *Academy of Management Review, 40*(4), 538–562. doi:10.5465/amr.2013.0101

Creed, W. E. D., Dejordy, R., & Lok, J. (2010). Being the change: Resolving institutional contradiction through identity work. *Academy of Management Journal, 53*(6), 1336–1364.

Crenshaw, K. (1991). Mapping the margins: Intersectionality, identity politics, and violence against women of color. *Stanford Law Review, 43*(6), 1241–1299.

DeVillis, R. F. (2003). *Scale development: Theory and applications* (2nd ed.). Thousand Oak, CA: SAGE Publications.

DeWitt, S. (2014). Single vs. married pastors: Take it from a guy who's been both. Retrieved July 2, 2018 from www.thegospelcoalition.org/article/single-vs-married-pastors-take-it-from-a-guy-whos-done-both-2/

Dresser, M. (2018). "Take that, Trump": Maryland Democrat Madaleno features same-sex kiss in ad for governor's campaign. Retrieved August 2, 2018 from www. baltimoresun.com/news/maryland/politics/bs-md-adwatch-madaleno-take-that-20180606-story.html

Duffy, N. (2017). Serbia's openly gay Prime Minister says her sexuality has never been an issue. Retrieved July 26, 2018 from www.pinknews.co.uk/2017/07/20/ser bias-openly-gay-prime-minister-says-her-sexuality-has-never-been-an-issue/

Eckes, S. E. (2007). The legal "rights" of LGBT educators in public and private schools. *Texas Journal on Civil Liberties & Civil Rights*, *23*(1), 29–54.

Egan, B. (2017). Leo Varadkar opens up about bridging the miles with FaceTime as his boyfriend Matt moves to the U.S. Retrieved July 26, 2018 from www.independ ent.ie/irish-news/leo-varadkar-opens-up-about-bridging-the-miles-with-facetime-as-his-boyfriend-matt-moves-to-the-us-36049324.html

Enroth, R. M. (1974). The homosexual church: An ecclesiastical extension of a subculture. *Social Compass*, *21*(3), 355–360. doi:10.1177/003776867402100310

Fassinger, R. E. (1991). The hidden minority: Issues and chalenges in working with lesbian women and gay men. *The Counseling Psychologist*, *19*(2), 157–176.

Gayle, D. (2015). Luxembourg's prime minister first EU leader to marry same-sex partner. Retrieved July 26, 2018 from www.theguardian.com/world/2015/may/15/luxembourg-prime-minister-eu-xavier-bettel-gauthier-destenay-gay-marriage

Grinberg, E. (2012). North Dakota paper reverses ban on same-sex wedding ads. Retrieved July 2, 2018 from www.cnn.com/2012/07/30/living/north-dakota-same-sex-wedding-ads/index.html

Haider-Markel, D. P. (2010). *Out and running: Gay and lesbian candidates, elections, and policy representation.* Washington, DC: Georgetown University Press.

Halbesleben, J. R. B., Harvey, J., & Bolino, M. C. (2009). Too engaged? A conservation of resources view of the relationship between work engagement and work interference with family. *Journal of Applied Psychology*, *94*(6), 1452–1465. doi:10.1037/a0017595

Herek, G. M. (1992). *The social context of hate crimes: Notes on cultural heterosexism* (pp. 89–104). Newbury Park, CA: Sage.

Holmes IV, O., Whitman, M. V., Campbell, K. S., & Johnson, D. E. (2016). Exploring the social identity threat response. *Equality, Diversity and Inclusion: An International Journal*, *35*(3), 205–220. doi:10.1108/EDI-08-2015-0068

Jost, J. T., & Hunyady, O. (2002). The psychology of system justification and the palliative function of ideology. *European Review of Social Psychology*, *13*(1), 111–153. doi:10.1080/10463280240000046

Judge, T. A., Ilies, R., & Scott, B. A. (2006). Work-family conflict and emotions: Effects at work and at home. *Personnel Psychology*, *59*(4), 779–814. doi:10.1111/j.1744-6570.2006.00054.x

King, E. B., Mohr, J. J., Peddie, C. I., Jones, K. P., & Kendra, M. (2017). Predictors of identity management: An exploratory experience-sampling study of lesbian, gay, and bisexual workers. *Journal of Management*, *43*(2), 476–502. doi:10.1177/0149206314539350

Korenman, S., & Newumark, D. (1992). Marriage, motherhood, and wages. *Journal of Human Resources*, *27*, 233–255.

Leipold, B. (2014). Navigating straight waters: The lived experiences of how out, white gay males have successfully naviagted the college presedential search process. *Journal of Psychological Issues in Organizational Culture*, *5*(3), 40–67. doi:10.1002/jpoc

Lundsteen, N. (2015). They aren't supposed to ask that. Retrieved June 2, 2018 from www.insidehighered.com/advice/2015/05/18/essay-how-handle-illegal-questions-during-academic-job-interviews

Lunkenheimer, A. (2018). Better. Retrieved August 2, 2018 from http://ashleylunken heimer.com/

MacManus, S. A., & Quecan, A. F. (2008). Spouses as campaign surrogates: Strategic appearances by presidential and vice presidential candidates' wives in the 2004 election. *PS: Political Science and Politics*, *41*(2), 337–348. doi:10.1017/S1049096508080529

Martin, G. (2018). Behind every great man there's a great woman. Retrieved June 12, 2018 from www.phrases.org.uk/meanings/60500.html

Molina, B. (2014). Apple CEO Tim Cook: "I'm proud to be gay." Retrieved February 22, 2017 from www.usatoday.com/story/tech/2014/10/30/tim-cook-comes-out/18165361/

Moore, D. L. (2008). Guilty of sin: African-American denominational churches and their exclusion of SGL sisters and brothers. *Black Theology: An International Journal*, *6*(1), 83–97. doi:10.1558/blth2008v6i1.83

Moore, D. L. (2012). Contested alliances: The black church, the right, and queer failure. *Black Theology: An International Journal*, *10*(3), 321–327.

Morrison, M. A., Morrison, T. G., & Franklin, R. (2009). Modern and old-fashioned homonegativity among samples of Canadian and American university students. *Journal of Cross-Cultural Psychology*, *40*(4), 523–542. doi:10.1177/0022022109335053

Morris-Young, D. (2014). Court greenlights fired gay teacher's lawsuit against Catholic school. Retrieved June 22, 2018 from www.ncronline.org/news/parish/court-greenlights-fired-gay-teachers-lawsuit-against-catholic-school

Offermann, L. R. (1986). Visibility and evaluation of female and male leaders. *Sex Roles*, *14*(9–10), 533–543.

Pergram, C. G. (2016). The new role of the modern political spouse. Retrieved from www.campaignsandelections.com/campaign-insider/the-new-role-of-the-modern-political-spouse

Petriglieri, J. L. (2011). Under threat: Responses and consequences of threats to individuals' identities. *Academy of Management Review*, *36*(4), 641–662.

Pharr, S. (1988). *Homophobia: A weapon of sexism*. Berkeley, CA: Chardon Press.

Pichler, S., & Holmes IV, O. (2017). An investigation of fit perceptions and promotability in sexual minority candidates. *Equality, Diversity and Inclusion: An International Journal*, *36*(7), 628–646. doi:10.1108/EDI-02-2017-0037

Pichler, S., & Ruggs, E. N. (2018). LGBT workers. In A. J. Colella, & E. B. King (Eds.), *The Oxford handbook of workplace discrimination* (pp. 177–195). New York, NY: Oxford University Press.

Pierce, C. A., & Aguinis, H. (2009). Moving beyond a legal-centric approach to managing workplace romances: organizationally sensible recommendations for HR leaders. *Human Resource Management*, *48*(3), 447–464. doi:10.1002/hrm.20289

Plaut, V. C., Thomas, K. M., & Goren, M. J. (2009). Is multiculturalism or color blindness better for minorities? *Psychological Science*, *20*(4), 444–446. doi:10.1111/j.1467-9280.2009.02318.x

Polis, J. (2018). Meet Jared Polis. Retrieved December 19, 2019 from www.colorado.gov/governor/gov-polis

Powell, G. N. (2001). Workplace romances between senior- level executives and lower-level employees: An issue of work disruption and gender. *Human Relations*, *54*(11), 1519–1544. doi:10.1177/00187267015411005

Ragins, B. R. (2004). Sexual orientation in the workplace: The unique work and career experiences of gay, lesbian, and bisexual workers. *Research in Personnel and Human Resources Management*, *23*, 35–120.

Reykjavik, M. D. (2014). Book pending on ex-Iceland PM's romantic life. Retrieved December 19, 2019 from https://www.icelandreview.com/news/book-pending-ex-iceland-pm%E2%80%99s-romantic-life/

Rivard, R. (2014). President's home or prison? Retrieved from www.insidehighered.com/news/2014/01/10/alabama-university-limits-presidents-love-life%0D

Rivera, L. A. (2017). When two bodies are (not) a problem: Gender and relationship status discrimination in academic hiring. *American Sociological Review, 82*(6), 1111–1138. doi:10.1177/0003122417739294

Roussanov, N., & Savor, P. G. (2014). Marriage and managers' attitudes to risk. *Management Science, 60*(10), 2496–2508.

Seibert, S. E., Kraimer, M. L., & Liden, R. C. (2001). A social capital theory of career success. *Academy of Management Journal, 44*(2), 219–237. doi:10.2307/3069452

Simoni, J. M., & Walters, K. L. (2001). Heterosexual identity and heterosexism. *Journal of Homosexuality, 41*(1), 157–172. doi:10.1300/J082v41n01

Smith, A., & Higgins, M. (2013). "My husband, my hero": Selling the political spouses in the 2010 general election. *Journal of Political Marketing, 12*(2–3), 197–210.

Solomon, B. C., & Jackson, J. J. (2014). The long reach of one's spouse. *Psychological Science, 25*(12), 2189–2198. doi:10.1177/0956797614551370

Sostre, A. (2015). Here's why the most important career choice you'll make is who you marry. Retrieved June 4, 2018 from http://sostrenews.com/heres-why-the-most-important-career-choice-youll-make-is-who-you-marry/

Sowell, T. (1975). *Affirmative action reconsidered: Was it necessary in academe?* Washington, DC: American Enterprise Institute for Public Policy Research.

St. Antoine, T. J. (1961). Color blindness but not myopia: A new look at state action, equal protection, and "private" racial discrimination. *Michigan Law Review, 59*(7), 993–1016.

Tilcsik, A., Anteby, M., & Knight, C. R. (2015). Concealable stigma and occupational segregation: Toward a theory of gay and lesbian occupations. *Administrative Science Quarterly, 60*(3), 446–481. doi:10.1177/0001839215576401

Toutkoushian, R. K. (1998). Racial and marital status differences in faculty pay. *The Journal of Higher Education, 69*(5), 513–541.

Toutkoushian, R. K., Bellas, M. L., & Moore, J. V. (2007). The interaction effects of gender, race, and marital status on faculty salaries. *The Journal of Higher Education, 78*(5), 572–601. doi:10.1353/jhe.2007.0031

Trimble, M. (2017). LGBT world leaders. Retrieved August 2, 2018 from www.usnews.com/news/best-countries/slideshows/openly-lgbt-world-leaders?slide=2

Urquhart, E. (2014). Meet the B siders: Celibate LGBTQ christians. Retrieved June 22, 2018 from www.slate.com/blogs/outward/2014/07/04/celibate_lgbtq_christians_the_mainstream_gay_community_should_be_more_welcoming.html

van Loggerenberg, M. (2015). The 2007 Dutch Reformed Church synod resolution: Impact on gay ministers. *HTS Teologiese Studies/Theological Studies, 71*(3), 1–9. doi:10.4102/hts.v71i3.2894

Walker, J. (2017). How LGBTQ politicians are winning elections in conservative areas. Retrieved February 27, 2017 from www.vice.com/en_us/article/pgxbqn/how-lgbtq-politicians-are-winning-elections-in-conservative-areas

Ward, E. G. (2005). Homophobia, hypermasculinity and the U.S. black church. *Culture, Health and Sexuality, 7*(5), 493–504. doi:10.1080/13691050500151248

Warren, T. S. (2015). University of Washington establishes several firsts by choosing woman president–who is also gay. Retrieved August 2, 2018 from www.oregonl ive.com/pacific-northwest-news/index.ssf/2015/10/uw_moves_longtime_profes sor_fr.html

Wilson, R. (2011). Gay academics find new paths to the top. Retrieved August 2, 2018 from www.chronicle.com/article/The-No-Big-Deal-Gay/126895? sid=at&utm_medium=en&utm_source=at

Wintour, P. (2017). Ana Brnabić: "I do not want to be branded Serbia's gay PM." Retrieved from www.theguardian.com/world/2017/jul/28/ana-brnabic-serbia-prime-minister-interview

Woods, J. D., & Lucas, J. H. (1994). *The corporate closet: The professional lives of gay men in America*. New York, NY: The Free Press.

4 Diversity Resistance and Gender Identity

How Far Have We Come and Where Do We Still Need to Go?

Katina Sawyer and Christian Thoroughgood

In recent years, transgender populations have garnered considerable coverage in the media. While prior court decisions have recognized gender expression as a protected category under Title VII's gender discrimination ban, gender identity is not a federally protected status (ACLU, 2018). Thus, the state of legal protections for gender identity minorities is precarious and vulnerable to new or alternative interpretations of the law. For example, recent debates have led lawmakers to question whether or not prior cases considering gender identity as a protected class for purposes of employment should be reversed (Transgender Law Center, 2017). Over the past several months, controversies have arisen regarding the current administration's attempt to roll back protections for transgender members of the military, transgender students who wish to use the bathroom of their choice and those who wish to have transition-related services covered by health insurance (ABC News, 2018). Thus, while some strides have been made toward gender identity inclusivity at work and beyond, we have a long way to go before we achieve equity.

Within the organizational sciences in particular, there has been very little published work that addresses the specific concerns of gender identity minorities at work. Most of the research that has been conducted within the LGBT community combines sexual orientation minority samples with gender identity minority samples, despite the fact that they may face unique challenges at work (Law, Martinez, Ruggs, Hebl, & Akers, 2011; Sawyer, Thoroughgood, & Webster, 2016). Further, many published studies use the term "LGBT" in the title and in describing their research sample, only to later note that transgender employees did not actually participate in the study (Beauregard, Arevshatian, Booth, & Whittle, 2018; Sawyer et al., 2016). Consequently, we know very little about transgender employees' experiences at work. Moreover, for those transgender employees who have participated in studies relevant to LGBT populations, the final product often focuses on explaining the experiences of sexual orientation minorities at work, evading an in-depth look at gender identity as a stand-alone construct of interest. In fact, there were only three articles (Drydakis, 2017;

Law, Martinez, Ruggs, Hebl, & Akers, 2011; Martinez, Sawyer, Thoroughgood, Ruggs, & Smith, 2017) that focused on this population published in the top 14 journals in organizational sciences from each of the journals' inception until 2017 (Martinez, Sawyer, & Wilson, 2017).

Looking beyond this select group of organizational science journals, scholars have attempted to address the dearth of research examining the work experiences of gender identity minorities by focusing on understanding the various aspects of their work lives. Foundational research has examined certain organizational antecedents (Ruggs, Martinez, Hebl, & Law, 2015) and psychological consequences of perceived discrimination (Brewster, Velez, DeBlaere, & Moradi, 2012) for gender identity minorities. Further, newer work has focused on the importance of authenticity and acceptance for transgender employees with respect to their job attitudes and well-being (Martinez et al., 2017; Thoroughgood, Sawyer, & Webster, 2017). However, despite the fact that researchers are starting to focus more heavily on the workplace experiences of gender identity minorities, these employees continue to report persistent discrimination at high levels. In fact, 30% of participants in a national U.S. survey of transgender individuals reported being fired, denied a promotion, or experiencing some other form of mistreatment in the workplace due to their gender identity or expression in 2015 (National Center for Transgender Equality, 2016).

Given the specific adversity that transgender employees may face in the workplace, we believe that comprehensive solutions for addressing resistance to inclusivity for these employees are necessary. In this chapter, we outline the unique challenges that gender identity minority employees continue to face from a legal and interpersonal perspective at work. We then provide suggestions for preventing discrimination, both overt and covert. However, we also believe that it is important to focus not just on eliminating discrimination, but on empowering employees who may not have extensive knowledge about gender identity minorities and who would be willing to serve as allies to these groups at work. Thus, our chapter will outline not only what needs to be done to resist discrimination toward gender identity minorities in organizations, but also what must be done to create more inclusive work environments. By focusing on eliminating negative attitudes and actively promoting positive attitudes toward gender identity minorities, we believe that our chapter will serve as a resource for researchers and practitioners who are striving to eliminate resistance to gender identity diversity. Finally, by creating a greater awareness of a wide range of possibilities for employees to express their gender, we hope to encourage a dialogue that calls into question our norms for gendered self-presentation, for the benefit of all employees – not just those who identify as gender identity minorities.

Gender Identity Minority Status: Defining Terms and Experiences

Gender is one of the most fundamental attributes by which individuals categorize and perceive others and by which they categorize and perceive themselves (Bem, 1983; Maccoby, 1988). However, what we often think of societally as "gender" is a far cry from the biologically determined sex that we are born with. In other words, biological sex is something individuals are born with, while gender is a way of expressing the self. The way that individuals dress, act, speak and groom themselves are all indicators of the societal construct of gender – as a result, gender is something we do and not something that is biologically imperative (Butler, 2002). Legally, rooted in the landmark Price Waterhouse v Hopkins case, courts have also interpreted discrimination based on gender expression as gender discrimination more broadly. In this and other similar cases, being penalized at work for expressing gender in non-normative ways was interpreted as creating unequal access to gender expressions for men and women (e.g., men are allowed to express gender in stereotypically masculine ways, but women are not).

Relevant to issues pertaining to gender identity expression more broadly, gender identity minorities are those who do not wish to conform to the gender expressive norms that are associated with their born sex. This term includes a variety of different groups that generally fall under the transgender umbrella, including but not limited to agender (genderless), FTM/MTF (born with female organs but engaging in male gender expressions or vice versa), or genderqueer (identifying outside of a gender binary as neither male nor female, both, or somewhere in between) (for more information about these and other terms, see Beemyn (n.d.) or Wentling, Windsor, Schilt, & Lucal (2008). Because gender norms are so strongly attended and conformed to in society, bucking these norms is rare. Moreover, when gender non-normative individuals do decide to express themselves authentically, there are often negative consequences for this behavior.

In fact, a recent report released by the Human Rights Campaign (2018) and the University of Connecticut, demonstrated that 51% of transgender youth are not using the bathroom of their choice, primarily because they are afraid for their safety. Further, on average, 37.6% of transgender, questioning, or genderqueer youth report having attempted suicide, compared to a rate of 13.7% of cisgender youth (Toomey, Syvertsen, & Shramko, 2018). Related, 87.5% of LGBTQ youth report hearing negative sentiments about transgender people while they are at school, potentially driving reports of more hostile school environments from transgender youth compared to their LGB counterparts (Kosciw, Greytak, Giga, Villenas, & Danischewski, 2016). As mentioned previously, transgender adults also face similar challenges, particularly in highly normative, group-focused spaces such as within organizations (Brewster et al., 2012; Law et al., 2011).

As a result, those who do not wish to adhere to gender norms in society do not frequently take the decision to transition or to express themselves in accordance with their authentic gender identity lightly. As demonstrated primarily within the context of LGB employees, the decision to disclose one's gender identity at work is often fraught with difficulty and anguish (Ragins, 2008; Ragins & Cornwell, 2001; Ragins, Singh, & Cornwell, 2007). The disclosure process may be even more complex for gender identity minorities, given they may wish to "pass" as a chosen gender identity (or the extent to which others "read" their gender appropriately; e.g., a male to female transgender individual who wants to be viewed and accepted by others as female, without having to explain their gender identity or provide proper pronouns to others) or they may not wish to be affiliated with any particular gender category, choosing instead to express themselves more acategorically (e.g., a genderqueer individual). Thus, when one is transitioning from one gender identity expression to another, they may need to inform coworkers of their wish to transition and make them aware of the proper pronouns to use after the transition occurs. Once that transition is made, however, disclosing one's transgender status may no longer be viewed as necessary, as some individuals wish to be known as transgender while others may wish to only align with their current gender identity (i.e., not transgender, but male or female, etc.). Some transgender individuals may also choose to have surgery, in order to change their biological sex to align with societal expectations for their gender identity. In general, when individuals transition from one societally accepted way of expressing identity to another (i.e., female to male or male to female), the decision to disclose may be largely based on passability. For those who are in constant transition or who view transitioning as irrelevant to their gender identity expression (because they are not moving from one expression to another, but rather are continuously expressing their gender in new and authentic ways), disclosure at work may be unnecessary, given the heightened visibility of their gender identity expression choices to others (for further reading on the ways in which acategorically identified individuals might navigate identity at work, see Clair, Humberd, Rouse, & Jones, In press).

Overall, gender identity is a complex construct, which can take on many different meanings and encompass a wide array of expressions. Each of these expressions has unique challenges and considerations associated with them, but they all share one thing in common – they are all forms of resistance against traditional norms for gender expression. Throughout this chapter, we will use the term "gender identity minority" to encompass terms such as transgender, questioning, queer, genderqueer, gender non-conforming, genderfluid, gender expansive, androgynous and additional terms related to gender expression (in addition to the readings listed above regarding terminology, please see the Human Rights Campaign (2018) Glossary of Terms for additional information on terminology). We use "gender identity minority" as an umbrella term given there are many different ways in which individuals may choose to non-normatively express their gender, but they all share the potential for societal marginalization as well as the ability to transgress and transform normative gender expression expectations. In the following sections, we will focus on the possibilities for organizational and societal transformation which may arise from both the practical need for employee protections againt discrimination and, more boldly, the idealistic goal of revolutionizing binary gender systems completely.

How Can Organizations Resist Discriminatory Systems and Attitudes at Work?

Gender identity minorities in the workplace have both legal and interpersonal concerns to contend with when determining how and when to express their identities authentically. Legally, in the U.S., transgender employees have not enjoyed the same protections against gender discrimination as cisgender women have. As mentioned above, past precedent has often included gender identity as an extension of Title VII's protections from gender discrimination under the law. For example, in a landmark case, Macy v. Holder, a transgender woman (who had been hired before transition), successfully won a case under Title VII's protections after she had been denied employment for authentically expressing her gender identity. However, given the precariousness with which past precedent can be determined, without a federal law that bars gender identity discrimination specifically, gender identity minorities remain at risk for discriminatory treatment without legal recourse. For example, in contrast to the Macy v. Holder case, the Goins v. West Group case upheld that being forced to use a bathroom that does not align with one's gender identity is not illegal. In this case, the court felt that assigning bathroom usage based on biological sex was a fair practice, given being unable to do so might disrupt organizations' ability to keep cisgender individuals from using any bathroom or locker facilities they saw fit – which the court saw as a potential threat to other employees. Indeed, at the time of this writing, the Supreme Court has agreed to hear a case which would clarify whether or not gender identity is covered under Title VII. This lack of clear workplace protections for gender identity minorities is important because discrimination remains prevalent. More than two million professionals leave their organizations each year due to unfairness, with an estimated cost to U.S. employers of roughly $64 billion annually (Level Playing Field, 2007). This experience is not at all uncommon to transgender individuals, of whom it is estimated 90% have experienced harassment, mistreatment, or discrimination on the job; 47% faced barriers to entry and promotion; and over 25% lost a job, because of their gender identity (Grant et al., 2008). Of additional legal concern, transgender individuals do not frequently have equal access to healthcare compared to their cisgender counterparts (Transgender Law Center 2004). It is unlikely that healthcare programs will cover transition-related procedures or treatments without employers going above the law to ensure that these benefits are offered.

Overall, the current legal arena within the United States is ambiguous. Without clear protections for gender identity minority employees and without consideration for their unique health-related needs, the burden of inclusivity falls on individual employers, judges and other decision makers to determine the fate of those who break with gender norms in society. Further, while beyond the scope of this chapter (see Sawyer et al., 2016 for a more comprehensive review), there are varying legal protections for gender identity minorities outside of the United States. Thus, globally, gender identity minorities and those who employ members of these populations may face even greater challenges in finding proper protections for their employment rights.

Even if the safety of federal protections is someday guaranteed, however, gender identity minorities may still be at risk for disrespectful treatment. While prior research has demonstrated that living authentically is related to positive psychological wellbeing (Goldman & Kernis, 2002; Ryan, LaGuardia, & Rawsthorne, 2005; Sheldon, Ryan, Rawsthorne, & Ilardi, 1997), for transgender employees, positive authenticity-related outcomes have been shown to depend on the reactions of coworkers, given the importance of others' verification of their identities in determining a positive sense of self (Martinez et al., 2017). For example, while it is important to be able to express oneself freely, if coworkers use improper pronouns in reference to gender identity minorities or "dead name" them (i.e., using one's prior name instead of a chosen name that aligns with one's gender identity), job attitudes may decrease. Further, transgender employees have been found to frequently experience microaggressions (Nadal, Davidoff, Davis, & Wong, 2014), which are difficult to legislate (Yoshino, 2007), and can only be ameliorated by addressing the underlying attitudes that drive them. As a result, acceptance of gender identity and inclusive behaviors from coworkers are necessary components in determining workplace attitudes and outcomes for gender identity minorities. When this is the case, gender identity minorities are more likely to feel comfortable at work and to perform better as a result (Grant et al., 2008). While improving the legal protections that transgender individuals have access to can certainly help in keeping employees and companies accountable for discriminatory behavior within organizations, the law can't change these underlying attitudes that may have negative effects on daily interactions between gender identity minorities and their coworkers. For these reasons, not only is it important that organizations enforce federal and state protections for gender identity minorities, but also that they attempt to spread awareness of and educate employees about gender identity inclusive behaviors at work. If organizations are truly to resist disrespect and discrimination perpetrated toward gender identity minorities, they must address both legal and social aspects driving these negative attitudes and behaviors.

How Can Gender Identity Minorities Resist the Effects of Discrimination in the Workplace?

While an ultimate solution to resisting discrimination should never rest with the recipient of discriminatory attitudes or behavior, there are practical reasons why those who are stigmatized might benefit from more effective coping with these negative experiences. First, while it would be ideal to be able to remove oneself from discriminatory or stigmatizing work environments, it is not always possible for individuals to exit their job or their organization. Discrediting the need for effective coping strategies for gender identity minority populations assumes that individuals always have complete agency over the environments they find themselves in, which is not economically or fiscally possible in many cases. Second, while it is imperative that gender identity majority group members take responsibility for creating inclusive workplaces, it is overly idealistic to imagine that this goal will be achieved easily or quickly. For these reasons, while gender identity minorities continue to face discrimination and negative attitudes and while it is impossible to assume that individuals might fully avoid these attitudes, it is necessary to determine how gender identity minorities might themselves resist their negative effects.

First, it might be useful for gender identity minorities to build resources that allow for greater resilience in the face of discriminatory behavior. Recent research has demonstrated that mindfulness, for example, may be important in buffering some of the negative effects of experiencing discrimination (Thoroughgood, Sawyer, & Webster, 2018a). Because mindfulness has been found to disrupt attentional processes that direct thought toward the past or future by allowing individuals to skillfully redirect attention to the present (Glomb, Duffy, Bono, & Yang, 2011), it is useful when appraising negative events. Instead of encouraging a focus on past negative encounters or the possibility for future discrimination, mindfulness may allow gender identity minorities (and other stigmatized group members) to disassociate from past negative events and to judge future events accurately. Thus, mindfulness may also allow gender identity minorities to preserve a more positive self-image in the face of negativity (e.g., "It's not about me, it's just something that is happening in my environment.") and to enter new situations without judgment (e.g., "Just because I have encountered negativity before, I will judge this situation only for what it is."). Overall, it may be important for gender identity minorities to develop their capacity for mindfulness by making mindfulness practice part of a daily routine. In this way, gender identity minorities might prepare themselves to face discrimination by developing mental resources that facilitate coping with negative life events. Even better, organizations might offer mindfulness programs in order to provide tools and resources for alleviating the symptoms of stigma at work. Again, while we want to be clear that we fully support eradicating bias instead of dealing with the aftermath, while bias still exists, understanding how to ameliorate its effects may be a useful short-term strategy for gender identity minorities to leverage.

Further, growing social support networks may help gender identity minorities to effectively cope with and resist the negative effects of discrimination. Social support networks have been found to help individuals cope with traumatic or distressing life events (Baumeister, Faber, & Wallace, 1999; Cohen & Wills, 1985). When individuals are able to participate in supportive conversations that remind them of their value or the value of the work they do, or when they are the recipient of empathic concern, social support has a particularly high impact (Zellars & Perrewé, 2001). Meta-analyses have demonstrated that social support can not only reduce the impact of strains, but can also reduce the perception of stressors (Viswesvaran, Sanchez, & Fisher, 1999). In this way, growing social support networks may help gender identity minorities to cope with perceived stress but also to appraise fewer situations as stressful. Of course, as may be the case in any non-normative group facing challenges being accepted in society, it may be harder for gender identity minorities to find and build support networks. However, while withdrawal may be a natural self-protective response to negative work or life environments (Podsakoff, LePine, & LePine, 2007), it may be important for gender identity minorities to resist this urge and to maintain positive relationships. In this way, it may be possible to draw on the support of others to get through difficult times at work. Further, to make this goal easier to achieve, we recommend that organizations offer affinity groups or networking sessions specifically for LGBT individuals at work. By providing convenient access to supportive communities, organizations might contribute to gender identity minorities' abilities to build and strengthen useful networks.

Finally, by advocating for the rights of gender identity minorities and by educating others in the face of discriminatory behavior, gender identity minorities may be able to change their own narrative. In fact, recent theorizing suggests that stigmatizing attitudes can be changed and shaped by those who experience stigma through ongoing discourse regarding their identity group (Lyons, Pek, & Wessel, 2017). While it can be exhausting to both experience stigma and to try to alleviate it, because contact is a key way to eliminate prejudicial attitudes (Pettigrew, 1998), gender identity minorities are in a unique position to push back on prejudicial behaviors and to promote an alternative, positive narrative. While it is never favorable to suggest that the burden of transforming societal attitudes toward vulnerable populations should rest with those who suffer due to stigma, for those who are able and interested, gender identity minorities in workplaces and elsewhere in society may be able to provoke and sustain valuable dialogue regarding appropriate ways to create inclusive organizations. In other words, while the responsibility for addressing negative attitudes regarding gender identity minorities at work should not rest on gender identity minorities themselves, group members may play a key role in changing societal perceptions and in promulgating inclusive workplace environments.

How Can Allies Resist Discrimination against Gender Identity Minorities?

As mentioned above, ideally, resistance against discrimination toward gender identity minorities would come from cisgender allies. By taking responsibility for negative workplace environments and by speaking out against poor treatment of gender identity minorities, majority group members have the ability to drive conversations about workplace equality regarding gender expression. Allyship occurs when "a person who is a member of the dominant or majority group ... works to end oppression in his or her personal and professional life through support of, and as an advocate for, the oppressed population" (Washington & Evans, 1991, p. 195). However, while allyship appears to be valued by those who are marginalized at work, not much is known about the specific behaviors that encompass allyship at work (Sabat, Martinez, & Wessel, 2013). In a preliminary study examining the ally behaviors that are preferred by marginalized groups (in this case, LGB employees), participants reported inclusion, safety and equity as the main reasons they appreciated allyship (Brooks & Edwards, 2009). The same study suggested that allies focus on the rewards instead of the risks inherent in allyship, by working to educate, advocate, drive organizational change and gather information about how to best support marginalized groups at work. Similarly, conceptual work has argued that allyship behaviors should encompass both social support and advocacy (Sabat et al., 2014). Thus, it may be important for cisgender individuals to take the time to listen and learn from gender identity minorities at work and then to work alongside them, in a capacity that is viewed as helpful.

Despite the need to focus on rewards rather than risks in order to enact ally behaviors, there may be risks inherent in taking action to support members of the gender identity minority community. Indeed, the extant literature demonstrates that confronting prejudice must be done persuasively in order to be effective (Crittle & Maddox, 2017), that confronters may experience backlash from observers when they are perceived as hostile (Martinez, Hebl, Smith, & Sabat, 2017) and that the prospect of interacting with dissimilar others in the context of discussing inequity is anxiety-provoking (Schultz, Gaither, Urry, & Maddox, 2015).Thus, given the personal and professional challenges involved enacting allyship, building allies' capacity for courage may be important. Acts of oppositional courage (Koerner, 2014) involve risk because they require the actor to confront powerful individuals at work, in order to correct a problematic situation. Confronting those who hold power in organizations may result in backlash, social and career related setbacks and even termination (Koerner, 2014). Thus, allies need to be ready to face these consequences and to do what is right, regardless of the circumstances. This may be more difficult for those who hold other stigmatized identities (e.g., women, racial minorities, sexual orientation minorities, religious minorities), given they may be less likely to find other alternative employment opportunities. For this reason, the onus falls on the least marginalized individuals at work to take up the mantle for resisting discrimination by enacting courageous behaviors at work. Research examining the conditions under which confronting prejudiced responses (Ashburn-Nardo, Morris, & Goodwin, 2008) is most effective demonstrates that confronting prejudice in a calm and direct manner is most effective (Martinez et al., 2017). By pointing out that the perpetrator is in the wrong but remaining calm while doing so, allies are most effective in changing the attitudes of others who are watching the confrontation occur. While oppositional courage represents a broader umbrella, which includes confrontation behavior (Thoroughgood, Sawyer, & Webster, 2018b), these findings may be extrapolated more broadly to suggest that majority group members should act with conviction, but remain in control over their emotions, when advocating for marginalized groups in the workplace. Overall, despite potential perils, majority group members should be ready to stand up for the rights of gender identity minorities at work, preferably by working with them to find the best solutions to the challenges they are facing and by addressing problems in a straightforward, rational manner.

Finally, organizational leaders can prioritize modeling gender identity minority inclusivity at work. Putting policies in place that support gender identity minority needs, while leading by example with regard to being supportive of gender identity minorities, can help to set the right tone for followers. Researchers have outlined best practices for gender identity minority inclusivity in the workplace (Sawyer & Thoroughgood, 2017), including driving inclusive policies and encouraging dialogue with gender identity minorities about gender expression equality at work. Human resources leaders may have a particularly important role to play in raising awareness regarding gender identity minority status at work and in educating other leaders about best practices for inclusivity (Brooks & Edwards, 2009). Overall, leaders have a great responsibility to drive equality more broadly. But, a particular focus on gender identity may be necessary, given employees may not be familiar with the unique challenges gender identity minorities may face. If employees perceive fewer risks in standing up for the rights of gender identity minorities at work, because leaders role model inclusive behaviors, barriers to equality will be broken down more easily.

Revolutionizing Gender Binaries at Work

As noted above, one of the primary reasons that gender identity minorities face discrimination in the workplace is because gender norms are so entrenched in society. By pushing up against our deeply entrenched norms for gender expression, gender identity minorities highlight alternative ways of thinking about gender and the limitations of our current gender conformist culture. For decades, queer theorists have been writing about the transformative power of questioning binaries (mostly focusing on hetero/homo-sexual binaries) and of reimagining sex and gender in an acategorical, fluid manner. By thinking about gender as a performance, instead of as a biological imperative (Butler, 2002), we can begin to deconstruct and reconstruct the ways in which we practice gender at work and in society more broadly.

While we often think of power in organizations as being driven by status or rank, making those who lead the most powerful individuals at work, there is also power in collective practices and rituals that are constituted by the masses (Butler, 2002). Gendered practices are performed in the workplace on a regular basis, through the general adherence to the "rules" of normative gender expression in the ways that individuals dress, communicate, relate and behave on the job. By maintaining and contributing to the constitution of gender norms at work, organizations serve as sites for the reproduction of gender binaries, making employees aware that stepping outside of gender bounds may have negative social and practical consequences. In this way, employees become self-surveilling (Foucault, 1977, 1980), holding themselves and others accountable for upholding gender norms at work. Indeed, recent legal battles have demonstrated employees' ongoing awareness of standards for grooming and self-presentation. For example, five former waitresses from the Borgata casino sued their former employer for gender discrimination in enforcing employee weight standards and, in the case of Jespersen v. Harrah Operations, female employees sued their employer because they were required to wear makeup, while their male counterparts were not. Thus, employees appear to be aware of the limitations of grooming requirements on self-expression and are looking for fair and equal application of these standards across genders.

By highlighting the ways in which gender binaries are made to feel natural due to their continuous replication in society (Butler, 2002), queer theorists point out the limiting and dangerous effects of normativity on individual practices and outcomes. Gender norms constrain the bounds of socially acceptable expressions of self and promote the punishment of those who step outside of those bounds. In this way, when we attempt to determine who is to "blame" for discrimination and negative attitudes toward gender identity minorities at work – the answer is that anyone who contributes to gender normativity through a strong personal commitment to enacting gender normative behavior is responsible for these attitudes and behaviors. This is not to say that individuals should not be able to adhere to stereotypical gender norms if they wish to – but a blind reliance on norms in determining our gender expressions on the whole is problematic. If individuals were encouraged to truly explore their personal gender expression desires at work, without fear of retribution for breaking with norms, positive change may result in organizations. By asking ourselves how the boundaries of gender expression are "created, regulated, and contested" (Namaste, 1994, p. 224) at work, we may be able to free employees from inauthentic expressions of gender and allow workers to enact genuine versions of the self within the workplace.

As previously highlighted, authenticity is important to employees, but acceptance of authentic self-expressions is even more paramount in determining workplace outcomes (Martinez et al., 2017). By relaxing, exploring and transgressing the boundaries of normative gender expression at work, organizations can create environments that strategically run counter to societal gender expectations. In other words, if we know that authenticity is important at work and we know that societal constraints may keep employees from freely expressing themselves, promoting the discovery and enactment of genuine gender expression for employees may have a positive impact on the way that people work and perceive their workplace overall. For example, even if an organization is legally allowed to uphold two separate dress codes for men and women, allowing all employees to adhere to either or both standards as they please may allow for a level of exploration of gender identity within workspaces that is currently impossible for many employees. Gender identity minorities highlight how the compulsory application of gender norms in organizations can limit employees and drive negative workplace outcomes. As Halberstam (2003, p. 314) notes, "Subcultures provide a vital critique of the seemingly organic nature of 'community' and make visible the forms of unbelonging and disconnection that are necessary to the creation of community."

The idea that subcultures are useful in organizations is counterintuitive because visible unbelonging in organizations is often thought of as an indicator of a lack of fit or a poor work attitude. However, when it comes to gender identity expression, gender non-normativity may be more indicative of overly rigid standards of expression on the part of the employer, as opposed to an indicator of employee misfit. By listening to and learning from employees who highlight the cracks in the normative, binary sex-gender systems (Rubin, 1984) upon which organizations have been built, we might create freeing, more innovative spaces, in which employees feel comfortable bringing the truest version of themselves to the table. In other words, by encouraging employee innovation on all fronts – including the innovative work of reimagining gender norms – organizations might become collectively more creative and productive, by way of instituting counter-normative forms of resistance and inclusivity.

Moving beyond categorical conceptualizations of identity in organizations and reimagining the ways in which employees might (re)define themselves at work may seem like a huge undertaking. However, researchers have proposed that organizations *need* to realign the way they measure and categorize identity at work, given current modes of capturing identity are outdated and misaligned with how employees view themselves (Clair et al., In press). Indeed, while bureaucratic institutions, such as government, create immense barriers to identity expression in the way that gender identities are documented and verified (Spade, 2008), workplaces don't have to wait for governments to change their systems and processes in order to introduce their own local changes. Sending the message that identity is given more nuanced and accurate consideration within companies may help to attract and retain employees who embrace non-normative gender expressions. By constraining employees to choose from a set of predetermined gender identity categories and by enforcing stereotypical expressions associated with each these categories, organizations inevitably erase natural variability in identity expression. By instead reimagining gender as a continuum and allowing employees to define their professional self-presentations for themselves, organizations can resist gender identity discrimination from the root up.

In essence, we are suggesting that a queer theory of management may be a necessary starting place for resisting gender identity discrimination. Applying a queer theory of management means that employees are no longer bound to stereotypical expressions of gender identity (or, ideally, other social identities), thereby cultivating organizational contexts that accurately reflect the diversity of gender identity conceptualizations already present within the working population. Further, a queer theory of management allows organizations to discontinue current, ongoing cycles of self-sabotage, in which companies attempt to actively combat discrimination (e.g., through trainings, diversity mission statements, etc.), while simultaneously supporting and reifying normative, binary structures that provide the very foundation for discriminatory attitudes and behaviors (e.g., requiring gendered grooming habits, forcing employees to "check" a box regarding their gender in relevant paperwork, etc.). By first deconstructing normative expectations for identity expression, organizations can break free of the current, counterproductive approach to inclusivity, since even organizations that attempt to create inclusive work environments find themselves caught in a conundrum. Without rethinking norms, organizations enhance the likelihood for systemic discrimination by allowing traditional conceptualizations of identity to remain unquestioned, but then concurrently attempting to combat the occurrence and aftermath of inequity that arises because of these norms. In other words, it's no wonder that organizations struggle with diversity and inclusion efforts when they were built on and continue to reproduce categorical, stereotypical ways of thinking about identity and identity expression – a beautiful flower will never blossom if the soil it grows in is poisoned. If gender identity inclusive environments are going to take root within organizations, we need to change the foundational assumptions that guide the normative, gendered structures and cultures of companies first. In sum, organizations should begin by asking questions about why formal and informal rules for gender expression exist and whether or not they are rooted in gender bias or stereotypes.

Then, upon answering these questions honestly, organizations should attempt to root out policies and practices that police gender identity, in addition to training employees on the importance of accepting and encouraging authentic gender expressions. In this way, companies might best prepare their workforces, and themselves, to become trailblazers for gender expression inclusivity.

Implications for Researchers and Practitioners Attempting to Resist Gender Identity Discrimination

With regard to implications for researchers, we urge readers to focus more heavily on examining the experiences of gender identity minorities at work. We know very little about the work experiences of those who express gender in non-normative ways. Understanding more about how gender identity minorities experience the workplace will help employers better understand how to create gender identity inclusive environments. Further, as discussed above, it's important that researchers document not only the negative workplace experiences of gender identity minorities, but also the positive characteristics that they bring to the workplace.

For example, do those who think non-normatively about their gender identity apply similar acategorical, innovative thought to their jobs or organizations? Are those who have truly thought hard about who they are and which gender expressions are authentic for them also higher in self-concept clarity, making it more likely that they will self-select into jobs or organizations that "fit" them? Are those who are used to thinking non-normatively also likely to emerge as leaders in ambiguous or unprecedented situations? In all, while it is important to document the negative work experiences that gender identity minorities might encounter, it is also important to highlight what organizations might learn from gender identity minority employees. Further, understanding the conditions under which organizations are most equipped to tackle bias against gender identity minorities, as well as how organizations might promote non-normative thinking more broadly, would help leaders and organizations to create gender inclusive environments from the top down and the bottom up.

With regard to practitioners, creating organizational safeguards through policies and protections for gender identity minorities is truly the minimum that can be done. Going beyond having policies "on the books" and creating environments that empower employees to be gender authentic is the key to revolutionizing gender expression at work. By prompting all employees to feel free to express their gender identity however they'd like (albeit within the confines of what constitutes professional self-expression), organizations may create more freeing and innovative environments in which non-normative and non-binary thinking becomes commonplace. As mentioned above, there may be business-related benefits involved in promoting non-normative thought. These benefits may include a greater level of inclusivity in the workplace, the ability to synergize diversity and inclusion efforts with foundational organizational assumptions regarding self-expression, and enhanced innovation and creativity.

If companies are going to resist discrimination against gender identity minorities, they also need to resist the reification of gender norms that lay the foundation for discriminatory behavior. A comprehensive approach to gender identity inclusivity requires a queer approach to management, an approach that resists the urge to categorize and label and promulgates self-definition and expression. A queer approach to management allows for the organic naming of "what is" within organizations instead of pushing a pre-determined set of options onto employees that dictate "what must be." For example, allowing employees to self-elect their gender terms or pronouns instead of having to pick from a pre-determined set of terms would allow for a more accurate representation of gender within organizations and send an inclusive signal that aligning with a "box" is not necessary. In these ways, organizations can comprehensively role model gender inclusive ideals, while discontinuing counterproductive cycles of doing and undoing gender inclusivity at work.

Conclusions

Overall, we hope that this chapter helps to provide some concrete solutions for allies, organizational leaders and gender identity minority employees themselves in resisting gender identity discrimination. It was our goal to promote gender identity inclusivity across stakeholder groups at work and to provoke conversation about how organizations can resist normative gender identity-based thought in general. While we have attempted to start this conversation, there are most certainly research questions and practical solutions that we have missed. As such, we also hope that researchers and practitioners work with gender identity minorities to create more comprehensive solutions to eliminating bias stemming from gender expression and in transforming organizational conceptualizations of gender identity. There is far more variability in gender identity and expression in the working population than is currently reflected in the way that organizations "do gender." Rethinking gender norms in organizations allows for a radical shift in foundational assumptions about gender, spurring a transformative reimagining of inclusionary possibilities for all employees, regardless of gender identity.

References

ABC News. (2018). How redefining gender under Trump could affect transgender people's health. Retrieved December 27, 2019 from https://abcnews.go.com/Health/redefining-gender-trump-affect-transgender-peoples-health/story?

id=58742630ACLU. (2018). Transgender people and the law. Retrieved December 27, 2019 from www.aclu.org/know-your-rights/lgbtq-rights/

Ashburn-Nardo, L., Morris, K. A., & Goodwin, S. A. (2008). The confronting prejudiced responses (CPR) model: Applying CPR in organizations. *Academy of Management Learning & Education, 7*, 332–342.

Baumeister, R. F., Faber, J. E., & Wallace, H. M. (1999). Coping: The psychology of what works. In C. R. Snyder (Ed.), *Coping and egodepletion* (pp. 50–69). Oxford, UK: Oxford University Press.

Beauregard, T. A., Arevshatian, L., Booth, J. E., & Whittle, S. (2018). Listen carefully: Transgender voices in the workplace. *The International Journal of Human Resource Management, 29*, 857–884.

Beemyn, G. (n.d.). Transgender Terminology. Retrieved June 26, 2019 from https://hr.cornell.edu/sites/default/files/trans%20terms.pdf

Bem, S. L. (1983). Gender schema theory and its implications for child development: Raising gender-aschematic children in a gender-schematic society. *Signs: Journal of Women in Culture and Society, 8*(4), 598–616.

Brewster, M. E., Velez, B., DeBlaere, C., & Moradi, B. (2012). Transgender individuals' workplace experiences: The applicability of sexual minority measures and models. *Journal of Counseling Psychology, 59*, 60–70.

Brooks, A. K., & Edwards, K. (2009). Allies in the workplace: Including LGBT in HRD. *Advances in Developing Human Resources, 11*(1), 136–149.

Butler, J. (2002). *Gender trouble*. New York, NY: Routledge.

Clair, J. A., Humberd, B. K., Rouse, E. D., & Jones, E. B. (2019). Loosening categorical thinking: Extending the terrain of theory and research on demographic identities in organizations. *Academy of Management Review, 44*, 592–617.

Cohen, S., & Wills, T. A. (1985). Stress, social support, and the buffering hypothesis. *Psychological Bulletin, 98*, 310–357.

Crittle, C., & Maddox, K. B. (2017). Confronting bias through teaching: Insights from social psychology. *Teaching of Psychology, 44*, 174–180.

Drydakis, N. (2017). Trans employees, transitioning, and job satisfaction. *Journal of Vocational Behavior, 98*, 1–16.

Foucault, M. (1977). *Discipline and punish*. (A. Sheridan, trans). New York, NY: Pantheon.

Foucault, M. (1980). *Power/knowledge: Selected interviews and other writings, 1972–1977*. New York, NY: Pantheon.

Glomb, T. M., Duffy, M. K., Bono, J. E., & Yang, T. (2011). Mindfulness at work. *Research in Personnel and Human Resources Management, 30*, 115–157.

Goldman, B. M., & Kernis, M. (2002). The role of authenticity in healthy psychological functioning and subjective well-being. *Annals of the American Psychotherapy Association, 5*, 18–20.

Grant, J. A., Mottet, L. A., Tanis, J., Harrison, J., Herman, J. L., & Kiesling, M. (2008). Injustice at every turn: A report of the National Transgender Discrimination Survey. Retrieved from http://transequality.org/sites/default/files/docs/resources/NTDS_Exec_Summary.pdf.

Halberstam, J. (2003). What's that smell? Queer temporalities and subcultural lives. *International Journal of Cultural Studies, 6*, 313–333.

Human Rights Campaign. (2018) LGBTQ youth report. Retrieved November 15, 2018 from www.hrc.org/resources/2018-lgbtq-youth-report

Koerner, M. M. (2014). Courage as identity work: Accounts of workplace courage. *Academy of Management Journal, 57*, 63–93.

Kosciw, J. G., Greytak, E. A., Giga, N. M., Villenas, C., & Danischewski, D. J. (2016). *The 2015 National School Climate Survey: The experiences of lesbian, gay, bisexual, transgender, and queer youth in our nation's schools.* New York, NY: GLSEN.

Law, C. L., Martinez, L. R., Ruggs, E. N., Hebl, M. R., & Akers, E. (2011). Transparency in the workplace: How the experiences of transsexual employees can be improved. *Journal of Vocational Behavior, 79*(3), 710–723.

Level Playing Field Institute. (2007). The cost of employee turnover due solely to unfairness in the workplace. Retrieved December 27, 2019 from www.smash.org/corporate-leavers-survey

Lyons, B. J., Pek, S., & Wessel, J. L. (2017). Toward a" sunlit path": Stigma identity management as a source of localized social change through interaction. *Academy of Management Review, 42*, 618–636.

Maccoby, E. E. (1988). Gender as a social category. *Developmental Psychology, 24* (6), 755–765.

Martinez, L. R., Hebl, M. R., Smith, N. A., & Sabat, I. E. (2017). Standing up and speaking out against prejudice toward gay men in the workplace. *Journal of Vocational Behavior, 103*, 71–85.

Martinez, L. R., Sawyer, K. B., Thoroughgood, C. N., Ruggs, E. N., & Smith, N. A. (2017). The importance of being "me": The relation between authentic identity expression and transgender employees' work-related attitudes and experiences. *Journal of Applied Psychology, 102*, 215–226.

Martinez, L. R., Sawyer, K. B., & Wilson, M. C. (2017). Understanding the experiences, attitudes, and behaviors of sexual orientation and gender identity minority employees. *Journal of Vocational Behavior, 103*(Part A), 1–6.

Nadal, K. L., Davidoff, K. C., Davis, L. S., & Wong, Y. (2014). Emotional, behavioral, and cognitive reactions to microaggressions: Transgender perspectives. *Psychology of Sexual Orientation and Gender Diversity, 1*, 72–81.

Namaste, K. (1994). The politics of inside/out: Queer theory, poststructuralism, and a sociological approach to sexuality. *Sociological Theory, 12*, 220–231.

National Center for Transgender Equality. (2016). 2015 U.S. transgender survey. Retrieved December 27, 2019 from https://transequality.org/sites/default/files/docs/usts/USTS-Executive-Summary-Dec17.pdf

Pettigrew, T. F. (1998). Intergroup contact theory. *Annual Review of Psychology, 49*, 65–85.

Podsakoff, N. P., LePine, J. A., & LePine, M. A. (2007). Differential challenge stressor- hindrance stressor relationships with job attitudes, turnover intentions, turnover, and withdrawal behavior: A meta-analysis. *Journal of Applied Psychology, 92*, 438–454.

Ragins, B. R. (2008). Disclosure disconnects: Antecedents and consequences of disclosing invisible stigmas across life domains. *Academy of Management Review, 33*, 194–215.

Ragins, B. R., & Cornwell, J. M. (2001). Pink triangles: Antecedents and consequences of perceived workplace discrimination against gay and lesbian employees. *Journal of Applied Psychology, 86*, 1244.

Ragins, B. R., Singh, R., & Cornwell, J. M. (2007). Making the invisible visible: Fear and disclosure of sexual orientation at work. *Journal of Applied Psychology, 92*, 1103.

Rubin, G. (1984). Thinking sex: Notes for a radical theory of the politics of sexuality. In C.S. Vance (Ed), *Pleasure and danger: Exploring female sexuality.* Boston, MA: Routledge.

Ruggs, E. N., Martinez, L. R., Hebl, M. R., & Law, C. L. (2015). Workplace "trans"-actions: How organizations, coworkers, and individual openness influence perceived gender identity discrimination. *Psychology of Sexual Orientation and Gender Diversity, 2,* 404–412.

Ryan, R. M., LaGuardia, J. G., & Rawsthorne, L. J. (2005). Self-complexity and the authenticity of self-aspects: Effects on well-being and resilience to stressful events. *North American Journal of Psychology, 7,* 431–448.

Sabat, I. E., Lindsey, A. P., Membere, A., Anderson, A., Ahmad, A., King, E., & Bolunmez, B. (2014). Invisible disabilities: Unique strategies for workplace allies. *Industrial and Organizational Psychology, 7,* 259–265.

Sabat, I. E., Martinez, L. R., & Wessel, J. L. (2013). Neo-activism: Engaging allies in modern workplace discrimination reduction. *Industrial and Organizational Psychology, 6,* 480–485.

Sawyer, K., & Thoroughgood, C. (2017). Gender non-conformity and the modern workplace. *Organizational Dynamics, 1,* 1–8.

Sawyer, K., Thoroughgood, C., & Webster, J. (2016). Queering the gender binary: Understanding transgender workplace experiences. In T. Köllen (Ed.), *Sexual orientation and transgender issues in organizations* (pp. 21–42). Switzerland: Springer.

Schultz, J. R., Gaither, S. E., Urry, H. L., & Maddox, K. B. (2015). Reframing anxiety to encourage interracial interactions. *Translational Issues in Psychological Science, 1,* 392–400.

Sheldon, K. M., Ryan, R. M., Rawsthorne, L. J., & Ilardi, B. (1997). Trait self and true self: Cross-role variation in the big five personality traits and its relations with psychological authenticity and subjective well-being. *Journal of Personality and Social Psychology, 73,* 1380–1393.

Spade, D. (2008). Documenting gender. *Hastings Law Journal,* 59, 731–841.

Thoroughgood, C. N., Sawyer, K. B., & Webster, J. R. (In press). Finding calm in the storm: A daily investigation of how trait mindfulness buffers against paranoid cognition and emotional exhaustion following perceived discrimination at work. *Organizational Behavior and Human Decision Processes.*

Thoroughgood, C. N., Sawyer, K. B., & Webster, J. (2018b, April). Social effects of oppositional courage on individuals with stigmatized identities. In K. Dray & I. Sabat (Chairs) (Ed.), *Workplace allies: Exploring the process of becoming an effective and vocal ally.* Symposium presented at the 33rd Annual Conference of the Society for Industrial and Organizational Psychology. Chicago, IL.

Thoroughgood, C. N., Sawyer, K. B., & Webster, J. R. (2017). What lies beneath: How paranoid cognition explains the relations between transgender employees' perceptions of discrimination at work and their job attitudes and wellbeing. *Journal of Vocational Behavior, 103,* 99–112.

Toomey, R. B., Syvertsen, A. K., & Shramko, M. (2018). Transgender adolescent suicide behavior. *Pediatrics,* 142.

Transgender Law Center. (2004). Transgender health and the law: Identifying and fighting healthcare discrimination. Retrieved from: http://transgenderlawcenter.org/wp-content/uploads/2012/07/99737410-Health-Law-Fact.pdf

Transgender Law Center. (2017). In "vicious" memo, DOJ encourages illegal discrimination against transgender workers. Retrieved December 27, 2019 from www.trans genderlawcenter.org/archives/1405

Viswesvaran, C., Sanchez, J. I., & Fisher, J. (1999). The role of social support in the process of work stress: A meta-analysis. *Journal of Vocational Behavior, 54,* 314–334.

Washington, J., & Evans, N. J. (1991). Becoming an ally. In N. J. Evans & V. A. Wall (Eds.), *Beyond tolerance: Gays, lesbians and bisexuals on campus* (pp. 195–204). Alexandria, VA: American College Personnel Association.

Wentling, T., Windsor, E., Schilt, K., & Lucal, B. (2008). Teaching transgender. *Teaching Sociology, 36,* 49–57.

Yoshino, K. (2007). *Covering: The hidden assault on our civil rights.* New York, NY: Random House Trade Paperbacks.

Zellars, K. L., & Perrewé, P. L. (2001). Affective personality and the content of emotional social support: Coping in organizations. *Journal of Applied Psychology, 86,* 459–467.

5 Stigma as Diversity Resistance to Employees with Mental Illness

Kayla Follmer and Kisha Jones

One in five U.S. adults – nearly 46 million people – experiences a mental illness each year (National Alliance on Mental Illness, NAMI, 2017) and many of these individuals are actively involved in the workforce. Not only is mental illness prevalent in the United States, it also can affect employees' productivity (Greenberg, Fournier, Sisitksy, Pike, & Kessler, 2015) and general workplace experiences (Follmer & Jones, 2018). Thus, understanding and supporting employees with mental illness is a timely workplace concern. Unfortunately, empirical evidence suggests that many organizations lack the information or preparedness to support employees with mental illness (Fairclough, Robinson, Nichols, & Cousley, 2013). At the same time, rhetoric at the national level has portrayed individuals with mental illness as dangerous to themselves and others in society (McGinty, Kennedy-Hendricks, Choksy, & Barry, 2016). This increased negative discourse indicates a regression toward outdated, negative social perceptions of those with mental illness which affects how these individuals are perceived and treated in society, including at work. Thus, we believe educating researchers and practitioners about this unique employee population is an important step in raising awareness of and reducing barriers for employees with mental illness in order to combat resistance towards this population.

In this chapter, we aim to depict stigma as a form of diversity resistance that impacts how people with mental illness experience the workplace. We begin by defining mental illness, differentiating it from other disabilities and detailing mental illness stigma. Next, we explain how stigma impacts the treatment of individuals with mental illness, operating as diversity resistance in both overt and subtle forms. We consider how resistance towards this group can make it difficult for employees with mental illness to disclose their condition and make accommodation requests. Finally, we conclude with a discussion of strategies organizations can use to reduce resistance towards employees with mental illness in order to create a more supportive and inclusive work environment.

Mental Illness Defined

The term mental illness encompasses more than 200 classified mental health disorders outlined in the fifth edition of the American Psychiatric Association's *Diagnostic and Statistical Manual of Mental Disorders* (DSM-V). Any type of mental, behavioral, or emotional disorder is considered a mental illness. On the other hand, serious mental illness is defined as "a mental, behavioral, or emotional disorder resulting in serious functional impairment, which substantially interferes with or limits one of more major life activities," including employment (NIMH, 2019). Mental illnesses range in prevalence and severity, with some creating more disruptions than others. Yet, many mental illnesses are manageable if treated by a trained professional (Seligman, 1995).

It is not feasible to review all of the disorders that are covered in the DSM-V. Therefore, the scope of this chapter is focused on the disorders that are most prevalent among the working population, including depression, bipolar disorder, anxiety, eating disorders, attention-deficit hyperactivity disorder (ADHD) and post-traumatic stress disorder (PTSD). Importantly, this chapter does not address individuals with developmental disabilities (e.g., autism) or substance abuse disorders (e.g., addiction), even though these are also encompassed within the DSM-V. Table 5.1 provides a brief description of these disorders as well as an estimate of the prevalence rates and average ages of onset. Although it is difficult to quantify the exact number of individuals living with mental illness, data from 2017 suggest that 46.6 million adults were living with any mental illness, and of those, 11.2 million were affected by a serious mental illness. On average, mental illnesses are more prevalent among women (compared to men), individuals between the ages of 18 and 25 and individuals who report more than two races (NIMH, 2019).

Although frequently used interchangeably, it is necessary to differentiate between mental illness and mental health. Both mental illness and mental health represent unique continuums and it is possible for an individual to be high or low on either (Keyes, 2005). Whereas mental illness represents a diagnosable, psychiatric disorder, mental health represents a state of positive functioning and well-being that is generally related to one's perceptions of their quality of life and happiness. In this way, it is possible for an individual who does not have a diagnosable mental illness to maintain poor mental health. On the contrary, an individual who has a diagnosable mental illness may report positive mental health and functioning. Throughout this chapter, the focus is on individuals with clinical, psychiatric disorders and the challenges they experience related to inclusion in the workplace.

Mental Illness as a Form of Disability

According to the Americans with Disabilities Act (ADA) of 1990, a disability is defined as a "physical or mental impairment that substantially limits one or more major life activities," (U.S. Department of Justice, 2018)

Table 5.1 Descriptions and prevalence of mental illnesses

Disorder Name	Description of Disorder	Prevalence in Adults	Average Onset Age
Generalized Anxiety Disorder	Excessive worry about everyday problems. Difficulty minimizing or controlling concerns.	3.1%	31
OCD	Intrusive thoughts that produce anxiety. Ritualized, repetitive behaviors are used to reduce anxiety.	1.0%	19
PTSD	Persistent frightening thoughts or memories of a traumatic event. Trouble sleeping, feeling detached and easily startled.	3.5%	23
ADHD	Difficulty staying focused, paying attention, or controlling behavior.	4.1%	7
Anorexia Nervosa	Relentless pursuit of thinness; a distortion of body image and intense fear of gaining weight.	N/A	19
Binge Eating Disorder	Recurrent binge-eating episodes in which a person feels a lack of control over their food consumption behaviors.	1.2%	25
Bulimia Nervosa	Recurrent episodes of eating large amounts of food followed by compensation behaviors (e.g., purging, exercise).	.3%	20
Bipolar Disorder	Dramatic changes in mood, energy and activity levels. Moods can persist for weeks at a time.	2.6%	25
Dysthymia	Chronic, low level depression.	1.5%	31
Major Depressive Disorder	Depressed mood, loss of interest, changes in daily functioning (e.g., sleep disruptions, inability to concentrate).	6.7%	N/A
Antisocial Personality Disorder	Disregard of social norms and laws, propensity for risk taking and frequent lying behavior. Also called sociopathy.	1.0%	N/A
Avoidant Personality Disorder	Extreme shyness, feelings of inadequacy and acute sensitivity to social rejection. Characterized by avoidance of social interactions.	2.5%	N/A
Borderline Personality Disorder	Instability in moods, interpersonal relationships, self-image and behavior. Moods last for a few hours to a day.	1.6%	N/A
Dissociative Disorders	Disconnection between thoughts, identity, consciousness and memory.	2.0%	N/A
Schizophrenia	Deficits in thought processes, perceptions and emotional responsiveness.	1.1%	N/A

including one's ability to work. In 2008, the ADA Amendments Act (ADAAA) was enacted to broaden the definition of disability as well as the conditions that would qualify for consideration so as to be more inclusive of conditions or disorders that were previously not covered by ADA (e.g., depression, cancer). Included within the ADAAA of 2008 are "rules of construction" that aim to create consistency in the standards used to evaluate whether or not an individual has a disability. These rules of construction state that 1) a condition does not have to prevent or severely restrict a major life activity to be considered substantially limiting, 2) the determination of the extent to which an impairment is substantially limiting requires individualized assessment, 3) limitations to life activities will be made without considering the effects of treatment on a given condition and 4) an episodic impairment is considered a disability if the condition could limit a major life activity when active.

These rules of construction are especially pertinent when considering the characteristics of mental illness that may disrupt one's ability to work. A mental illness may not preclude an individual from working at all, though it could create limitations in the type of work or working conditions related to a given job. In addition, a mental illness should qualify as a disability regardless if the individual is receiving treatment for his or her disorder and should qualify even if the disorder occurs at irregular intervals (e.g., depression, eating disorder). Finally, it is important to consider that each employee's situation will be unique and that the extent to which a mental illness creates a significant impairment will vary across individual experiences.

Although mental illness is legally considered a disability, it is often overlooked in discussions or research of employees with disabilities. One reason for this omission may be due to inherent differences between characteristics of mental illness and physical disabilities. First, a mental illness is not always visible to other individuals. Consequently, the symptoms associated with mental illness may not be easily understood by others as a manifestation of a serious mental health disorder. While some physical disabilities (e.g., using a wheelchair, blindness) may be apparent by looking at an individual, it is often difficult to discern whether or not an individual has a mental illness simply by looking at or even interacting with him or her. Because many mental illnesses can be concealed from others in the workplace, a disorder may only become salient if the individual chooses to disclose his or her condition to others. In this way, the condition must be disclosed in order for other individuals to recognize that it exists and for the provision of accommodations.

Unfortunately, the lack of visibility of many mental illnesses can influence the extent to which a given disorder is perceived as a legitimate disability. Common perceptions of mental illnesses are that they are "all in the heads" of those who suffer from them, rather than representing legitimate health disorders (Rüsch, Todd, Bodenhausen, & Corrigan, 2010). A physical disability, on the other hand, may be perceived as a more valid condition than mental illness, based in part on its assumed anatomical and physiological foundations (Aftab, 2014). For this reason, an employer may perceive accommodation requests for physical disabilities as more legitimate than those for psychological disabilities.

Finally, mental illnesses may be differentiated from other physical disabilities[1] in that the symptoms associated with a mental illness often fluctuate across time periods. For example, an employee with depression may not consistently appear depressed to others. Rather, the employee may go several weeks or months without exhibiting depressive symptoms and then suddenly experience a prolonged depressive episode. These changes in an employee's attitudes or behaviors may affect the extent to which others perceive the disorder as serious or legitimate. Further, variability in the manifestation of symptoms also affects work-related accommodations for employees with mental illness. For example, an employee with depression may require increased scheduling breaks, flexible work hours, or time away from work for treatment-related reasons, but only during a depressive episode. This experience can be contrasted with that of an employee with hearing or visual impairments, for example, who may consistently – rather than intermittently – rely upon accommodations. Given the relatively reduced legitimacy of some mental illnesses compared to physical illnesses, this inconsistent need for accommodations may be interpreted as a poor work ethic and run counter to expectations of an ideal employee.

In summary, mental illnesses are not only serious, legitimate psychiatric conditions that can affect employees' work experiences, but they are also protected as disabilities under federal law. Despite these protections, employees with disabilities often experience mistreatment in the workplace, much of which is related to the stigma associated with having a mental illness.

Mental Illness Stigma

Stigma is a pervasive, unavoidable aspect of social life in which individuals are labeled as either "normal" or "abnormal." Individuals who fail to adhere to conceptualizations of normalcy may be perceived to have an identity that is deviant, flawed, or spoiled (Goffman, 1963); thus, these individuals bear a social "mark" that subjects them to stigmatization. A social mark does not, however, need to be visible in order for it to be stigmatizing.

The degree of stigma attached to a given identity varies based upon six dimensions that underlie stigma content: concealability, course, disruptiveness, aesthetic qualities, origin and peril (Jones et al., 1984). *Concealability* is the extent to which a stigmatizing condition is visible to others; *course* concerns the pattern of a stigmatizing condition over time; *disruptiveness* is the extent to which a stigmatizing condition impedes social interaction and communication; *aesthetic qualities* highlight the extent to which a stigmatizing characteristic makes an individual "repellent, ugly or upsetting" (p. 24); *origin* concerns the circumstances under which the individual acquired the stigmatized identity and *peril* is the extent to which the stigmatizing characteristic is dangerous to the self or others.

Each specific mental illness varies in the extent to which it is considered stigmatizing based on perceived alignment between a disorder's symptoms and the six dimensions of stigma. For example, with more severe forms of mental illness (e.g., schizophrenia, bipolar disorder), individuals may be perceived as dangerous to others (e.g., high perceptions of peril) yet such perceptions are unlikely to be applied to individuals with mental illnesses such as ADHD or anxiety. On the other hand, the origin of disorders may be perceived differently, with some mental illnesses perceived to be "caused" by or in the control of the individual (e.g., eating disorders) while others may be perceived as out of someone's control (e.g., obsessive-compulsive disorder). Indeed, prior research has demonstrated that attitudes toward individuals with mental illness vary based on the type of disorder being evaluated (Follmer & Jones, 2017); thus, to understand mental illness and its related stigma, it is important to first consider the extent to which it is generally considered stigmatizing by others.

The stigma of mental illness is a pervasive social phenomenon that has persisted across time (Angermeyer & Dietrich, 2006; Follmer & Jones, 2017). Everyday rhetoric includes words like "psycho" and "crazy" to describe individuals with mental illness, which can thwart progress toward acceptance and inclusion of these individuals in society at large, as well as in the workplace. Indeed, the relationship between mental illness and violence is greatly exaggerated in the media (Stuart, 2006). For instance, the current president of the United States has, on multiple occasions, disparaged individuals with mental illness, thereby contributing to the stigma that these persons are dangerous and unstable. In response to two school shooting incidences, Donald Trump used the words "wacky" and "mentally disturbed" to describe the shooters. Even more, he publicly recounted previous eras in which "[the U.S.] had mental institutions. We had a lot of them. And you could nab somebody like this, because they knew something was off" (Shear, 2018). Comments such as these at the national level exacerbate and perpetuate negative perceptions of individuals with mental illness. Further, the continued negative attention toward individuals with mental illness facilitates the alienation of these individuals in society and detracts from efforts needed to enact positive social change.

Public vs. Internalized Stigma

Stigma represents a double-edged sword for those with mental illness. The first type of stigma – and the type prominently discussed in this chapter – is called *public stigma*, which refers to the reaction that the general population has to people with mental illness (Corrigan, 2000). According to the social-cognitive model of stigma (Corrigan, 2000; Crocker & Lutsky, 1986), stigma develops from the ways in which individuals process information. Signals are taken from the environment and these are given meaning through attitudes and stereotypes, which in turn influence discrimination. *Signals* may include information about an individual's symptoms, labels that have been applied to an individual (e.g., schizophrenic, mentally ill) and a person's physical appearance. When these signals are observed, individuals attempt to make sense of them by way of *stereotypes*. Common stereotypes of individuals with mental illness include perceptions that these persons are dangerous, dependent on others, lacking in competence and unstable (Corrigan et al., 2000). Stereotypes, in turn, contribute to *discriminatory treatment* of those with mental illness. Taken together, public stigma influences how society, as a whole, views and treats those with mental illness.

The consequence of public stigma is that individuals living with mental illness may feel devalued or excluded from society. Those who experience mental illness are likely aware of the stereotypes associated with mental illness, but they may also apply these stereotypes to themselves. This process is called *internalized stigma* or *self-stigma*. Ritsher, Otilingam and Grajales (2003) identified four dimensions of internalized stigma that explain how stigma shapes the ways in which individuals think about themselves and their place in society. *Alienation* refers to the extent to which individuals with mental illness feel like "less than a full member of society" (Ritsher et al., 2003, p. 36); *stereotype endorsement* involves individuals applying common stereotypes about mental illness to themselves (e.g., mentally ill people are dangerous, therefore I am dangerous); *discrimination experience* includes individuals' perceptions of mistreatment due to their mental illness; and *social withdrawal* represents an individual's attempt to limit social interaction with others due to having a mental illness. A meta-analysis of the consequences of internalized stigma (Livingston & Boyd, 2010) revealed that higher levels of internalized stigma were related to decreased perceptions of hope, empowerment, self-esteem, quality of life and social support. Even more, internalized stigma was associated with reduced treatment adherence, suggesting that internalized stigma may preclude individuals from seeking care that they need to manage their mental illness.

Taken together, stigma is a pernicious social phenomenon that can impart great harm for its targets. Although public stigma affects how individuals with mental illness are treated by others, internalized stigma can result in self-inflicted psychological harm. Regardless of the form it takes, stigma creates unnecessary obstacles for those living with mental illness and can make it harder for these persons to thrive in broader society, as well as the workplace.

Stigma as Diversity Resistance

Stigma is harmful to individuals with mental illness because it taints perceptions of and treatment toward these persons. On balance, stigma can perpetuate inaccuracies about individuals with mental illness, which can impede their ability to obtain and maintain employment (Ren, Paetzold, & Colella, 2008). In this way, stigma represents a unique form of diversity resistance – "a range of practices and behaviors within and by organizations that interfere, intentionally or unintentionally, with the use of diversity as an opportunity for learning and effectiveness" (Thomas & Plaut, 2012, p. 5). In this section, we review conceptual models created by Stone and Colella (1996) and Thomas and Plaut (2012) that aim to explain how stigma influences mistreatment of those with mental illness as well as illustrate what that mistreatment may look like in organizations. We also consider the impact of intersectionality on one's experience of diversity resistance related to having a mental illness.

Stone and Colella's Model of Treatment of Individuals with Disabilities

In 1996, Stone and Colella proposed a theoretical model that outlined the legal, organizational and individual factors that influence treatment of individuals with disabilities. This model defined "disability" broadly as to be more inclusive of individuals with both physical and mental disabilities. More specifically, the model posits that specific stigma characteristics (e.g., peril, visibility) of a given disorder can result in categorization, stereotyping, inaccurate expectations (e.g., person is unqualified or unable to perform) and negative affective states among observers of the individual with a mental illness. Essentially, the knowledge that an individual has a stigmatized identity can result in an "us vs. them" mentality (Rüsch, Angermeyer, & Corrigan, 2005), in which those with stigmatized identities are cognitively categorized as separate from non-stigmatized individuals. In this way, awareness that one is working with or around an individual with a stigmatized mental illness can result in psychological consequences for the individuals who observe them. These general propositions have been supported in the research literature. Using a vignette study, Angermeyer and Matschinger (2003) demonstrated that perceptions of individuals with depression and schizophrenia as dangerous resulted in increased feelings of both fear and anger. In another vignette study conducted by Feldman and Crandall (2007), perceptions of dangerousness (i.e., peril) and personal responsibility for condition (i.e., origin) were the greatest predictors of participants' desires to socially distance themselves from individuals with psychiatric disorders. Likewise, Corrigan and colleagues (2000) found that controllability (i.e., origin) and stability (e.g., course) of the disorder were most strongly related to perceptions of stigma for mental illness. Specifically, psychosis and depression were both perceived to be more controllable than either intellectual impairments or cancer. On balance, these studies illustrate that individuals with mental illness were often assumed to be dangerous and as responsible for their own condition, which in turn shaped observers' negative affective responses, such as discomfort being near these individuals.

Unfortunately, stigma not only shapes how individuals with mental illness are cognitively perceived by others, it also influences the ways in which they are treated. In the Stone and Colella (1996) model, negative perceptions of those with mental illness yield serious implications for their job performance ratings, job assignments, pay raises and inclusion in their workgroup. Several qualitative and quantitative research studies have demonstrated that individuals with mental illness are often subjected to discriminatory treatment in the workplace. Employees with mental illness are unemployed or underemployed relative to their skills and abilities, especially among individuals with the most severe forms of mental illness (Baldwin & Marcus, 2007). Two qualitative studies involving employees with serious mental illness reported that participants frequently experienced discriminatory treatment at work, such that coworkers questioned their work ability and competence, employees felt ostracized from others in the workplace and, in some extreme cases, employees were terminated after diagnoses were disclosed (Baker & Procter, 2014; Russinova, Griffin, Bloch, Wewiorski, & Rosoklija, 2011). Similar results have been obtained in vignette-based and survey studies, in which managers indicated a decreased likelihood of hiring an individual with depression as compared to individuals with physical disabilities (Corrigan, Larson, & Kuwabara, 2007) as well as general concerns about hiring individuals with mental illness (Hauck & Chard, 2009). These studies illustrate the harmful effects of stigma on the work outcomes of employees with mental illness.

Thomas and Plaut's (2012) Taxonomy of Diversity Resistance

The Stone and Colella (1996) model provided a foundation for conceptualizing why individuals with disabilities may be mistreated at work. However, the study of discriminatory behaviors targeted at employees with mental illness may be further enriched by considering these behaviors as specific acts of diversity resistance. According to Thomas and Plaut's (2012) taxonomy of diversity resistance, resistance manifests via subtle or overt behaviors and can occur at both the individual and organizational level. We next discuss how stigma influences the ways in which individuals with mental illness are treated in the workplace, thereby creating barriers to their inclusion.

Overt Diversity Resistance to Mental Illness

Overt forms of diversity resistance are often blatant, hard-to-miss discriminatory and harassing behaviors targeted towards a stigmatized group (Hebl, Foster, Mannix, & Dovidio, 2002; Thomas & Plaut, 2012). Although this type of discrimination is considered illegal towards members of protected classes, including those with disabilities like mental illness, it still occurs in modern organizations. Overt discrimination can be perpetrated by individual employees or by the broader organization via discriminatory employment practices or retaliation. In one study of employees with mental illness (Russinova et al., 2011) participants recounted being passed over for promotions, being discredited in the hiring process and even being verbally harassed at work due to their mental illness. One participant stated that their boss made public comments such as, "Let me know if you're going to blow us all away," while another recounted that they were discouraged from applying for a job that had previously been promised to them after receiving in-patient psychiatric treatment (Russinova et al., 2011, p. 232). In a qualitative study of employees with depression, one participant shared a story in which her boss discussed the need to hire a new paralegal after the prior paralegal (who had depression) left the organization. During a company meeting the boss stated, "Oh, we're looking for a paralegal but not one that has manic depression" (Follmer, 2017). In other instances, employees have reported being denied reasonable accommodations at work related to their mental illness (Gold, Oire, Fabian, & Wewiorski, 2012).

These studies and others (for a review see Follmer & Jones, 2018) exemplify how individuals with mental illness are subjected to overt forms of discrimination in the workplace. When coworkers and supervisors make blatant disparaging comments to those with mental illness, it can create a hostile work environment and decrease the quality of work life for those who have been targeted. In addition, organizations that explicitly deny opportunities or fail to provide employees with accommodations not only contribute to the continued resistance towards individuals with mental illness but are also breaking the law. Mistreatment of employees with mental illness may be explained, in part, by discriminatory organizational routines that focus on how the "majority of workers view the world or approach tasks" (Wooten & James, 2005, p.130). These discriminatory routines manifest in organizations that allow harassment of individuals with disabilities, lack the infrastructure to support employees with disabilities and are unwilling to provide reasonable accommodations to employees who may need them. In this way, organizational routines that fail to consider the unique experiences of employees with mental illness may create unspoken norms of mistreatment that consciously perpetuate discriminatory treatment toward these individuals.

Subtle Diversity Resistance to Mental Illness

Subtle diversity resistance assumes a more passive form than overt resistance; however, it contributes to diversity resistance via behaviors that alienate stigmatized individuals or failures to intervene when mistreatment occurs. For instance, in a study of job termination among those with severe mental illnesses, over half of the cases where people were fired or left without new employment stemmed from interpersonal issues with others in their workplace (Becker et al., 1998; Stuart, 2006). These interpersonal difficulties can include coworkers who withhold help and assistance as well as actively avoid and socially distance themselves from those with mental illness (Hebl, Madera, & King, 2008). Subtle discrimination is harmful because it allows for mistreatment to be perpetuated in a way that often occurs "under the radar." That is, rather than manifesting overtly, negative attitudes toward those with mental illness manifest in ways that are hard to detect (Angermeyer & Matschinger, 2003). Such subtle mistreatment represents a form of diversity resistance because it contributes to exclusionary social behavior as well as allows witnessed mistreatment of employees with mental illness to go unaddressed. For instance, Follmer and Jones (2017) found that warmth and competence stereotypes of employees with anxiety, depression and bipolar disorder predicted social distancing intentions towards employees with those disorders. In this study, the desire to socially distance oneself included, as examples, discomfort sharing an office with an individual with mental illness as well as discomfort in supervising or being supervised by an employee with mental illness. If others in the workplace do not want to be associated or affiliated with coworkers who are known to have a mental illness, it is likely that they will not intervene or speak up on the behalf of a person who has mental illness.

Intersectionality and Diversity Resistance to Mental Illness

Intersectionality theory considers the joint effects of various identities on the extent to which one experiences advantage and disadvantage. Given the implications that the intersections of identities such as class, race and gender have for employment discrimination, as well as access to healthcare and resources that could support the successful management of one's mental illness (Rosenfield, 2012), it is important to consider whether the intersections of certain identity groups lead to unique circumstances that put them at risk to experience diversity resistance. Double jeopardy theory (Almquist, 1975) suggests that the extent to which individuals maintain multiple minority group memberships heightens the amount of disadvantage they experience in society. There is already evidence that this theory plays out when considering the experience of mental illness among people with multiple marginalized identities. For instance, LGBT people of color with mental illness experience increased stigmatization and lack of satisfaction with the mental health treatment they receive (Holley, Tavassoli, & Stromwall, 2016).

More research is needed to explore ways populations at certain intersections experience diversity resistance based on mental illness. Existing work has considered how intersections of marginalized identities impact one's experience with employment discrimination due to disability. Shaw, Chan and McMahon (2012) examined the interactive effects of demographic characteristics on the likelihood of experiencing disability discrimination as documented by Equal Employment Opportunity Commission (EEOC) discrimination claims and found that being female, being older, having a behavioral disability, being a racial minority and working for either a very large or very small company put one at a higher chance of experiencing discrimination based on disability. In a later study, McMahon and colleagues (2017) found that individuals with learning disabilities were more likely to experience discrimination such as harassment, withholding of training and disciplinary actions compared to individuals with other disabilities. Although this work did not look specifically at mental illness, these results suggest that discrimination may vary across identity groups such that the type of disability, including various forms of mental illness, shapes the way mistreatment manifests in organizations. Future research should further explore this as well as whether strategies for increasing inclusion are equally effective across intersecting identities.

As detailed by our consideration of Stone and Colella's (1996) and Thomas and Plaut's (2008) models, we see that stigma has severe implications for the work outcomes of people with mental illness, impacting not only hiring outcomes and career mobility, but producing both overt and subtle mental illness resistance. Further, one's status as a member of multiple minority groups may heighten one's negative workplace experiences due to their mental illness. This not only has general negative effects on employees with mental illness but can also impact the extent to which they feel comfortable seeking accommodations to assist them in performing their job duties. In fact, because the accommodations are provided by law as a method of alleviating the management of the conditions of those with disabilities, we consider factors that impede one's ability to seek accommodations as a form of diversity resistance.

Resistance as a Barrier to Seeking Accommodations

The Americans with Disabilities Act of 1991 and the Americans with Disabilities Act Amendments Act of 2008 had the purpose of providing those with disabilities an opportunity to receive assistance in functioning across daily life activities, including work. Importantly, these laws require that employees disclose their condition at work and initiate conversations around what reasonable accommodation they think will be of assistance. In most cases, this conversation is had with the employee's direct supervisor (Follmer & Beatty, 2018) and discussions take place until an agreed upon accommodation plan is achieved. Although the intent of this legislation is to enable people with disabilities to receive help in performing to the best of their abilities, subtle and overt diversity resistance towards those with mental illness can make it difficult for them to take advantage of it. Specifically, individual and organizational diversity resistance may preclude individuals from self-disclosing their mental illness out of concern for subsequent mistreatment.

There are several models that provide insight into factors that impact the disclosure of concealable identities in the workplace to coworkers and supervisors, including perceived support, anticipated acceptance and stigma characteristics (e.g., K. P. Jones & King, 2014; Ragins, 2008). In the accommodation request likelihood framework, Baldridge and Veiga (2001) proposed a conceptual model of the factors that would impact whether employees with disabilities would make a request for accommodations. Several factors were outlined, including the extent to which the employee believes the accommodation would address his/her need, how s/he would be perceived after making the request (e.g., as a troublemaker or incompetent), perceived resistance from the supervisor, the appropriateness of the request, the accommodation and inclusion culture, attributes of the accommodation being requested and attributes of the disability. In this way, individuals assess their environment to determine whether the advantages of receiving an accommodation outweigh the possible social consequences that may ensue.

Each of the factors presented in Baldridge and Veiga's (2001) model represent possible ways that diversity resistance can deter individuals with mental illness from seeking accommodations. First, employees' beliefs about the extent to which the accommodation would be helpful to their work performance may be influenced by the employees' knowledge of accommodations. If organizational leaders show resistance to mental illness (e.g., stereotype individuals with mental illness as incompetent or dangerous), they will be less proactive in providing examples of accommodations that employees with mental illness could request. Further, employees who believes that they will be viewed negatively after disclosing a mental illness during the accommodation request process may be reluctant to request accommodations at all. Further, employees may feel the need to protect their image in the workplace and concerns about future career mobility or perceptions of receiving an unfair advantage could preclude them from seeking needed accommodations. For instance, accommodations for people with mental illness can include modifying work schedules and time, work rules or procedures, performance expectations, or job tasks (Fabian, Waterworth, & Ripke, 1993) – all activities that could be viewed negatively by coworkers if noticed. Indeed, research has shown that coworkers have negative perceptions of individuals with disabilities who outperform others after receiving an accommodation (Paetzold et al., 2008).

Negative evaluations of employees with mental illness who request accommodations may be even more likely in competitive versus cooperative environments. In contrast to a work environment that promotes ethical treatment of employees with mental illness (Barclay & Markel, 2009), a work environment with high diversity resistance to mental illness would make accommodation requests more difficult. Of concern, continuous negative stereotyping and stigmatization of people with mental illnesses can produce institutional discrimination where people with mental illness are excluded from certain job levels and positions (Corrigan et al., 2005), thereby illustrating the importance of promoting access to accommodations for those who need them. Despite the ADA's intentions to provide more opportunities for individuals with mental illness, many organizations demonstrate resistance towards employees with mental illness. Such resistance is not only unethical it is also illegal and cost prohibitive.

Notable Court Cases Involving Employees with Mental Illness

Further evidence for the presence of diversity resistance towards individuals with mental illness exists in records of the legal action taken against organizations through lawsuits and claims filed with the EEOC. The EEOC reported that their office resolved approximately 5,000 charges of mental health-based discrimination in the 2016 fiscal year. Even more, the office recouped $20 million in damages for those employees who had been unlawfully denied employment because of their mental illness (EEOC, 2016). Below, we review recent court cases and settlements that involved discrimination toward employees with mental illness.

The EEOC filed a lawsuit against AccentCare, Inc. on behalf of an IT analyst who was fired after disclosing her bipolar disorder to request time off from work for treatment. The courts found AccentCare's actions to violate ADA law, ruled in favor of the EEOC and in 2017 required the organization to pay $25,000 in monetary relief to the former employee (EEOC, 2017). In 2018, Mine Rite Technologies settled a discrimination lawsuit brought forth by an employee with PTSD. The employee reported that his supervisor harassed him at work and often called him disparaging names such as "psycho." On Thursdays, the employee attended therapy sessions for his PTSD, to which his boss referred to these days as "Psycho Thursday." The abuse escalated to the point that the employee eventually quit his job. In the settlement of the case, Mine Rite Technologies agreed to pay the employee $75,000 (EEOC, 2018). As another example, Crain Automotive Holdings agreed to pay $27,100 to settle a discrimination lawsuit with an employee who experienced depression and anxiety in 2019. In this case, the employer refused to provide a medical leave of absence to the employee and subsequently fired her, a violation of ADA law (EEOC, 2019).

These cases illustrate how employees with mental illness are often victims of workplace discrimination and mistreatment. We argue that organizations have both an ethical and legal imperative to make their organizations more welcoming of persons with mental illness. Below, we discuss strategies that organizations can implement to minimize resistance towards employees with mental illness thereby increasing support and inclusion of these employees.

Strategies for Overcoming Resistance to Individuals with Mental Illness

Because all employees deserve the opportunity to obtain and maintain fulfilling employment in a safe environment, it is imperative that organizations actively create an inclusive work atmosphere. While organizations should take steps to actively minimize stigma (and its effects) in the workplace, they also can strategically implement policies and procedures that signal to those with mental illness that they are welcome.

Overarching Strategies for Reducing Stigma

By and large, there are steps that organizations can take to help reduce the stigma of mental illness and, subsequently, resistance to individuals with these disorders. To date, researchers have identified three broad strategies for minimizing the stigma associated with mental illness – education, contact and protest (Corrigan, Morris, Michaels, Rafacz, & Rüsch, 2012).

Education

Educational programs aimed at increasing general knowledge about mental illness may correct the common misconceptions and stereotypes applied to individuals with mental illness, thereby reducing the stigma attached to these individuals (Corrigan & Gelb, 2006). Results from a systematic review of stigma reduction programs demonstrated that educational initiatives generally provide short-term changes in attitudes (Thornicroft et al., 2016), though small meta-analytic effect sizes provide modest evidence for the long-term persistence of these changes (Corrigan, Michaels, & Morris, 2015). Research also suggests that information-based programs may be most effective for individuals who had previous knowledge of mental illness prior to the training (Rüsch et al., 2005). In addition, the content of the educational materials can significantly affect attitudes related to individuals with mental illness. Specifically, studies have shown that focusing on biological or genetic explanations for mental illness can actually increase negative attitudes about individuals with mental illness (Holzinger, Dietrich, Heitmann, & Angermeyer, 2008).

Although training is beneficial for all employees in an organization, it is especially critical for individuals in managerial positions. Martin, Woods and Dawkins (2013) suggested that organizations provide managers with both conceptual (e.g., what is done in a given situation) and procedural (e.g., how to execute actions) knowledge about mental illness. Prior research has shown that training programs that focus on improving conceptual knowledge are beneficial for increasing mental health literacy (Stuart, Thornicroft, & Arboleda-Florez, 2005), decreasing stigma (Rusch et al., 2005) and improving self-efficacy in applying learned concepts within the workplace context (Dimoff, Kelloway, & Burnstein, 2016). Educational programs targeting procedural knowledge can help leaders to execute the appropriate actions when managing an employee with mental illness such as accurately and sensitively discussing mental health issues, monitoring workplace behavior and engaging in reflective learning (Martin et al., 2013).

Contact

Providing opportunities for interpersonal contact between individuals with and without mental illness is another way to minimize stigma. A meta-analysis of the effects of contact as an anti-stigma method demonstrated that it was more effective at reducing perceptions of danger, incompetence and blame as compared to education methods (Corrigan et al., 2015). Contact can help replace inaccurate and negative stereotypes with more positive perceptions of minority group members. However, the type of contact is important – research has shown that face-to-face contact was more effective in reducing stigma than video contact (Corrigan et al., 2015). Although this may happen as employees with mental illness feel more comfortable disclosing their condition to others at work, there are also active methods organizations can take to provide all employees (both neurotypical and those with mental illness) the opportunity to interact with individuals with mental illness. For instance, an organization could incorporate individuals with mental illness into training and corporate social responsibility initiatives through partnering with local mental health organizations.

Protest

Finally, stigma may be minimized through social activism, by way of protest. Protest strategies target the social injustices experienced by individuals with mental illness (Corrigan et al., 2012). One way that protests minimize stigma is by redressing the negative stereotypes of individuals with mental illness often portrayed in the media. Further, collective social action can target the public stigma attached with mental illness (Corrigan, 2004).

In organizations, mental health allies are also needed to help to promote social activism. Allies are non-stigmatized others who provide support and advocacy on behalf of stigmatized individuals (Washington & Evans, 1991). When allies speak out for individuals with mental illness, they help to minimize the stigma associated with these disorders. In the workplace diversity literature, allies have traditionally been discussed in the context of lesbian, gay and bisexual employees (Sabat, Martinez, & Wessel, 2013), though extensions of ally research to the work experiences of employees with mental illness may yield fruitful contributions to the field.

Strategies to Support Individuals with Mental Illness

I/O psychology and management scholars have long advocated for fostering workplace inclusion of marginalized employees via the implementation of organizational policies, procedures and resources (King & Cortina, 2010) as well as fostering a climate for diversity.

Policies and Procedures

Federal law prohibits discrimination against employees with mental illness and also entitles employees to reasonable accommodations to enable them to fully meet the demands of the job. Providing employees with accommodations is important not only to adhere to governmental regulations but also to help employees to achieve their full potential on the job. In this way, employees will be able to perform their jobs more effectively, which can improve their productivity and workplace contributions (Fabian et al., 1993).

Further, many of the accommodations required by employees with mental illness are reasonable to implement. Previous estimates have suggested that 90% of accommodations, cost less than $100 to implement (Granger, Baron, & Robinson, 1996) and, in some cases, organizations do not incur any direct costs (Fabian et al., 1993). Commonly reported accommodations include weekly meetings with a supervisor, exchanging minor tasks among employees, creating a quieter workplace and allocating extra time to learn on the job (Wang et al., 2011).

Accommodations not only help employees to meet the requirements of their jobs, but also are related to improvements in employees' mental health over time (Bolo et al., 2013). A recent study by Follmer and Beatty (2018) demonstrated that receiving accommodations can also influence employees' job satisfaction and turnover intentions. Specifically, having a need for accommodations at work was related to decreased job satisfaction and increased turnover intentions; however, these effects were ameliorated when employees actually received accommodations. In other words, when employees' needs for accommodations were actually met, they experienced increased levels of job satisfaction and decreased desires to leave their organizations. Some research findings have suggested employees with mental illness know very little about their rights to request accommodations at work (Goldberg, Killeen, & O'day, 2005). In this way, organizations may benefit from educating employees about their rights to accommodations, especially since these accommodations can increase job performance and tenure (Wåhlin, Ekberg, Persson, Bernfort, & Öberg, 2013).

In addition to accommodations, organizations may also implement open door policies which encourage employees to reach out to organizational personnel with concerns or to obtain support. An open-door policy provides employees with a personnel contact for discussing mental health concerns, thereby signaling a safe environment in which to disclose, seek available resources, or request accommodations. Previous research has suggested that open door policies can aid employees' return to work after injury (Shaw, Robertson, Pransky, & McLellan, 2003), though additional research is needed to understand the effectiveness of these policies for those with mental illness.

Accommodations and open door policies assist employees with mental illness while at work. There are, however, times when employees are not able to fully meet the requirements of their job due to the conditions of their mental illness and may require time away from work as a result. Organizations can implement clear short-term (e.g., sick leave) and long-term (e.g., Family Medical Leave Act; FMLA) policies to make provisions for employees who need time off from work in order to manage their mental illness. Under the FMLA guidelines, employees with mental illness may be entitled to 12 weeks of unpaid leave such that employees can focus on recovery while maintaining assurance that their jobs are secure. It is imperative that employees have knowledge of the parameters associated with these leave policies in order to facilitate conversations about invoking them. In fact, research suggests that open communication between employees and supervisors can aid in reintegrating employees into the workplace after taking time off (Noordik, Nieuwenhuijsen, Varekamp, van der Klink, & van Dijk, 2011).

Resources

Although policies and procedures aim to outline employees' rights within the workplace, it may be the case that an employee with mental illness requires support outside of accommodations or time off work. Examples of workplace resources include providing access to health insurance and employee assistance programs (EAPs).

Many employees with mental illness rely on health insurance to obtain treatment for their disorders (Goetzel, Ozminkowski, Sederer, & Mark, 2002). Research has suggested that providing employees with healthcare yields both personal and organizational benefits. For example, employees who have access to treatment and psychological services experience increased well-being and work productivity (Mechanic & Olfson, 2016). Unfortunately, stigma can often prevent individuals from seeking the treatment they need (Corrigan, 2004), indicating a need for organizations to employ multiple strategies for assisting employees with mental illness.

A second resource that organizations may offer are EAPs, which provide "confidential assessment, counseling and therapeutic services" for employees experiencing personal, emotional, or psychological problems (Arthur, 2000, p. 550). These programs generally maintain a hotline into which employees may call to obtain advice regarding medical, legal, domestic and financial concerns and, at times, may include referring employees to external mental health counselors for additional support. These programs allow employees to acquire needed psychological or medical services related to their mental illness, which may otherwise go unrecognized or untreated. There are several advantages associated with EAPs including an emphasis on confidentiality in the help-seeking process as well as access to a predetermined number of free or cost-effective services (Arthur, 2000). Despite the benefits of EAPs, research suggests that employees are apprehensive about seeking organizational-sponsored support (Butterworth, 2001) due to fears of confidentiality loss and lack of organizational trust. Studies on the use of EAPs have shown that trust in the organization, program confidence and managerial support for the program significantly predict employees' actual or intended EAP use (French, Roman, Dunlap, & Steele, 1997; Milne, Blum, & Roman, 1994). The stigma surrounding mental illness is also likely to deter employees from seeking help through organization-sponsored programs and prior research has shown that EAP use is stigmatized by peers (Butterworth, 2001). Thus, EAPs have the potential to provide employees access to mental health care but may be underutilized due to misunderstanding the purpose and bounds of care provided to employees.

Organizational Diversity Climate

Diversity climate refers to individual and collective perceptions about an organization's stance on diversity (Mor Barak, Cherin, & Berkman, 1998). Establishing a climate for diversity helps to incorporate minority employees into the broader work environment, thereby improving their desire to remain with the organization. To date, most research on diversity climate has focused on inclusion strategies as they relate to women and racial/ethnic minorities. Although less attention has been paid to developing strong diversity climates for employees with concealable identities (such as mental illness), Huffman, Watrous-Rodriguez, and King (2008) identified strategies that organizations can implement to build supportive cultures. Examples of strategies for creating a strong diversity climate include using inclusive language and eliminating offensive slurs as they relate to mental illness (e.g., psycho, crazy), surveying employees about their perceptions of diversity climate, promoting inclusive practices to job applicants and providing employees with mentoring opportunities.

Importantly, organizations should consider mental illness as a unique type of diversity and actively include these employees in the organization's diversity materials. For example, organizations can utilize diversity statements to demonstrate their consideration and inclusivity of employees with mental illness, which may increase employees' trust in the organization, encourage safe disclosure and even decrease stigmatization of these individuals. Previous research has shown that organizational diversity policies positively contribute to impressions of the organization (Williams & Bauer, 1994) and are related to decreased discrimination directed toward groups of individuals included in the policy (Button, 2001). Of course, organizations must do more than provide inclusive statements, they must also put these statements into action – in essence, they must "walk the talk" in order to establish a strong diversity climate.

Promoting Mental Health in Organizations

The previous section described strategies that organizations could use to offer support to employees with mental illness. However, there are also strategies that organizations could employ to promote flourishing mental health among all employees. According to Keyes (2007), there are three factors that are related to flourishing mental health: positive emotions, positive psychological functioning and positive social functioning.

Positive emotions include the extent to which an individual feels cheerful, in good spirits and is mostly satisfied in with his/her life domains. Indeed, organizations can rely on affective events theory (Weiss & Cropanzano, 1996) to help promote positive emotions in the workplace through daily uplift activities such as giving praise to an employee for work well done or hosting daily work rallies to engender excitement among employees. Further, research suggests that leaders' emotions are "contagious" in the workplace, suggesting that when leaders display and feel positive emotions, employees will also be more likely to display and feel such feelings (Bono & Ilies, 2006). Organizations that proactively attempt to foster positive moods among employees can aid in promoting more positive emotions.

Positive psychological functioning relates to an employee's perceptions of engagement with existential challenges in life (Keyes, Shmotkin, & Ruff, 2002). Employees with flourishing mental health will demonstrate self-acceptance, autonomy and environmental mastery as well as seek personal growth, purpose in life and positive relations with others. Drawing upon work characteristics theory (Hackman & Oldham, 1975) and theories of psychological empowerment (Spreitzer, 1995), organizations can help to craft employees' jobs such that they are meaningful and provide autonomy and continued opportunity for skill mastery. Further, organizations can clearly explain the contributions that employees make both to the organization and society at large in order to fulfill desires for maintaining a purpose in life.

Positive social functioning concerns an employee's social well-being. Employees will be more likely to maintain flourishing mental health when they hold positive attitudes toward human differences, believe society can grow and evolve positively, see their own daily activities as valuable to society, are interested in society and social life and feel a sense of belonging to a community. Creating a strong climate for inclusion and engaging employees in diversity training can help them to accept and positively regard human differences in the workplace. Social functioning can also be promoted when the organization provides feedback to employees to help them understand the contributions they make to society through their work, as well as foster a supportive work environment. For example, hosting informal social gatherings inside and outside of the workplace can help employees to develop social bonds with others at work, which may increase their perceptions of community.

Conclusion

Recent news reports have suggested that mental illness is on the rise globally. Despite the increase in the number of employees with mental illness, most organizations are underprepared to support these individuals in the workplace. Alarmingly, the policies, procedures and climates of many organizations may actually create hostile environments that, in effect, act as a form of diversity resistance toward those with mental illness. Stigma remains a critical contributor to diversity resistance toward individuals with mental illness. Thus, organizations would benefit from implementing interventions aimed at minimizing this stigma, which in turn can help eliminate both subtle and overt forms of mistreatment as well as improve the overall workplace experiences of those with mental illness.

Note

1 We contrast mental illnesses with physical disabilities not to suggest that one form of disability is more or less difficult to manage in the workplace, but to illustrate the unique challenges that come with recognizing mental illnesses as worthy of support and understanding.

References

Aftab, A. (2014). Mental illness vs. brain disorders: from Szasz to DSM-5. *Psychiatric Times*, *31*(2), 20G.

Almquist, E. M. (1975). Untangling the effects of race and sex: The disadvantaged status of Black women. *Social Science Quarterly*, *56*, 129–142.

Angermeyer, M. C., & Dietrich, S. (2006). Public beliefs about and attitudes towards people with mental illness: a review of population studies. *Acta Psychiatrica Scandinavica*, *113*(3), 163–179.

Angermeyer, M. C., & Matschinger, H. (2003). The stigma of mental illness: Effects of labelling on public attitudes towards people with mental disorder. *Acta Psychiatrica Scandinavica*, *108*(4), 304–309.

Arthur, A. R. (2000). Employee assistance programmes: The emperor's new clothes of stress management? *British Journal of Guidance & Counselling*, *28*(4), 549–559.

Baker, A. E. Z., & Procter, N. G. (2014). Losses related to everyday occupations for adults affected by mental illness. *Scandinavian Journal of Occupational Therapy*, *21*, 287–294.

Baldridge, D. C., & Veiga, J. F. (2001). Toward a greater understanding of the willingness to request an accommodation: Can requesters' beliefs disable the Americans with Disabilities Act? *Academy of Management Review*, *26*, 85–99.

Baldwin, M. L., & Marcus, S. C. (2007). Labor market outcomes of persons with mental disorders. *Industrial Relations: A Journal of Economy and Society*, *46*(3), 481–510.

Barclay, L. A., & Markel, K. S. (2009). Ethical fairness and human rights: The treatment of employees with psychiatric disabilities. *Journal of Business Ethics*, *85*, 333.

Becker, D. R., Drake, R. E., Bond, G. R., Xie, H., Dain, B. J., & Harrison, K. (1998). Job terminations among persons with severe mental illness participating in supported employment. *Community Mental Health Journal, 34*(1), 71–82.

Bolo, C., Sareen, J., Patten, S., Schmitz, N., Currie, S., & Wang, J. (2013). Receiving workplace mental health accommodations and the outcome of mental disorders in employees with a depressive and/or anxiety disorder. *Journal of Occupational and Environmental Medicine, 55*(11), 1293–1299.

Bono, J. E., & Ilies, R. (2006). Charisma, positive emotions and mood contagion. *The Leadership Quarterly, 17*(4), 317–334.

Butterworth, I. E. (2001). The components and impact of stigma associated with EAP counseling. *Employee Assistance Quarterly, 16*(3), 1–8.

Button, S. B. (2001). Organizational efforts to affirm sexual diversity: A cross-level examination. *Journal of Applied psychology, 86*(1), 17–28.

Corrigan, P. (2004). How stigma interferes with mental health care. *American Psychologist, 59*(7), 614–625.

Corrigan, P. W., Larson, J. E., & Kuwabara, S. A. (2007). Mental illness stigma and the fundamental components of supported employment. *Rehabilitation Psychology, 52*(4), 451.

Corrigan, P. W., Morris, S. B., Michaels, P. J., Rafacz, J. D., & Rüsch, N. (2012). Challenging the public stigma of mental illness: a meta-analysis of outcome studies. *Psychiatric Services, 63*(10), 963–973.

Corrigan, P. W., River, L. P., Lundin, R. K., Wasowski, K. U., Campion, J., Mathisen, J., … Kubiak, M. A. (2000). Stigmatizing attributions about mental illness. *Journal of Community Psychology, 28*(1), 91–102.

Corrigan, P., & Gelb, B. (2006). Three programs that use mass approaches to challenge the stigma of mental illness. *Psychiatric Services, 57*(3), 393–398.

Corrigan, P., Michaels, P. J., & Morris, S. (2015). Do the effects of anti-stigma programs persist over time? Findings from a meta-analysis. *Psychiatric Services, 66* (5), 543–546.

Corrigan, P.W., Kerr, A., & Knudsen, L. (2005). The stigma of mental illness: Explanatory models and methods for change. *Applied and Preventive Psychology, 11*, 179.190.

Crocker J., Lutsky N. (1986) Stigma and the dynamics of social cognition. In S. C. Ainlay, G. Becker, & L.M. Coleman (Eds.), *The dilemma of difference: Perspectives in social psychology*. Boston, MA: Springer.

Dimoff, J. K., Kelloway, E. K., & Burnstein, M. D. (2016). Mental health awareness training (MHAT): The development and evaluation of an intervention for workplace leaders. *International Journal of Stress Management, 23*(2), 167.

Equal Employment Opportunity Commission (April 2019). Crain Automotive Holdings to pay $27,100 to settle EEOC disability discrimination lawsuit. Retrieved from www.eeoc.gov/eeoc/newsroom/release/4-30-19a.cfm

Equal Employment Opportunity Commission (December 2016). EEOC issues publication on the rights of job applicants and employees with mental health conditions. Retrieved from www.eeoc.gov/eeoc/newsroom/release/12-12-16a.cfm

Equal Employment Opportunity Commission (December 2017). AccentCare to pay $25,000 to settle EEOC disability discrimination suit. Retrieved from www.eeoc.gov/eeoc/newsroom/release/12-1-17.cfm

Equal Employment Opportunity Commission (March 2018). Mine Rite Technologies to pay $75,000 to settle EEOC disability suit. Retrieved from www.eeoc.gov/eeoc/newsroom/release/3-23-18.cfm

Fabian, E. S., Waterworth, A., & Ripke, B. (1993). Reasonable accommodations for workers with serious mental illness: Type, frequency, and associated outcomes. *Psychosocial Rehabilitation Journal, 17*(2), 163–172.

Fairclough, S., Robinson, R. K., Nichols, D. L., & Cousley, S. (2013). In sickness and in health: Implications for employers when bipolar disorders are protected disabilities. *Employee Responsibilities and Rights Journal, 25*(4), 277–292.

Feldman, D. B., & Crandall, C. S. (2007). Dimensions of mental illness stigma: What about mental illness causes social rejection? *Journal of Social and Clinical Psychology, 26*(2), 137–154.

Follmer, K. B., & Beatty, J. E. (2018, August). The roles of perceived need and stigma in the decision to request accommodations at work. In D. C. Baldridge (Chair) (Ed.), *New directions in disability research: work contexts, inclusivity, and wellbeing interactions*. Chicago, IL: Symposium conducted at the annual Academy of Management Conference.

Follmer, K. B., & Jones, K. S. (2017). Stereotype content and social distancing from employees with mental illness: The moderating roles of gender and social dominance orientation. *Journal of Applied Social Psychology, 47*(9), 492–504.

Follmer, K. B., & Jones, K. S. (2018). Mental illness in the workplace: An interdisciplinary review and organizational research agenda. *Journal of Management, 44*(1), 325–351.

French, M. T., Roman, P. M., Dunlap, L. J., & Steele, P. D. (1997). Factors that influence the use and perceptions of employee assistance programs at six worksites. *Journal of Occupational Health Psychology, 2*(4), 312–324.

Goetzel, R. Z., Ozminkowski, R. J., Sederer, L. I., & Mark, T. L. (2002). The business case for quality mental health services: why employers should care about the mental health and well-being of their employees. *Journal of Occupational and Environmental Medicine, 44*(4), 320–330.

Goffman, E. (1963). *Stigma: Notes on the management of spoiled identity*. Englewood Cliffs, NJ: Prentice-Hall.

Gold, P. B., Oire, S. N., Fabian, E. S., & Wewiorski, N. J. (2012). Negotiating reasonable workplace accommodations: Perspectives of employers, employees with disabilities, and rehabilitation service providers. *Journal of Vocational Rehabilitation, 37*(1), 25–37.

Goldberg, S. G., Killeen, M. B., & O'day, B. (2005). The disclosure conundrum: How people with psychiatric disabilities navigate employment. *Psychology, Public Policy, and Law, 11*(3), 463–500.

Granger, B., Baron, R. C., & Robinson, S. (1996). *A national study on job accommodations for people with psychiatric disabilities*. Philadelphia, PA: Matrix Research Institute.

Greenberg, P. E., Fournier, A., Sisitsky, T., Pike, C. T., & Kessler, R. C. (2015). The economic burden of adults with major depressive disorder in the United States (2005 and 2010). *Journal of Clinical Psychiatry, 76*(2), 155–162. doi:10.1002/da.20580

Hackman, J. R., Oldham, G., Janson, R., & Purdy, K. (1975). A new strategy for job enrichment. *California Management Review, 17*(4), 57–71.

Hauck, K., & Chard, G. (2009). How do employees and managers perceive depression: A worksite case study. *Work, 33*(1), 13–22.

Hebl, M. R., Foster, J. B., Mannix, L. M., & Dovidio, J. F. (2002). Formal and interpersonal discrimination: A field study of bias toward homosexual applicants. *Personality and Social Psychology Bulletin, 28*(6), 815–825.

Hebl, M., Madera, J. M., & King, E. (2008). Exclusion, avoidance and social distancing.In K.M. Thomas (Ed.), *Diversity resistance in organizations* (pp. 127–150). New York, NY: Taylor Francis.

Holley, L. C., Tavassoli, K. Y., & Stromwall, L. K. (2016). Mental illness discrimination in mental health treatment programs: Intersections of race, ethnicity, and sexual orientation. *Community Mental Health Journal, 52*(3), 311–322.

Holzinger, A., Dietrich, S., Heitmann, S., & Angermeyer, M. (2008). Evaluation of target-group oriented interventions aimed at reducing the stigma surrounding mental illness. *Psychiatrische Praxis, 35*(8), 376–386.

Huffman, A. H., Watrous-Rodriguez, K. M., & King, E. B. (2008). Supporting a diverse workforce: What type of support is most meaningful for lesbian and gay employees? *Human Resource Management, 47*(2), 237–253.

Jones, E. E., Farina, A., Hastorf, A. H., Markus, H., Miller, D. T., & Scott, R. A. (1984). *Social stigma: The psychology of marked relationships.* New York, NY: W. H. Freeman and Company.

Jones, K. P., & King, E. B. (2014). Managing concealable stigmas at work: A review and multilevel model. *Journal of Management, 40*(5), 1466–1494.

Keyes, C. L. (2007). Promoting and protecting mental health as flourishing: A complementary strategy for improving national mental health. *American Psychologist, 62*(2), 95.

Keyes, C. L. M. (2005). Mental illness and/or mental health? Investigating axioms of the complete state model of health. *Journal of Consulting and Clinical Psychology, 73*(3), 539–548.

Keyes, C. L., Shmotkin, D., & Ryff, C. D. (2002). Optimizing well-being: the empirical encounter of two traditions. *Journal of Personality and Social Psychology, 82* (6), 1007–1022.

King, E. B., & Cortina, J. M. (2010). The social and economic imperative of lesbian, gay, bisexual, and transgendered supportive organizational policies. *Industrial and Organizational Psychology, 3*(1), 69–78.

Livingston, J. D., & Boyd, J. E. (2010). Correlates and consequences of internalized stigma for people living with mental illness: A systematic review and meta-analysis. *Social Science & Medicine, 71*(12), 2150–2161.

Martin, A., Woods, M., & Dawkins, S. (2015). Managing employees with mental health issues: Identification of conceptual and procedural knowledge for development within management education curricula. *Academy of Management Learning & Education, 14*(1), 50–68.

McGinty, E. E., Kennedy-Hendricks, A., Choksy, S., & Barry, C. L. (2016). Trends in news media coverage of mental illness in the United States: 1995–2014. *Health Affairs, 35*(6), 1121–1129.

McMahon, B. T., McMahon, M. C., West, S. L., Conway, J. P., & Lemieux, M. (2017). The nature of allegations of workplace discrimination for Americans with learning disabilities. *Journal of Vocational Rehabilitation, 46*(1), 31–37.

Mechanic, D., & Olfson, M. (2016). The relevance of the Affordable Care Act for improving mental health care. *Annual Review of Clinical Psychology, 12*, 515–542.

Milne, S. H., Blum, T. C., & Roman, P. M. (1994). Factors influencing employees' propensity to use an employee assistance program. *Personnel Psychology, 47*(1), 123–145.

Mor Barak, M. E., Cherin, D. A., & Berkman, S. (1998). Organizational and personal dimensions in diversity climate: Ethnic and gender differences in employee perceptions. *The Journal of Applied Behavioral Science, 34*(1), 82–104.

National Alliance of Mental Illness (2017, May 4). Personal stories: Continuing the conversation. Retrieved from: www.nami.org/Personal-Stories/Continuing-the-Conversation

National Institute of Mental Health (February 2019). Mental illness. Retrieved from www.nimh.nih.gov/health/statistics/mental-illness.shtml

Noordik, E., Nieuwenhuijsen, K., Varekamp, I., van der Klink, J. J., & van Dijk, F. J. (2011). Exploring the return-to-work process for workers partially returned to work and partially on long-term sick leave due to common mental disorders: a qualitative study. *Disability and Rehabilitation, 33*(17–18), 1625–1635.

Paetzold, R. L., García, M. F., Colella, A., Ren, L. R., Triana, M. D. C., & Ziebro, M. (2008). Perceptions of people with disabilities: When is accommodation fair? *Basic and Applied Social Psychology, 30*(1), 27–35.

Pinfold, V., Stuart, H., Thornicroft, G., & Arboleda-Flórez, J. (2005). Working with young people: The impact of mental health awareness programs in schools in the UK and Canada. *World Psychiatry, 4*, 48–52.

Ragins, B. R. (2008). Disclosure disconnects: Antecedents and consequences of disclosing invisible stigmas across life domains. *Academy of Management Review, 33*(1), 194–215.

Ren, L. R., Paetzold, R. L., & Colella, A. (2008). A meta-analysis of experimental studies on the effects of disability on human resource judgments. *Human Resource Management Review, 18*(3), 191–203.

Ritsher, J. B., Otilingam, P. G., & Grajales, M. (2003). Internalized stigma of mental illness: psychometric properties of a new measure. *Psychiatry Research, 121*(1), 31–49.

Rosenfield, S. (2012). Triple jeopardy? Mental health at the intersection of gender, race, and class. *Social Science & Medicine, 74*(11), 1791–1801.

Rüsch, N., Angermeyer, M. C., & Corrigan, P. W. (2005). Mental illness stigma: Concepts, consequences, and initiatives to reduce stigma. *European Psychiatry, 20*, 529–539.

Rüsch, N., Todd, A. R., Bodenhausen, G. V., & Corrigan, P. W. (2010). Do people with mental illness deserve what they get? Links between meritocratic worldviews and implicit versus explicit stigma. *European Archives of Psychiatry and Clinical Neuroscience, 260*(8), 617–625.

Russinova, Z., Griffin, S., Bloch, P., Wewiorski, N. J., & Rosoklija, I. (2011). Workplace prejudice and discrimination toward individuals with mental illnesses. *Journal of Vocational Rehabilitation, 35*(3), 227–241.

Sabat, I. E., Martinez, L. R., & Wessel, J. L. (2013). Neo-activism: Engaging allies in modern workplace discrimination reduction. *Industrial and Organizational Psychology, 6*(4), 480–485.

Seligman, M. E. P. (1995). The effectiveness of psychotherapy: The consumer reports study. *American Psychologist, 50*(12), 965–974.

Shaw, L. R., Chan, F., & McMahon, B. T. (2012). Intersectionality and disability harassment: The interactive effects of disability, race, age, and gender. *Rehabilitation Counseling Bulletin, 55*(2), 82–91.

Shaw, W. S., Robertson, M. M., Pransky, G., & McLellan, R. K. (2003). Employee perspectives on the role of supervisors to prevent workplace disability after injuries. *Journal of Occupational Rehabilitation, 13*(3), 129–142.

Shear, M. D. (2018, February 26). Trump says he would have rushed in unarmed to stop school shooting. *The New York Times*, Retrieved from www.nytimes.com/2018/02/26/us/politics/trump-school-shooter-florida.html

Spreitzer, G. M. (1995). Psychological empowerment in the workplace: Dimensions, measurement, and validation. *Academy of Management Journal, 38*(5), 1442–1465.

Stone, D. L., & Colella, A. (1996). A model of factors affecting the treatment of disabled individuals in organizations. *Academy of Management Review, 21*(2), 352–401.

Stuart, H. (2006). Mental illness and employment discrimination. *Current Opinion in Psychiatry, 19*(5), 522–526.

Thomas, K. M. & Plaut, V.C. (2012). The many faces of diversity resistance in the workplace. In K.M. Thomas (Ed.), *Diversity resistance in organizations* (pp. 5–22). New York, NY: Taylor & Francis.

Thornicroft, G., Mehta, N., Clement, S., Evans-Lacko, S., Doherty, M., Rose, D., & Henderson, C. (2016). Evidence for effective interventions to reduce mental-health-related stigma and discrimination. *The Lancet, 387*(10023), 1123–1132.

United States Department of Justice (2018). Information and technical assistance on the Americans with Disabilities Act. Retrieved from www.ada.gov/ada_intro.htm

Wåhlin, C., Ekberg, K., Persson, J., Bernfort, L., & Öberg, B. (2013). Evaluation of self-reported work ability and usefulness of interventions among sick-listed patients. *Journal of Occupational Rehabilitation, 23*(1), 32–43.

Wang, J., Patten, S., Currie, S., Sareen, J., & Schmitz, N. (2011). Perceived needs for and use of workplace accommodations by individuals with a depressive and/or anxiety disorder. *Journal of Occupational and Environmental Medicine, 53*(11), 1268–1272.

Washington, J., & Evans, N. J. (1991). In Evans, N.J., & V.A. Wall (Eds.), *Becoming an ally. Beyond tolerance: Gays, lesbians, and bisexuals on campus* (pp. 195–204). Virginia: American College Personnel Association.

Weaver, K.B. (2017). Full disclosure: An examination of disclosure decisions and identity management motives among employees with depression (Unpublished doctoral dissertation). State College, PA: The Pennsylvania State University.

Weiss, H. M., & Cropanzano, R. (1996), Affective events theory: A theoretical discussion of the structure, causes and consequences of affective experiences at work. *Research in Organizational Behavior, 18*, 1–74.

Williams, M. L., & Bauer, T. N. (1994). The effect of a managing diversity policy on organizational attractiveness. *Group & Organization Management, 19*(3), 295–308.

Wooten, L. P., & James, E. H. (2005). Challenges of organizational learning: Perpetuation of discrimination against employees with disabilities. *Behavioral Sciences & the Law, 23*(1), 123–141.

6 Diversity Resistance Redux

The Nature and Implications of Dominant Group Threat for Diversity And Inclusion

*Victoria C. Plaut, Celina A. Romano,
Kyneshawau Hurd and Emily Goldstein*

... to achieve a more equal gender and race representation, Google has created several discriminatory practices: Programs, mentoring, and classes only
for people with a certain gender or race; a high priority queue and special
treatment for "diversity" candidates; hiring practices which can effectively
lower the bar for "diversity" candidates ...

(Damore, 2017, "The Harm of Google's biases," para. 1)

"The lack of clear, communicated policies and actions to advance diversity and
inclusion with concrete accountability and leadership from senior executives has
left many of us feeling unsafe and unable to do our work," Knapp [software
engineer at Google] said.

(Fiegerman & O'Brien, 2018, para. 6)

The first quote, from Google employee James Damore's 2017 memo, which
took issue with the company's diversity practices and progressive leanings,
captures aspects of the resistance facing diversity initiatives geared at
improving the representation and inclusion of women and people of color in
organizations. At the same time that Damore and others have railed against
certain diversity initiatives, voices from non-dominant groups[1] have reflected
a very different reality: that they experience bias and harassment, not favorit-
ism, and as reflected in the quote above, that they feel a lack of safety. Moreover, the very types of initiatives to which Damore objects in his memo have
not yielded much improvement, and despite many efforts, leading technology
companies remain woefully lacking in their representation of women across
racial groups and of people of color (Evans & Rangarajan, 2017).

We focus on the co-existence of these dueling realities – the view that
diversity initiatives give special treatment to women and people of color
versus the sentiment that not enough has been done to ensure these groups'
equitable treatment and inclusion – and explore their interconnection.
Research has identified a host of threats that some dominant group members (e.g., White people, men) perceive when they encounter changing

demographics or policies and practices that address the experiences and historical exclusion of non-dominant group members (e.g., people of color, women across racial groups) and improve their hiring, promotion and retention. Furthermore, these threats can fuel a variety of anti-diversity responses. At the same time, diversity initiatives have shifted from an initial focus on the experiences and representation of non-dominant groups to broader, more diffuse and less morally-related goals. These shifts may reflect an implicit capitulation to the preferences and comfort of dominant groups in decisions about the selection and design of diversity initiatives. These choices have consequences for the experiences of non-dominant groups. Many cues elicited by diversity programs do not convey safety to non-dominant groups. And, ironically, the very types of programs that some dominant group members resist are themselves ineffectual in part because they have been diluted, possibly in catering to non-dominant groups.

Of course, reactions such as those encapsulated in the opening quotes exist at other companies besides Google. However, although prior research has provided evidence of resistance to or wariness of diversity and inclusion initiatives (see, e.g., Dobbin & Kalev, 2016), we know of no work that suggests the precise scope of the problem (and "the problem" itself is not easily defined). This is the case whether considering the reactions of dominant group or non-dominant group members. Nevertheless, we suggest that there is value in understanding these processes.

Our analysis proceeds as follows. First, we map the various types of threats perceived by dominant groups and their connection to diversity resistance. Much of this work is grounded in the intergroup threat literature. Second, we explain the ways in which diversity initiatives have shifted over time in a way that reflects accommodation of the interests and preferences of dominant groups. Third, we examine the experiences of non-dominant groups with the resulting organizational approaches to diversity. We then explore some implications of research across these three areas for best practices.

Diversity, Threat and Resistance among Dominant Groups

Organizational efforts to foster diversity and inclusion can be met with resistance. For example, in discussing backlash to company diversity and inclusion work, former Intel CEO Brian Krzanich suggested that White men could feel "under siege" and that the leadership team had even experienced threats around their "position on diversity and inclusion" (Dickey, n.d.). Diversity resistance within organizations can take many forms, including opposition toward policies such as affirmative action, less support toward broad diversity initiatives, concern over unfair treatment amidst pro-diversity values, less support toward recruiting and hiring individuals from marginalized backgrounds and so on. This section will explore and explain some of the psychological factors that undergird diversity resistance among dominant group members, including a range of perceived threats, preferences for hierarchy, White identity forms and management and perceived anti-White discrimination.

A Multiplicity of Threats

Multiple types of threats, including those related to prototypicality, group status, culture, resources, and existentiality, have been identified as psychological processes that increase resistance towards diversity (see Table 6.1 for summary of threats). It is important to note that though the literature, and this section, treat these threats as distinct, some overlap conceptually and produce similar effects. Yet, examining them separately and illuminating their subtle differences could be not only theoretically but also practically useful, highlighting their applicability to diversity resistance in a variety of contexts.

One threat that fuels resistance toward diversity is a prototypicality threat, or the fear that one's subgroup will no longer be the quintessential representative of a group. Danbold and Huo (2015) found that exposing Whites to information projecting their loss of status as prototypical Americans led them to endorse diversity at lower levels. Within organizations, employees may be inclined to resist diversity efforts if they believe that

Table 6.1 Perceived Threats to Dominant Groups and Related Concepts

Threat	Definition	Citations
Prototypicality Threat	Fear that one's subgroup will no longer be the quintessential representative of a group	e.g., Danbold and Huo (2015)
Group-status Threat	Perceived threat to one's group's societal status	e.g., Craig and Richeson (2014a)
Symbolic Threat	Perceived threat to one's group's culture, values, or beliefs	e.g., Stephan, Ybarra and Bachman (1999)
Realistic Threat	Threat (real or perceived) to one's group's resources (e.g., jobs, property, food, etc.)	e.g., LeVine and Campbell (1972); Sherif (1966)
Existential Threat	Perceived threat to one's group's existence	e.g., Greenberg et al. (1990)
Meritocratic Threat	The possibility that one's accomplishments are not due to personal merit	e.g., Knowles, Lowery, Chow and Unzueta (2014)
Group-image Threat	The threat to feeling positively about one's group when the group has benefitted from unfair social advantage	e.g., Knowles et al. (2014)
Perceived Exclusion	Concern over whether multicultural diversity efforts include one's group	e.g., Plaut, Garnett, Buffardi and Sanchez-Burks (2011)

their group will no longer remain the prototypical representative of their field or organization. Such resistance may be especially likely to occur within fields traditionally occupied by dominant group members, such as law, business, or STEM (Science, Technology, Engineering and Math). For example, in one study, men who believed that men represented the prototypical members of the STEM community showed greater resistance toward a diversity initiative seeking to increase women's representation in STEM (Danbold & Huo, 2017). The fields upholding the most prominent stereotypical representations of who represents such a field, and thus, the fields in most need of diversifying, may be the ones that experience the most backlash.

Another threat that can lead dominant group members to resist diversity is group-status threat, or perceiving that one's group's societal status is in jeopardy. For example, exposing White Americans to demographic shift information where White Americans are projected to become the "minority" population leads them to express more negative attitudes toward minority groups (Craig & Richeson, 2014a), endorse more conservative policy positions on race- and non-race related issues (Craig & Richeson, 2014b) and anticipate heightened discrimination against their group in the future (Craig & Richeson, 2017). Therefore, employees belonging to high status groups, such as men and Whites, may be particularly susceptible to resistance if they perceive diversity efforts as threatening their high social standing. Dominant group members, however, are not alone in displaying such reactions. When Hispanic population growth was made salient to African Americans and Asian Americans, they too exhibited a conservative shift in policy attitudes (Craig & Richeson, 2018).

Diversity resistance may also emerge if individuals perceive their group's culture, values, or beliefs to be threatened, a phenomenon known as symbolic threat (W. G. Stephan et al., 1999). Yogeeswaran and Dasgupta (2014) found that White Americans construed multiculturalism as more threatening to American values and culture when it was described in concrete as opposed to abstract terms. Morrison, Plaut and Ybarra (2010) found that symbolic threat helped to explain why, when exposed to multiculturalism, Whites highly identified with their racial identity allocated fewer funds to diversity-related organizations. Symbolic threats also predict prejudice toward non-White immigrants and racial minorities (Stephan et al., 2002, 1999). Diversity efforts within organizations can entail diversifying its employees or adapting its values and culture to promote a more inclusive workplace. Therefore, the very changes organizations undergo in an attempt to be more diverse and inclusive may threaten those who see their workplace as changing its beliefs, or recruiting, retaining and promoting those with beliefs different than their own.

Realistic threats, or threats to resources (e.g., jobs, property, food, etc.), whether real or perceived, may also spark resistance toward diversity (Sherif, Harvey, White, Hood, & Sherif, 1961). Studies have found that realistic threats can catalyze intergroup conflict as well as prejudice against and stereotypes of outgroups (Stephan & Stephan, 2000; Stephan et al., 1999; Zárate, Garcia, Garza, & Hitlan, 2004) and predict attitudes toward affirmative action (Renfro, Duran, Stephan, & Clason, 2006). If, as part of a diversity initiative, an organization makes an effort to recruit, retain and promote members of marginalized groups at higher rates, dominant group members may perceive such efforts as threatening their own job security and other resources such as pay and promotion opportunities, fueling backlash.

Concerns that one's group's existence is in danger or in risk of extinction may also illicit opposition toward diversity. In a study conducted by Greenberg and colleagues (1990), when mortality was made salient, individuals espoused negative reactions to those with dissimilar cultural worldviews. Recently, such threats have manifested, for example, as concerns among White nationalists. While marching in Charlottesville, White nationalists chanted, "You will not replace us!", with supporters claiming that the growing diversity and demographic changes within the nation pose "an existential threat to the country" (Serwer, 2018). In a political climate in which Whites have mobilized around their group's existential threat, organizations may experience backlash explicitly linked to a fear of marginalized group members displacing Whites. Such backlash has the potential to severely hinder organizations' efforts in diversifying their workplace laterally and hierarchically.

White Identity and the Racial Hierarchy

Implementing diversity initiatives and policies may not only threaten dominant group members, but could also disrupt White employees' comfort within the racial hierarchy. As members of the racially dominant group, White individuals reap numerous benefits based on their group membership, and diversity initiatives may signal to Whites that their racial privilege is at risk.

To begin, both the form and strength of White identity can implicate Whites' attitudes toward diversity. Goren and Plaut (2012) found that Whites who felt prideful of their White identity showed more diversity opposition and more racial bias than weakly identified or power-cognizant Whites. Another study conducted by Lowery and colleagues (2006) found that the more Whites identified with their racial group, the greater opposition they exhibited to affirmative action policy. White employees who identify strongly or pridefully with their White identity, then, may be particularly likely to resist efforts to recruit, retain and promote employees of color.

Whites who prefer to maintain the racial hierarchy may also resist diversity efforts. Studies have shown that high levels of social dominance orientation lead to more resistance towards diversity policies, such as affirmative action (Sidanius & Pratto, 1999). Additionally, social dominance orientation has been found to relate to workplace bullying and discrimination (Parkins, Fishbein, & Ritchey, 2006). White employees high in social dominance may be more likely to explicitly exclude, harass and target employees from racial and ethnic minority backgrounds, which may have implications not only for diversity initiative implementation, but also for the wellbeing and workplace satisfaction of employees of color.

Coping with Racial Privilege and Challenges to Racial Privilege

American society has been organized in a way that largely shields Whites from being directly confronted with the realities of their racial privilege. Therefore, for many White people, organizational efforts to foster diversity, such as de-biasing or multicultural trainings, may be the first time their racial understandings of the world are directly challenged (DiAngelo, 2011). Whites may experience "race-based stress" in these situations and react in defensive, emotional and hostile ways, a phenomenon known as "white fragility" (DiAngelo, 2011), which may impede diversity efforts. Knowles et al. (2014) explain how Whites use identity-management strategies to cope with psychological threats related to their racial privilege. Meritocratic threat, for example, threatens Whites' understanding of their accomplishments as due to personal merit, thereby requiring them to relinquish credit for their successes. Additionally, Whites may face group-image threat if they come to understand their group as benefitting from unfair social advantage and as the culprit of historical wrongs (Knowles et al., 2014). Whites may manage their White identity in self-protective ways in response to these threats. They may, for example, deny the existence of White privilege or distance themselves from whiteness (Knowles et al., 2014). In doing so, they may endorse a more colorblind worldview and decrease support for diversity-related policies (Knowles et al., 2014).

In addition to psychologically coping with racial privilege, Whites also react to threats to their privilege. For example, research suggests that addressing issues of diversity and inclusion may itself lead to dominant groups' perception that they are being discriminated against. Norton and Sommers (2011) found that Whites perceive racial discrimination as a zero-sum game: as they perceive a decline in anti-Black bias, they perceive an increase in anti-White bias, such that it now exceeds anti-Black bias. Similarly, White people high in SDO code progress towards equality for people of color as a loss for Whites (Eibach & Keegan, 2006). Relatedly, Dover and colleagues (2016) found that White men were more concerned about unfair treatment and anti-White discrimination when exposed to a company's pro-diversity (vs. neutral) values. Even a company's expression of diversity values then, absent any concrete policies and procedures, may lead Whites to believe that they will be discriminated against. This zero-sum mentality may lead Whites to reject diversity policies and practices focused on the inclusion of marginalized groups to the extent they perceive these efforts as discriminatory.

Whites may also resist diversity efforts to the extent that they feel excluded from them. Research conducted by Plaut and colleagues (2011) found that

White men's feelings of inclusion in organizational diversity predicted their support for diversity. Additionally, White men higher in the "need to belong" were more attracted to an organization espousing a colorblind ideology than to one espousing a multicultural ideology (Plaut et al., 2011).

In conclusion, diversity resistance stems at least in part from dominant group members perceiving diversity efforts as threatening their prototypicality, group status, culture, resources, or existence. Additionally, prideful construals of White identity, as well as preferences for hierarchy may also fuel diversity resistance. Lastly, forms of White identity management, such as denial and distancing, as well as perceiving diversity as increasing anti-White discrimination may also spark resistance.

Equity vs. Diversity: Shifting Foci of Diversity Initiatives and the Incorporation of Dominant Group Preferences

Some scholars have speculated that minimizing backlash among more privileged group members has driven the design of diversity programs (see e.g., Konrad & Linnehan, 2003). While this is possible, we do not know of any studies that have explicitly connected the adoption of initiatives with intentional avoidance of dominant group backlash. However, we do know that programs originally focused on equal opportunity have given way to "diversity management" (Kelly & Dobbin, 1998). In addition, there are subtle ways in which Whites' preferences have become incorporated into the design of diversity programs regardless of overt intentions.

A Shift in Focus

There is evidence that corporate programs originally aimed at countering discrimination and promoting equality have morphed, becoming diluted over time. Scholars have traced a trajectory of these programs from antidiscrimination practices to diversity management (Kelly & Dobbin, 1998) and have identified a shift from a focus on affirmative action, equal opportunity and the experiences of historically excluded groups to a nebulous and broad conceptualization of diversity (Edelman, Fuller, & Mara-Drita, 2001; Konrad & Linnehan, 2003). Edelman et al. (2001) documented, for example, how non-legally protected dimensions that dissociate diversity from civil rights (e.g., geography, attitudes, communication style) have been added to the diversity discourse over time in the management literature. These additions manifest on current diversity and inclusion websites. Even a cursory examination of companies at the very top of the Fortune 500 list reveals a commitment to broad definitions of diversity. For example, Walmart (n.d.) defines diversity as "unique styles, experiences, identities, ideas and opinions" (para. 1); ExxonMobil (n.d.) endeavors to "promote the inclusion of thought, skill, knowledge and culture" (para. 1); and United-Health Group (n.d.) claims to "value unique points of view" (para. 1). Indeed another development is the focus on "diversity of thought," touted by Deloitte as the new frontier of inclusion (see Diaz-Uda, Medina, & Schill, 2013). Diversity of thought has been characterized as an organizational resource that forms the foundation for reaping the benefits of diversity (e.g., Woods, 2008).

Cross-sectional studies also suggest the existence of broad and nebulous conceptions of diversity. A study of White college students' diversity conceptions found that while they commonly associated diversity with race and ethnicity, many simply identified "differences" or "backgrounds" or groups generally interacting (Banks, 2009). A study of U.S. adults found that many stuck to general descriptions when asked what diversity means to them, such as "being exposed to many different people from many different cultures" (Bell & Hartmann, 2007). And although respondents were able to discuss the challenges of diversity, they (especially majority-group respondents) evaded the topic of inequality with respect to diversity. In addition, group-based motives can contribute to broadening construals of diversity in certain individuals in particular situations. In one experiment, when racial diversity in an organization was characterized as low but occupational diversity was high, anti-egalitarians (those scoring high on a measure of social dominance orientation) broadened their definition of diversity to include non-racial factors such as occupational diversity in order to justify their opposition to affirmative action policies (Unzueta, Knowles, & Ho, 2012). Presumably, incorporating occupational diversity in their representation of diversity helped them to see the organization as more diverse.

The picture that emerges of conceptions of diversity from the management, psychology and sociology literatures is one of a diluted focus on inequality. Notably, these broad conceptions themselves may beget an obliviousness to inequality. For example, recent research suggests that broad definitions of diversity – those that include characteristics such as background/culture, perspectives, skills and ideologies – may make it more difficult to detect racial underrepresentation as a problem relative to a narrow definition that focuses on legally protected characteristics (Akinola et al., 2019).

How White Preferences Get Instantiated into Diversity Initiatives

In addition, if they simply focus on making dominant group members feel comfortable, these initiatives risk missing the mark. Research suggests that dominant groups prefer to focus on commonalities between them and non-dominant groups, rather than discuss group-based power (Saguy, Dovidio, & Pratto, 2008). Other studies show that group identity concerns (such as identification with nation) encourage sanitized historical narratives and preference for positive, celebratory narratives (Kurtiş, Adams, & Yellow Bird, 2010). In a fascinating study of Black History Month representations in schools, White-majority schools tended to celebrate diversity and individual achievement rather than acknowledge historical barriers, which was more prominent in Black-majority schools (Salter & Adams, 2016). In addition,

White students tended to like the displays from White-majority schools more, and exposure to the White-majority school displays fostered the lowest recognition of racism and lowest support for anti-racism policies in contrast with the Black-majority schools. These studies suggest that focusing on majority group members pushes initiatives in the direction of celebrating diversity and away from recognizing and addressing inequities.

Another way in which the dominant group's preferences may be incorporated into the design of diversity initiatives and move them further from a focus on inequality is through the business case rationale for diversity (see Williams, 2017). For example, the website of McKinsey & Company, author of an oft-cited study on this topic (Hunt, Prince, Dixon-Fyle, & Yee, 2018) states, "We know intuitively that diversity matters. It's also increasingly clear that it makes sense in purely business terms" (Hunt, Layton, & Prince, 2015, para. 1). A 2019 Google search for the phrase "business case for diversity" yielded 161,000 hits, highlighting its ubiquity. As equal opportunity has given way to diversity management (Kelly & Dobbin, 1998), the business case has increasingly been deployed in favor of the legal or moral one to foster support for diversity efforts (Williams, 2017).

Indeed, many diversity initiatives today are guided by and advertised with the idea that diversity benefits the organization, as opposed to the idea that diversity is the right thing to do (see Trawalter, Driskell, & Davidson, 2016). And there is empirical support for the idea that Whites – the group that tends to occupy positions of power in organizations – prefer the instrumental framing (Kidder, Lankau, Chrobot-Mason, Mollica, & Friedman, 2004; Starck, 2018; Trawalter et al., 2016). For example, one study found that Whites prefer the message that "diversity is good" (e.g., it "enables creative problem-solving") to the message that "diversity is fair" (e.g., it "promotes group equity"; Trawalter et al., 2016, p. 74). In another study, White MBA students held more positive attitudes toward a company's instrumental justification (i.e., a diversity initiative to reflect the customer base) than one stressing affirmative action to meet EEO goals (Kidder et al., 2004). Another study found not only that Whites felt more positively toward a university's message that emphasized the benefits of diversity to students than one that emphasized diversity as a moral commitment to fairness and equity, but also that the instrumental framing produced more psychological comfort (Starck, 2018). Specifically, it connoted more value and more sense of belonging to Whites, while the moral message produced greater feelings of threat of being perceived through the lens of the stereotype that Whites are prejudiced.

In addition, an instrumental "diversity benefits" model can work counter to the goal of equal opportunity (see also Hurd & Plaut, 2018; Trawalter et al., 2016; Williams, 2017). For example, the "diversity is good" message leads Whites to broaden their conception of diversity to include other, non-racial dimensions and to de-prioritize hiring Black job candidates (in favor of, for example, an applicant from another country; Trawalter et al., 2016). In another study, White participants who viewed a business case rationale for diversity were, ironically, less likely to appoint a Black teammate as leader of the team than were those who saw an anti-discrimination rationale for diversity (Williams, 2017). In addition, participants exposed to the business case for diversity were less likely to promote a Black job candidate and less likely to agree that striving for diversity is morally right or fair to non-dominant groups.

Therefore, whether intentional or not, the design of diversity initiatives appears to have gravitated toward the preferences of dominant groups (e.g., a broader conception of diversity; an instrumental versus moral rationale for diversity). Furthermore, these developments have negative consequences for equity in that they can lead to diminished identification and remedying of underrepresentation of non-dominant groups.

Diversity Approaches and the Experiences of Employees of Color

According to our analysis above, whether intentionally or unintentionally, dominant group preferences can easily become incorporated into the choices that are made about how to motivate and structure diversity initiatives. Importantly, the decisions organizations make about how to frame and approach diversity can fundamentally alter the way people experience a space, especially people from non-dominant social identity groups. Therefore, diversity approaches can have significant consequences for the success and well-being of the very groups whose representation and inclusion they are purportedly intended to improve. Moreover, backlash to diversity initiatives and the presence of diversity structures themselves can also derail organizations' progress on diversity, equity and inclusion.

Reading Diversity Cues

Because historically dominant groups have created and maintained institutions that functioned to keep out non-dominant groups, non-dominant groups have developed mechanisms of vigilance (for a review, see Emerson & Murphy, 2014). As organizations attempt to distance themselves from this history of exclusion and signal to non-dominant groups a commitment to diversity, they construct diversity cues that non-dominant groups use to diagnose how they will be treated and perceived. Non-dominant groups use a variety of cues to evaluate a given setting, ranging from cues in the physical environment (e.g., Cheryan, Plaut, Davies, & Steele, 2009), to depictions of people from non-dominant groups in corporate advertising (e.g. Avery, 2003), to diversity ideologies or framings such as multiculturalism and colorblindness (Purdie-Vaughns, Steele, Davies, Ditlmann, & Crosby, 2008; Wilton, Good, Moss-Racusin, & Sanchez, 2015). Cues such as these activate what are called Social Identity Contingencies, or expectations about how the setting will respond to one's social identity (Purdie-Vaughns et al., 2008). That is, for those who possess social identities at risk of social devaluation or other harm (e.g., Black people, women, people who the environment disables, etc.), certain cues in the environment can either confirm that their group will be devalued and induce threat (i.e., negative social identity contingencies) or indicate that their social identity will be valued or at the very least a non-issue and portray safety (i.e., positive and/or neutral social identity contingencies). Ultimately, the cues emitted by organizations' diversity initiatives can signal to these groups whether they can belong in and trust the organization and can affect recruitment, engagement and retention.

Cues that center the diversity preferences of dominant group members (e.g., Whites) can negatively impact non-dominant groups. For instance, as previously mentioned, Whites who feel threatened or excluded by diversity may be particularly apt to deploy or be attracted to colorblind diversity ideologies (Knowles et al., 2014; Knowles, Lowery, Hogan, & Chow, 2009; Plaut et al., 2011). Yet, adopting this diversity ideology as a means of mitigating diversity resistance among Whites may not engender positive reactions from non-dominant groups. For instance, in one study researchers exposed students to fictitious university diversity messages that were either colorblind or multicultural. They found that compared to White women and men, women of color exposed to colorblind diversity messaging had lower expectations of diversity at the university, higher expectations of encountering bias at the university and lower anticipated performance on a group task, as well as lower actual performance on a math test (Wilton et al., 2015). Moreover, when combined with other institutional cues, colorblind framings of diversity may engender distrust and threat among non-dominant groups. In one study, participants viewed a brochure from a fictitious company that varied in both the level of non-dominant group representation in the photos (low or high) and the type of diversity philosophy the company explicitly endorsed (colorblind or value-diversity). Participants evaluating the brochure with a colorblind diversity philosophy and low representation expressed relatively more concern that their racial identity would be devalued at that company and trusted the company less compared to participants in the value-diversity condition (Purdie-Vaughns et al., 2008).

Yet, adopting a more multicultural (e.g., valuing difference) philosophy does not immunize an organization from emitting threatening cues to non-dominant groups, especially if they do not trust the institution's motivation, implementation and engagement of the philosophy (Purdie-Vaughns & Walton, 2011). When non-dominant groups have concerns about how they will be treated within an institution, they may be skeptical about its claims of valuing difference. For instance, in the Purdie-Vaughns et al. (2008) study, participants who saw a company brochure that endorsed a value-diversity philosophy and high non-dominant group representation reported just as many threatening identity contingencies (e.g., expecting to be passed over for promotions because of race, feeling excluded from social events because of race, feeling that their race would be relevant to how others view them, etc.) as participants who saw an organizational brochure with low representation and a colorblind philosophy. These ostensibly positive cues may nonetheless raise non-dominant groups' concerns, for instance, about being valued solely on the basis of one's identity (Purdie-Vaughns et al., 2008). Consistent with this possibility, Apfelbaum and colleagues found that compared to Black participants who viewed a "value in equality" philosophy, Black participants who viewed a "value in difference" philosophy (e.g., multiculturalism) performed worse on a challenging cognitive task (Apfelbaum, Stephens, & Reagans, 2016).

Are people of color justified in being wary of a more multicultural philosophy? Organizations could unintentionally pigeonhole people even with well-intentioned multicultural messaging. Gutiérrez and Unzueta (2010) examined the effects of exposing people (a predominantly Asian and White sample) to a multicultural ideology stressing the usefulness of recognizing and valuing group differences for fostering intergroup harmony, as opposed to a colorblind ideology. Participants exposed to multiculturalism expressed greater liking for a man of color who showed stereotype-consistent as opposed to stereotype-inconsistent extracurricular interests (e.g., a Black man who likes basketball vs. surfing). This raises concerns about whether focusing on group differences constrains identity expression for racial and ethnic minorities. Other research has shown that Whites exposed to this same multiculturalism message stereotype ethnically-different groups more (Wolsko, Park, Judd, & Wittenbrink, 2000). Another study exposed people from various racial and ethnic groups (though the samples were plurality or majority White) to the same multiculturalism or colorblindness essays (Wilton, Apfelbaum, & Good, 2018). They found that multiculturalism increased the extent to which participants thought in racially essentialist ways (i.e., thought of race as biological). Underscoring these psychological tendencies, Ely and Thomas (2001) found that when an organization is guided by an "access and legitimacy" diversity paradigm (i.e., they are focused on diversity among their employees in so far as they want to gain access to and legitimacy with previously inaccessible, diverse markets), they reproduce racial divisions and stereotypes.

Moreover, justifying the value of diversity through its instrumental benefit to the organization may lead to anti-egalitarian consequences. As described in the previous section, dominant group members prefer and feel more comfortable with an instrumental rationale for diversity, which seems to have replaced the moral one (Williams, 2017). Yet, this operationalization forces diversity's value to hinge on the benefits it provides to organizations rather than on an intrinsic appreciation of the realities of difference or a commitment to eliminating harassment and discrimination. Not only does the instrumental framing potentially lead to ultimately less equitable decision-making (Hurd & Plaut, 2018; Starck, 2018; Williams, 2017), but it may also be met with negative reactions from non-dominant groups. For example, Starck (2018) shows that, in evaluating a college for their children, non-White parents do not show the same preference for the instrumental model as Whites and anticipate better outcomes from the moral frame, including belonging and threat.

As institutions try to balance White threat and resistance to diversity with the inclusion needs of non-dominant groups, their adoption of certain diversity ideologies and rationales over others often prioritizes an approach towards diversity that is more palatable to Whites while yielding potentially negative outcomes for non-dominant groups. But it is not just the insidious instantiation of white preferences into the structure of diversity initiatives that can signal a threat to non-dominant groups. The backlash to diversity itself may be enough of a cue to non-dominant groups that an organization is not safe for them. For instance, according to the first Google Diversity Annual Report issued after Damore's infamous anti-diversity diatribe went viral, Google continued to struggle with retention of employees of color, with the 2017 rate of attrition being highest among Black and Latinx employees relative to other groups (Google, 2018). Of course, we cannot prove a causal link between Damore's professed resistance to diversity and the exodus of Google employees of color (and some of the data points are presumably pre-memo), but this example does challenge organizations and intuitions to consider how backlash to their diversity programs affects non-dominant groups and can unravel progress. Previous research suggests that dominant groups' attitudes toward diversity in a work unit predict non-dominant groups' perceptions of bias (Plaut, Thomas, & Goren, 2009), and perceptions of diversity climate link to outcomes such as turnover intentions (Buttner, Lowe, & Billings-Harris, 2010; McKay et al., 2007). Moreover, even when there is high representation of non-dominant group members, it is difficult to overcome a negative climate (Vargas, Westmoreland, Robotham, & Lee, 2018).

In sum, this work cautions organizations from instituting diversity initiatives without considering the consequences for non-dominant groups' experience and well-being. Diversity ideologies and rationales for adopting these initiatives bear meaning for non-dominant groups and can have a profound impact on their expected and actual experiences. This point becomes particularly relevant when determining how to address resistance to diversity. As organizations attempt to create identity-safe spaces to replace infrastructures of exclusion, they may attempt to do so while not also alienating or having to contend with prevailing opposition to diversity. Yet, framings of diversity that are, even partially, motivated by trying to mitigate diversity resistance among dominant groups (as opposed to focusing on tailored identity-safe applications of inclusion) have implications for non-dominant groups and their institutional outcomes, experiences and expectations. They can sanitize diversity and divorce it from remedying structural inequality (Purdie-Vaughns & Walton, 2011). Moreover, they can maintain threatening environments for non-dominant groups.

Implications for Best Practices

Given the analysis above, diversity and inclusion can easily become a tightrope walk in which organizations attempt to balance threats: dominant groups' threat of diversity and non-dominant groups' threat of not existing – or not being safe – in a space. The design of diversity and inclusion initiatives needs to move beyond this. In particular, organizations might ask themselves what their diversity and inclusion initiatives could look like if they were not trying to (intentionally or unintentionally) incorporate dominant group interests.

First, organizations should reflect on how their diversity programs may, ironically, reflect the preferences of the dominant group. For example, do they focus more on celebrating diversity than on tackling systemic problems? Does celebrating diversity come at the expense of recognizing historically-related obstacles, harms and disparities?

Second, organizations should consider and understand the consequences of their motivations and choices for a diversity initiative. For example, what are the effects of bringing dimensions such as "diversity of thought" to the diversity and inclusion table? How might the business case for diversity alienate the very employees it is designed to recruit?

Third, in promoting diversity and inclusion initiatives, organizations should be thoughtful about their communication strategies with all stakeholders. The opportunity for explanation and voice could be helpful for dominant and non-dominant groups. Explanation-based, transparent and responsive communication could help, for example, dispel misguided zero-sum interpretations of diversity programs. Transparency may also aid communication with non-dominant groups, for example, helping to preserve organizational trust among these groups (McKay & Avery, 2005).

Fourth, organizations should hear and center the voices of non-dominant groups but also involve dominant group members in identifying and implementing solutions. Involving dominant group members in this way may minimize resistance (Kalev, Dobbin, & Kelly, 2006) without catering to the dominant group. Moreover, this involvement could increase perceptions of autonomy, agency and ownership of ideas.

Fifth, and relatedly, organizations should promote and guide the involvement of allies. For example, they could provide guidance on amplifying the voices of non-dominant groups, bystander training, mentoring and sponsorship.

Finally, guiding all of these efforts, organizations should focus first and foremost on creating bias- and harassment-free environments in which people will have the opportunity to succeed. This involves at least two approaches: 1) fostering a positive climate for success, by, for example, diagnosing belonging, norm-setting and accountability and 2) providing pathways to success, such as visible, developmental opportunities and providing constructive feedback.

It is important to keep in mind that research does not point to a silver bullet or secret sauce for creating an environment in which all organizational members feel equally dedicated to diversity and inclusion, though we have tried to sketch some strategies for bringing more people to the table. For some organizations that have not built diversity, equity and inclusion into their values, practices and demographics from the start, building programs that work and that produce minimal resistance may require long-term and multi-pronged cultural change. And even then, it might not be possible to get every member on board. The principle we are trying to bring to the forefront is protecting non-dominant employees from harm and exclusion and ensuring pathways for their success. Of course, this focus need not be antithetical to building support among dominant group members.

Conclusion

The quotes introducing this chapter offer a glimpse into dueling realities regarding diversity initiatives. For Damore, diversity initiatives at Google represent heavy-handed "discriminatory practices," (Damore, 2017, The Harm of Google's biases, para. 1) that negatively impact the company and employees from dominant groups. In contrast, Knapp draws attention to "The lack of clear, communicated policies … accountability and leadership…" (Fiegerman & O'Brien, 2018, para. 6) on issues related to diversity and inclusion and highlights the toll that this takes on employees from non-dominant groups. This speaks to contrasting views of diversity initiatives: for some they represent overreach, while for many others they fall short. As we have seen, a myriad of possible threats experienced by dominant group members can produce backlash. While broadening notions of diversity may assuage threat for dominant group members, it comes at a steep cost. Catering – even unintentionally – to backlash has the potential to alter the landscape of how non-dominant groups experience diversity and could encourage institutions to frame diversity in ways that may actually undermine inclusion and harm non-dominant groups. Thus, rather than promoting the interests and protecting the experiences of non-dominant group members, as intended, these initiatives could instead serve as another form of privilege for dominant members. Therefore, organizations must critically examine their motives, framing and programs around diversity and inclusion and watch for cases in which diversity efforts have become unmoored from their intended focus. Simply put, the centrality of non-dominant voices is integral to the success of any initiative aimed at promoting diversity and inclusion.

Note

1 We use the term non-dominant groups to refer to a variety of identity dimensions associated with less dominance in the social hierarchy. Our examination of resistance and diversity and inclusion initiatives focuses first and foremost on race and ethnicity, although we also draw from examples and evidence related to gender and, where possible, race*gender. While some experiences of other non-dominant group members (e.g., gender identity, sexual orientation, disability, religion and

others) will resemble those of women across racial groups and those of people of color, we hesitate to claim that the dynamics will be the same across dimensions.

References

Akinola, M., Opie, T. R., Ho, G. C., Unzueta, M., Castel, S., Kristal, A., Stevens, F., & Brief, A. (2019). Diversity isn't what it used to be: The consequences of the broadening of diversity. Working Paper.

Apfelbaum, E. P., Stephens, N. M., & Reagans, R. E. (2016). Beyond one-size-fits-all: Tailoring diversity approaches to the representation of social groups. *Journal of Personality and Social Psychology, 111*(4), 547–566. doi:10.1037/pspi0000071

Avery, D. R. (2003). Reactions to diversity in recruitment advertising–are differences black and white? *Journal of Applied Psychology, 88*(4), 672–679. doi:10.1037/0021-9010.88.4.672

Banks, K. H. (2009). A qualitative investigation of white students' perceptions of diversity. *Journal of Diversity in Higher Education, 2*(3), 149–155. doi:10.1037/a0016292

Bell, J. M., & Hartmann, D. (2007). Diversity in everyday discourse: The cultural ambiguities and consequences of "happy talk." *American Sociological Review, 72* (6), 895–914. doi:10.1177/000312240707200603

Buttner, H., Lowe, K. B., & Billings-Harris, L. (2010). Diversity climate impact on employee of color outcomes: Does justice matter? *Career Development International, 15*(3), 239–258. doi:10.1108/13620431011053721

Cheryan, S., Plaut, V. C., Davies, P. G., & Steele, C. M. (2009). Ambient belonging: How stereotypical cues impact gender participation in computer science. *Journal of Personality and Social Psychology, 97*(6), 1045–1060. doi:10.1037/a0016239

Craig, M. A., & Richeson, J. A. (2014a). More diverse yet less tolerant? How the increasingly diverse racial landscape affects White Americans' racial attitudes. *Personality and Social Psychology Bulletin, 40*(6), 750–761. doi:10.1177/0146167214524993

Craig, M. A., & Richeson, J. A. (2014b). On the precipice of a "majority-minority" America: Perceived status threat from the racial demographic shift affects White Americans' political ideology. *Psychological Science, 25*(6), 1189–1197. doi:10.1177/0956797614527113

Craig, M. A., & Richeson, J. A. (2017). Information about the U.S. racial demographic shift triggers concerns about anti-White discrimination among the prospective White "minority." *PLos One, 12*(9), e0185389. doi:10.1371/journal.pone.0185389

Craig, M. A., & Richeson, J. A. (2018). Hispanic population growth engenders conservative shift among non-Hispanic racial minorities. *Social Psychological and Personality Science, 9*(4), 383–392. doi:10.1177/1948550617712029

Damore, J. (2017, August 5). Google's ideological echo chamber. Retrieved from https://gizmodo.com/exclusive-heres-the-full-10-page-anti-diversity-screed-1797564320

Danbold, F., & Huo, Y. J. (2015). No longer "All-American"? Whites' defensive reactions to their numerical decline. *Social Psychological and Personality Science, 6*(2), 210–218. doi:10.1177/1948550614546355

Danbold, F., & Huo, Y. J. (2017). Men's defense of their prototypicality undermines the success of women in STEM initiatives. *Journal of Experimental Social Psychology, 72*, 57–66. doi:10.1016/j.jesp.2016.12.014

DiAngelo, R. (2011). White fragility. *The International Journal of Critical Pedagogy, 3*(3), 54–70. Retrieved from http://libjournal.uncg.edu/ijcp/article/view/249

Diaz-Uda, A., Medina, C., & Schill, B. (2013, July 23). Diversity's new frontier | Deloitte Insights. Retrieved January 28, 2019 from www2.deloitte.com/insights/us/en/topics/talent/diversitys-new-frontier.html

Dickey, M. R. (n.d.). Intel CEO says leadership team has received threats for company's stance on diversity. Retrieved from http://social.techcrunch.com/2016/04/22/intel-ceo-says-leadership-team-has-received-threats-for-companys-stance-on-diversity/

Dobbin, F., & Kalev, A. (2016). Why diversity programs fail and what works better. *Harvard Business Review, 94*, 7.

Dover, T. L., Major, B., & Kaiser, C. R. (2016). Members of high-status groups are threatened by pro-diversity organizational messages. *Journal of Experimental Social Psychology, 62*, 58–67. doi:10.1016/j.jesp.2015.10.006

Edelman, L. B., Fuller, S. R., & Mara-Drita, I. (2001). Diversity rhetoric and the managerialization of law. *American Journal of Sociology, 106*(6), 1589–1641. doi:10.1086/321303

Eibach, R. P., & Keegan, T. (2006). Free at last? Social dominance, loss aversion, and white and black Americans' differing assessments of racial progress. *Journal of Personality and Social Psychology, 90*(3), 453–467. doi.org/10.1037/0022-3514.90.3.453

Ely, R. J., & Thomas, D. A. (2001). Cultural diversity at work: The effects of diversity perspectives on work group processes and outcomes. *Administrative Science Quarterly, 46*(2), 229–275. doi:10.2307/2667087

Emerson, K. T. U., & Murphy, M. C. (2014). Identity threat at work: How social identity threat and situational cues contribute to racial and ethnic disparities in the workplace. *Cultural Diversity and Ethnic Minority Psychology, 20*(4), 508–520. doi:10.1037/a0035403

Evans, W., & Rangarajan, S. (2017, October 19). Hidden figures: How Silicon Valley keeps diversity data secret. Retrieved from www.revealnews.org/article/hidden-figures-how-silicon-valley-keeps-diversity-data-secret/

ExxonMobil. (n.d.). Diversity and inclusion. Retrieved February 8, 2019 from http://corporate.exxonmobil.com/Community-engagement/Sustainability-Report/Safety-health-and-the-workplace/Diversity-and-inclusion

Fiegerman, S., & O'Brien, S. A. (2018, June 6). Google employee confronts execs over diversity: Many of us feel "unsafe." *CNN Business*. Retrieved from https://money.cnn.com/2018/06/06/technology/alphabet-shareholder-meeting/index.html

Google. (2018). *Google diversity annual report 2018*. Retrieved from https://static.googleusercontent.com/media/diversity.google/en//static/pdf/Google_Diversity_annual_report_2018.pdf

Goren, M. J., & Plaut, V. C. (2012). Identity form matters: White racial identity and attitudes toward diversity. *Self and Identity, 11*(2), 237–254. doi:10.1080/15298868.2011.556804

Greenberg, J., Solomon, S., Pyszczynski, T., Rosenblatt, A., Veeder, M., Kirkland, S., & Lyon, D. (1990). Evidence for terror management theory II: The effects of mortality salience on reactions to those who threaten or bolster the cultural worldview. *Journal of personality and social psychology, 58*(2), 308–318.

Gutiérrez, A. S., & Unzueta, M. M. (2010). The effect of interethnic ideologies on the likability of stereotypic vs. counterstereotypic minority targets. *Journal of Experimental Social Psychology, 46*(5), 775–784. doi:10.1016/j.jesp.2010.03.010

Hunt, V., Layton, D., & Prince, S. (2015, January). Why diversity matters | McKinsey. Retrieved February 5, 2019 from www.mckinsey.com/business-functions/organiza tion/our-insights/why-diversity-matters

Hunt, V., Prince, S., Dixon-Fyle, S., & Yee, L. (2018). *Delivering through diversity.* Retrieved from www.mckinsey.com/%7E/media/McKinsey/Business%20Func tions/Organization/Our%20Insights/Delivering%20through%20diversity/Deliver ing-through-diversity_full-report.ashx

Hurd, K., & Plaut, V. C. (2018). Diversity entitlement: Does diversity-benefits ideology undermine inclusion? *Northwestern University Law Review, 112*(6), 1605–1636.

Kalev, A., Dobbin, F., & Kelly, E. (2006). Best practices or best guesses? Assessing the efficacy of corporate affirmative action and diversity policies. *American Sociological Review, 71*(4), 589–617. doi:10.1177/000312240607100404

Kelly, E., & Dobbin, F. (1998). How affirmative action became diversity management: Employer response to antidiscrimination law, 1961 to 1996. *American Behavioral Scientist, 41*(7), 960–984. doi:10.1177/0002764298041007008

Kidder, D. L., Lankau, M. J., Chrobot-Mason, D., Mollica, K. A., & Friedman, R. A. (2004). Backlash toward diversity initiatives: Examining the impact of diversity program justification, personal and group outcomes. *International Journal of Conflict Management, 15*(1), 77–102. doi:10.1108/eb022908

Knowles, E. D., Lowery, B. S., Chow, R. M., & Unzueta, M. M. (2014). Deny, distance, or dismantle? How White Americans manage a privileged identity. *Perspectives on Psychological Science, 9*(6), 594–609. doi:10.1177/1745691614554658

Knowles, E. D., Lowery, B. S., Hogan, C. M., & Chow, R. M. (2009). On the malleability of ideology: Motivated construals of color blindness. *Journal of Personality and Social Psychology, 96*(4), 857–869.

Konrad, A. M., & Linnehan, F. (2003). Affirmative action as a means of increasing workforce diversity. In M. J. Davidson & S. L. Fielden (Eds.), *Individual diversity and psychology in organizations* (Vol. 1, pp. 95–112). West Sussex, UK: John Wiley & Sons.

Kurtiş, T., Adams, G., & Yellow Bird, M. (2010). Generosity or genocide? Identity implications of silence in American Thanksgiving commemorations. *Memory, 18* (2), 208–224. doi:10.1080/09658210903176478

LeVine, R. A., & Campbell, D. T. (1972). *Ethnocentrism: Theories of conflict, ethnic attitudes, and group behavior.* Oxford, UK: John Wiley & Sons.

Lowery, B. S., Unzueta, M. M., Knowles, E. D., & Goff, P. A. (2006). Concern for the in-group and opposition to affirmative action. *Journal of Personality and Social Psychology, 90*(6), 961–974. doi:10.1037/0022-3514.90.6.961

McKay, P. F., & Avery, D. R. (2005). Warning! Diversity recruitment could backfire. *Journal of Management Inquiry, 14*(4), 330–336.

McKay, P. F., Avery, D. R., Tonidandel, S., Morris, M. A., Hernandez, M., & Hebl, M. R. (2007). Racial differences in employee retention: Are diversity climate perceptions the key? *Personnel Psychology, 60*(1), 35–62. doi:10.1111/j.1744-6570.2007.00064.x

Morrison, K. R., Plaut, V. C., & Ybarra, O. (2010). Predicting whether multiculturalism positively or negatively influences White Americans' intergroup attitudes: The role of ethnic identification. *Personality and Social Psychology Bulletin, 36*(12), 1648–1661. doi:10.1177/0146167210386118

Norton, M. I., & Sommers, S. R. (2011). Whites see racism as a zero-sum game that they are now losing. *Perspectives on Psychological Science, 6*(3), 215–218. doi:10.1177/1745691611406922

Parkins, I. S., Fishbein, H. D., & Ritchey, P. N. (2006). The influence of personality on workplace bullying and discrimination. *Journal of Applied Social Psychology, 36*(10), 2554–2577. doi:10.1111/j.0021-9029.2006.00117.x

Plaut, V. C., Garnett, F. G., Buffardi, L. E., & Sanchez-Burks, J. (2011). "What about me?" Perceptions of exclusion and Whites' reactions to multiculturalism. *Journal of Personality and Social Psychology, 101*(2), 337–353. doi:10.1037/a0022832

Plaut, V. C., Thomas, K. M., & Goren, M. J. (2009). Is multiculturalism or color blindness better for minorities? *Psychological Science, 20*(4), 444–446. doi:10.1111/j.1467-9280.2009.02318.x

Purdie-Vaughns, V., Steele, C. M., Davies, P. G., Ditlmann, R., & Crosby, J. R. (2008). Social identity contingencies: How diversity cues signal threat or safety for African Americans in mainstream institutions. *Journal of Personality and Social Psychology, 94*(4), 615–630. doi:10.1037/0022-3514.94.4.615

Purdie-Vaughns, V., & Walton, G. M. (2011). Is multiculturalism bad for African Americans? Redefining inclusion through the lens of identity safety. In L. R. Tropp, & R. K. Mallett (Eds.), *Moving beyond prejudice reduction: Pathways to positive intergroup relations* (pp. 159–177). Washington, DC: American Psychological Association. doi:10.1037/12319-008

Renfro, C. L., Duran, A., Stephan, W. G., & Clason, D. L. (2006). The role of threat in attitudes toward affirmative action and Its beneficiaries. *Journal of Applied Social Psychology, 36*(1), 41–74. doi:10.1111/j.0021-9029.2006.00003.x

Saguy, T., Dovidio, J. F., & Pratto, F. (2008). Beyond contact: Intergroup contact in the context of power relations. *Personality and Social Psychology Bulletin, 34*(3), 432–445. doi:10.1177/0146167207311200

Salter, P. S., & Adams, G. (2016). On the intentionality of cultural products: Representations of Black history as psychological affordances. *Frontiers in Psychology, 7*, 1166. doi:10.3389/fpsyg.2016.01166

Serwer, A. (2018, August 10). The White nationalists are winning. *The Atlantic.* Retrieved from www.theatlantic.com/ideas/archive/2018/08/the-battle-that-erupted-in-charlottesville-is-far-from-over/567167/

Sherif, M. (1966). *In common predicament: Social psychology of intergroup conflict and cooperation.* Boston: Houghton Mifflin.

Sherif, M., Harvey, O. J., White, B. J., Hood, W. R., & Sherif, C. W. (1961). *Intergroup cooperation and competition: The Robbers Cave experiment.* Norman, OK: University Book Exchange.

Sidanius, J., & Pratto, F. (1999). *Social dominance: An intergroup theory of social hierarchy and oppression.* New York, NY: Cambridge University Press. doi:10.1017/CBO9781139175043

Starck, J. G. (2018). *The Best case for inclusivity: Value or values?* (Unpublished master's thesis). Princeton, NJ: Princeton University.

Stephan, C. W., & Stephan, W. G. (2000). The measurement of racial and ethnic identity. *International Journal of Intercultural Relations, 24*(5), 541–552. doi:10.1016/S0147-1767(00)00016-X

Stephan, W. G., Boniecki, K. A., Ybarra, O., Bettencourt, A., Ervin, K. S., Jackson, L. A., ... Renfro, C. L. (2002). The role of threats in the racial attitudes of Blacks and Whites. *Personality and Social Psychology Bulletin, 28*(9), 1242–1254. doi:10.1177/01461672022812009

Stephan, W. G., Ybarra, O., & Bachman, G. (1999). Prejudice toward immigrants. *Journal of Applied Social Psychology, 29*(11), 2221–2237. doi:10.1111/j.1559-1816.1999.tb00107.x

Trawalter, S., Driskell, S., & Davidson, M. N. (2016). What is good isn't always fair: On the unintended effects of framing diversity as good. *Analyses of Social Issues and Public Policy, 16*(1), 69–99. doi:10.1111/asap.12103

UnitedHealth Group. (n.d.). Diversity & inclusion. Retrieved February 8, 2019 from www.unitedhealthgroup.com/about/diversity.html

Unzueta, M. M., Knowles, E. D., & Ho, G. C. (2012). Diversity is what you want it to be: How social-dominance motives affect construals of diversity. *Psychological Science, 23*(3), 303–309. doi:10.1177/0956797611426727

Vargas, E., Westmoreland, A. S., Robotham, K., & Lee, F. (2018). Counting heads vs. making heads count: Impact of numeric diversity and diversity climate on psychological outcomes for faculty of color. *Equality, Diversity and Inclusion: An International Journal, 37*(8), 780–798.

Walmart. (n.d.). Diversity & inclusion. Retrieved February 8, 2019 from https://corporate.walmart.com/global-responsibility/opportunity/diversity-and-inclusion

Williams, J. B. (2017). Breaking down bias: Legal mandates vs. corporate interests. *Washington Law Review, 92*, 1473–1514.

Wilton, L. S., Apfelbaum, E. P., & Good, J. J. (2018). Valuing differences and reinforcing them: Multiculturalism increases race essentialism. *Social Psychological and Personality Science*, 194855061878072. doi:10.1177/1948550618780728

Wilton, L. S., Good, J. J., Moss-Racusin, C. A., & Sanchez, D. T. (2015). Communicating more than diversity: The effect of institutional diversity statements on expectations and performance as a function of race and gender. *Cultural Diversity and Ethnic Minority Psychology, 21*(3), 315–325. doi:10.1037/a0037883

Wolsko, C., Park, B., Judd, C. M., & Wittenbrink, B. (2000). Framing interethnic ideology: Effects of multicultural and color-blind perspectives on judgments of groups and individuals. *Journal of Personality and Social Psychology, 78*(4), 635–654. doi:10.1037/0022-3514.78.4.635

Woods, S. (2008). Thinking about diversity of thought. Retrieved January 28, 2019 from https://digitalcommons.ilr.cornell.edu/cgi/viewcontent.cgi?article=1106&context=workingpapers

Yogeeswaran, K., & Dasgupta, N. (2014). The devil is in the details: Abstract versus concrete construals of multiculturalism differentially impact intergroup relations. *Journal of Personality and Social Psychology, 106*(5), 772–789. doi:10.1037/a0035830

Zárate, M. A., Garcia, B., Garza, A. A., & Hitlan, R. T. (2004). Cultural threat and perceived realistic group conflict as dual predictors of prejudice. *Journal of Experimental Social Psychology, 40*(1), 99–105. doi:10.1016/S0022-1031(03)00067-2

7 The Response to Social Justice Issues in Organizations as a Form of Diversity Resistance

Enrica N. Ruggs, Karoline M. Summerville and Christopher K. Marshburn

As increased momentum is building around social justice issues in the U.S., organizations have found themselves in positions in which they are making decisions about whether to publicly support social justice movements and how to do so. In this chapter, we discuss how organizational response to social justice crises may signal resistance to diversity. Over the past ten years, the United States has seen a resurgence in cries for social justice for marginalized groups resulting from pivotal social events that have and continue to occur. For instance, the 2013 acquittal of George Zimmerman, a man who shot and killed a 17-year old Black teenager named Trayvon Martin because he felt the teen looked suspicious led to the uprising of the #BlackLivesMatter movement. This movement, a network of nationwide chapters which rally against violence and systemic racism toward Black people and help build local power for Black communities, intensified in 2014 after a police officer shot Michael Brown, an unarmed Black man, in Ferguson, MO. Since then, we have seen increased attention given to scrutinizing instances of police shooting unarmed Black civilians. In particular, there have been protests and calls to action to address racialized policing and injustice in our criminal justice system.

In addition to racial justice, movements around increased equality regarding sexual orientation led to the U.S. Supreme Court case, Obergefell v. Hodges (2015) to legalize same-sex marriage, which was ultimately successful. Sexual orientation and gender identity rights movements were spurred by state legislation eliminating or seeking to eliminate anti-discrimination protection for individuals on the basis of sexual orientation and gender identity, such as the House Bill 2 (commonly known as the Bathroom Bill) passed in North Carolina in 2016. Increased awareness of gender inequality and sexual harassment has also been revived. For instance, in 2017, #MeToo became a viral social media message that built on the MeToo movement, which was founded by Tarana Burke in 2006 and calls for greater gender equality in employment and sanctions against people who engage in sexual harassment.

Although some of these issues may not directly affect employees as targets of injustice, they no doubt influence people from all walks of life as such events often spark collective trauma. Rooted in sociology, collective trauma occurs when society witnesses a traumatic event that spurs some collective sentiment or a *group-level* psychological reaction (Alexander, Eyerman, Giesen, Smelser, & Sztompka, 2004; Hirschberger, 2018). The event becomes ingrained in the minds of a group of people, rather than a single-person's memory, commonly referred to as collective memory (Olick & Robbins, 1998). The Holocaust, slavery and the 9/11 terrorist attacks are all examples of collective trauma events that are embedded in collective memories. Such traumatic events become interwoven in the moral and political fabric of a society and threaten the collective identity within a society. In other words, collective trauma events shock the social system and signal social disintegration due to a number of psychological reactions ranging from insecurity and fear to anger and political intolerance (Skitka, Bauman, & Mullen, 2004). This trauma is likely not contained to a single moment or domain in one's life, but rather it may spill over into various domains of life, including the work domain. Support for this assertion is seen in survey data where employees reported feelings of anger toward employers and fear three months after the 9/11 attacks (Mainiero & Gibson, 2003).

The experience of collective trauma at work can lead to negative consequences such as increased negative affect and distraction (Busso, McLaughlin, & Sheridan, 2014; Holman, Garfin, & Silver, 2014; Schuster et al., 2001). Recently, Leigh and Melwani (in press) theorized that such collective trauma events are *mega-threats*, or large-scale negative diversity-related events that receive significant media attention. They proposed that mega-threats, such as highly publicized police killings of unarmed Black men and sexual harassment scandals are events that seep into the psyche of individuals and affect people personally and relationally when they are at work. Specifically, Leigh and Melwani posit that mega-threats can lead to negative emotions that people attempt to make sense of by processing and engaging with others at work. Additionally, McCluney, Bryant, King, and Ali (2017) note that racially traumatic social events serve as racial identity threats for Black employees and can lead to negative outcomes at work if these employees do not receive organizational support and resources to help cope with the trauma. The influence of trauma-inducing mega-threats during work may be particularly likely given the continued increasing diversity of the workforce. Namely, in 2017, 47% of the U.S. labor force was comprised of women, 22% of the labor force was comprised of racial minorities and 17% of the labor force identified as Hispanic or Latino (Bureau of Labor Statistics, 2018). Despite negative consequences related to collective trauma for employees, organizations do not always take steps to address the potential effects of societal-level trauma or critical social justice events on employees' lives. The broader discussions about social justice issues, which are frequently related to individuals from marginalized backgrounds, are often left out or silenced in organizational contexts.

The goal of this chapter is to examine the ways in which organizations' response to (or lack thereof) social justice issues or mega-threats may serve as a form of resistance to diversity issues affecting organizations. We use the term organization to refer to any place of employment including corporate firms, education institutions and non-profit organizations. In this chapter, we begin by briefly discussing the ways in which collective trauma can negatively affect individuals and may spillover to the workplace. Then, we discuss how organizations respond to such issues and illustrate how these responses may signal where the organization stands on diversity on a spectrum ranging from absolute resistance to embracing diversity issues. Drawing from the management and social psychological literatures, we integrate a model of organizational response to social and political issues (Greening & Gray, 1994) with a theoretical framework on White identity management (Knowles, Lowery, Chow, & Unzueta, 2014) to develop a model (see Figure 7.1) that outlines organizational responses to social justice issues that range from resisting to embracing diversity. Our model also examines external and organizational antecedents that influence organizational responses, as well as employee outcomes associated with the selected organizational response. Further, we present examples of real

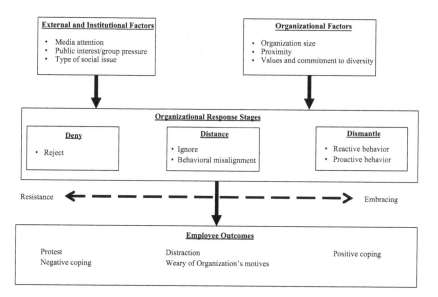

Figure 7.1 Model of Organizational Responses to Social Justice as a form of Resistance to Diversity

organizational responses to social justice issues and discuss how these decisions fit into the various stages of resistance. Finally, we provide recommendations for organizations to engage in proactive behaviors that illustrate a shift from resistance to embracing social justice issues as a way to embrace broader diversity and inclusion issues.

Social Injustice and Collective Trauma

Several events in the U.S. have shown that trauma can extend to those who did not experience the event firsthand and can result in collective psychological distress. Examples include terrorist attacks – such as the attacks on September 11, 2001 in New York (Boscarino, Figley, & Adams, 2004; Creamer & Liddle, 2005; Schuster et al., 2001; Silver, Holman, McIntosh, Poulin, & Gil-Rivas, 2002) and the Boston Marathon bombing in 2013 (Busso et al., 2014; Holman et al., 2014). Indeed, a study of over 1,000 people who kept online journals before and after the September 11th attacks showed that following the attacks, people expressed greater negative emotions and increased psychological distancing in their journal entries, which can indicate an avoidance of the present tense (Cohn, Mehl, & Pennebaker, 2004). Hurricane Katrina, which caused mass devastation and the displacement of over 400,000 people from New Orleans and surrounding areas has also been linked to collective distress such as increases in post-traumatic distress for a large group of people (Galea, Tracy, Norris, & Coffey, 2008; Xiong et al., 2008).

Additionally, historical faults and social injustices, like slavery and the aftermath of racial discrimination and mistreatment have led to distress and trauma experienced by many in a similar way (Alexander et al., 2004). In 2016, sociology professor Neil Gross wrote an article in *The New York Times* to argue that America is still experiencing collective trauma which can occur in the form of environmental catastrophes, wars, genocide, financial crises and large scale social movements. Research provides evidence that points to the link between social injustice, like racism and negative mental and physical health outcomes (Pascoe & Smart Richman, 2009). For instance, in a longitudinal panel study, a National Survey of Black Americans shows that participants who reported higher rates of racial mistreatment also reported higher levels of psychological distress (Williams, Yu, Jackson, & Anderson, 1997). Some work has documented how conversations about individual experiences of racism are added to the collective memory of the targeted group (May, 2000). Additionally, a survey found that in the months following the Pulse nightclub shooting in Orlando, FL, a mass shooting targeting LGBT individuals, people from the LGBT community expressed concerns for their safety and the safety of LGBT peers (Stults, Kupprat, Krause, Kapadia, & Halkitis, 2017). Taken together, there is evidence that mega-threats and other social justice incidents can be experienced as collective trauma, or an outpouring of similar negative reactions to the same event or series of events. Furthermore, because such trauma may negatively affect various facets of people's lives, including their work lives, it is vital to understand which organizational responses are harmful or helpful for employees who are directly and indirectly affected by social justice-related events.

Organizational Responses to Social Issues: Stages of Resistance to Diversity

The conceptualization of organizational response to social issues is not new. In the mid-1990s, Greening and Gray (Greening & Gray, 1994) developed and tested a model examining how organizations respond to social and political issues within society. In their model, Greening and Gray (1994) use an issues management approach to explain factors that influence when and how organizations respond to social and political problems. Issues management refers to processes and procedures for handling and responding to social and political issues. Greening and Gray (1994) outline five components of issues management development organizations may use: formalization of issues management activities (formal functional departments such as public affairs that deal with such issues), resource commitment (resources allocated from top management to deal with issues management), dedicated committees (standing or ad hoc committees developed to handle different types of issues), integration with strategic planning (conducting issues analyses and building this into the strategic plan of the organization) and integration with line functions (line managers taking steps to address issues). Drawing from institutional theory (DiMaggio & Powell, 1983) and resource dependence theory (Pfeffer & Salancik, 1978), they state that both institutional and organizational variables influence the structural development of issues management and approaches that are used by organizations. Institutional variables that influence issues management development include pressure from external interest groups, media exposure about the issues and crises or disasters that occur. Organizational variables include the size of the organization and the level of commitment from top management to address issues. In testing their model, Greening and Gray focused primarily on institutional factors that directly influenced the firm such as media exposure about the organization as it relates to the issue at hand or crises (e.g., accidents, scandal, disasters) that directly led to negative outcomes such as loss of profits for the organization. Such attention to direct relations between institutional factors likely encourages greater formalized responses to issues.

Building from Greening and Gray's (1994) model, we posit that organizations may respond to mega-threats and other social justice issues even though these issues may or may not have direct links to their profits. Further, we posit that some organizational responses to such issues can serve as a form of resistance to diversity. To more closely examine this notion, we draw from the social psychology literature. Specifically, we integrate

Knowles et al.'s (2014) 3D (deny, distance and dismantle) model on White Americans' identity management strategies to cope with threats associated with White privilege (i.e., the benefits afforded to White individuals based on their race; Healey & O'Brien, 2007; McIntosh, 1989). The model suggests that White Americans may fear that their accomplishments will be credited to White privilege and thus others will perceive their status as not fully earned. The authors suggest that in order to cope with these feelings, White individuals either *deny* that White privilege exists, *distance* their own identity from the identity of their social group, or *dismantle* privilege by actively working to reduce it. The first two strategies are helpful in restoring one's self-concept; however, they can have the unintended consequence of leading to greater racial inequality. Dismantling on the other hand, can be effective in reducing inequality because it acknowledges the existence of privilege head on and seeks a solution to minimize privilege and its consequences. Although dismantling appears to be a noble strategy, Knowles and colleagues (2014) emphasize that the motivation for such action is still self-motivated. Specifically, dismantling is a coping strategy that wards off group-image threat – or perceptions that one's ingroup is perceived negatively by outgroups.

An organization's response toward diversity may signal the organization's identity development regarding diversity issues. For instance, Chrobot-Mason and Thomas (2002) suggest that a company where organizational racial identity is high (i.e., an organization that values racial differences in employees) is more likely to include diversity in their business strategies compared to a company that is low on organizational racial identity. Organizations that exhibit low levels of diversity identity, such as those that ignore social justice issues, may help create a negative climate toward diversity within the organization. A negative climate for diversity can result in increased exclusion, stress and burnout for employees who value diversity (Thomas & Plaut, 2008), as well as financial costs due to workplace conflict and legal issues with regard to discrimination (Thomas & Plaut, 2008; Tsui & Gutek, 1999).

Both Greening and Gray's (1994) issues management model and Knowles and colleagues' (2014) 3D model of managing White identity are useful for understanding organizational responses to social justice issues, which goes beyond day-to-day organizational practices. The issues management model allows us to understand the factors external to and within organizations that influence organizational responses. The 3D model recognizes self-prioritization as a determinant for how people respond to inequity. And, because organizations likely prioritize self-preservation over social justice, the 3D model accounts for how such an orientation factors into an organization's decision to resist (deny or distance) or embrace (dismantle) diversity issues. For instance, dismantling may require diversity issues to be framed in ways that motivate organizations to protect facets of their group-image. Therefore, our model in Figure 7.1 integrates the issues management model and the 3D model to provide a useful framework to examine why and how organizations respond to trauma-inducing mega-threats. We posit that organizations use response strategies that map onto those outlined by Knowles et al. (2014) to manage their reputations and organizational identity with regard to social justice issues and further, diversity.

As seen in Figure 7.1, we believe that the ways in which organizations respond to social justice issues such as mega-threats fall within various stages that move on a continuum from signaling resistance to diversity issues to proactively embracing diversity issues. Responses that signal resistance fit within Thomas and Plaut's (2008) taxonomy of diversity resistance as a form of subtle organizational resistance. This taxonomy highlights how an organization's silence on diversity issues or mixed messages about diversity can both serve as subtle signals of organizational resistance. We believe that organizations may send similar signals when they choose to respond (or not) to mega-threats in society. However, we also show how some responses further down the continuum move toward a more positive illustration of embracing diversity.

Following Greening and Gray (1994), we believe that institutional and organizational factors influence the response strategy selected by organizations. Specifically, when crises receive greater media exposure and when there is strong pressure from interest groups in place, organizations are more likely to engage in a more active and public response that signals a higher degree of embracing (as opposed to resisting) diversity-related issues. For instance, following the uprising of the #MeToo movement in 2017, which received high levels of national media attention, several companies such as Facebook and the Screen Actors Guild made public statements on things the organization was doing to address and prevent sexual harassment (Kantor, 2018).

Organizational factors such as organizational size and proximity to social justice issues are also likely to influence the response strategy selected. In our model, we extend beyond Greening and Gray's (1994) and Knowles et al. (2014) models to examine outcomes of different organizational responses. We focus specifically on individual-level outcomes related to employees. In the next section, we discuss the phases of our model and provide real world examples of organizational responses that fit into each response stage. We start by discussing different organizational responses to social justice events, then discuss the external and organizational factors that influence the type of response used and the employee outcomes that may result from different responses.

Denying Diversity

Knowles et al. (2014) discussed the use of denial as a way to defend one's self-concept against threats related to competence and ability. Organizations can deny diversity when they choose to reject social justice issues. We believe that organizations use denial as a means of either defending values the organization holds dear or maintaining the status quo. In this way, denial can be seen as an organizational response strategy that mirrors cultural ideologies such as colorblindness. Such ideologies have been shown to have negative consequences for individuals, in particular minorities (Plaut, Thomas, & Goren, 2009). Such cultural ideologies that minimize group differences, and the practices derived from these ideologies, can create more harm than good when used to respond to social justice issues, as responses built on denial stifle the needs of minority employees. Organizations exhibit denial through two forms of rejecting social justice issues: rejection of the values of the issue and reframing the issue away from social justice.

Stage 1: Reject Social Justice Issues

Rejection of social justice issues from organizations can take different forms. In some cases, rejection may be the result of conflicting values between the social justice issue and the values of organizational leaders. An example of this is the stance of Chick-fil-A CEO Dan Cathy in 2012 in response to social pressures surrounding same-sex marriage. In an interview, Cathy stated that the company is "supportive of the family – the biblical definition of the family unit" (Blume, 2012). He made similar comments on a radio show in the same year, which sparked backlash from consumers (McGregor, 2012). In this case, Chick-fil-a refused to show support for same-sex marriage equality due to the Christian-based values of the organization's leadership.

As seen in Figure 7.1, organizational denial may have negative implications for employees. In this example, such a rejection of diversity may have implications for employees who identify with the Lesbian, Gay, Bisexual and Transgender (LGBT) community as anti-LGBT politics have been shown to be associated with negative long-term consequences for LGBT identified people (Russell, Bohan, McCarroll, & Smith, 2011). Furthermore, comments condemning homosexuality may foster anti-LGBT sentiments in the workplace and create norms that could lead to increased prejudice and discrimination against LGBT employees (McDonald & Crandall, 2015), which would negatively affect their mental health (Almeida, Johnson, Corliss, Molnar, & Azrael, 2009; Pizer, Sears, Mallory, & Hunter, 2012).

Rejection of social justice can also manifest when organizations attempt to reframe social justice issues as a way to divert attention away from or undermine the issue. One example is the National Football League's (NFL) response to football players taking a knee during the national anthem. In 2016, Colin Kaepernick, a former quarterback for the San Francisco 49ers placed the NFL at the center of media attention when he decided not to stand for the national anthem to make a statement about police brutality against Black people in the U.S. before the team's first preseason game. Other NFL players followed suit as a way to support the movement in the wake of several cases of police violence toward Black civilians shown across national news platforms during this time. In his initial statement, NFL commissioner Roger Goodell reframed the meaning of the protest from one meant to bring attention to social injustice to focusing on patriotism (or lack thereof). In part, Goodell stated that he did not agree with Kaepernick's protest because "… we believe very strongly in patriotism in the NFL … I think it's important to have respect for our country, for our flag, for the people who make our country better …." Although he somewhat changed his verbal message in a later statement in 2018, saying that the protests were not a sign of unpatriotic players, he did so when announcing a new NFL policy voted on by team owners to require players to stand during the national anthem or remain in the locker room; otherwise players would be sanctioned for violating the policy (National Football League, 2016). The establishment of the policy along with his statement that standing during the anthem is equivalent with respect for the flag (which was repeated in his 2018 statement) signaled a rejection of the purpose of the movement. The NFL's response of the original issue led to displeasure for many employees, particularly Black players. Many NFL players (i.e., employees) expressed displeasure with the policy and the NFL Players Association filed a grievance on behalf of the players in response to the policy (Dwyer, 2018). The NFL's response to the social justice issue via policy change can be viewed in part as a strategic attempt to stifle the protests and distract audiences from the root of the issue, which is police brutality against Black civilians.

Distancing from Diversity

Knowles et al. (2014) define distancing as a way for White people to detach their own identity from that of White privilege. In our model, we examine the ways in which organizations attempt to distance themselves from social justice issues. Distancing from diversity and social justice issues is a common strategy among organizations because of the political nature of these issues. The prevailing philosophy is that business and politics should remain separate as to not alienate customers and employees from the organization. By distancing the organization from social justice issues, organizations may be attempting to insulate themselves from the political nature of such issues and continue business as usual. However, when an organization engages in distancing strategies, they often affect their employees and customers directly or indirectly. As seen in Figure 7.1, organizational distancing from social justice issues often manifests in two forms: ignoring the issue and behavioral misalignment.

Stage 2: Ignore Social Justice Issues

The traditional response for many companies is to remain silent or neutral on diversity-related issues. Such a tactic is in line with economist and

Nobel Laureate Milton Friedman's (1970) philosophy that corporations are responsible for increasing profit and should not become involved in social issues beyond this goal. Although this division sounds good in theory, in practice the complete separation of business and politics is often hard to distinguish. For instance, shortly after Donald Trump was inaugurated in 2016, he announced a temporary executive order to ban immigration from seven Muslim countries – a move that prompted response from many CEO's because of the effect that the Executive Order had on business operation. Many organizations hire, contract and serve people who are immigrants, and therefore, immigrants are a vital part of business operation. However, a large number of companies stayed silent (at least publicly) and refused to comment on the ban (Cain, 2017; Korschun, 2017). It is likely these organizations refused to speak out for or against the travel ban to avoid upsetting and possibly losing customers from either side of the political spectrum (Korschun & Smith, 2018). Yet, the silence on social issues deemed important by many, can be seen as a form of rejecting the implications and importance of the social justice for diverse individuals affected by the events.

Stage 3: Behavioral Misalignment

In some cases, organizations will respond in contradictory ways. For instance, they may issue a statement supporting a social justice issue or staying neutral, then later engage in a behavior or create a policy that does not match this stance. In these instances, organizational responses fit into the stage of behavioral misalignment. This type of response may lead some employees and customers to view the organization as hypocritical.

One example of behavioral misalignment can be seen in Target's response to marriage equality issues in 2012. Target, a retail company, conveyed a neutral position on the subject in a public statement saying they recognize the "broad range of views" on the subject (Bhasin, 2014). Despite this statement, the company donated funds to support Republican candidate Tom Emmer, who opposed same-sex marriage. Upon learning this information, many shoppers boycotted Target because they perceived their actions to be hypocritical of the values Target proclaimed. People may perceive behavioral misalignment as a form of organizational resistance and this attribution may intensify negative reactions from consumers.

Another example of behavioral misalignment is Uber's changing stance on the travel ban issue raised by Donald Trump in 2017. Following the announcement of the travel ban, yellow cab drivers announced a protest of the ban in which they would not work for a period of time on a specific day. Uber publicly announced their drivers would continue to operate during the protest and surge pricing would be turned off. Many customers were outraged, as they saw the initial response as an attempt to profit from the taxi driver protest and ignore diversity-related issues in the midst of a collective trauma. In response, several Uber customers deleted the Uber app from their phones and boycotted the company. Eventually, the company declared the executive order was wrong and Uber CEO Travis Kalanick refused to participate in Trump's economic council (Segall & Zeleny, 2017); however, the damage was done. Many customers saw the initial response as a true indication of where Uber stood on the issue and viewed the later declaration as a marketing ploy more than a stance for social justice.

Dismantling Social Inequality and Injustice

The final strategy discussed in Knowles et al.'s (2014) theory of responding to White privilege is dismantling systems of privilege. This involves acknowledging that privilege exists and taking steps to reduce or remove it. Many organizations take action to dismantle social inequality by engaging in behavior that directly acknowledges injustice and attempts to correct for it when possible and fully embrace diversity. Some companies engage in dismantling as a way to address issues occurring within the company; whereas others take a proactive approach to dismantling injustice. Below we discuss both reactive and proactive responses as stages of dismantling injustice.

Stage 4: Reactive Behavior

At times, diversity-related events may occur within the company that relate to larger social justice issues and therefore, exacerbate collective traumas. In these instances, companies may feel compelled to act in response to the events in order to maintain a positive reputation with external stakeholders. In this way, the motivation behind reactive responses may be seen as self-interested on the part of the organization because it is in part used to protect organizational image. This motivation aligns with the motivation of dismantling White privilege discussed by Knowles et al. (2014). Many times, reactive responses take the form of public statements, diversity trainings, or some other form of action geared toward reaffirming company values toward diversity.

One example of reactive behavior is Starbucks's reaction to a racially biased incident that occurred at a Philadelphia Starbucks in 2018. On April 12, 2018, a store manager called the police on two 23-year-old Black men, Rashon Nelson and Donte Robinson, who were sitting in the coffee shop waiting for another friend to arrive. One of the men asked the store manager to use the shop's restroom and the manager told him the restrooms were for "customers only." According to news reports, the manager called the police on the men less than two minutes later (Siegel, 2018). Starbucks CEO, Kevin Johnson, met with the two men to apologize for the way they were treated and also made a public apology. The store manager who instigated the incident no longer works for Starbucks (Isidore, 2018). This incident gained national media coverage and became part of the national conversation about race in America. In response, Starbucks shut down 8,000 stores to deliver a store-wide diversity training to combat unconscious bias (Donnelly, 2018).

Stage 5: Proactive Response to Social Justice Issues

An organization is proactive in their response to social justice issues when they take initiative to respond to larger societal level social justice events even when their organization is not directly affected by the event. For example, after Trump was elected, many CEOs felt it necessary to acknowledge the election and reaffirm their values toward diversity. Ben and Jerry's, an ice cream company, wrote an open letter to President Trump to "challenge" the president to "hear the voices of all Americans" (Hauser, 2016). Additionally, Matt Maloney, the chief executive of Grubhub, an online food delivery corporation, sent an email to employees affirming the organization's commitment to protect the rights of those most vulnerable to experiences of harassment and discrimination in the U.S. In this email, Maloney was sending a message that Grubhub was against the divisive commentary and platforms Trump discussed that threatened the rights of individuals from marginalized groups such as the admitted sexual harassment and the negative sentiment toward immigrants.

Another way companies can be proactive with regard to social justice issues is to recognize and respond to problems within their own organization before someone else points it out. Some organizations have started to do this in regards to creating parity to men's and women's pay. For example, Adobe hired an external company to conduct a review of their pay practices in 2016 and found that women were paid $.099 per every $1.00 men were paid. Although the gap was not nearly as large as many organizations, Adobe took steps to ensure complete parity between men and women's pay by the following year (Calfas, 2018). Other companies such as Salesforce and Starbucks have taken similar steps to reach gender pay equity (Calfas, 2018).

Response Stage Mobility

It is important to note that an organization's response to a social justice issue or event is not necessarily a static declaration of where they will always stand on that particular issue. Many organizations evolve over time and their stance may change from one that is more resistant to one that is more embracing as the social climate and pressures for justice related to a particular group change or intensify. For instance, the Boy Scout's of America's response to sexual orientation and gender identity-related issues have evolved from a stance of rejecting openly gay and transgender youth from joining the organization to reactive policies allowing openly gay youth (passed in 2013), openly gay adult leaders (passed in 2015) and transgender youth (passed in 2017) to be a part of the organization (Chokshi, 2017). Additionally, an organization's response to one social justice issue may vary from their response to a different social justice issue. For instance, Grubhub executives sent emails to employees confirming commitment to diversity after Donald Trump was elected, but the company appeared to be silent (at least publicly) following other trauma-inducing events such as the Pulse nightclub shooting, where LGBT individuals were targets in a mass shooting that left 49 people dead. As we discuss in the next section, there are several reasons that influence why organizations select different responses that are more or less resistant to diversity. These reasons can also explain why a single organization may change its response on an issue or respond differently to different issues.

Variables that Contribute to Organizational Response

Several contextual variables may influence the way in which an organization chooses to respond to social justice events. As noted in Stage 2, many choose to remain silent about these issues. This response is often the path that may result in the least amount of disruption from any stakeholders; however, this non-response can signal resistance to diversity because it is passive and may suggest that organizations do not care about an issue that may negatively affect individuals, particularly those from marginalized groups. Greening and Gray (1994) discuss the institutional and organizational variables that influence organizations' issues management development process. Following this model, we also note the distinction between institutional (external) context and organizational characteristics that can influence organizational response to social justice issues. We believe that the variables noted by Greening and Gray also apply to our model; however, we discuss additional context variables and organizational characteristics that may play a role in organizational response choice.

Institutional Context

As seen in Figure 7.1, we believe institutional context variables influence the stage an organization selects to respond to social justice issues. Greening and Gray (1994) found support showing that pressure from social interest groups and media exposure about an issue in which the organization was involved were both positively related to issues management development. Following the premise set by Greening and Gray (1994), we believe that generally when institutional pressures are stronger, organizations are more likely to respond in ways that exhibit higher levels of embracing diversity relative to when these pressures are low. Some support for this has been seen with pressure from climate change interest groups. One study examined the response to climate change interest groups from organizations in the oil and gas industry over an almost 30-year period and found that increased protests and pressure from interest groups on organizations was positively related to increased public statements and affiliation with interest groups to attempt to address concerns from organizations (Hiatt, Grandy, & Lee, 2015). When examining social justice issues, an example of external pressure leading to a response that is less resistant to diversity is seen in the grocery store Publix announcing its plan to suspend donations to political campaigns in May 2018 after protests against the organization. Publix had been contributing to a republican campaign of a candidate who was a supporter of the National Rifle Association (Arnold, Brinkmann, & Shoneki, 2018). Anti-gun activists, led by survivors of the Parkland High School mass shooting, spoke out in protests and held a "die-in" where protesters laid on the floor of a Florida Publix grocery store and put out a national call for people to boycott the grocery chain.

We believe that media attention also influences the response that an organization has toward a social justice issue. Greening and Gray (1994) found evidence that higher levels of media exposure are positively related to issues management development. They specifically examined media exposure about the firm and its relation issues management. As such, in instances where there was more media coverage about the firm, organizations were more likely to develop issue management practices. When examining social justice issues, media attention also likely plays a role. We believe that greater media attention toward the organization and also toward the social justice issue in general will prompt responses that are more embracing of diversity versus resistant. The relation between interest group pressure and media attention is likely high given the increase in media, in particular social media, as a way to disseminate information about social justice issues and interest groups. An example of media attention influencing response is seen through the examples of Publix and Starbucks (provided above). In both cases, there was a high level of media coverage surrounding the social issue at hand, which likely played a role in the reactive responses selected by both organizations.

We note that increased media attention may not always lead to embracing responses. For instance the national anthem protests by NFL players received a large amount of media attention, yet the organization's response was not an immediate embracing of the issue. As we note earlier, in line with Knowles et al. (2014), we believe that organizational responses are also likely motivated by organizational self-interest. In the case of the NFL, the protests were rejected by many consumers and may have had a negative effect on the television consumption of games. As media attention both praising and condemning the protests increased, the NFL shifted its response in an attempt to appeal to both sides. As noted earlier, the commissioner stayed consistent with the message of respecting the flag, implying that the protests were a sign of disrespect; however, the NFL pledged money to support community initiatives related to social justice. As such, we believe increased media attention sparks a more active response (i.e., something beyond ignoring the issue) and these responses often fall closer toward embracing versus resisting diversity.

Organizational Characteristics

Using a resource-based perspective, Greening and Gray (1994) hypothesized that organizational size and top management commitment should positively affect an organization's issues management development. They found some support for the relationship between organizational size (operationally defined by average firm sales over a three-year period) and issues management development, specifically as it relates to resource commitment and formalization of units. The findings for top management commitment were more conflicting with stronger support seen based on top managements' self-reported commitment measure versus an analysis of commitment via annual reports to stakeholders. In line with Greening and Gray (1994), we believe that organizational size may be positively related to selecting responses that are more indicative of embracing diversity because larger organizations are more likely to have the resources to support reactive and proactive behaviors around these issues. Furthermore, larger organizations have more resources to sustain potential blow back from consumers who may be unhappy with their stance of embracing social justice issues. We also believe that an organization's values and commitment to diversity influence organizational response. Specifically, we believe that organizations that espouse diversity values and have policies and practices that align with these values are more likely to provide organizational responses to social justice issues that embrace diversity.

In addition to size and values, the proximity of an organization to a social justice event is likely to influence the extent to which they respond to social justice issues in ways that are less resistant to diversity. We refer to proximity in terms of both geography/physical location (e.g., Starbucks' response to a racial bias event that happened at their organization versus responses to events that happen at other locations) and closeness of the event to individuals within the organization (e.g., the extent to which the demographic makeup of individuals within the organization reflects that of individuals directly affected by the social justice issue). Research has shown that at the individual-level, closer proximity and contact with people with marginalized identities can help to reduce prejudice and increase embracing diversity (for example see Pettigrew & Tropp, 2008). We believe that similar effects of proximity are possible at the organizational-level.

Organizations closer in geographical proximity to a mega-threat or other social justice event are more likely to respond in a manner that embraces diversity compared to those that are more distal to the issue or event because closer organizations are more likely to face more pressure from the local community to respond. Organizations want to appear legitimate to key external stakeholders because social action is believed to increase financial performance (Webb & Farmer, 1996). Thus, organizations within the local community where a traumatic event occurs may confront increased pressure to react than companies who are further away because emotional response within the community will be more intense overall, which will bolster external pressures for local organizations after a tragedy or in the midst of social justice issues. Although people who are more distal to a traumatic mega-threat can experience negative psychological effects, some evidence shows people who are closer to the event may suffer greater consequences. For instance, a study on employee trauma after the 9/11 attacks, found that the closer an individual lived and worked to the location of the attacks, the more likely they were to experience higher levels of fear and stress from the attacks than people who were more distanced from the event (Mainiero & Gibson, 2003). Another example of geographical proximity influencing organizational response is seen through PayPal's behavior following the passing of House Bill 2 in North Carolina, a law which made it legal for organizations to discriminate against people on the basis of sexual orientation and gender identity. Following the passing of this state law, the CEO of PayPal released a statement withdrawing plans to open a new operations center within the state of North Carolina (Schulman, 2016). While it is possible that PayPal may have responded to this legislation even if it had not announced plans to conduct business in the state two weeks prior, the geographic proximity of the trauma to the organization's planned business likely sparked a public response.

Organizational response may also be spurred by the proximity of employees within the organization to the incident. The ties and/or perceived similarity that employees within the organization have to those involved in the traumatic event can spark emotional responses such as empathy, which can motivate an organization to respond. This can be illustrated through the response of Price Waterhouse Cooper (PwC) to the shooting death of a Black employee, Botham Jean, by his neighbor, a White police officer in September 2018. Following the incident, a chairman of PwC sent an email to employees encouraging them to be empathetic toward colleagues, and work to become allies with Black coworkers and other underrepresented minorities (Holman, 2018).

Employee Outcomes Related to Denial, Distancing and Dismantling Stages

Most research on organizational response to social justice issues focuses on the effect on external stakeholders. Recently, organizational researchers have shifted to examining the effect of organizational response on existing employees as well as prospective employees. Broadly, these studies provide evidence that organizational response influences employees' perceptions of the organization's reputation, which influences employee attitudes toward the organization and behaviors on the job (Albinger & Freeman, 2000; Chaudhary, 2018; Riordan, Gatewood, & Bill, 1997; Turban & Greening, 1997). As seen in Figure 7.1, we believe that organizational response to trauma-inducing social justice issues can lead to a variety of employee outcomes. Overall, we posit that organizational responses that signal resistance to diversity will have a negative effect on employees' job performance, engagement and organizational commitment, particularly for employees who are affected by the social justice issue at hand. For employees who are adversely affected by social justice issues, having their organizations embrace diversity may lead them to stay engaged and committed to the organization, be more satisfied in their jobs and cope in a more positive manner when social justice issues arise.

We believe employees who relate more to the group targeted by an organizational response that embraces diversity will receive the most benefits (e.g., women will likely experience more positive benefits than other minority groups if an organization embraces gender equality); however, we do believe that many employees outside of the referenced marginalized group will still benefit from organizational responses that signal embracing any diversity-related issues. Indeed, there is evidence that employees from marginalized groups respond positively to diversity cues in recruitment materials even when their group is not represented in such materials (for example, see Avery, Hernandez, & Hebl, 2004). Additionally, one study found that majority group members who feel more connected with minority groups are more likely to view organizational responses that signal resistance to diversity negatively (Dahling, Wiley, Fishman, & Loihle, 2016). Specifically, in an experiment using real organizational statements about marriage equality for same-sex couples, Dahling et al. (2016) found that heterosexual individuals who read anti-equality messages from organizations (e.g., Chick-fil-a's response to marriage equality) expressed greater intentions to decrease work effort, protest the organization's position and quit than individuals who read messages that moved further away from resistance (neutral on marriage equality and pro-marriage equality). The negative outcomes of organization responses that resisted diversity were stronger for majority-group members who expressed high (vs. low) identification with the LGBT community.

It is possible that some individuals from non-marginalized groups will perceive organizational responses that embrace social justice issues negatively. However, if organizations can illustrate how these responses fit within an overarching diversity policy that is multicultural and inclusive of all employees, then majority group members are more likely to view responses to specific social justice more positively (Galinsky et al., 2015) and in line with the organization's values.

Drawing from the corporate social responsibility (CSR) literature, which states that corporations that are socially responsible make decisions and actions for reasons other than economical or technical gain (Carroll, 1991), research has shown that organizational response to social issues is related to job performance (Chaudhary, 2018; Korschun, Bhattacharya, & Swain, 2014) and organizational commitment (Brammer, Millington, & Rayton, 2007). For instance, Chaudhary (2018) found that employees who held positive perceptions of their organization's CSR initiatives performed had higher job performance and engaged in more organizational commitment behaviors (OCBs). Response to social justice issues is related to CSR in the sense that people believe that organizations have a responsibility to respond to social and political issues (Greening & Gray, 1994). Thus, it is likely that as organizations move to distance themselves from politics in the rejection and behavioral misalignment stages, employees may become less engaged at work and possibly participate more in activism outside of the organization in the form of protests or boycotts, as seen in the Uber case study (Williams, 2017).

Organizations who reject or attempt to distract attention away from social justice issues risk employee disengagement and dissatisfaction. Employees likely become wary of organization's motives behind reframing social justice issues and may take action to disassociate from their organizations. Additionally, rejecting social justice issues may negatively influence the organization's reputation, which in turn, can result in more negative job attitudes and behavior. For instance, one study found that employees with negative perceptions of corporate image were less satisfied with their jobs and were more likely to have higher turnover intentions (Riordan et al., 1997).

Organizations who take action, whether reactive or proactive, have a better chance of encouraging positive coping behaviors in their employees – proactive companies more so than reactive companies. Employees may still question the motives behind the organization's reactive response because organizational action is often thought to be profit-driven; however, the behavior should ultimately lead to positive outcomes for employees who are invested in the social justice issue. Organizations who are more proactive in embracing diversity are likely to foster employees who are more committed and engaged in the organization (Brammer et al., 2007; Chaudhary, 2018). Additionally, they are more likely to have a competitive advantage when it comes to hiring because prospective employees, particularly those from minority backgrounds and those who value diversity, may be more attracted to the organization due to their diversity efforts (Albinger & Freeman, 2000; Turban & Greening, 1997).

Organizational Responses: Moving from Resistance to Embracing Social Justice Issues

As we argue, ignoring or denying social justice issues that lead to collective trauma can be seen as a form of diversity resistance in organizations because it directly discounts the distress that employees feel and bring to work with them as a result of societal events. Although it is not an organization's responsibility to take sides on every political or social issue, or to try to resolve all social injustice issues, organizations should be concerned with the ways in which societal issues influence their employees. Further, in cases where social issues and events challenge values that an organization espouses, such as the value of diversity, organizations should feel compelled to acknowledge the potential for feelings of trauma at least internally within the organization. Below we offer recommendations for behaviors organizations can engage in to move from resistance to proactive responses related to mega-threats and collective trauma in society.

One action that organizations can take is to acknowledge large-scale societal events that may elicit collective trauma, particularly if such events occur in close geographical proximity to the organization's location (e.g., same region). Such acknowledgements can come in the form of an internal statement or a public statement. We recognize that some mega-threats involve subjects that may be sensitive or involve politics, and we are not suggesting that companies must take a stance on all issues. However, we are suggesting that it can be beneficial for organizations to recognize collective trauma when it occurs and acknowledge that such trauma has led to distress for many people in society. A public (internal or public facing) recognition of collective trauma demonstrates awareness of how such events create discord among social groups and can often leave those in marginalized groups feeling vulnerable. This can aid employee coping as it indicates a sense of empathy from the organization regarding feelings of collective trauma.

An acknowledgment of collective trauma related to social justice, such as the trauma seen by changes made to increase deportation of immigrant and migrant individuals with little regard to background factors, may be viewed particularly favorably for organizations that have stated values related to diversity. It can send a signal that the organization lives its values of diversity by showing care and concern for how the events are affecting people, particularly those who may be more likely to experience negative outcomes related to the issues. Public acknowledgment also allows organizations to reaffirm their values to diversity and inclusion by stating their recognition of how mega-threats and collective trauma hinder broader strides toward social justice in society.

A second action that organizations can take to help employees cope with collective trauma is to dedicate a safe space for employees to collect their thoughts privately, share their emotions with others, receive social support from each other and gather to demonstrate collectively as a sign of communion. Organization-sponsored safe spaces can take on different forms. For some organizations this may be a dedicated physical space, such as a break room or conference room, where employees are able to congregate together for the purpose of coping with the collective trauma. Providing dedicated space may not be feasible for all organizations; therefore, safe space can also be represented through temporal space, or dedicated time where employees are allowed to take a brief break to gather with others for the purpose of collective coping.

All employees may not need this opportunity, as people experience events differently (i.e., some people may not experience trauma related to a mega-threat); however, for those who experience societal-level events as collective trauma, this can be a useful way to help them cope and refocus on work tasks. In these spaces, employees can engage in social sharing and communal coping, which have been shown to aid in helping people positively cope with trauma (Páez, Basabe, Ubillos, & Gonzalez-Castro, 2007). Social sharing, which involves discussing one's feelings about the collective trauma event (Rimé, 2009), can shape the emotional climate within a space. Although negative emotions are likely to be shared early on, there is evidence that social sharing ultimately leads to positive feelings and positive emotional climates (Páez et al., 2007). Similarly, safe spaces can allow people to engage in communal coping where they actively seek support from people who are experiencing the event in a similar manner. For instance, in a study of law students, students of color noted that they relied on safe spaces in the form of multicultural centers and groups at school to buffer hostility they experienced in the larger, predominantly white environment within their respective law schools (Deo, 2013). Communal coping allows people to seek greater understanding about the event and sympathy from others (Páez et al., 2007) and also provides a support system of individuals who empathize with emotions experienced as a result of collective trauma. In the days and weeks following a mega-threat, employees may need opportunities to process the trauma during times that fall within work hours. Employees may use already allowed break times (such as lunch or other breaks) to take time in a safe space to seek shared communion among others or even gather themselves in solitude. Indeed, in addition to seeking support from others, employees can experience positive benefits of coping independently through meditation or other activities that allow them to temporarily relax cognitively and physically (Littleton, Horsley, John, & Nelson, 2007; Petchsawang & Duchon, 2012).

Organizations should also think about proactive strategies that can have a longer lasting effect on embracing diversity and inclusion at a broader level, which includes understanding how societal events affect employees. This can be achieved by applying some of the issue management components identified by Greening and Gray (1994) to social justice issues in a dedicated and strategic manner. For instance, organizations may develop committees that focus on tracking social justice issues that may be relevant to the organization and its employees and understanding why the issue may be important to employees. Such functions may be built into previously existing committees (e.g., diversity committees, employee assistance programs) or may be assigned to ad hoc committees as the need arises.

One way to understand how social justice issues may be relevant to employees is to think about the ways in which social justice events may influence employees indirectly or may parallel the experiences that employees (particularly those from traditionally marginalized groups) experience in their organizations. For instance, the recent revival of calls for gender equality has placed a sharper focus on sexual harassment experienced in workplace settings. As noted earlier, the #MeToo movement has led to feelings of collective trauma as women (and some men) reflect on their experiences with sexual harassment. This is an experience that many people are currently still facing in organizations and some manifestations of it may go ignored or unrecognized due in part to following the status quo. If an organization takes a stance to support the principles of the #MeToo movement (increased gender equality, end to sexual harassment), this is also an opportunity to examine how the events that led to the crisis may be playing out within one's own organization. There are several ways an organization can take such a stance including providing a public or internal (i.e., organization only) announcement discussing the consequences of sexual harassment and the organizational policies regarding sexual harassment. Once this stance is proclaimed, it can be used as a catalyst to examine how organizational policies and practices may support gender harassment or inequality. For instance, are there subtle forms of hostile environments that are ignored, or are there structural organizational systems set in place that provide advantages to men that are not afforded to women? Developing committees or creating time for already existing committees to examine social justice issues can also help organizations identify their own blind spots regarding increasing diversity and inclusion.

Conclusion

Social justice events can spark collective trauma among people who experience the event in a similar negative way. When individuals go to work the grief or negative emotions they feel from traumatic social events may still be present. We present a conceptual model that highlights how the ways in which employers respond to social justice events can send different signals to employees about how the organization truly views diversity. If organizations deny or distance themselves from social justice issues this may send a message that they do not have vested interest in the well-being of individuals from diverse backgrounds, including their employees. Such responses may lead to negative employee attitudes and behavior toward their organization. However, if organizations engage in strategies that seek to dismantle social injustice, such behaviors can signal embracing diversity that has positive effects on employees and other stakeholders. Thinking about the ways in which organizational responses to social justice events align with an organization's values toward diversity can serve as a way for organizations to turn trauma into an opportunity for diversity and inclusion.

References

Albinger, H. S., & Freeman, S. J. (2000). Corporate social performance and attractiveness as an employer to different job seeking populations. *Journal of Business Ethics, 28*(3), 243–253. doi:10.1023/A:1006289817941

Alexander, J. C., Eyerman, R., Giesen, B., Smelser, N. J., & Sztompka, P. (2004). *Cultural trauma and collective identity.* Berkeley, CA: University of California Press.

Almeida, J., Johnson, R. M., Corliss, H. L., Molnar, B. E., & Azrael, D. (2009). Emotional distress among LGBT youth: The influence of perceived discrimination based on sexual orientation. *Journal of Youth and Adolescence, 38*(7), 1001–1014. doi:10.1007/s10964-009-9397-9

Arnold, K., Brinkmann, P., & Shoneki, M. (2018, May 25). Publix suspends political contributions as David Hogg "die-ins" begin. *Orlando Sentinel.* Retrieved August 26, 2018 from www.orlandosentinel.com

Avery, D. R., Hernandez, M., & Hebl, M. R. (2004). Who's watching the race? Racial salience in recruitment advertising. *Journal of Applied Social Psychology, 34,* 146–161.

Bhasin, K. (2014, August 6). Target finally comes out in support of gay marriage. *Huffington Post.* Retrieved from www.huffingtonpost.com

Blume, A. K. (2012, July 16). "Guilty as charged," Cathy says of Chick-fil-A's stand on biblical & family values. Retrieved August 26, 2018 from www.bpnews.net

Boscarino, J. A., Figley, C. R., & Adams, R. E. (2004). Compassion fatigue following the September 11 terrorist attacks: A study of secondary trauma among New York City social workers. *International Journal of Emergency Mental Health, 6*(2), 57–66. Retrieved from www.ncbi.nlm.nih.gov

Brammer, S., Millington, A., & Rayton, B. (2007). The contribution of corporate social responsibility to organizational commitment. *The International Journal of Human Resource Management, 18*(10), 1701–1719. doi:10.1080/09585190701570866

Bureau of Labor Statistics. (2018). Retrieved March 17, 2019 from www.bls.gov/cps/cpsaat11.htm

Busso, D. S., McLaughlin, K. A., & Sheridan, M. A. (2014). Media exposure and sympathetic nervous system reactivity predict PTSD symptoms after the Boston marathon bombings. *Depression and Anxiety, 31*(7), 551–558. doi:10.1002/da.22282

Cain, Á. (2017, September 18). As America becomes more polarized, private companies are getting pushed into the political spotlight. *Business Insider.* Retrieved December 21, 2019 from www.businessinsider.com/private-companies-politics-2017-8

Calfas, J. (2018, April 9). How HowHow these major companies are getting equal pay right. *Fortune.* Retrieved August 26, 2018 from http://fortune.com

Carroll, A. B. (1991). The pyramid of corporate social responsibility: Toward the moral management of organizational stakeholders. *Business Horizons, 34*(4), 39–49.

Chaudhary, R. (2018). Corporate social responsibility and employee performance: A study among Indian business executives. *The International Journal of Human Resource Management,* 1–24. doi:10.1080/09585192.2018.1469159

Chokshi, N. (2017, December 22). Boy scouts, reversing century-old stance, will allow transgender boys. *The New York Times*. Retrieved from www.nytimes.com

Chrobot-Mason, D., & Thomas, K. M. (2002). Minority employees in majority organizations: The Intersection of individual and organizational racial identity in the workplace. *Human Resource Development Review, 1*(3), 323–344. doi:10.1177/1534484302013004

Cohn, M. A., Mehl, M. R., & Pennebaker, J. W. (2004). Linguistic markers of psychological change surrounding September 11, 2001. *Psychological Science, 15*(10), 687–693. doi:10.1111/j.0956-7976.2004.00741.x

Creamer, T. L., & Liddle, B. J. (2005). Secondary traumatic stress among disaster mental health workers responding to the September 11 attacks. *Journal of Traumatic Stress, 18*(1), 89–96. doi:10.1002/jts.20008

Dahling, J. J., Wiley, S., Fishman, Z. A., & Loihle, A. (2016). A stake in the fight: When do heterosexual employees resist organizational policies that deny marriage equality to LGB peers? *Organizational Behavior and Human Decision Processes, 132*, 1–15. doi:10.1016/j.obhdp.2015.11.003

Deo, M. E. (2013). Two sides of a coin: Safe space & segregation in race/ethnic-specific law student organizations. *Washington University Journal of Law & Policy, 42*, 83–129.

DiMaggio, P. J., & Powell, W. W. (1983). The iron cage revisited: Institutional isomorphism and collective rationality in organizational fields. *American Sociological Review, 48*(2), 147–160. doi:10.2307/2095101

Donnelly, G. (2018, May 24). Starbucks released part of its diversity training curriculum. *Fortune*. Retrieved August 26, 2018 from http://fortune.com/

Dwyer, C. (2018, July 10). NFL players union files grievance over league's new national anthem policy. Retrieved March 17, 2019 from www.npr.org/2018/07/10/627717067/nfl-players-union-files-grievance-over-leagues-new-national-anthem-policy

Friedman, F. (1970, September 13). A Friednzan doctrine-. *The New York Times*. Retrieved from www.nytimes.com

Galea, S., Tracy, M., Norris, F., & Coffey, S. F. (2008). Financial and social circumstances and the incidence and course of PTSD in Mississippi during the first two years after Hurricane Katrina. *Journal of Traumatic Stress, 21*(4), 357–368. doi:10.1002/jts.20355

Galinsky, A. D., Todd, A. R., Homan, A. C., Phillips, K. W., Apfelbaum, E. P., Sasaki, S. J., … Maddux, W. W. (2015). Maximizing the gains and minimizing the pains of diversity: A policy perspective. *Perspectives on Psychological Science, 10*, 742–748.

Greening, D. W., & Gray, B. (1994). Testing a model of organizational response to social and political issues. *Academy of Management Journal, 37*(3), 467–498. Retrieved from December 21, 2019 www.jstor.org/stable/256697

Hauser, C. (2016, January 20). American companies tailor responses to Trump election. *The New York Times*. Retrieved from www.nytimes.com

Healey, J. F., & O'Brien, E. (2007). *Race, ethnicity, and gender: Selected readings*. Thousand Oaks, CA: Pine Forge Press.

Hiatt, S. R., Grandy, J. B., & Lee, B. H. (2015). Organizational responses to public and private politics: An analysis of climate change activists and U.S. oil and gas firms. *Organization Science, 26*(6), 1769–1786. doi:10.1287/orsc.2015.1008

Hirschberger, G. (2018). Collective trauma and the social construction of meaning. *Frontiers in Psychology, 9*, 1–14.

Holman, E. A., Garfin, D. R., & Silver, R. C. (2014). Media's role in broadcasting acute stress following the Boston Marathon bombings. *Proceedings of the National Academy of Sciences, 111*(1), 93–98. doi:10.1073/pnas.1316265110

Holman, J. (2018, September 11). PwC chairman encourages talks on race after Dallas shooting. Retrieved March 17, 2019 from www.bloomberg.com/news/art icles/2018-09-11/pwc-chairman-encourages-dialogue-on-race-after-dallas-shooting

Isidore, C. (2018, April 17). Starbucks CEO meets with two Black men arrested at Philadelphia store. *Cable News Network*. Retrieved June 19, 2018 from http:// money.cnn.com

Kantor, J. (2018, March 23). #MeToo called for an overhaul: Are workplaces really changing? *New York Times*. Retrieved March 17, 2019 from www.nytimes.com/ 2018/03/23/us/sexual-harassment-workplace-response.html

Knowles, E. D., Lowery, B. S., Chow, R. M., & Unzueta, M. M. (2014). Deny, distance, or dismantle? How White Americans manage a privileged identity. *Perspectives on Psychological Science, 9*(6), 594–609. doi:10.1177/1745691614554658

Korschun, D. (2017, February 7). Column: Should companies stay politically neutral? *Public Broadcasting Service*. Retrieved July 27, 2018 from www.pbs.org

Korschun, D., Bhattacharya, C. B., & Swain, S. D. (2014). Corporate social responsibility, customer orientation, and the job performance of frontline employees. *Journal of Marketing, 78*(3), 20–37. doi:10.1509/jm.11.0245

Korschun, D., & Smith, N. C. (2018, March 7). Companies can't avoid politics — and shouldn't try to. *Harvard Business Review*. Retrieved August 1, 2018 from https://hbr.org

Leigh, A., & Melwani, S. (2019). BlackEmployeesMatter: Mega-threats, identity fusion, and enacting positive deviance in organizations. *Academy of Management Review, 44*, 564–591.

Littleton, H., Horsley, S., John, S., & Nelson, D. V. (2007). Trauma coping strategies and psychological distress: A meta-analysis. *Journal of Traumatic Stress, 20*(6), 977–988. doi:10.1002/jts.20276

Mainiero, L. A., & Gibson, D. E. (2003). Managing employee trauma: Dealing with the emotional fallout from 9-11. *Academy of Management Perspectives, 17*(3), 130–143.

May, R. A. B. (2000). Race talk and local collective memory among African American men in a neighborhood tavern. *Qualitative Sociology, 23*, 201–214. doi:10.1023/A:1005482816598

McCluney, C. L., Bryant, C. M., King, D. D., & Ali, A. A. (2017). Calling in Black: A dynamic model of racially traumatic events, resourcing, and safety. *Equality, Diversity and Inclusion: An International Journal, 36*(8), 767–786.

McDonald, R. I., & Crandall, C. S. (2015). Social norms and social influence. *Current Opinion in Behavioral Sciences, 3*, 147–151. doi:10.1016/j.cobeha.2015.04.006

McGregor, J. (2012, July 19). Chick-fil-A CEO Dan Cathy steps into gay-marriage debate. *The Washington Post*. Retrieved August 26, 2018 from www.washington post.com

McIntosh, P. (1989). White privilege: Unpacking the invisible knapsack. Retrieved August 27, 2018 from https://nationalseedproject.org

National Football League. (2016, September 7September 7). Goodell recognizes Kap's right to protest, disagrees with action. Retrieved December 21, 2019 from www.nfl.com/news/story/0ap3000000696136/article/goodell-recognizes-kaps-right-to-protest-disagrees-with-action

Obergefell v. Hodges, 135 U.S. 1039 (2015).

Olick, J. K., & Robbins, J. (1998). Social memory studies: From "collective memory" to the historical sociology of mnemonic practices. *Annual Review of Sociology*, *24*, 105–140.

Páez, D., Basabe, N., Ubillos, S., & Gonźalez-Castro, J. L. (2007). Social sharing, participation in demonstrations, emotional climate, and coping with collective violence after the March 11th Madrid bombings. *Journal of Social Issues*, *63*(2), 323–337. doi:10.1111/j.1540-4560.2007.00511.x

Pascoe, E. A., & Smart Richman, L. S. (2009). Perceived discrimination and health: A meta-analytic review. *Psychological Bulletin*, *135*, 531–554. doi:10.1037/a0016059

Petchsawang, P., & Duchon, D. (2012). Workplace spirituality, meditation, and work performance. *Journal of Management, Spirituality & Religion*, *9*(2), 189–208. doi:10.1080/14766086.2012.688623

Pettigrew, T. F., & Tropp, L. R. (2008). How does intergroup contact reduce prejudice? Meta-analytic tests of three mediators. *European Journal of Social Psychology*, *38*(6), 922–934.

Pfeffer, J., & Salancik, G. (1978). *The external control of organizations: A resource dependence perspective.* New York, NY: Harper & Row.

Pizer, J. C., Sears, B., Mallory, C., & Hunter, N. D. (2012). Evidence of persistent and pervasive workplace discrimination against LGBT people: The need for federal legislation prohibiting discrimination and providing for equal employment benefits. *Loyola of Los Angeles Law Review*, *45*, 715–779. http://digitalcommons.lmu.edu/llr/vol45/iss3/3

Plaut, V. C., Thomas, K. M., & Goren, M. J. (2009). Is multiculturalism or color blindness better for minorities?. *Psychological Science*, *20*, 444–446.

Rimé, B. (2009). Emotion elicits the social sharing of emotion: Theory and empirical review. *Emotion Review*, *1*(1), 60–85. doi:10.1177/1754073908097189

Riordan, C. M., Gatewood, R. D., & Bill, J. B. (1997). Corporate image: Employee reactions and implications for managing corporate social performance. *Journal of Business Ethics*, *16*(4), 401–412. Retrieved from: https://link.springer.com/content/pdf/10.1023/A:1017989205184.pdf

Russell, G. M., Bohan, J. S., McCarroll, M. C., & Smith, N. G. (2011). Trauma, recovery, and community: Perspectives on the long-term impact of anti-LGBT politics. *Traumatology*, *17*(2), 14–23. doi:10.1177/1534765610362799

Schulman, D. (2016, April 5). PayPal withdraws plans for Charlotte expansion. Retrieved March 17, 2019 from www.paypal.com/stories/us/paypal-withdraws-plan-for-charlotte-expansion

Schuster, M. A., Stein, B. D., Jaycox, L. H., Collins, R. L., Marshall, G. N., Elliott, M. N., & Berry, S. H. (2001). A national survey of stress reactions after the September 11, 2001, terrorist attacks. *The New England Journal of Medicine; Boston*, *345*(20), 1507–1512. doi:10.1056/NEJM200111153452024

Segall, L., & Zeleny, J. (2017, February 2). Uber CEO drops out of Trump's business advisory council. *Cable News Network*. Retrieved July 28, 2018 from https://money.cnn.com

Siegel, R. (2018, May 3). Two Black men arrested at Starbucks settle with Philadelphia for $1 each. *Washington Post*. Retrieved August 26, 2018 from www.washingtonpost.com.

Silver, R. C., Holman, E. A., McIntosh, D. N., Poulin, M., & Gil-Rivas, V. (2002). Nationwide longitudinal study of psychological responses to September 11. *JAMA: Journal of the American Medical Association, 288*(10), 1235–1244. doi:10.1001/jama.288.10.1235

Skitka, L. J., Bauman, C. W., & Mullen, E. (2004). Political tolerance and coming to psychological closure following the September 11, 2001, terrorist attacks: An integrative approach. *Personality and Social Psychology Bulletin, 30*, 743–756.

Stults, C. B., Kupprat, S. A., Krause, K. D., Kapadia, F., & Halkitis, P. N. (2017). Perceptions of safety among LGBTQ people following the 2016 Pulse nightclub shooting. *Psychology of Sexual Orientation and Gender Diversity, 4* (3), 251–256.

Thomas, K. M., & Plaut, V. C. (2008). The many faces of diversity resistance in the workplace. In K. M. Thomas (Ed.), *Diversity resistance in organizations* (pp. 1–22). New York, NY: Lawrence Erlbaum Associates.

Tsui, A. S., & Gutek, B. A. (1999). *Demographic differences in organizations: Current research and future directions*. New York, NY: Lexington Books.

Turban, D. B., & Greening, D. W. (1997). Corporate social performance and organizational attractiveness to prospective employees. *Academy of Management Journal, 40*(3), 658–672. www.jstor.org/stable/257057

Webb, N. J., & Farmer, A. (1996). Corporate goodwill: A game theoretic approach to the effect of corporate charitable expenditures on Firm behaviour. *Annals of Public and Cooperative Economics, 67*(1), 29–50. doi:10.1111/j.1467-8292.1996.tb01946.x

Williams, D. R., Yan, Y., Jackson, J. S., & Anderson, N. B. (1997). Racial differences in physical and mental health: Socio-economic status, stress and discrimination. *Journal of Health Psychology, 2*(3), 335–351. doi:10.1177/135910539700200305

Williams, L. (2017, February 3). How Uber drivers are dealing with — and protesting — Trump's immigration bans. *ThinkProgress*. Retrieved August 27, 2018 from https://thinkprogress.org

Xiong, X., Harville, E. W., Mattison, D. R., Elkind-Hirsch, K., Pridjian, G., & Buekens, P. (2008). Exposure to Hurricane Katrina, post-traumatic stress disorder and birth outcomes. *The American Journal of the Medical Sciences, 336*(2), 111–115. doi:10.1097/MAJ.0b013e318180f21c

8 Artful Avoidance

Initial Considerations for Measuring Diversity Resistance in Cultural Organizations

Brea M. Heidelberg

Introduction

Nonprofit arts organizations in the United States exist at a special intersection of the artistic disciplines, business management and nonprofit management (Heidelberg, 2019). Operating at this intersection creates field and organizational nuances that can alter the impact of theories and practices from other fields. These organizations (hereafter referred to as cultural organizations) are responsible for sharing artistic practices and products with the general public in order to maintain their 501(c)(3) nonprofit status. They must also demonstrate knowledge of business practices that have come to permeate the consciousness of nonprofit management (Redaelli, 2012). Systemic inequalities that disproportionately impact people of color, women and the LGBTQ community are a significant part of the foundational fabric of the nonprofit arts world. White, male, heteronormative privilege has permeated dance companies, museums, symphony orchestras and theater companies – impacting board recruitment and retention (Buse, Bernstein, & Bilimoria, 2016) and creating systemic inequities in the dissemination and monitoring of public funding (Sidford, 2011). Additional demonstrations of the aforementioned privilege include biased programming choices (e.g. what is considered valid or respectable forms of artistic expression) and inequitable hiring practices that restrict the diversity of decision-makers – all of which directly contribute to the systemic oppression of marginalized communities within the nonprofit arts field.

Concerted attention from individuals in positions of power and accomplices who want to create positive change have added to the collective power of individuals from marginalized communities to make powerful strides toward creating more equitable, diverse and inclusive cultural organizations. Equity, diversity and inclusion (EDI) initiatives include plans to further diversify the arts management workforce, diversify the boards of arts and cultural institutions and diversify the types of people served by cultural institutions. Many of these initiatives also include some aspect of education designed to help staff and boards understand theories connected to EDI work such as colorblind racism, aversion racism and systemic oppression. Within the past five years, cultural equity statements have emerged as one of the main ways an organization can signal their commitment to EDI work.

Cultural equity statements are articulations of an organization's definition of cultural equity and a declaration of their intent to work toward or uphold that definition. Statements usually include a brief summary of what cultural equity work looks like in each organization's specific context. This usually includes programmatic examples (e.g. artistic choices) as well as functional perspectives that discuss the organization's funding structures, leadership roles and decision-making processes. Organizations take a considerable amount of time, anywhere from 12–24 months, to develop their cultural equity statements. For many organizations, the process is similar to strategic planning – which includes the development of ad hoc committees consisting of members at various levels within the organization, listening tours, draft reviews and feedback sessions with various stakeholder groups (Americans for the Arts, 2016; Greater Philadelphia Cultural Alliance, n.d.). I chose to focus on cultural organizations with an equity statement because the development and dissemination of an equity statement signify that an organization is one that is working toward more equitable practices. The presence of an equity statement is an indicator that the organization, in some part or as a whole, has identified EDI as an important initiative and has committed in some way to letting EDI guide their work, even if that commitment is superficial. Because equity statements are not a necessary component of equity work, part of the process of developing an equity statement is performative. I am interested in the connection between performance and practice in organizations who create equity statements.

One phenomenon related to this work that is often present, but not explicitly discussed at the organizational level in cultural organizations is diversity resistance, as individuals who have historically held power knowingly and unknowingly seek to maintain the status quo of society and their organization (Hite & McDonald, 2006). Generally speaking, resistance to change is detrimental as organizational contexts are ever evolving (Georgalis, Samaratunge, Kimberley, & Lu, 2015). Diversity resistance is acutely problematic in organizations seeking to create and maintain equitable practices. Overt bias, discrimination and violence against underrepresented individuals occurs every day in the United States. While expressions of diversity resistance can include workplace violence, it is just as likely to manifest in milder, seemingly neutral behaviors such as inaction and subtle undermining of diversity change efforts (Thomas & Plaut, 2012). Diversity resistance can be particularly troubling and difficult to address when it stems from individuals and organizations who have explicitly stated a desire to increase their personal and organizational capacity for EDI work. It is imperative that organizations familiarize themselves with the many subtle forms of diversity resistance and how these forms of resistance can erode work towards equity in their organizations. This knowledge is an integral part of helping organizations learn how to identify and address diversity resistance within their own institutions as well as within the field as a whole. Without the work of identifying and naming diversity resistance, equity efforts will ultimately fail despite the best of intentions.

One main reason organizations seeking equity struggle to further their EDI goals is the lack of a direct link between the ideas espoused in an equity state-ment and the skills and practices needed to achieve those goals (Hays-Thomas, Bowen, & Boudreaux, 2012). Bezrukova, Spell, Perry and Jehn (2016) have demonstrated that training alone is not nearly as effective as training that occurs alongside other initiatives. At best, it can provide superficial, short-term changes in attitudes that usually result in the hiring of a few staff members of color (Was-serman, Gallegos, & Ferdman, 2008). However, "there is a critical difference between merely having diversity in an organization's workforce and developing the organizational capacity to leverage diversity as a resource" (Roberson, 2006, p. 234). This chapter helps advance EDI practice in cultural organizations by providing information that connects values to specific practices and pitfalls in EDI work. Linking theorization of EDI to conceptualization and operationaliza-tion is an important step toward collective positive action toward creating an equitable and inclusive arts ecosystem.

This chapter seeks to open a field-wide dialogue about how to assess diversity resistance in cultural organizations. The field of nonprofit arts management is often required to synthesize work from other fields in order to generate theories, practices and assessment processes that can account for the nuances of the field (Heidelberg, 2019). In acknowledgement of this process, I provide a narrow syn-thesis of the diversity resistance literature that is ripe for adaptation and adoption by cultural organizations seeking to assess their EDI efforts. I apply Lee and Nowell's (2015) nonprofit performance measurement framework to that litera-ture and offer initial considerations for how cultural organizations can begin to assess gaps in the stated equity intent of cultural organizations and their actual impact with their organizations and communities. This work is timely because efforts to purposefully and thoughtfully engage in EDI practices are not only needed but have been called for throughout the cultural ecosystem (Cuyler, 2017; Heidelberg, n.d.). Additionally, funding institutions are beginning to strongly encourage or even require cultural organizations to have an equity state-ment, plan, or policies in place in order to be eligible for funding (NASAA, 2016). Despite the aforementioned acknowledgement for the need to develop diverse audiences, staff and boards, stalled efforts continue to plague the field (Le, 2015). The lack of a cohesive and field-specific discussion of diversity resistance may be contributing to organizations' inability to understand the gap that may exist between intentions and impact. I conclude with a discussion of how this work can be used to further diversity resistance and management research in the field of nonprofit arts management.

Defining Diversity and Diversity Management

The way diversity has been defined within diversity management literature has evolved over time to become more inclusive and nuanced. Roberson, Holmes and Perry (2017) offers the most comprehensive definition of diversity that includes any "compositional difference" among individuals within the workplace (p. 72). This covers both categories of difference discussed throughout literature regarding diversity in the workplace: attributes of difference that are readily apparent or easily detected (e.g. gender, race, age) (Jackson, May, & Whitney, 1995) and those that are less easily identified (e.g. education and job-related attributes) (Simons, Pelled, & Smith, 1999). Roberson (2018) presents an integrated approach that distinguishes between, yet still encompasses, both surface-level diversity and deep-level diversity, reflecting attributes like values and beliefs which are subject to change, unlike surface-level attributes (Harrison, Price, & Bell, 1998; Harrison, Price, Gavin, & Florey, 2002).

Just as definitions of diversity have undergone a number of revisions, so too has the term diversity management. Diversity management can be narrowly defined as the management of individuals with diverse surface-level attributes (Tatli & Özbilgin, 2012), often to gain a competitive edge over other organizations (Robinson & Dechant, 1997). It can also be more broadly conceptualized as managing the workplace environment to encourage and empower a multitude of voices, reaching deep-level attributes in addition to surface-level ones (Trittin & Schoeneborn, 2017). For the purposes of this chapter, I am following the lead of the organizations that have developed equity statements. Within that context there is an explicit nod to the narrow definition of diversity management. This is based on cultural organizations' longstanding silencing and marginalization of individuals based on surface-level attributes. However, in order to avoid tokenism and the erroneous assumption that individuals who share surface-level attributes must also share deep-level attributes, the broader definition must be included. It is also important to note that the broader definition better aligns with calls for social and creative justice within the cultural sector (Banks, 2017).

Diversity management has received increased attention from scholars as the population continues to diversify. Researchers from fields such as human resource development (Alcazar, Fernandez, & Gardey, 2013; Nishii, Khattab, Shemla, & Paluch, 2018), organization and general management studies (Cox & Blake, 1991; Roberson, Ryan, & Ragins, 2017) and organizational psychology (Roberson, 2018) have looked at the ways organizations manage diversity.

The Field of Nonprofit Arts Management

The field of nonprofit arts management has its intellectual roots in a number of fields including business management, nonprofit management, public administration and public policy. These fields all contribute something to field norms and practices, but each fail to fully encompass the uniqueness of operating a nonprofit organization focused primarily on an artistic discipline or the artistic process (Wyszomirski, 2013). Ways of generating knowledge and identifying best practices have come from the aforementioned ancestral disciplines with varied results as the field tries to find a balance in how to adapt and adopt the practices of other fields. A push for more organizational structure and fiscal responsibility helped usher in a new era for arts managers in the 1960s with a desire for more business-minded organizational practices (Redaelli, 2012). However, an overreliance on business practices can negatively impact nonprofit organizations' public service ethos, while nonprofit management's dual focus on serving the general public can sometimes seem like its working against the goal of artistic excellence (Wyszomirski, 2013). The need to have a triple focus on fiscal responsibility, public value and artistic excellence can create an organizational context that makes knowledge transfer from other fields difficult.

With regard to EDI work, the field of nonprofit arts management has fallen in line with nonprofit management's attempts to address diversity and equity at the board, funder and staff levels (Matlon, Van Haastrecht, & Wittig Menguc, 2014; Ostrower, 2005). EDI knowledge transfer from the general nonprofit management field has resulted in significant similarities in both initiatives and blind spots across the two fields. General nonprofit management has grappled with inequity and diversity resistance in a number of forms. Organizations who state equitable intent, while continuing to engage in inequitable practices have been the focus of online communities such as fake quity.com and nonprofitaf.com who discuss these issues through a nonprofit lens. This, alongside stalled equity efforts (Young, 2010), which have also emerged in the field of nonprofit arts management, indicate a need for cultural organizations to begin systematically assessing their EDI work in order to lessen or remove the gap between stated intent and impact.

Conceptual Framework: Nonprofit Performance Management

Resources to support nonprofit organizations continue to be scarce (Malatesta, 2014). As a result, funders, independent citizens and policymakers have increased calls for public accountability among nonprofits to ensure that organizations seeking public funding are run effectively and efficiently (Benjamin, 2012). In response, these organizations have turned to performance measurement (Modell, 2001, 2009; Yang, 2008). Performance management can also increase organizational transparency and support organizational decision-making (Mannarini, Talo, D'Aprile, & Ingusci, 2018; Poister, 2008; Russ-Eft & Preskill, 2009). While nonprofit organizations have learned a lot from their traditional private organizational counterparts with regard to management and organizational behavior (Sawhill & Williamson, 2001), the difficulty of adapting performance measurement practices to a mission-driven organization have been noted throughout the nonprofit management literature (Bryson, 2011; Drucker, 2010; Sawhill & Williamson, 2001).

Performance measurement in mission-driven organizations cannot rely on the same measures of profit and loss as private enterprises (Sawhill &

Williamson, 2011), as profit generation is not their primary purpose. Increasing scholarly attention in performance measurement in nonprofits has resulted in a number of approaches such as the balanced scorecard for nonprofits (Kaplan, 2001), the public value scorecard (Moore, 2003) and Sowa et al.'s (2004) multidimensional integrated model of nonprofit organizational effectiveness (MIMNOE). However, each approach only focuses on a subset of factors (e.g. resource efficiency) from the overall realm of performance measurement areas.

In response to the limited scope of performance measurement frameworks, Lee and Nowell (2015) presented an integrated framework which includes seven aspects of nonprofit organizational performance that operate at various points in the value-generation process: inputs, organizational capacity, outputs, outcomes: behavioral and environmental changes, outcomes: constituent satisfaction[1], public value accomplishment and network/institutional legitimacy. These seven aspects encompass the previously disparate performance measurement traditions in the nonprofit literature.

Laying the Foundation

This investigation highlights and synthesizes the work of Turnbull, Greenwood, Tworoger and Golden (2010), Hays-Thomas et al. (2012) and Wiggins-Romesburg and Githens (2018), who all worked to further articulate the skills needed to manage diversity and inclusion in the workplace. I use Lee and Nowell's (2015) performance management framework to demonstrate the utility of each piece toward use in cultural organizations and demonstrate how this work can be layered to offer a set of considerations for future work in assessing and navigating diversity resistance within cultural organizations that have created cultural equity statements.

Turnbull et al. (2010) created the Inclusion Skills Measurement Profile (ISM) which identifies skills necessary to implement EDI practices and manage diversity resistance at the intra- and inter-personal, group and organizational levels across seven different areas: diversity sensitivity, integrity with difference, interacting with differences, valuing difference, team inclusion, managing conflict over difference and embedding inclusion. Turnbull et al. (2010) identify key competencies associated with each category that are largely internal. This work was designed specifically to address the gap that exists between EDI intent and impact. Turnbull et al.'s (2010) competency categories stem from the tacit knowledge of experts yet is "firmly grounded in relevant theory" (p. 1). This work is focused on the for-profit context, with frequent mention of how successful EDI work can increase competitive advantage. However, the process of gathering information about and working to rectify values, behaviors and attitudes that may impede or erode EDI progress is applicable and adaptable across organizational context. Finally, the desired competencies in the ISM are explicitly named and there is no need for practitioners to make a leap from abstract theory to practical application. This alleviates potential issues of adaptation and knowledge sharing that can arise when theory-based assumptions are not shared across disciplines.

Hays-Thomas et al. (2012) present values, knowledge and skills "for effectiveness in diverse environments" (p. 138) that help provide a conceptual through line from Turnbull et al.'s (2010) theoretical and competency-based work toward demonstrable skills. A significant contribution of Hays-Thomas et al.'s work (2012) is the differentiation of skills needed at various organizational levels. The skills needed by the executives who set the tone for the organization is different from those needed by middle managers who implement policies. Both of those types of work are distinct from that of the line staff, who are responsible for executing the daily tasks associated with the organization's core work (Hays-Thomas et al., 2012, p. 137). Hays-Thomas et al.'s (2012) articulation of outwardly observable indicators, grounded in theories from counseling psychology, make this work accessible to arts managers without a theory-based background as well as those who struggle to operationalize more theoretical articulations of diversity management strategies. The authors use critical incident methodology – having respondents from a variety of fields recount incidents where they witnessed or experienced positive or negative events related to diversity. This methodology is not unique to, nor does it require grounding in, a specific discipline.

Wiggins-Romesburg and Githens (2018) created a Resistance-Integration Continuum based on the premise that "diversity resistance is inversely related to integration" (p. 184). They identify five organizational diversity perspectives on a continuum that range from *high resistance/low integration* to *low resistance/high integration*: resistance, discrimination prevention, access and legitimacy, inclusion and integration and learning. This nuanced and descriptive continuum synthesizes the diversity resistance literature and provides a tool for classifying organizational approaches to diversity. Resistance is when an organization actively refuses to engage in organizational change related to diversity, instead working to maintain inequality (Davidson & Proudford, 2008). Discrimination prevention approaches comply with legal anti-discrimination requirements and often include diversity training, career development and advancement opportunities specifically for minorities and women. However, this approach is known for championing "color- or gender-neutral" strategies designed to "treat everyone the same" (Wiggins-Romesburg & Githens, 2018, p. 185). This strips people of their individuality and fails to truly acknowledge systemic barriers to equity. The access and legitimacy approach aligns the demographics of the organization with those of the organization's constituents (Ely & Thomas, 1996) and champions multiculturalism, usually for profit or productivity purposes. A common pitfall of this approach is adoption without fully "understanding or appreciating the cultural differences at play" (Wiggins-Romesburg & Githens, 2018, p. 186). This lack of understanding may cause friction within the organization and between the organization and its constituents. Additionally, the focus on using equity as a way to increase profits may result in superficial equity practices that are not authentic nor sustainable. Authentic and holistic inclusion approaches celebrate difference and work to remove systemic barriers to full organizational participation among traditionally marginalized groups. Therefore, the access and legitimacy approach may fall short with regard to organization-wide inclusion and equity in "decision making, power and influence structures" due to the fact that longstanding mental models that promote power inequities may remain intact (Wiggins-Romesburg & Githens, 2018, p. 187). Finally, integration and learning celebrate diversity and champion continued learning and transformation. Integration and learning work to dismantle and examine "tacit assumptions and underlying processes" that underpin inequitable practices (Wiggins-Romesburg & Githens, 2018, p. 187). Without this examination and dismantling process, change resistance will continue despite the removal of larger explicit indicators of inequity because the organizational culture remains unchanged. A significant component of this approach is organizational work to increase equity and equitable practices throughout the broader society. Wiggins-Romesburg and Githens' (2018) continuum of diversity resistance applies social cognition theory to clearly define steps along the pathway toward creating an equitable organization. This combination of theory and practical application does not require practitioners to fully understand the theoretical underpinnings in order to successfully navigate the knowledge transfer process. Additionally, this work is not bound by discipline – although it does stem from a human resource development perspective. A budding understanding and appreciation for human resource development within the field of nonprofit arts management is occurring (Heidelberg, 2016), making this work attractive to practitioners who often lack formal training in this management area (Varela, 2013).

Considerations for Identifying and Assessing Diversity Resistance

Combining the intellectual contributions of Turnbull et al. (2010), Hays-Thomas et al. (2012) and Wiggins-Romesburg and Githens (2018) allows for consideration of the full spectrum of diversity resistance at the organizational level while also accounting for the individually-based values and skills that are necessary to create sustainable organizational change. Layering the three approaches also provides information useful in the pursuit of personal- and organizational-level indicators of diversity resistance. For example, Hays-Thomas et al. (2012) provide a list of personal-level values for effectively operating in diverse environments (p. 138). *Flexibility*, or the ability to adapt organizational policies and procedures to achieve equity in changing organizational contexts, exists opposite an organizational *rigidity* that finds organizations unwilling to change sources of inequity. In diversity resistance settings, *humility* would be open acknowledgment of an exclusionary policy or practice – while the opposite, *ostentation*, would be refusing to acknowledge any wrongdoing. The humility – ostentation continuum, which addresses values present at the managerial and executive level, also encompasses "openness to new ideas" which emerges at the staff level and "openness to try new things" which is a factor at the executive level (Hays-Thomas et al., 2012, p. 138).

Hays-Thomas et al.'s (2012) traits, when considered alongside Wiggins-Romesburg and Githens' (2018) diversity resistance spectrum, allow for a fuller understanding of how personal-level values can directly impact organizational-level diversity resistance. In Table 8.1 I combine Hays-Thomas et al.'s (2012) personal-level values and Wiggins-Romesburg and Githens' (2018) organization-level resistance stages. This amalgamation provides a clearer understanding of the personal-level values most likely present in organizations operating at different levels along the resistance-integration continuum.

Having a conceptual framework for identifying diversity resistance in this way can help researchers and practitioners begin operationalizing these values – a necessary precursor to assessing organizational equity efforts.

In addition to laying the foundation for operationalizing values associated with diversity resistance, it is also important to consider how performance measurement tools can be adapted in order to assess equity work. In Table 8.2, I expand upon Lee and Nowell's (2015) performance management framework by providing definitions for each performance measurement area that are based on the diversity management literature. I also add positive (+) and negative (-) organizational indicators, adapted from Wiggins-Romesburg and Githens' (2018), which help ground the area and definition within the organizational context.

Lee and Nowell note that the seven aspects and core perspectives "do not operate independently of one another" (2015, p. 304). This acknowledgement of the interdependence of performance measures is particularly useful for the purpose of assessing equity efforts in cultural organizations, which requires

Table 8.1 Diversity Resistance at the Organizational and Personal Levels

Organizational Resistance Level	Personal-level Values
Resistance	Maximum diversity, rigidity and ostentation Minimum integration, flexibility and humility
Diversity Prevention	High diversity, rigidity and ostentation Low integration, flexibility and humility
Access and Legitimacy	Moderate diversity, rigidity and ostentation Moderate integration, flexibility and humility
Inclusion	Low diversity, rigidity and ostentation High integration, flexibility and humility
Integration and Learning	Minimum diversity, rigidity and ostentation Maximum integration, flexibility and humility

Table 8.2 Diversity Management Performance Measurement Areas

Performance Management Area	Diversity Management Definition	Organizational Indicators
Inputs	Acquiring resources and efficiently using them to increase equity throughout the organization	Diversity and equity mandates without funding structures to support associated initiatives (-) Diversity programs with no structure or employee support (-)
Organizational Capacity	Creating environments where the people and structures in place facilitate an organization's ability to become equitable	Diffuse power structures (+) Tokenism (-) Welcoming and largely inclusive environment for staff and board members (+)
Outputs	Designing equity initiatives with consideration of scale, scope and quality Creating organizational output targets that connect directly to the organizational mission and equity statement	Claims that discrimination no longer exists or has been "solved" within the organization (-) Equity programming that is unsustainable (-) Programming based on tokenism (-)
Outcomes: Behavioral and Environmental Changes	Assessing the changes to the target population or condition to be addressed by equity initiatives	Learning and equity commitment that is reflected throughout the organization (+) Uneven changes to organizational power structures (-)
Outcomes: Constituent Satisfaction	Identifying and addressing the equity needs of the organization's constituents	Aversive bias (-) A staff whose demographics reflect those of the organization's constituents (+) Changes in organizational demographics that do not impact existing power structures (-) Disjointed or spotty diversity strategies (-) Modern bias that champions assimilation by minority groups (-) Symbolic bias that appears to champion meritocracy, but really supports inequality (-)
Public Value Accomplishment	Identifying and assessing the ultimate impact the organization hopes to create for the community/society	Organizational silence on societal inequities – even those that are business-related (-)

(Continued)

Table 8.2 (Cont.)

Performance Management Area	Diversity Management Definition	Organizational Indicators
		Public advocacy to create change throughout the broader society (+)
Network/Institutional Legitimacy	Assessing reputation legitimacy within the community and the field	Inequitable inter-organizational partnerships (-) Organizational impact on the community (-/+) Organizational transparency about their equity journey (+)

consideration of intra- and inter-organizational, as well as interpersonal performance measures. Diversity resistance can emerge at any organizational level and have wider organizational and community impact. Therefore, a narrow organizational focus could potentially miss elements of diversity resistance that may emerge in areas outside of such a narrow scope. Layering approaches to identifying and assessing diversity resistance allows me to present initial considerations of ways organizations can identify how individual employees' values can impact where their organization lies on the diversity resistance continuum. It also allows me to offer organizational indicators that arts managers may look at to begin to determine the relative internal and external impact of their equity work.

Conclusion

The first purpose of this chapter was to identify diversity resistance work that could speak specifically to the field context of nonprofit arts organizations. Turnbull et al. (2010), Hays-Thomas et al. (2012) and Wiggins-Romesburg and Githens (2018) were used alongside Lee and Nowell's (2015) nonprofit performance measurement framework to present initial considerations for ways to assess equity efforts and diversity resistance in cultural organizations. This chapter has practical implications for assessing the efforts of cultural organizations who have signaled a commitment to equity work by developing and disseminating an equity statement. First, it provides a starting point for designing customized assessment tools within the specialized and nuanced context of cultural organizations. For those that have already created evaluation methods for assessing their equity work, this study provides a means of critically reflecting upon their evaluation choices. It also lays the foundation for the consideration and creation of evaluation protocols that may be utilized by policymakers and funders whose focus is community- or field-wide cultural equity. Finally, this work furthers the nonprofit arts management literature by providing field-specific consideration of how diversity management theories and practices from other fields can inform equity work in cultural organizations.

Potential next steps in this line of research include the development of an assessment tool based on the considerations discussed in this chapter. It is important to note that performance measurement is not the end of the assessment process. Further research into what can be done with diversity resistance assessment results is necessary in order to avoid leaving organizations without a map of what comes after the assessment process. Also, practice-oriented inquiries into how equity work is done and assessed throughout the field will better position both researchers and practitioners to undertake work that informs more equitable practices throughout the cultural sector.

This chapter synthesizes literature from diversity management, diversity resistance and nonprofit performance management to provide preliminary considerations for identifying and assessing diversity resistance management in cultural organizations. My work serves the field of arts management by providing field-specific application of the resistance-integration continuum. In doing so, I assert that an approach to examining equity work within cultural institutions that is based on both organizational context and individual-level values is ideal. As more organizations develop equity statements and otherwise begin or continue equity work, it is important to further the dialogue about practical means for assessing the relative success of their efforts.

Note

1 In the original source this aspect is referred to as "outcomes: client/consumer satisfaction. I have substituted "constituent" for "client" to better reflect language used in nonprofit arts organizations.

References

Alcazar, F. M., Fernandez, P. M. R., Gardey, G. S. (2013). Workforce diversity in strategic human resource management models: A critical review of the literature and implications for future research. *Cross Cultural Management: An International Journal, 20*(1), 39–49.

Americans for the Arts. (2016). 10 Steps we took to create the Americans for the Arts statement on cultural equity. Retrieved November 18, 2018 from www.americansforthearts.org/about-americans-for-the-arts/cultural-equity/steps-we-took-to-create-the-americans-for-the-arts-statement-on-cultural-equity

Banks, M. (2017). *Creative justice: Cultural industries, work and inequality.* Abingdon, UK: Pickering & Chatto Publishers.

Bartolini, S. (2014). Relational goods. In A. C. Michalos (Ed.), *Encyclopedia of quality of life and well-being research* (pp. 5428–5429). Dordrecht, Netherlands: Springer.

Benjamin, L. M. (2012). Nonprofit organizations and outcome measurement: From tracking program activities to focusing on frontline work. *American Journal of Evaluation, 33*, 431–447.

Bezrukova, K., Spell, C. S., Perry, J. L., & Jehn, K. A. (2016). A meta-analytical integration of over 40 years of research on diversity training evaluation. *Psychological Bulletin, 142*, 11.

Bryson, J. M. (2011). *Strategic planning for public and nonprofit organizations: A guide to strengthening and sustaining organizational achievement* (Vol. 1). San Francisco, CA: John Wiley.

Buse, K., Bernstein, R. S., & Bilimoria, D. (2016). The influence of board diversity, board diversity policies and practices, and board inclusion behaviors on nonprofit governance practices. *Journal of Business Ethics, 133*(1), 179–191.

Clark, C., Rosenzweig, W., Long, D., & Olsen, S. (2004). Double bottom line project report: Assessing social impact in double bottom line ventures. *The Rockefeller Foundation*, 1–73.

Cox, T. H., & Blake, S. (1991). Managing cultural diversity: Implications for organizational competitiveness. *Academy of Management Perspectives, 5*(3), 45–56.

Cuyler, A. C. (2017). Diversity, equity, & inclusion (DEI) in the cultural sector: What's next? *CultureWork, 21*(3), 2–6.

Davidson, M. N., & Proudford, K. L. (2008). Cycles of resistance: How dominants and subordinates collude to undermine diversity efforts in organizations. In K. M. Thomas (Ed.), *Diversity resistance in organizations* (pp. 279–302). New York, NY: Taylor & Francis Group.

Davis, D. E., Hook, J. N., Worthington, E. L., Jr, Van Tongeren, D. R., Gartner, A. L., Jennings, D. J., & Emmons, R. A. (2011). Relational humility: Conceptualizing and measuring humility as a personality judgment. *Journal of Personality Assessment, 93*(3), 225–234.

Drucker, P. F. (2010). *Managing the nonprofit organization: Principles and practices.* New York, NY: Harper Collins.

Ely, R. J., & Thomas, D. A. (1996). Making differences matter: A new paradigm for managing diversity. *Harvard Business Review, 74*(5), 79–90.

Georgalis, J., Samaratunge, R., Kimberley, N., & Lu, Y. (2015). Change process characteristics and resistance to organisational change: The role of employee perceptions of justice. *Australian Journal of Management, 40*(1), 89–113.

Greater Philadelphia Cultural Alliance. (n.d.). Diversity, equity, and inclusion. Retrieved from www.philaculture.org/dei

Harrison, D. A., Price, K. H., & Bell, M. P. (1998). Beyond relational demography: Time and the effects of surface-and deep-level diversity on work group cohesion. *Academy of Management Journal, 41*(1).

Harrison, D. A., Price, K. H., Gavin, J. H., & Florey, A. T. (2002). Time, teams, and task performance: Changing effects of surface-and deep-level diversity on group functioning. *Academy of Management Journal, 45*(5), 1029–1045.

Hays-Thomas, R., Bowen, A., & Boudreaux, M. (2012). Skills for diversity and inclusion in organizations: A review and preliminary investigation. *The Psychologist-Manager Journal, 15*(2), 128–141.

Heidelberg, B. (2016 October). *A Human resources research agenda for the nonprofit arts.* Paper presented at the 42nd International Conference on Social Theory, Politics & the Arts Montreal, Quebec.

Heidelberg, B. (2019). The Professionalization of arts management in the United States: Are we there yet? *Cultural Science Management and Education, 3*(1), 53–66.

Heidelberg, B. (Forthcoming). Teaching culturally responsive performing arts management in higher education. In T. Stein (Ed.), *Workforce diversity and the arts*, (pp. 176–186). New York, NY: Routledge.

Hite, L. M., & McDonald, K. S. (2006). Diversity training pitfalls and possibilities: An exploration of small and mid-size U.S. organizations. *Human Resource Development International, 9*(3), 365–377.

Jackson, S. E., May, K. A., & Whitney, K. (1995). Understanding the dynamics of diversity in decision making teams. In R. A. Guzzo & E. Salas, *Team decision making effectiveness in organizations* (p. 261). San Francisco, CA: Jossey-Bass.

Kaplan, R. S. (2001). Strategic performance measurement and management in nonprofit organizations. *Nonprofit Management & Leadership, 11*, 353–370.

Lee, C., & Nowell, B. (2015). A framework for assessing the performance of nonprofit organizations. *American Journal of Evaluation, 36*(3), 299–319.

Los Angeles County Arts Commission. (2017). Strengthening diversity, equity and inclusion in the arts and culture sector for all Los Angeles county residents.

Malatesta, D., & Smith, C. R. (2014). Lessons from resource dependence theory for contemporary public and nonprofit management. *Public Administration Review, 74*(1), 14–25.

Mannarini, T., Talo, C., D'Aprile, G., & Ingusci, E. (2018). A psychosocial measure of social added value in non-profit and voluntary organizations: Findings from a study in the South of Italy. *VOLUNTAS: International Journal of Voluntary and Nonprofit Organizations, 29*(6), 1315–1329.

Matlon, M. P., Van Haastrecht, I., & Wittig Menguc, K. (2014). *Figuring the plural: Needs and supports of Canadian and U.S. ethnocultural arts organizations*. Chicago, IL: School of the Arts Institute of Chicago/Art Institute of Chicago.

Modell, S. (2001). Performance measurement and institutional processes: a study of managerial responses to public sector reform. *Management Accounting Research, 12*(4), 437–464.

Modell, S. (2009). Institutional research on performance measurement and management in the public sector accounting literature: a review and assessment. *Financial Accountability & Management, 25*(3), 277–303.

Moore, M. H. (2003). The public value scorecard: a rejoinder and an alternative to 'strategic performance measurement and management in non-profit organizations' by Robert Kaplan. *Hauser Center for Nonprofit Organizations Working Paper, 18*, 1–27.

National Assembly of State Arts Agencies (NASAA). August 2016. State policies & programs addressing diversity.

Nishii, L. H., Khattab, J., Shemla, M., & Paluch, R. M. (2018). A multi-level process model for understanding diversity practice effectiveness. *Academy of Management Annals, 12*(1), 37–82.

Ostrower, F. (2005). Diversity on cultural boards: Implications for organizational value and impact. Working paper: University of Texas at Austin.

Poister, T. H. (2008). *Measuring performance in public and nonprofit organizations*. Hoboken, NJ: John Wiley & Sons.

Redaelli, E. (2012). American cultural policy and the rise of arts management programs: The creation of a new professional identity. In J. Paquette (Ed.), *Work and identity: The creation, renewal and negotiation of professional subjectivities* (pp. 159–174). Routledge.

Roberson, Q., Holmes, O., IV, & Perry, J. L. (2017). Transforming research on diversity and firm performance: A dynamic capabilities perspective. *Academy of Management Annals*, *11*(1), 189–216.

Roberson, Q., Ryan, A. M., & Ragins, B. R. (2017). The evolution and future of diversity at work. *Journal of Applied Psychology*, *102*(3), 483.

Roberson, Q. M. (2006). Disentangling the meanings of diversity and inclusion in organizations. *Group and Organization Management*, *13*(2), 212–236.

Roberson, Q. M. (2018). Diversity and inclusion in the workplace: A Review, synthesis, and future research agenda. *Annual Review of Organizational Psychology and Organizational Behavior*.

Robinson, G., & Dechant, K. (1997). Building a business case for diversity. *Academy of Management Perspectives*, *11*(3), 21–31.

Russ-Eft, D., & Preskill, H. (2009). *Evaluation in organizations: A systematic approach to enhancing learning, performance, and change*. New York, NY: Basic Books.

Sawhill, J. C., & Williamson, D. (2001). Mission impossible? Measuring success in nonprofit organizations. *Nonprofit management and leadership*, *11*(3), 371–386.

Sidford, H. (2011). *Fusing arts, culture and social change*. Washington, DC: National Committee for Responsive Philanthropy.

Simons, T., Pelled, L. H., & Smith, K. A. (1999). Making use of difference: Diversity, debate, and decision comprehensiveness in top management teams. *Academy of management journal*, *42*(6), 662–673.

Sowa, J. E., Selden, S. C., & Sandfort, J. R. (2004). No longer unmeasurable? A multidimensional integrated model of nonprofit organizational effectiveness. *Nonprofit and Voluntary Sector Quarterly*, *33*(4), 711–728.

Tatli, A., & Özbilgin, M. F. (2012). An emic approach to intersectional study of diversity at work: A Bourdieuan framing. *International Journal of Management Reviews*, *14*(2), 189–200.

Thomas, K. M., & Plaut, V. C. (2012). The many faces of diversity resistance in the workplace. In K. Thomas (Ed.), *Diversity resistance in organizations* (pp. 1–22). East Sussex, UK: Psychology Press.

Trittin, H., & Schoeneborn, D. (2017). Diversity as polyphony: Reconceptualizing diversity management from a communication-centered perspective. *Journal of Business Ethics*, *144*(2), 305–322.

Turnbull, H., Greenwood, R., Tworoger, L., & Golden, C. (2010). Skill deficiencies in diversity and inclusion in organizations: Developing an inclusion skills measurement. *Academy of Strategic Management Journal*, *9*(1), 1444.

Van Dijk, H., van Engen, M., & Paauwe, J. (2012). Reframing the business case for diversity: A values and virtues perspective. *Journal of Business Ethics*, *111*(1), 73–84.

Varela, X. (2013). Core consensus, strategic variations: Mapping arts management graduate education in the United States. *Journal of Arts Management, Law and Society*, *43*(2), 74–87.

Waiting for Unicorns: The Supply and Demand of Diversity and Inclusion.

Vu Le, Nonprofit AF Blog: https://nonprofitaf.com/2015/03/the-supply-and-demand-of-diversity-and-inclusion/

Waiting for Unicorns: The Supply and Demand of Diversity and Inclusion.

Nonprofit AF. Retrieved from https://nonprofitaf.com/2015/03/the-supply-and-demand-of-diversity-and-inclusion/

Wasserman, I. C., Gallegos, P. V., & Ferdman, B. M. (2008). Dancing with resistance: Leadership challenges in fostering a culture of inclusion. In K. Thomas (Ed.), *Diversity resistance in organizations* (pp. 205–230). East Sussex, UK: Psychology Press.

Wiggins-Romesburg, C. A., & Githens, R. P. (2018). The Psychology of Diversity Resistance and Integration. *Human Resource Development Review, 17*(2), 179–198.

Wyszomirski, M. J. (2013). Shaping a triple-bottom line for nonprofit arts organizations: Micro-, macro-, and meta-policy influences. *Cultural Trends, 22*(3-4), 156–166.

Yang, K. (2008). Examining perceived honest performance reporting by public organizations: Bureaucratic politics and organizational practice. *Journal of Public Administration Research and Theory, 19*(1), 81–105.

Young, J. O. (2010). *Cultural appropriation and the arts*. Hoboken, NJ: John Wiley & Sons.

9 The Dance of Inclusion

New Ways of Moving With Resistance

Plácida V. Gallegos, Ilene C. Wasserman and Bernardo M. Ferdman

Dancing with Resistance from the Minuet to the Tango

> You asked me to dance and
> I wondered what music would be playing.
> The palace orchestra continued the official minuet.
>
> We started slow with the samba and dreamed of salsa,
> dancing to our own rhythms.
> The crown's enthusiasm cooled as we upped our tango tempo.
> The orchestra tried to drown us out.
>
> I did not want to let go of your hand
> as we danced with the resistance.
> New partners emerged
> teaching the steps to new dancers.
>
> Anita Perez Ferguson

With the benefit of over 10 years of hindsight, we appreciate the opportunity to look back and revisit the insights we shared in "Dancing with Resistance: Leadership Challenges in Fostering a Culture of Inclusion" (Wasserman, Gallegos, & Ferdman, 2008), our chapter in the first edition of this book. As we explored new patterns, issues and challenges that have emerged since we wrote our original chapter, we found that what we wrote over a decade ago continues to apply. Yet, changes in the social context and conditions provoke us to reflect on the implications of our arguments and our perspective for new ways of coordinating and being in rhythm together in the process of fostering inclusive organizations. The nature of what we previously likened to dancing to continuously changing music now appears more akin to dancing on a constantly shifting dance floor, with shape-shifting partners and in an unpredictable space that is hard to describe and that may not even feel fully safe. No longer do we need to consider primarily *what* is changing; rather, how we do what we do, who we are together and the values in which we are grounded seem fundamentally unstable. We also feel strongly that it is no longer a question of what to do "if" resistance occurs but rather how we prepare ourselves to be intentional and mindful with our responses "when" it happens. We have found that significant change invariably involves generating resistance and that its absence often indicates that only incremental or minimal real change is happening in the organization.

In Wasserman et al. (2008), we took the perspective that organizational culture, and thereby inclusion, are a consequence of social engagement at the interpersonal, intergroup and systems levels. We grounded our description and analysis of inclusive leadership in Heifetz and colleagues' (Heifetz & Laurie, 1997; Heifetz & Linsky, 2002) focus on the importance of adaptive work, and in that context see each of these levels of analysis as providing a different view of the dance floor – some require us to be out there moving while others require us to be above the dance floor, perhaps on multiple balconies, observing the action and framing what may need to be done to foster inclusion. Building on these multiple and co-occurring levels of system, we offer here (see Table 9.1) specific examples and strategies of "the dance" to guide the moves that support leadership for inclusion, including disruption, alignment, noticing patterns, coordinating and engaging individuals and groups across the organization. These stances and actions, taken together and carried out across the levels of system, collectively constitute an approach to leading for inclusion that considers and engages with what would otherwise be framed as negative resistance in a more productive and inclusive way. Our approach to inclusion (Ferdman, 2014; Ferdman & Deane, 2014; Gallegos, 2014; Wasserman, 2014; Wasserman et al., 2008) frames it as a positive and empowering approach to groups and organizations that is geared toward building patterns of behavior, interaction, leadership and organizational culture that bring out the best in people, provide opportunities for authenticity, safety and growth – both individually and collectively – and use differences as a source of advantage and mutual benefit. This is accompanied by work to address historical and current patterns of intergroup inequity and systematic exclusion and oppression. In this view, inclusion brings benefits for everyone, in all groups, and therefore what is experienced as resistance to diversity can therefore be productively reframed in the process of what we refer to as "the dance" – the elements of which we outline in Table 9.1.

In line with our original analysis, we take a relational communication (Parker, Hall, Kram, & Wasserman, 2015) approach that uses inquiry, response, reflection and meaning-making to generate and frame our comments. The idea of "dancing with resistance" that we introduced earlier (Wasserman et al., 2008) referred to the possibility of taking the energy contained in what could be experienced as a challenge to diversity and transforming it to a source of learning, change and engagement. By focusing on *dancing with*, we sought to highlight the relational aspects of the experience of resistance and leaders' possible responses to it, as well as to encourage a more systemic view of that process, in the context of fostering inclusion in

Table 9.1 The Dance of Resistance Across Levels of System and Strategies for Engaging Toward Inclusion

		LEVELS OF SYSTEM		
		Individual/ Interpersonal	*Group/ Intergroup*	*System/ Organization*
"The Dance": **Elements and Strategies**	*D*isrupt	Notice – then question assumptions and embrace conflict	Heighten awareness of differences and amplify or engage	Name and change policies and practices to support learning
	*A*lign	Identify shared values and concerns	Promote dialogue about differences and deep listening across groups	Be explicit about norms and change those that no longer align (e.g., business meetings in bars or at golf courses)
	*N*otice	Ask: Who is in and who is out?	Pay attention to how groups and their members interact across boundaries	Discern the unintended impact of traditional practices (e.g., performance appraisal systems)
	*C*oordinate	Provide opportunities to hear other/different stories	Bring in history of group membership and educate across differences	Promotions and awards to include all groups because optics and symbolic change matter
	*E*ngage	Participate in one-to-one dialogue (e.g., different learning styles of introverts vs. extroverts)	Provide intentional and mutual mentoring across differences	Offer opportunities for many/all voices to emerge and to be heard

organizations and work groups. When conflict occurs at individual and intergroup levels, it is often the work of leadership to expand the frame to the largest level of organization so as to create deeper understanding and a coherent vision that allows the resistance to be understood, engaged and constructively utilized. The metaphor of *dancing with* permits and encourages leaders to seek out and to see the productive and positive energy in what would otherwise be experienced as problematic and unacceptable and to work to translate what is typically experienced as a complaint into a commitment toward valued goals (Wasserman et al., 2008).

In the current social and political climate, this process has only become more challenging and complex, with more steps to learn, more partners to follow and lead and shifting – and often polarizing – societal and global influences framing choices and experience within organizations. Given the cacophony provided by myriad channels and platforms, it becomes less likely that different people are hearing the same music. High levels of polarization create fewer opportunities for constructive engagement across differences and increase the likelihood of misunderstanding among people who hold views significantly different from each other.

Note, for example, the controversy that erupted in the U.S. and even worldwide in 2017 after James Damore, a Google engineer, released a memo challenging the company's diversity policies and practices and accusing it of creating an "ideological echo chamber" (www.nbcnews.com/news/us-news/google-engineer-fired-writing-manifesto-women-s-neuroticism-sues-company-n835836). Damore was fired by the company, but it is not clear how much people in the organization learned in the process, nor what the ultimate impact was on inclusion. Globalization and social media have added more complexity to forms and sequences of communicating. This increase in complexity has made what on one level might be seen as a simple interpersonal exchange – for example between a supervisor and an employee – into an event that has the potential to be disseminated worldwide or to impact the intergroup, political and social environment. Moreover, it is now even more likely that the workforce consists of people from a variety of cultures, ages, sexual orientations, gender identities, perspectives, life experiences and biases. These many layers of complexity require even more attention, presence, intention, resilience, courage and relational wisdom (Ferdman, 2017; Gallegos, 2014; Wasserman, 2014; Wasserman et al., 2008) than ever before.

There are yet more factors intensifying the complexity of shared meaning. Often today, facts are not agreed upon and truth is not generally seen as verifiable or reliable. On such a dance floor, what happens when we inevitably step on someone's toes? Do we apologize and commit to improving in the future? Ideally, we give each other grace and recognize our good intentions while continuing to challenge ourselves and each other to become more inclusive. We commit to giving each other clear and direct feedback with explicit requests for changed behavior. Patience and kindness, along with firm boundaries and accountability, are critical to our authentic engagement across differences. And so is the capacity to understand and live with the contradictions and paradoxes of diversity, inclusion and life itself (Ferdman, 2016, 2017). Only then can we continue to grow toward our highest aspirations, walking toward our talk and demonstrating our commitment to lifelong learning and to practicing inclusive values.

Cultural humility can provide a helpful frame for this. Cultural humility (Waters & Asbill, 2013) is a collaborative process that includes the ability to take an interpersonal stance that is "other-oriented" (or being open to understanding the unknown other). The essential components of cultural humility are a commitment to lifelong learning, a commitment to making social inequality visible (Tervalon & Murray-García, 1998) and advocacy for others through community partnerships. Cultural humility, by definition, is larger than our individual selves. As diverse groups meet in the workplace, intergroup humility is also called for; we must ask and learn how our group contributes to problematic dynamics, recognizing how these are co-created as well as the need for us to take greater responsibility not only for what has happened, but for creating something different and better. It is a process that we must advocate for collaboratively and systemically (Waters & Asbill, 2013). A recent example in the U.S. elections has to do with a White candidate being accused of racism responding defensively, touting his record as a long-term civil rights advocate to explain why his remarks that were considered insensitive by African American leaders were not actually problematic. When under attack, the easiest stance to take is to defend ourselves with our record of positive action rather than owning the unintended impact of our comments or actions in the present (as in former Vice President Joe Biden's defensiveness regarding Senator Corey Booker's accusation). Approaching our work to foster inclusion with cultural humility turns our attention to how we are enacting systemic patterns and while we can strive to shift those patterns we will never be perfect at practicing inclusion. Nonetheless, it is important to stay the course and to continue to advocate for inclusion (Ferdman, 2018).

In reflecting on the current state of dancing with resistance to foster inclusion in organizations, we generated the following questions, some new and some similar to those that framed our original chapter (Wasserman et al., 2008). As you consider our ideas and suggestions and apply your learning, we invite you to bring your own experiences and reflections to bear on your own humble inquiry (Schein, 2013) and application.

1) What recent shifts in the social context have implications for key leadership challenges in fostering cultures of inclusion? (Some examples: #MeToo, Whiteness and White fragility (DiAngelo, 2018), backlash against voting and civil rights, admissions and financial scandals at elite colleges, generational leadership transitions in government and other sectors.)

2) How have the meanings and connotations of our words changed as the world around us has changed? What diversity and inclusion issues are we seeing and dealing with now that are different than those that were most salient a decade or more ago? (For example, terminology and self-naming by groups require continuous attention to emerging trends, such as how *LatinX* has become the prominent and ubiquitous name among many for Hispanics in the U.S. context in a relatively short time span.)

3) What is the continuing role of resistance in transforming organizational cultures from less to more inclusive? How can leaders work with their people in the space between the old and new cultures, even as they push for change? How do leaders' thinking and action in response to resistance influence the process and outcomes of change efforts, both

positively and negatively? The importance and danger of speaking truth to power needs to be supported by organizational and cultural norms that reward bravery and protect those who bring contrary perspectives. Formal and informal leaders model inclusive behaviors in the service of shifting toward cultures that can tolerate the discomfort of learning new moves (Ferdman, 2016; Gallegos, 2014). Here there is no substitute for practice. Just as good dancers spend countless hours in preparation, leaders need to find safe spaces to learn where and how constructive feedback is welcomed and mistakes are seen as progress.

4) What does a truly inclusive culture look and feel like? Has that changed in the last decade? How does having a clear and compelling vision of the future allow leaders to dance with organizational resistance? What is the cost of not having such a view of the future and how might that limit achieving our best hopes? Are we bound to repeat the errors of the past and recreate systems that oppress and dominate certain groups while unfairly elevating others by bestowing unearned power and privilege? Rather than focusing on problem areas and the most difficult issues, inclusive leaders play a key role in elevating best practices and demonstrating their commitment to innovation and experimentation. Moving from performing, competitive and individualistic cultures toward those that are more collaborative, collective and inclusive does not happen without vision, intentionality and persistence. The eventual comfort and sense of increasing mastery allows groups to experience the rewards of inclusion and organizations to truly benefit from the richness of diversity within their boundaries. But such benefits do not occur overnight and require patience and strategic vision to be achieved.

Wasserman et al. (2008) was based on various assumptions. In particular, we assumed that a commitment to fostering diversity, equity and inclusion aligns with strengthening the processes, practices and performance of the organization and its people. We assumed that leaders were interested in learning how to become more inclusive in their leadership and how to work with resistance in a constructive, generative way. As we reconsider these assumptions now, we need to explore changes in the contextual forces that affect workplace diversity and organizations that are looking to become more inclusive and equitable. In the following paragraphs, we highlight some of these forces.

Increasingly Polarized Public Discourse Manifesting in and Fueled by Media

Many people in the U.S. and around the world seem to have a stronger sense of continuous outrage about "them" (the "others") than 10 or 15 years ago. It can be more challenging than ever to remain committed to curiosity about others' perspectives when differences are increasingly framed in terms of moral and ethical issues and human rights. At times there seems to be no solid ground on which to stand as these issues invite controversy, provoke debate, or raise questions of leadership competence. There is distinctly more telling and less listening in the public sphere, which spreads into organizational life with the consequence of greater cautiousness and more taboo subjects to be avoided at work (cf. Winters, 2017).

The Conflating of Individual Rights of Expression with the Health and Safety of Our Community

What are the boundaries of inclusion, particularly in light of the increasing threats and acts of violence against religious minorities, LGBTQ+ communities and people of color? (For one discussion of this challenge, see Ferdman, 2018; Ng & Stamper, 2018). Competing interests in society make it impossible to remain neutral when core values are at stake. Indeed, on some issues there needs to be a clear and unambiguous stance that aligns with our highest values, where not all sides are good and deserving of a full hearing (e.g., organizations and their leaders must have clear positions about and responses to hate speech, anti-Semitism, violence and abuse). Although the tension between freedom of speech and group or organizational imperatives is a real one, it can be managed in ways that support and increase inclusion (cf. Ferdman, 2017, 2018). Clarity of organizational mission and values serves as a touchpoint and rationale for taking controversial stances. Engaging resistance directly and being willing to have difficult conversations becomes more than rhetoric or optional; these are crucial and valued leadership competencies, now more than ever.

Expanding Naming and Framing of Identity and Identity Groups

It is more common now to see ourselves and each other as simultaneously inhabiting multiple identities and negotiating complex power dynamics that shift in unpredictable and nuanced ways. *Intersectionality* has almost become part of popular parlance. Intersectionality is an analytical framework that incorporates the multiple locations of individual identities within broader social and historical contexts for the explicit purpose of taking responsibility for understanding the effects of social power. Black feminists have raised awareness about multiple, interlocking oppressions by challenging mainstream feminist practices that have not dealt with how racism, sexism and classism frequently "intersect in the lives of real people" (Crenshaw, 1991, p. 1242). We hear this in our language as we expand identity markers to include more widespread acknowledgement and use of pronouns representing gender fluidity and non-binary experiences of gender (e.g., he, she, they, ze, etc.) and other emerging ways of self-definition. Yet, for many people, this can be experienced as confusing or disorienting, especially in the process of change. Our differences and the dances we learn as a result of engaging across them can strongly influence how we meet the world, how it perceives us and what happens in those encounters. Dancers are expected to master multiple genres of music and dance (e.g. western, cha-cha and waltzing, all in one day) while being ready to shift rapidly with little warning or practice. At stake is more than individual worth and self-esteem; it also matters for full inclusion of people from marginalized groups, including immigrants, trans and intersex people, gender-fluid and genderqueer people, biracial and multiracial people and many others who do not neatly fit into old or traditional categories.

An important and timely emphasis on dominant identities has emerged that shifts attention from only considering subordinated and marginalized groups to include the often-invisible privileged group memberships and dynamics. In particular, the rise of White nationalism and resurgence of anti-Semitism require progressive White people to consider the impact of their not noticing their own racially dominant identities and to become more intentional about how they engage with intergroup power relations. Whereas in the past, White, or cisgender, or able-bodied, or heterosexual, was seen as the norm and therefore unremarkable, these days there is greater awareness of the need for those of us with unearned power and privilege to recognize how inclusion benefits us as well and to speak clearly from those social locations.

This recognition comes with heightened expectations that White and other leaders with dominant identities be conscious and "woke" about their own identities and able to engage across differences in a more robust and comprehensive way, understanding the history that created current systems and patterns and their collective complicity in maintaining inequality. In terms of our prominent dance metaphor, greater attention to intergroup power relations demands that we find better ways to negotiate our steps; those in subordinated groups are unwilling to uncritically follow the lead of those in dominant groups. Where is the practice floor where we all can learn how to move without inflicting further harm on those whose toes have been stepped upon for generations? How can learning happen if either side of the intergroup dance is afraid to make even a single misstep and the consequences of imperfections can be broadcast across the world in a matter of seconds? Groups need robust yet permeable boundaries that do not collapse into the comfort of individual differences.

Increasing Acts of Hate- and Discrimination-Based Violence Against Members of More Groups

In Wasserman et al. (2008), we noted that it often took blatant acts of overt discrimination or violence for intergroup aggression to be negatively sanctioned, and we saw those acts as being relatively rare (p. 183). According to more recent statistics from the FBI (www.fbi.gov/news/stories/2017-hate-crime-statistics-released-111318), the number of reported hate crime incidents in 2017 was up 17% over 2016 totals, representing the first consecutive three-year annual increase and the largest single-year increase since 2001, when hate crimes targeting Arab Americans, American Muslims and those perceived to be Arab or Muslim surged in the aftermath of 9/11. This increase – coinciding with burgeoning racism, xenophobia and anti-Semitism – shows that crimes motivated by race or ethnicity (which rose 18% in 2017), accounted for a majority of all incidents reported, with hate crimes specifically targeting African Americans representing the greatest share and anti-Arab hate crimes increasing 100%. Anti-Jewish incidents surged over 37% in 2017, while anti-Muslim incidents, which had stabilized after two years of increases, remained well above historical averages from before the 2016 election cycle. Indeed, these statistics suggest that we are going backwards. What keeps us from gaining traction in our change efforts? What would more direct engagement with resistance allow if we developed greater capacities to be in contact and respectfully but passionately engaged? How do we get beyond fragility and learn to jump in and improvise rather than waiting to get it all figured out and focusing on avoiding any mistakes?

In sum, these forces challenge us not only to continue to dance with resistance, but to do so in new ways and with heightened alertness, intent and wisdom. Part of this requires more attention to the meaning and context of behavior and to the connotations and meanings of our words. Cultural humility teaches us that there are many competing and multiple perspectives on any issue, including our own. Engaging in more vibrant and productive conflict requires courage and authenticity (Gallegos, 2014). Leading with questions and curiosity versus telling reframes leadership and collaboration in critical ways. For example, increased attention is being placed on recognizing and disrupting unconscious bias in the workplace as well as addressing the prevalence of microaggressions in our everyday interactions. Small behaviors with potentially huge and unintended consequences demand our constant vigilance and willingness to change behavior that may have been normal or acceptable in the past.

In particular, the systemic level of context needs more amplification. When people say, "I'm not explicitly behaving badly towards others" or "I am not responsible for the past," accountability is debatable. Prejudice and oppression are systemic, which means that they do not require conscious individual behavior to continue perpetuating injustice. The personal/self and interpersonal are embedded in complex systems and cannot be considered in isolation. Using our dancing metaphor, context in social interactions is the music we dance to (even when we are not fully listening), therefore it behooves us to pay close attention to what we are hearing (and not hearing) and to what others hear (and do not hear), as well as to how the systems and structures in which we are embedded channel and affect our behavior and experiences and those of others. A recent example in our experience was being at a wedding where generations had very different expectations for the type of music and dance that would be included. Event planners had to deal with the challenge of what to do when satisfying the preferences of one generation has the potential to alienate the others.

How Have the Connotations and Meanings of Our Words Changed as the World has Changed Around Us?

Language is part of culture, a living and growing thing. The words *resistance* and *inclusion* have more nuanced meanings than they did a decade ago. Today, when we put a hashtag in front of resistance it becomes #Resistance and takes on a new meaning for some (for example, resistance to the U.S. President, Donald Trump, and his policies and practices). And as we stretch ourselves toward increasingly critical understandings of privilege and oppression, sometimes the word *inclusion* can feel ambiguous or limited. Inclusion on whose terms and to what end, for example? So, what do we mean by these words?

Resistance implies exerting force against something, so *diversity resistance* implies pushing back against diversity (and by extension) inclusion. In Wasserman et al. (2008), we advocated "dancing with resistance," in the sense of engaging with it in a positive way. We wrote: "When resistance is ignored or addressed ineffectively, it becomes a negative force that can threaten change. When leaders expect, acknowledge and embrace resistance, it becomes a powerful instrument for change" (Wasserman et al., 2008, p. 177). Now, in some recent contexts, *resistance* has taken on a reference that is more about protecting diversity and inclusion. This more positive connotation is consistent with our assertion that resistance can indeed be a constructive force when engaged directly. The very pace of change within organizations is controversial; those in subordinated groups have lost patience and want action yesterday. Dominant group members who are just beginning to see injustice in a new light ask for patience and worry that there is too much change too fast. These colliding forces are predictable, especially when long-term culture change is the desired outcome and leaders are committed to staying the course at a pace that draws criticism from one subgroup or another. How can leaders anticipate and provide support during these turbulent though necessary learning processes, knowing that any compromise leaves all parties dissatisfied?

Inclusion

Paradoxically, on the one hand, inclusion involves creating openness to more ideas and possibilities as well as to the ability of people to express themselves more authentically, without assimilating, while on the other hand, it also involves setting firmer boundaries with regard to the behaviors and mindsets that make these types of expression possible, safe and both individually and collectively beneficial (Ferdman, 2017). In the context of the current U.S. administration, inclusion requires strong advocacy for inclusive values and norms (Ferdman, 2018). We believe that leaders and practitioners should strive for a certain type of inclusion, which one of us (Ferdman, 2018) recently described as follows:

> [The] type of inclusion that does not require conformity, but that is nonetheless characterized by strong, unifying, and egalitarian bases for ongoing engagement that allow both maintaining and assessing the meaning and limits of collective principles and values, even as these evolve and are adapted to new conditions and members. Although inclusion and diversity involve continued exploration of both boundaries and what is in the boundaries, they also require standing firm for certain principles: diversity and inclusion should and do not mean that everything goes, and there must be consensus and alignment around core values and norms, even as these are tested in practice.
>
> (p. 101)

Indeed, sometimes resistance comes in the form of either arguing for inclusion as assimilation ("we are happy to include anyone who becomes sufficiently like us") or using the language and concepts of inclusion to undermine it (Ferdman, 2018). Neither of these should be acceptable. Naming these challenges and calling out exclusive behavior in service of the larger goal is a hallmark of inclusive practice. Yet, subtlety and grace must also be present in order to strengthen communities even as they undergo significant growth pains.

The emergence and evolution of new language and concepts are telltale signals that our thinking about diversity, equity and inclusion is changing. The use of the acronym "DE&I" has been a relatively swift and recent turn in our lexicon. The central role of equity in promoting inclusion has gained considerable traction and can be seen as a clear message that systems and structures of exclusion are being challenged. Other examples include the terms *LatinX, LGBTQIIA, intersexual* and *intersectionality*.

Many of these examples call our attention to seismic shifts in our social landscape.

Conclusion

Engaging resistance is no longer an option for organizations that want to be relevant in the future. Yet, for those whose worldview privileges some groups and oppresses others, fostering a society and organizations that embrace cultural differences is problematic (Adams, 2000). As the world faces increasing challenges to inclusive communication across differences, there is a paradoxical need to foster greater safety and greater risk-taking, often simultaneously. In our consulting practices, we continue to explore different ways of thinking about and engaging diversity and fostering inclusion that better fit the changing environments our clients face. Change initiatives challenge our taken-for-granted way of seeing and engaging the world. This almost always provokes challenges and tensions that must be skillfully engaged on the journey toward cultural competence and humility. Our repertoire of moves on the dance floor must span many cultures, genres, harmonies and histories to flow in rhythm and balance.

As leaders invest in building their repertoire of moves, they are recognizing the need and importance for ongoing conversations with their counterparts in other organizations – a form of peer coaching at the highest levels (Parker, Wasserman, Kram, & Hall, 2015). In the spring of 2017, a group of CEOs spearheaded by Tim Ryan, U.S. Chairman and Senior Partner from PwC, formed a steering committee to address diversity and inclusion in the workplace. This steering committee resulted in the creation of the CEO Pledge (www.ceoaction.com/pledge/ceo-pledge/) committed to by 350 members from 85 industries (www.forbes.com/sites/bonniemarcus/2018/01/19/what-happens-when-ceos-take-a-pledge-to-improve-diversity-and-inclusion/#3c2eaf075627). This group continues to invest in their own learning to be more inclusive leaders and in speaking out on related issues.

We believe, hope and trust that the ideas and examples in this chapter are relevant and provide grounding for maintaining balance while welcoming and engaging with differences and ultimately for fostering and leading cultures of inclusion. We increasingly recognize missteps and discordant music as opportunities to expand how we coordinate, learn and dance on!

References

Adams, M. (2000). Conceptual frameworks: Introduction In M. Adams, W. J. Blumenfeld, R. Castañeda, H. W. Hackman, M. L. Peters, & X. Zúñiga (Eds.), *Readings for diversity and social justice: An anthology on racism, anti-Semitism, sexism, heterosexism, ableism, and classism* (pp. 5–8). New York, NY: Routledge.

Crenshaw, K. (1991). Mapping the margins: Intersectionality, identity politics, and violence against women of color. *Stanford Law Review, 43*(6), 1241–1299.

DiAngelo, R. (2018). *White fragility: Why it's so hard for White people to talk about racism`*. Boston, MA: Beacon Press.

Ferdman, B. M. (2014). The practice of inclusion in diverse organizations: Toward a systemic and inclusive framework. In B. M. Ferdman & B. R. Deane (Eds.), *Diversity at work: The practice of inclusion* (pp. 3–54). San Francisco, CA: Jossey-Bass.

Ferdman, B. M. (2016). If I'm comfortable does that mean I'm included? And if I'm included, will I now be comfortable? In L. M. Roberts, L. P. Wooten, & M. N. Davidson (Eds.), *Positive organizing in a global society: Understanding and engaging differences for capacity-building and inclusion* (pp. 65–70). New York, NY: Routledge.

Ferdman, B. M. (2017). Paradoxes of inclusion: Understanding and managing the tensions of diversity and multiculturalism. *The Journal of Applied Behavioral Science, 53*(2), 235–263. doi:10.1177/0021886317702608

Ferdman, B. M. (2018). In Trump's shadow: Questioning and testing the boundaries of inclusion. *Equality, Diversity, and Inclusion: An International Journal, 37*(1), 96–107. doi:10.1108/EDI-09-2017-0177

Ferdman, B. M., & Deane, B. R. (Eds.). (2014). *Diversity at work: The practice of inclusion*. San Francisco, CA: Jossey-Bass.

Gallegos, P. V. (2014). The work of inclusive leadership: Fostering authentic relationships, modeling courage and humility. In B. M. Ferdman & B. R. Deane (Eds.), *Diversity at work: The practice of inclusion* (pp. 177–202). San Francisco, CA: Jossey-Bass.

Heifetz, R. A., & Laurie, D. L. (1997). The work of leadership. *Harvard Business Review, 75*(1), 124–134.

Heifetz, R. A., & Linsky, M. (2002). *Leadership on the line: Staying alive through the dangers of leading*. Boston, MA: Harvard Business School Press.

Ng, E., & Stamper, C. (Eds.). (2018). Special issue: Equality and diversity in the time of Trump. *Equality, Diversity, and Inclusion: An International Journal, 37*(1).

Parker, P., Hall, D. T., Kram, K., & Wasserman, I. (2015). *Peer coaching at work: Principles and practices*. Stanford, CA: Stanford University Press.

Parker, P., Wasserman, I., Kram, K., & Hall, D. T. (2015). A relational communication approach to peer coaching. *The Journal of Applied Behavioral Science, 51*(2), 231–252. doi:10.1177/0021886315573270

Schein, E. (2013). *Humble inquiry: The gentle art of asking instead of telling*. San Francisco, CA: Berrett-Koehler.

Tervalon, M., & Murray-García, J. (1998). Cultural humility versus cultural competence: A critical distinction in defining physician training outcomes in multicultural education. *Journal of Health Care for the Poor and Underserved, 9*(2), 117–125.

Wasserman, I. C. (2014). Strengthening interpersonal awareness and fostering relational eloquence. In B. M. Ferdman & B. R. Deane (Eds.), *Diversity at work: The practice of inclusion* (pp. 128–154). San Francisco, CA: Jossey-Bass.

Wasserman, I. C., Gallegos, P. V., & Ferdman, B. M. (2008). Dancing with resistance: Leadership challenges in fostering a culture of inclusion. In K. M. Thomas (Ed.), *Diversity resistance in organizations* (pp. 175–200). Mahwah, NJ: Lawrence Erlbaum.

Waters, A., & Asbill, L. (2013, August). Reflections on cultural humility. *CYF News*, American Psychological Association., Retrieved February 24, 2014 from www.apa.org/pi/families/resources/newsletter/2013/08/cultural-humility

Winters, M. (2017). *We can't talk about that at work! How to talk about race, religion, and other polarizing topics*. Oakland, CA: Berrett-Koehler.

10 African-American Professionals in Public Relations and the Greater Impacts

Candace P. Parrish and Janice Z. Gassam

Status of Diversity and Resistance in Public Relations

The Public Relations Society of America (2018) defines public relations as "… a strategic communication process that builds mutually beneficial relationships between organizations and their publics." One of the tenants of the field is the ability to come up with unique and original ideas; creativity is a valuable asset in PR. With the increased importance and focus on creativity and innovation, the assumption would be that this industry values, celebrates and champions the inclusion of individuals from diverse backgrounds. However, extent literature on diversification in PR reveals the industry's further need for racial diversity (Eibach & Keegan, 2006; Jones, 2011; McGirt, 2018; Sha, 2013; Tindall, 2012). Research by Bell, Villado and Lukasik (2010) indicated that with diversity comes more creativity and innovation. In an industry like PR that thrives off innovation and creativity, it is of ever-increasing importance to fully understand why there is a lack of racial diversity, how this can impact the industry, and what measures can be taken to recruit and retain diverse talent within this industry.

Discussions about diversity are necessary for an organization to thrive and progress but simply uttering the word "diversity" draws ire, mixed reactions, and resistance. The word "diversity" is plagued with negative connotations. One of the primary reasons why there is resistance to diversity is the belief that celebrating multiculturalism equates to exclusion of those in the dominant group. The desire to be included is a basic human instinct. A focus on other cultures may cause White Americans (as the majority race/culture) to believe that diversity is "only for minorities" and impedes on the basic human need to belong (Plaut, Garnett, Buffardi, & Sanchez-Burks, 2011). In addition, there is a perception among many White Americans that a focus on diversity, equity, and inclusion will lead to a loss of power and status within society (Eibach & Keegan, 2006). In order to make diversity and inclusion initiatives successful, it is necessary to first understand why there is resistance as well as creating strategies to combat this resistance.

En route, this chapter will review diversity efforts, as well as the need for diversity resistance literature around the public relations (PR) industry.

PR is an industry comprised of mostly Whites, with African-Americans (AA)*, LGBTQ and disabled practitioners being largely underrepresented (Pomper, 2007; Vardeman-Winter & Place, 2017). With the amount of racial and political dialogue taking place in this country, it is critical that PR professionals are able to recognize each side of the discussion. The recent gaffes of companies such as Pepsi, Shea Moisture, Dove, Nivea, H&M, Prada, and Gucci, reveal why diversity in the PR industry is so critical. With a homogenous group of individuals crafting and managing a message or image of an organization for public consumption, unique and varied voices and ideas are necessary (Tindall, 2012). Contextually, this chapter references the term "diversity resistance," as defined by Thomas and Plaut (2008), as "a range of practices and behaviors within or by organizations that interfere, intentionally or unintentionally, with the use of diversity as an opportunity for learning and effectiveness" (p. 5).

PR's Role in an Organizational Context

The practice of PR entails the act of managing communication and relationships between an organization and their publics. The field was created during the 1950s when big and industrial business began to boom and reputation and brand management was needed to increase communication between businesses and their surrounding communities (Edwards, 2010; Logan, 2016). Since, PR has progressed largely and is practiced in fields from entertainment to health. In an organization, PR employees/departments work closely with fields like advertising and marketing that also communicate and promote an organization's brand. Yet, PR uniquely promotes an organization's brand and focuses on building and maintaining relationships between an organization and its stakeholders (customers, employees, community partners and investors).

As social communication successfully adapted to societies digital progression, PR largely includes practice and strategies implemented through social media. Social media has greatly enhanced the opportunity for PR in digital communication because it allows direct two-way communication where a consumer can share feedback and interact with an organization right on their social media channels. Likewise, the growth of organizations using social media to communicate with their stakeholders has also allowed for negative opportunities. This unfortunate outcome of social media use for organizational communication has exposed some of the shortcomings of many companies, including those that reveal directly and indirectly their lack of organizational diversity.

---

* African American and Black will be used interchangeably throughout the chapter.

Cases Reflecting the Impact of Diversity Resistance in Public Relations

In April of 2017 Pepsi pulled a controversial advertisement featuring Kendall Jenner. In the one-and-a-half-minute ad, a protest with a group of young adults was featured. The next scene featured Jenner walking into a crowd of protestors and grabbing a can of Pepsi and handing it to a police officer nearby. The officer accepts the can of Pepsi from Jenner, which seemingly diffuses the protest. After the officer accepts the can, the crowd erupts into laughter and cheers. Criticism of the ad was swift, with many commenters remarking that the ad minimized the Black Lives Matter movement and the recent anti-police violence protests that had been taking place in the U.S. at that time. Pepsi, who acknowledged their misstep, issued an apology. The general consensus was that if Pepsi had more people of color working on their PR, this type of situation could have been avoided (D'addario, 2017; Victor, 2017).

Skin care company Shea Moisture also caused controversy following an advertisement that the company put out in 2017. The company drew ire from their core customer base following an ad about "hair hate." The ad featured a blonde, two redheads and one curly-haired woman, discussing their negative experiences with their hair being labeled as different. Since the company's inception, their core customer base has been Black women with kinky and coarse hair textures (Nussbaum, 2017). When the ad came out, many were shocked to see that there was seemingly only one Black woman featured in the ad. The general consensus on social media indicated that many customers felt that the company was using marketing tactics to cater to a different demographic while disregarding the core customers that made the company successful (Payne & Duster, 2017). Following the backlash, the company did issue an apology and subsequently pulled the ad. In addition, Twitter users claimed to have found the marketing team responsible for the ad, which was seemingly comprised of no persons of color (Coffee, 2017). Again, the lack of diversity on their communication team was theorized to be the main reason for the tone-deaf ad.

Dove is another company that experienced a PR gaffe after the release of a Facebook ad in 2017. The company created an ad featuring a Black woman who turned into a White woman. The meaning behind the ad was supposed to be that Dove will help one go from dirty to clean, but the racial undertones of the ad did not bode well with the public. Many Twitter users expressed their concern over the connotations of the ad. The visual of a Black woman representing a dirty person and a White woman representing a clean person was the message that many derived from Dove's ad. Online users complained that the ad was not only offensive but was also racist, and Dove later issued a statement expressing their regret for putting out such an offensive ad that "missed the mark." A spokesperson for Dove declined to comment on how many Black people were involved in the creation of the ad, or if there were any at all (Astor, 2017). This wasn't the first time that Dove found itself in hot water. In 2011, the company sparked controversy following an ad for their Visible Care bodywash, which showcased a Black woman changing into a Latina woman and then changing into a White woman. The purpose of the ad was to demonstrate how the usage of the bodywash could improve skin and make it more beautiful, but again their ad missed the mark and was seen as racially insensitive (Held, 2017).

Another skin care company that found itself in a similar situation as Dove was Nivea. In 2017, Nivea caused a public uproar after the creation of an ad showcasing the words "White is purity" alongside a photo of a woman with dark hair dressed in all white (Tsang, 2017). Critics said that the ad, which was posted on Nivea's Middle East Facebook page, was racist and insensitive. Nivea pulled the ad and Beiersdorf, the company that owns Nivea, issued an apology and stated that the reason the ad was taken down was over concerns that it was an example of ethnic discrimination. (Kottasova, 2017).

H&M is another example of a company that found itself in turmoil after posting an image of a sweatshirt that was found to be offensive. In 2018, the clothing retailer posted an image of a young Black boy wearing a hoodie that read "coolest monkey in the jungle." After the image was posted online, there was severe backlash with many criticizing the company for racial insensitivity and calling for a boycott. Given the racist history of the word "monkey" and how it has been used as a slur against people of African descent, the company was heavily criticized for its decision to have a Black boy wear a sweater with a monkey on it (Stack, 2018). Many celebrities including Snoop Dogg, R&B singer The Weeknd and rapper G-Eazy all chimed in, expressing their disdain and disappointment with the clothing brand. Both The Weeknd and G-Eazy, who at the time had collaborations with H&M, announced that they would be severing ties with the company following the incident. As an attempt to change the narrative, a modified image of the young boy in the sweatshirt was created by rapper Chris Classic, who posted the modified image on his Instagram. The modified image featured the same boy wearing a sweatshirt, but instead of a monkey, the sweatshirt featured a crown with the words "King of the World" written above the boy. Many celebrities, including Lebron James, reposted this modified image (Stump, 2018). Comedian Marlon Wayans shared his thoughts about the H&M incident and stated that the company desperately needed more Black people to oversee things, which may prevent situations like this from happening in the future. H&M did issue an apology following the backlash, expressing regret for the egregious error (Ziv, 2018).

In recent years, cases of well-known fashion brands that have had similar missteps as H&M have persisted. Prada was forced to pull products from their Pradamalia line after images of the products from the line were criticized due to their eerily similar resemblance to Blackface. New York-based attorney Chinyere Ezie posted images of the products from a Manhattan storefront on to her Facebook page and heavy backlash ensued. Prada ended up pulling products from the line and issued a statement on Twitter indicating that the company "abhors racist imagery." In regard to the product line, Prada stated, "They are imaginary creatures not intended to have any reference to the real world and certainly not Blackface" (Ly, 2018). There is a possibility that during the creation of the product line, there was a lack of understanding regarding the history of Blackface. It is unclear what the demographic makeup of the creator(s) of this line were. One could argue that having more people of color involved in the creation process could have prevented a blunder like this from happening.

Gucci was also guilty of a similar and equally egregious error in February of 2019. Gucci experienced severe backlash following the release of their sweaters, which many felt looked eerily similar to Blackface. The sweater was featured as part of Gucci's Fall Winter 2018 line. Washington Post columnist Michelle Singletary shared her thoughts on how disrespectful that sweater was, especially given that it was Black History Month when the image went viral (Held, 2019). After the image went viral, many called for a boycott of Gucci, including director Spike Lee and rapper T.I. Gucci's response to the situation was swift. The luxury fashion brand apologized for the design and stated their commitment to diversity (Griffith, 2019). Gucci also took their commitment to change a step further when the company announced the launch of Gucci Changemakers, a global initiative to "create more opportunities for talented young people of diverse backgrounds." The components of the global initiative include hiring more global leaders that focus on diversity, creating a multicultural design scholarship program and improving cultural awareness and sensitivity (Gassam, 2019).

What are the consequences of these PR missteps? According to Clutch, a website that collects research and data on different companies, companies that have a strong PR strategy and loyal customer base were better able to bounce back following public scandals. Clutch surveyed Pepsi consumers following the backlash of their aforementioned advertisement with Kendall Jenner. What their survey found was that most consumers have not changed their buying habits following the backlash. Only 1% of those surveyed changed their Pepsi buying behavior following the PR blunder. According to the survey, the consumer's willingness to buy Pepsi dropped from 55% immediately after the PR crisis to 56% seven months after the PR crisis (Seter, 2017).

While it is unclear what the direct results of PR blunders like a Blackface hoodie or a racially insensitive ad actually are, loss of profit may be the most tangible outcome. It was reported in March of 2018 that H&M was experiencing a drop in sales and was having difficulty selling merchandise (Payton, 2018). It is unclear whether the controversy exacerbated the issues that H&M was already experiencing but it's not outside the realm of possibilities. While the aftermath of Gucci's PR mistake is still being assessed, it was reported in April 2019 Gucci actually experienced sales growth and outperformed their peers (Kar-Gupta & White, 2019).

Directly following a PR crisis, there are often calls for boycotts, and companies can experience reputational damage. Assessing the long-term effects are more difficult to evaluate. For Papa John's, comments made by the former CEO John Schnatter may have directly hit the company's profits (Wiener-Bronner, 2019). Schnatter has publicly criticized NFL leadership regarding the player protests. To add insult to injury, in July of 2018, Forbes reported that the former CEO had used a racial epithet during a company conference call (Kirsch, 2018). The restaurant chain has since experienced a consistent decline in sales. In March of 2019, it was reported that the company's North American sales were declining for the fifth straight quarter (Meyersohn, 2018; Wiener-Bronner, 2019).

While some larger companies may not suffer negative consequences in the long term due to effective PR crisis management, small and mid-sized organizations may not enjoy that same luxury. Lacking the funds to pay for a diverse and qualified crisis management team can impact how well a company bounces back following public scandals and disasters. Consequences following a misstep may come in the form of negative public perception, lowered employee morale or reputational damage. Some of these variables are more difficult to quantify than others. Examining a company's social media pages can provide an indication of public and customer perception, although sometimes it's difficult to evaluate the veracity of online feedback. It's in an organization's best interest to avoid these sorts of issues from the beginning because crisis management following a public incident can be costly.

As exemplified in the recent media crises mentioned above, diversity resistance can have deleterious effects on the organizational brand, image and bottom line due to lack of awareness regarding sanative cultural and racial matters (Reader, 2017; Tindall, 2012). Smith (2017) highlighted expressions from industry insiders that alluded to the executive and "C-Suite" level of communication in boardrooms being mostly or all White and male. Considering so, this chapter is a scholarly review that will uniquely focus on diversity in PR, regarding 1) the lack of literature on AA experiences with diversity resistance, 2) AA roles in promotion to executive or senior-level management positions, and 3) providing strategies toward effective coverage of AA experiences with diversity and resistance. In doing so, the chapter will first dive into the history of AA in the workforce, the history of diversity work and research in the PR industry, and then discuss some reasons and solutions as to the lack of diversity within the field. For the purposes of this chapter, *diversity resistance* is defined as "the dynamic interplay of individual and collective behaviors, with individual resistance rooted in unconscious motivation and organizational resistance rooted in the collective behavior of individuals" (Wiggins-Romesburg & Githens, 2018, p. 179).

African Americans in the Workplace

Over the past couple of decades, there has been an increasing amount of research studies and reports that highlight the many challenges AA face in workplace environments. Extant research has affirmed that AA, as a minority racial group, face more challenges (than the majority racial group) while looking for employment, while working on jobs and while seeking upward mobility within organizations. A study by Lyons, Velez, Mehta, and Neill (2014) found that AA are more likely to be detached from the organizations in which they work if the organization does little to affirm diversity among its employees. Adding to this notion, the study suggests that AA who are more likely to identify with their job feel that their organization affirms racial diversity. Take the global company of Starbucks for example. In 2014, the organization launched a national campaign that aimed to open dialogue on race relations both in the U.S. and within their organization (Logan, 2016). Based upon the prior research study, it is more likely that AA employees at Starbucks will identify with their organization than other leading companies who don't even acknowledge the plight and/ or struggle of AA in society. Research surrounding AA in workplace environments also highlights that factors like education and socioeconomic status play a role in job satisfaction and workplace experiences. AA with higher socioeconomic status are less likely to experience overt discrimination and harassment than AA with lower socioeconomic status (Lyons et al., 2014).

Although the existence of struggles AA face in the workforce have been documented, more research is needed to further explore more specific lived experiences and barriers AA encounter while on the job. When discussing the specific experiences of AA in the workplace, it is imperative to examine the relationship between diversity and microaggressions. Microaggressions refer to seemingly benign speech or behavior towards an AA that is unconscious and unintentionally offensive and racist. Microaggressions occur on a smaller scale basis and often occur in everyday life (Solorzano, Ceja, Yosso, 2001). Microaggressions can often be a form of diversity resistance. When examining the impacts of racial microaggressions on AA, Sue, Torino, Capodilupo, & Rivera (2009) found that despite their seemingly benign nature, microaggressions can have a profound negative impact on AA lives. An even more challenging barrier to overcome when it comes to microaggressions is the believability factor. Often times when AA express feelings of being slighted, their feelings are not recognized or validated (Sue et al., 2009). Microaggressions represent the more insidious form of racism that, for some, is more difficult to conceptualize. Deitch et al. (2003) noted that the nature of racism has morphed over recent years. Racism has become more covert, making it harder to identify. Resistance to integration and diversity is something that is more concealed and can create barriers to inclusivity in the workplace. Even with the popularity of diversity and inclusion programs in the workplace, workplace discrimination continues (Thomas & Plaut, 2008).

One study conducted by Rose & Bielby (2011) found that board members of American corporations based the racial diversity opportunity of company-wide leadership groups on societal norms. Further, racial diversity in these leadership groups was not persuaded by natural ability to accomplish the job, rather by attempt to appease stakeholders by keeping diverse representation on the board to a minimum (Rose & Bielby, 2011). Usually AA were stated to be represented as one or two (maximum) members of a leadership board so not entirely excluded, yet, not to racially dominate the board, which goes against societal norms (Rose & Bielby, 2011). This type of stereotyping and resistance to diversity can have a great impact on AA who work hard to exemplify their work and worth in an organization. Oftentimes, their expertise or help isn't needed and they are merely in the position to serve as a face of diversity. These occurrences are classified under the notion of *cosmetic diversity*, which is defined for the purposes of this chapter as pseudo-diverse efforts in an attempt to appease various stakeholder groups (investors, consumers and/or employees) while not actually effectively implementing diversity initiatives. Cosmetic diversity is a paradigm that is gaining discussion and use within digital rhetoric concerning diversity, however, the term is rarely defined.

Diversity in Public Relations

Diversity has been a popular topic of discussion in national academic and practice research and digital discourse for the past three decades. As research highlighted the need for multicultural and diverse people and viewpoints in workplace and corporate environments, (Jones, 2011; Tindall, 2012; Vardeman-Winter, 2011; Vardeman-Winter & Place, 2017), organizations seemingly clamored to increase diversity and multicultural efforts. However, with the influx of efforts to increase diverse faces and voices came resistance to these efforts from racial majority organizational leaders, workers and stakeholders. While resistance to these types of diversity-enhancing initiatives has increased over the past decade, more research is needed in a vast majority of practice-based areas that highlights direct experiences and consequences of diversity resistance in the workplace. The field of PR qualifies as one of those fields lacking scholarly review of the immediate and long-term effects of diversity resistance within the field.

Mirroring the practical field of PR, the academic side of the communication practice has taken grave interest in the functioning (or lack thereof) of diversity efforts in the workplace (Brown, White, & Waymer, 2011; Logan, 2016; Stellar, 2014; Vardeman-Winter, 2011; Vardeman-Winter & Place, 2017). Not only limited to race, however, researchers have covered diversity in public relations from a plethora of gender, sexual orientation, culture and generational groups (Vardeman-Winter & Place, 2017). PR research has made great strides in communicating and affirming the need for diversity, yet little in efforts to highlight specific experiences of and/or solutions for minority group members while they navigate diversity resistance in the field. Even slimmer is the amount of literature that highlights the disadvantages, resistance and barriers AA face when they seek upper-level and/or management positions within the field of public relations.

In one representational study regarding the lived experiences of AA in PR settings/careers, Tindall (2009) interviewed 12 AA practitioners and highlighted their portrayals of support and racism in their work environments. Regarding support within the industry, the AA practitioners shared that they have to intentionally build their own networks for support (Tindall, 2009). The lack of AA in upper-level and management positions also creates a void in like-minded mentorship for young AA professionals. As a result, many of the study's participants found white mentors (Tindall, 2009), which potentially denies them of mentoring equipped with strategies specific to race and diversity from a minority perspective. AA in this study also reported casual and micro-aggressive encounters with racism on a daily basis via remarks and voiced stereotyping from non-minority peers and supervisors (Tindall, 2009). This often creates a tense environment for AA who feel they have to be self-aware and make sure to avoid associating with projected stereotypes of the Black community (Boulton, 2016; Tindall, 2009).

Cosmetic Diversity's Adverse Effects

The same method of research conducted by Rose and Bielby (2010) regarding inclusion of AA in leadership group positions should be applied in PR practice research. PR researchers have specific opportunities to make sure that diversity isn't just covered as a popular trend or limited to cosmetic attempts. Deeper research, specifically qualitative research, is needed to highlight the actual lived experiences of AA who have been through or are going through circumstances in the PR workforce relating to diversity resistance and cosmetic diversity. Vardeman-Winter and Place (2017) discovered that the amount of information known about the presence and status of AA in PR is virtually zero. This leads to a question directed toward the field and the research that examines the field: If diversity efforts in practice and research have been on the rise in PR, why is there so little information on the status, barriers to success and promotion of lived experiences of AA in the field?

Circumstances like this, where there are voids in scholarly literature, are not the first occasions where AA voices, accomplishments, and experiences were left out of the documented history of PR. For years, various members of diversity initiatives in the academic PR field have advocated for the inclusion of AA practitioners and scholars in PR history and textbooks. Pioneers such as Dr. Inez Kaiser, the first AA to own an international PR firm; Ofield Dukes, who often represented civil rights activists and entertainers; and Moss Kendrix, who advised The Coca-Cola Company on how to communicate to the Black community in the 1950's, are currently not represented in mainstream textbooks (Fraser, 2017; Greer, 2013; Schudel, 2011). All of these pioneers deserve their rightful place in history via representation in textbooks and history literature next to Ivy Lee and Edward Bernays – white practitioners who are most credited and highlighted in PR history literature.

The oversight of AA experiences in and contributions to PR in extant scholarly literature can be better explained through the exploration of whiteness and how its performance regarding majority groups often disregards and omits the viewpoints of minority groups. A study by Vardeman-Winter (2011) encourages white practitioners to accept what whiteness means in society, therefore opening up space for dialogue on how to dismantle workplace norms that are created within White culture. In previous field-wide research, White executives represent approximately 79.5% of the PR workforce (Murphy and Public Relations Coalition, 2015), further demonstrating how imbalanced the diverse playing fields are within the profession. Regarding data on PR professionals, there was a long streak of underrepresented data pertaining to the quantifiable make-up of AA in the PR profession. The National Black Public Relations Society (NBPRS) has helped to close this gap through conductance of a national survey to help advance statistics in this area (Foote, 2017). The survey found that most AA in the profession of PR are working in corporate (20.6%) and agency (23.8%) settings (NBPRS, 2017). NBPRS was created in 1997 as an organization to support AA PR professionals and further advocate for diversity in the field (NBPRS, 2017).

It is also important to note that the origins of the PR field are closely aligned with the cultural needs of the majority White race as the field gained definition during the nineteenth century in a time when big business began to emerge (Edwards, 2010; Logan, 2016). In this era, the voices of minorities and the efforts for inclusion were not of particular interest to practitioners (Edwards, 2010; Logan, 2016). At the same time, many professions, like PR, excluded AA from participation in the field through notions of mental inferiority and laws created to maintain segregation (Logan, 2016). Although the industry is evolving and aiming toward inclusivity, these complicated race relations in American history are still the root of communication focus in PR practice today. This history gives insight as to why some executives only engage in cosmetic diversity or resist diversity altogether, as the scope of inclusivity is not a major element to the progression of their organization or the field as they have come to perceive.

Considering the history of race and diversity in the U.S., it is important to note that cosmetic diversity efforts can work directly against the actual aim of implementing diversity strategies. Historical viewpoints have expressed that previous tactics to hire diverse audiences only to use them to communicate to "like" audiences or be the face of diversity for the company can have adverse effects on the employee's experience and likelihood to remain in the position. This is further explained via Thomas and Ely's (1996) Access & Legitimacy paradigm, which has been adapted by organizations that use the resources and positioning of diverse employees only to gain attention and buy-in from diverse populations. The issue with this paradigm and viewpoint is that diverse employees hired to gain access to these populations can become and feel exploited and excluded from other organizational opportunities (Thomas & Ely, 1996). Organizations can avoid exploiting diversity practices and diverse employees by looking past the bottom line for motivations to include diversity into the company's culture. Organizations should also allow diverse populations equal opportunities for growth and allow them to apply their full skill-sets to the position, not just to the diversity-related work.

Additionally, many minority group members (including Asian and Latino practitioners) entering the PR field have to subscribe to and practice whiteness to a certain extent in order to maintain good standing with supervisors and clients. One framework relating to this type of workplace performance is called "code switching." Code switching has become prominent in the discussion of conformity to whiteness across all professions but has also been studied specifically in the area of communications. The term and academic framework refer to the varied types of performance in which minorities engage to assimilate their voice (pitch and/or volume) and vernacular (rhythm and/or tone quality) to appease and communicate to those in dominant cultures (Boulton, 2016). In advertising, a communication field often held in close regard to PR, Boulton (2016) highlighted the extent to which AA felt they needed to code switch in order to maintain good standing in their profession. In the study, some participants voiced "mirroring White culture in this way [code switching] helped to dissociate them from negative stereotypes of Black identity" (Boulton, 2016, p. 132).

Solutions to Decipher the True Impact of Diversity Resistance in PR

Given the current state of both diversity and resistance to the inclusive framework, it is more crucial than ever to shape and create strategies for enhancing positive and inclusive experiences for AA in PR. As previously affirmed, the extent of inclusive and reflexive PR practice hinges on effective implementation of diverse hiring, equity, and inclusion within the field (Jones, 2011; Tindall, 2012; Vardeman-Winter & Place, 2017). Thus, this chapter provides the following strategies for increasing effective diversity efforts based on the voids that exist in the coverage of diversity and resistance in PR practice and scholarly research:

- *Research the Lived Experience* – As stated earlier, it is essential to learn more about the everyday work experiences of AA pursuing entry-level and upper-level/management careers in PR. Without these qualitative benchmarks, the actual voices and plights of AA in PR will continue to be disproportionately represented compared to the amount of diversity research that hints at the possibility of resistance to diversity. This chapter is unique in that it is the first of its kind to highlight the lack of diversity resistance literature in the PR field. Given this void, in-depth interviews, focus groups and case studies should all be immediate directions for future research in order to build the literature on and strategies for more effective diversity education and implementation in the PR field. In the past 10 years, there has only been one study that interviewed 12 AA practitioners (Tindall, 2009). While certainly this study's efforts are innovative and certainly commendable, it should not exist as the sole representation for AA voices and experiences in diversity research within PR. Research should also be conducted with PR agencies or organizations and PR teams to learn more about their internal practices for diversity in the workplace.

• *Generate Relative and Accurate Quantifiable Data* – The representation of solid, relative and accurate quantitative data is also needed is this area of scholarly and practical research. The review of extant literature on diversity as it pertains to AA in the field of PR was lacking overall. There were a few studies and reports that provided minimal statistics regarding the approximate make-up of AA professionals in PR (Group, 2005; NBPRS, 2017). However, there is a need for more updated and multi-dimensional information regarding the approximate number of AA working in the field, what types of positions they are holding and the national organizations/agencies in which they work. This type of continual, statistical and quantifiable information can be used to build upon qualitative outcomes to create a more representational body of literature regarding AA in PR – taking into account both their voices/experiences and contribution within the field.

• *Build and/or Adapt Conceptual Frameworks and Programs* – Suggestions for the enhancement of diversity, especially among AA in PR is scattered and scarce. Efforts to unify the field in terms of diversity research and implementation are both necessary. There are different frameworks that have been used in diversity literature in PR, such as code switching, whiteness and stereotype threat. This particular review of literature yields that many of these frameworks often overlap during the description of AA experiences in PR. Thus, conceptual frameworks and programs aimed toward effective diversity, equity and inclusion should be created, validated and applied. These multidimensional frameworks and programs will focus on the specific yet compound experience of AA in the PR workforce by providing potential solutions and strategies for decreasing racism and discrimination, placing whiteness under a critical lens and addressing tokenism and cosmetic diversity.

• *Implement Organizational Campaigns and Trainings* – Campaign strategies are needed to help implement diversity programs and cultivate environments of education and understanding among the majority race in PR. To merely mention frameworks and programs in scholarly literature is not enough. This research makes a clear need for researchers and diversity practitioners to work together in creating specific trainings and campaigns around diversity programs and frameworks. Onsite trainings and campaigns could put the frameworks that we mention in scholarly research into practice, and which have shown to decrease the potential for minorities to experience discrimination on the job (King, Dawson, Kravitz, & Gulick, 2010). Educational trainings for majority races could be carefully crafted and offered to provide viewpoints of persons of color in workplace environments and in communication with clients/consumer markets. These onsite trainings and campaigns are necessary to show majority race executives why cosmetic diversity is unjust and ultimately toxic to gaining strides and trust of minority populations nationally (King et al., 2010). Higher-level executives and board members stand to benefit from being educated on how diverse employee populations in PR can positively affect the organization's bottom line – in terms of increasing effective communication with and the gaining of organizational buy-in from diverse publics.

- *Evaluate Diversity Practice* – The importance of evaluation as a method of communication practice enhancement is a well-known strategy in the field of PR. A program or campaign that lacks evaluation can be costly to the host organization as the aims and objectives may not have been actually reached (Dianova, 2016; Janoske, 2011). Specific to this research, if programs and campaigns are created (based upon updated qualitative and quantitative literature) for more effective efforts in diversity implementation and increased efforts to combat diversity resistance, they should certainly be evaluated for outcomes that could be enhanced or changed altogether. Evaluation is a key strategy in continually assessing the direction and need for diversity literature and practical solutions in the field of PR. If new and updated research and programs are created, they should be evaluated to ascertain whether or not they are essential to the promotion of more diverse populations, and AA specifically, in PR. This type of continual or process evaluation helps to protect the field against redundant research and practical endeavors (Dianova, 2016) and helps ensure the growth of diversity within the profession.

Conclusion

In summary, the need for diversity in workplace environments has been continually affirmed in both research and unfortunate crises in PR practice (Jones, 2011; Tindall, 2012; Vardeman-Winter, 2011; Vardeman-Winter & Place, 2017). Many of the high profile PR guffaws that have taken place in well-known companies have been attributed to the lack of diverse thinking and individuals in their PR and advertising departments (Reader, 2017; Smith, 2017; Tindall, 2012). If executives and professionals in the field can work toward rebuilding the foundation of the PR field, where the focus was not directed toward social justice and diverse population inclusion, great strides can be achieved. Internal and external communication efforts by PR professionals can be maximized and effectively revolutionized with true inclusion of diverse cultures, faces and voices in the field. The strategies provided within this chapter contribute to a thin body of literature focused on the detriment of both the resistance to diversity and cosmetic diversity attempts in the PR field. However, it further provides a foundation for future directions in terms of promoting diversity, equity and inclusion among practitioners in the field of PR.

References

Astor, M. (2017). Dove drops an ad accused of racism. Retrieved January 8, 2020 from www.nytimes.com/2017/10/08/business/dove-ad-racist.html

Bell, S. T., Villado, A. J., Lukasik, M. A., Belau, L., & Briggs, A. L. (2011). Getting specific about demographic diversity variable and team performance relationships: A meta-analysis. *Journal of Management, 37*(3), 709–743. https://doi.org/10.1177/0149206310365001

Boulton, C. (2016). Black identities inside advertising: Race inequality, code switching, and stereotype threat. *Howard Journal of Communication, 27*(2), 130–144.

Brown, K. A., White, C., & Waymer, D. (2011). African-American students' perceptions of public relations education and practice: Implications for minority recruitment. *Public Relations Review, 37*(5), 522–529.

Coffee, P. (2017). Shea Moisture has been direct messaging black influencers unhappy with its recent Vaynermedia ads. www.adweek.com/agencyspy/shea-moisture-has-been-direct-messaging-Black-influencers-unhappy-with-its-recent-vaynermedia-ads/130460/

D'addario, D. (2017). Why the Kendall Jenner Pepsi ad was such a glaring misstep. Retrieved January 8, 2020 from https://time.com/4726500/pepsi-ad-kendall-jenner/

Deitch, E. A., Barsky, A., Butz, R. M., Chan, S., Brief, A. P., & Bradley, J. C. (2003). Subtle yet significant: The existence and impact of everyday racial discrimination in the workplace. *Human Relations, 56*(11), 1299–1324.

Dianova, Y. (2016). Why measure PR efforts? www.axiapr.com/blog/why-measure-pr-efforts

Edwards, L. (2010). 'Race' in public relations. In Heath, R. L. (Ed.), *The SAGE handbook of public relations* (pp. 205–221). Los Angeles, CA: Sage.

Eibach, R. P., & Keegan, T. (2006). Free at last? Social dominance, loss aversion, and white and Black Americans' differing assessments of racial progress. *Journal of Personality and Social Psychology, 90*(3), 453.

Foote, N. (2017, November). NBPRS State of the Industry. Retrieved January 8, 2020 from http://nbprs.org/nbprs-state-of-the-industry/.

Fraser, L. (2017). 5 Lessons to learn about black public relations pioneers. www.Blackenterprise.com/5-lessons-to-learn-about-Black-public-relations-pioneers/

Gassam, J. (2019). Will Gucci's comprehensive diversity and inclusion plan repair the company's image? Retrieved January 8, 2020 from www.forbes.com/sites/janicegassam/2019/02/16/will-guccis-comprehensive-diversity-and-inclusion-plan-repair-the-companys-image/#7d72d74d716a

Greer, B. W. (2013). Consuming America: Moss Kendrix, Coca-Cola and the identity of the black American consumer. Retrieved January 8, 2020 from www.coca-cola

company.com/stories/consuming-america-moss-kendrix-coca-cola-and-the-making-of-the-Black-american-consumer

Griffith, J. (2019). Spike Lee, T.I. boycott Gucci, Prada over 'blackface' fashion. Retrieved January 8, 2020 from www.nbcnews.com/news/us-news/spike-lee-t-i-boy cott-gucci-prada-over-Blackface-fashion-n969821

Murphy, J. E., & Public Relations Coalition. (2015). Diversity Tracking Survey 2005. Retrieved January 8, 2020 from https://instituteforpr.org/diversity-tracking-survey-2005/.

Held, A. (2017). Dove expresses 'regret' for racially insensitive ad. Retrieved January 8, 2020 from www.npr.org/sections/thetwo-way/2017/10/08/556523422/dove-expresses-regret-for-ad-that-missed-the-mark

Held, A. (2019) Gucci apologizes and removes sweater following 'blackface' backlash. Retrieved January 8, 2020 from www.npr.org/2019/02/07/692314950/gucci-apologizes-and-removes-sweater-following-Blackface-backlash

Janoske, M. (2011). Public relations metrics: Research and evaluation. *Journal of Communication, 61*(4), E26–E30.

Jones, L. (2011). Public relations needs more diverse voices. *Public Relations Tactics, 18*(3), 6.

Kar-Gupta, S., & White, S. (2019). Kering shares slide as Gucci's growth slows. www.reuters.com/article/us-kering-results/kering-shares-slide-as-guccis-growth-slows-idUSKCN1RU0P7

King, E. B., Dawson, J. F., Kravitz, D. A., & Gulick, L. M. (2010). A multilevel study of the relationships between diversity training, ethnic discrimination and satisfaction in organizations. *Journal of Organizational Behavior, 33*(1), 5–20.

Kirsch, N. (2018). Papa John's founder used N-word on conference call. Retrieved January 8, 2020 from www.forbes.com/sites/noahkirsch/2018/07/11/papa-johns-founder-john-schnatter-allegedly-used-n-word-on-conference-call/#7a0891154cfc

Kottasova, I. (2017). Nivea pulls 'white is purity' ad after outcry. Retrieved January 8, 2020 from https://money.cnn.com/2017/04/05/news/companies/nivea-white-is-purity-racist-ad/index.html

Logan, N. (2016). The Starbucks race together initiative: Analyzing a public relations campaign with critical race theory. *Public Relations Inquiry, 5*(1), 93–113. doi:10.1177/2046147X15626969

Ly, L. (2018). Prada pulls products after accusations of Blackface imagery. Retrieved January 8, 2020 from www.cnn.com/style/article/prada-pulls-products-Blackface-imagery/index.html

Lyons, H. Z., Velez, B. L., Mehta, M., & Neill, N. (2014). Tests of the theory of work adjustment with economically distressed African Americans. *Journal of Counseling Psychology, 61*(3), 473–483.

McGirt, E. (2018). RaceAhead: Why is public relations so white? *Leadership.* Retrieved January 8, 2020 from http://fortune.com/2018/02/08/raceahead-why-is-public-relations-so-white/

Meyersohn, N. (2018). Papa John's founder resigns as chairman after using N-word on conference call. Retrieved January 8, 2020 from https://money.cnn.com/2018/07/11/news/companies/papa-johns-pizza-john-schnatter/index.html

Nussbaum, R. (2017). Shea moisture has issued an apology and pulled its controversial "hair hate" ad. Retrieved January 8, 2020 from www.glamour.com/story/shea-moisture-hair-hate-ad

Payne, A., & Duster, C. (2017). Shea Moisture ad falls flat under backlash. Retrieved January 8, 2020 from www.nbcnews.com/news/nbcblk/shea-moisture-ad-falls-flat-after-backlash-n750421

Payton, E. (2018). H&M, a fashion giant, has a problem: $4.3 billion in unsold clothes. Retrieved January 8, 2020 from www.nytimes.com/2018/03/27/business/hm-clothes-stock-sales.html

Plaut, V. C., Garnett, F. G., Buffardi, L. E., & Sanchez-Burks, J. (2011). "What about me?" Perceptions of exclusion and whites' reactions to multiculturalism. *Journal of Personality and Social Psychology, 101*(2), 337.

Pomper, D. (2007). Multiculturalism in the public relations curriculum: Female African American practitioners' perceptions of effects. *Howard Journal of Communication, 16*(4), 295–316.

PRSA. (2018). All about PR. Retrieved January 8, 2020 from www.prsa.org/all-about-n-worpr/

Reader, B. (2017). Dove ad screw up: Lack of diversity, inclusion in the 'decision room' still a huge problem in America. Retrieved January 8, 2020 from www.bkreader.com/2017/10/dove-ad-eff-lack-diversity-inclusion-decision-room-huge-problem-still/

Rose, C. S., & Bielby, W. T. (2011). Race at the top: How companies shape the inclusion of African Americans on their boards in response to institutional pressures. *Social Science Research, 40*(3), 841–859.

Schudel, M. (2011). Ofield Dukes, prominent D.C. public relations figure, dies at 79. Retrieved January 8, 2020 from www.washingtonpost.com/local/obituaries/ofield-dukes-prominent-dc-public-relations-figure-dies-at-79/2011/12/08/gIQA38jxgO_story.html?noredirect=on&utm_term=.38e55fbecb3d

Seter, J. (2017). How PR crises impact brand reputation. Retrieved January 8, 2020 from https://clutch.co/pr-firms/resources/how-pr-crises-impact-brand-reputation

Sha, B.-L. (2013). Diversity in public relations special issue editor's note. *Public Relations Journal, 7*(2), 1–7.

Smith, R. (2017). Pepsi, McDonald's, Dove (twice): Can the PR industry stop adland's disastrous campaigns? Retrieved January 8, 2020 from www.prweek.com/article/1447097/pepsi-mcdonalds-dove-twice-pr-industry-stop-adlands-disastrous-campaigns

Stack, L. (2018). H&M apologizes for 'monkey' image featuring black child. Retrieved January 8, 2020 from www.nytimes.com/2018/01/08/business/hm-monkey.html

Stellar, A. (2014). Diversity communications toolkit: A guide to diversity communications/engagement in education, by the National School Public Relations Association. *Journal of School Public Relations, 35*(1), 150.

Stump, S. (2018). H&M apologizes following outrage over 'monkey' sweatshirt ad seen as racist. Retrieved January 8, 2020 from www.today.com/style/h-m-apologizes-following-outrage-over-monkey-sweatshirt-ad-seen-t120979

Sue, D. W., Lin, A. I., Torino, G. C., Capodilupo, C. M., & Rivera, D. P. (2009). Racial microaggressions and difficult dialogues on race in the classroom. *Cultural Diversity and Ethnic Minority Psychology, 15*(2), 183. doi.org/10.1037/a0014191

Thomas, D. A., & Ely, R. J. (1996). Making differences matter: A new paradigm for managing diversity. *Harvard Business Review, 74*(5), 79. Retrieved January 8, 2020 from https://search-ebscohost-com.sacredheart.idm.oclc.org/login.aspx?direct=true&db=edb&AN=9609167709&site=eds-live&scope=site

Thomas, K. M. (Ed.). (2008). *Diversity resistance in organizations*. New York, NY: Taylor & Francis Group/Lawrence Erlbaum Associates.

Thomas, K. M., & Plaut, V. C. (2008). The many faces of diversity resistance in the workplace. In K. M. Thomas (Ed.), *Series in applied psychology. Diversity resistance in organizations* (pp. 1–22). Taylor & Francis Group/Lawrence Erlbaum Associates.

Tindall, N. T. J. (2009). In search of career satisfaction: African-American public relations practitioners, pigeonholing, and the workplace. *Public Relations Review, 35`* (4), 443–445. doi:10.1016/j.pubrev.2009.06.007

Tindall, N. T. J. (2012). The effective, multicultural practice of public relations. *Public Relations Tactic, 19*(2), 6.

Tsang, A. (2017). Nivea pulls 'white is purity' ad after online uproar. Retrieved January 8, 2020 from www.nytimes.com/2017/04/04/business/media/nivea-ad-online-uproar-racism.html?module=inline

Vardeman-Winter, J. (2011). Confronting whiteness in public relations campaigns and research with women. *Journal of Public Relations Research, 23*(4), 412–441. doi:10.1080/1062726X.2011.605973

Vardeman-Winter, J., & Place, K. R. (2017). Still a lily-white field of women: The state of workforce diversity in public relations practice and research. *Public Relations Review, 43*(2), 326–336.

Victor, D. (2017). Pepsi pulls ad accused of trivializing black lives matter. Retrieved January 8, 2020 from www.nytimes.com/2017/04/05/business/kendall-jenner-pepsi-ad.html

Wiener-Bronner, D. (2019). Papa John's reports fifth straight quarter of declining sales. Retrieved January 8, 2020 from www.cnn.com/2019/02/26/business/papa-johns-earnings/index.html

Wiggins-Romesburg, C. A., & Githens, R. P. (2018). The psychology of diversity resistance and integration. *Human Resource Development Review, 17*(2), 179–198.

Ziv, S. (2018). H&M 'needs some Black people in their company to look over shit': Marlon Wayans. Retrieved January 8, 2020 from www.newsweek.com/hm-needs-some-Black-people-their-company-look-over-shit-marlon-wayans-776720

Index

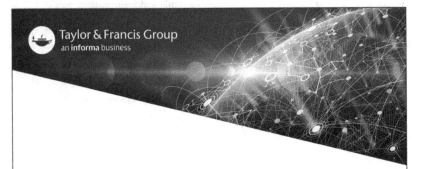

Printed in the United States
by Baker & Taylor Publisher Services

Sub- and Supercritical Hydrothermal Technology

Sub- and Supercritical Hydrothermal Technology

Industrial Applications

Edited by
Sandeep Kumar and Florin Barla

CRC Press
Taylor & Francis Group
Boca Raton London New York

CRC Press is an imprint of the
Taylor & Francis Group, an **informa** business

CRC Press
Taylor & Francis Group
6000 Broken Sound Parkway NW, Suite 300
Boca Raton, FL 33487-2742

First issued in paperback 2021

ISBN-13: 978-1-03-208554-8 (pbk)
ISBN-13: 978-1-138-08509-1 (hbk)

Publisher's Note
The publisher has gone to great lengths to ensure the quality of this reprint but points out that some imperfections in the original copies may be apparent.

Visit the Taylor & Francis Web site at
http://www.taylorandfrancis.com

and the CRC Press Web site at
http://www.crcpress.com

Contents

Preface

Chemical processes have been a cornerstone of industrial revolution and societal progress. The development of sustainable chemical processes using green chemistry concepts can potentially minimize resource use and environmental impacts. The use of sub- and supercritical water (collectively called hydrothermal) as reaction media and reactant provides a path for reducing the use of organic solvents. The technology can be integrated in industrial processing for safer, more flexible, economical, and ecological production processes in the context of green and sustainable engineering.

In recent years, there has been much research focused on the use of subcritical water for a wide range of applications including extraction, liquefaction, carbonization, gasification, coolant in nuclear power plants, aqua-thermolysis, waste oxidation, inorganic synthesis, and mineralization. Hydrothermal processing of lignocellulosic biomass, biosolids, and microalgae for bioproducts, biochar, and biofuels are among the rapidly growing sectors which are on the cusp of using this technology at industrial scales. At the same time, supercritical water oxidation (SCWO) for managing hazardous waste, hydrothermal processing of coal, heavy oil and bitumen, and subcritical water use as a coolant in nuclear reactors have already seen the industrial-scale operation of hydrothermal technology.

Even though hydrothermal processing, an environmentally benign method, has a wide range of applications, most of the publications or previous books are focused on narrow applications in biofuels or materials synthesis. There is a need to develop new chemical and environmental processes and products using a system approach that results in zero waste, high energy efficiency, zero toxicity, and minimal impact over life cycle. This book is focused on the industrial applications of hydrothermal technology with an objective to help move toward green chemistry and engineering.

Supercritical water oxidation is used to treat a wide variety of industrial wastewaters and hazardous/non-hazardous wastes. The SCWO reactions take place above the critical point of water when water behaves like a nonpolar solvent causing almost complete solubility of organics and gases such as oxygen. The products of SCWO are nontoxic and free from air pollutants such as NO_x, SO_x, and dioxins. The SCWO chapter discusses the technology, applications, and challenges of commercial plants.

Lignocellulosic biomass and algae are being considered for producing biofuels, bioproducts, and nanocellulose using an integrated biorefinery system. The biopolymers (e.g., starch, cellulose, hemicellulose, lignin, proteins, and lipids) present in different biomass/algae are the source of biofuels for transportation use and many bioproducts such as organic acids, plastics, fertilizers, lubricants, industrial chemicals, and many other products. A successful deployment of industrial-scale biorefineries will help, partly replacing the fossil fuels and reducing the greenhouse gas emissions. A total of five chapters have been devoted to discussing hydrothermal processing of biomass/algae focused on extraction, pretreatment, liquefaction, carbonization, and gasification processes excluding SCWO. Besides the extraction of different bioproducts from biomass/algae, these chapters describe the application and current status of sub- and supercritical water technology for liquid fuels (bio-ethanol, biocrude, liquid hydrocarbons), gaseous fuels (methane, hydrogen, synthesis gas), and solid fuels (biochar, other functional carbonaceous materials) production.

As we move toward biomass resources for producing biofuels and different bio-products, the need to manage nutrients (nitrogen and phosphorous) is becoming increasingly important. The recovery and recycling of nutrients are important for sustainability as well as reducing the non-point pollutions in air and waterways. The need for nutrients in algae cultivation is substantial. However, these nutrients can be fractionated and mineralized to recover in the form of solid nitrogen- and phosphorous-based minerals (e.g. struvite, hydroxyapatite) which can be used for slow-release fertilizer and biomedical applications. The chapter on hydrothermal mineralization (HTM) discusses the process and products from processing nutri-ents-rich feedstock or wastewater.

Supercritical water has a liquid-like density, which gives a large capacity for solvation. It has high molecular diffusivity and low viscosity, making it an ideal medium for efficient mass transfer. Further, its high compressibility gives large den-sity variations with very small pressure changes, yielding extraordinary selectivity characteristics, which are most important in the removal of nitrogen from coal or coal liquids. SCW is miscible with light gases such as H_2, CO, and O_2. As water temperature approaches the critical point, it becomes miscible with oils and aromat-ics. This provides a unique, homogeneous reaction medium for coal, heavy oil, and bitumen processing.

This book recognizes that some SCW reactors have reported significant opera-tional problems. These mainly concern the deposition of salts at the point in the reactor or feed delivery system where the feed goes above the critical temperature and inorganic salts become insoluble. Salt buildup has plugged reactors and forced unplanned shutdowns, and every coal-water mixture will contain inorganic salts. In the event of salt deposition or other issues caused by SCW, operation becomes an obstacle in scale up.

A novel SCW reactor design called the "deep-well reactor" is presented in this book. In this concept, the reactor consists of a long pipe, positioned vertically within a cased, cemented, well bore, possibly as deep as 8,000 feet. The presented design could reduce both the cost of construction and the operating costs. The deep-well reactor concept may present an economic advantage large enough to make the method commercially viable or provide a greater return than processing in conven-tional reactors.

Massive accumulation of plastic wastes has caused serious environmental prob-lems worldwide. There are well-established systems for material recycling of glass, metals, and paper but not for more complex materials such as textiles that are burned or discarded in landfills. In addition, synthetic dyes and other substrates such as paper, leather, fur, hair, waxes, greases, fiber lubricants, finishing agents, and plas-tics are extensively used in the textiles manufacturing and dyeing process, making the recycling process exceptionally challenging. Currently, polyethylene terephthal-ate (PET), a condensation polymer abundantly used in production of soft drink bot-tles, although its main application is by far in the textile industry, recycling is one of the most successful and widespread examples of polymer recycling. However, new methods were developed, adapted, and employed such as sub- and supercritical water treatment to find feasible alternatives for polymer fibers recycling. The last chapter discusses various methods that have shown potential in polymer fiber recycling.

Sandeep Kumar
Florin Barla

Acknowledgments

My gratitude goes to my wife, Sweta Kumari, my parents, and family who believed in me and supported me in all my endeavors. The love and sweet smile of my son Shashank is always a source of rejuvenation after exhaustive work. I am thankful to the National Science Foundation (NSF) CAREER Award #CBET-1351413.

Sandeep Kumar

I would like to express my sincere gratitude to Dr. Kumar for his invaluable guidance and continuous support throughout the course of this project. My deepest appreciation belongs to my family; my wife, Hang, my sons, Andrew and Alex, and to my lovely daughter, Anna, for their patience and understanding.

Florin G. Barla

Editors

Sandeep Kumar, PhD, is a chemical engineer and since July 2010 is Associate Professor in the Department of Civil & Environmental Engineering, Old Dominion University, Norfolk, Virginia. As Director of Energy Cluster at Frank Batten College of Engineering, Old Dominion University, Dr. Kumar is working on interdisciplinary research, energy-related courses, and outreach activities. Dr. Kumar has nearly 20 years of professional experience in industry, R&D, and academic research in the area of carbon black, nuclear fuels, and biofuels with responsibilities in new process development, process engineering, and project management. Dr. Kumar started his professional career in 1994 as a process engineer and was a certified project management professional (PMP®) between 2007 and 2011.

Dr. Kumar earned his PhD in Chemical Engineering from Auburn University, Auburn, Alabama in May 2010. His current research focuses on the application of sub- and supercritical water technology for the conversion of biomass/algae to advanced biofuels and bioproducts. Dr. Kumar's areas of interest are supercritical fluids (CO_2 and water), biofuels, hydrodeoxygenation of lipids, and the applications of chemical and environmental engineering concepts to produce renewable fuels. His expertise is in high temperature and high-pressure reactions involving biomass components such as proteins, lipids, cellulose, hemicelluloses, and lignin. Dr. Kumar has served as reviewer/panelist for several scientific international journals and grant agencies. He has successfully secured more than $1.5M of external research grants including the prestigious NSF CAREER award. Dr. Kumar has been awarded two patents, and has published five book chapters, 30 peer-reviewed journal publications, and more than 50 conference presentations.

Florin Barla, PhD, is a Director of Process Development at Tyton Biosciences, where he is solving real-world problems by implementing Tyton Biotechnology at the lab- and pilot-scales. With expertise in chemical engineering, extraction techniques including sub- and supercritical fluid technology, and a hands-on approach to development, Dr. Barla is the lead operator of Tyton Bio's proprietary digesters and associated water-based technologies. Experienced in applying subcritical fluids technology, he is a member of the reactors design development team that designed the first pilot-scale subcritical water reactor in the US, built to address lingo-cellulosic hydrolysis; non-tree pulping; textile recycling, and food packaging recycling. He won the Award of Excellence—Tyton Biosciences (2017).

Prior to Tyton Bio, he was the plant manager of a dairy facility located in Romania. He also was an Assistant Professor of Food Science at Stefan cel Mare University,

Romania. Dr. Barla earned his PhD in Applied Life Sciences from Osaka Prefecture University, Osaka, Japan (2010) and completed his postdoctoral training at Ishikawa University, Japan and Old Dominion University, Norfolk, Virginia. Dr. Barla has been awarded two patents and has published two book chapters, more than ten peer-reviewed journal publications, and more than ten conference presentations.

Contributors

Elena Barbera
Department of Industrial
  Engineering DII
University of Padova
Padova, Italy

Maoqi Feng
Polykala Technologies LLC
San Antonio, Texas

Chen Li
Department of Civil and Environmental
  Engineering
Old Dominion University
Norfolk, Virginia

Eleazer P. Resurreccion
Department of Civil Engineering
  Technology
Montana State University-Northern
Havre, Montana

Anuj Thakkar
Department of Civil and Environmental
  Engineering
Old Dominion University
Norfolk, Virginia

1 Subcritical Water Technology in Bioproducts Extraction and Nanocellulose Production

Florin Barla and Sandeep Kumar

CONTENTS

1.1 INTRODUCTION

The industry is facing one of the major challenges nowadays in its effort to transition from an economic system based primarily on nonrenewable resources to one based on renewable ones, a sustainable production and conversion of biomass to provide food, health, fiber, industrial products, and energy (Venkata et al., 2016; Yamamoto et al., 2014). The vast amount of available biomass represents an attractive potential feedstock for many energy and chemical processes. From an economic standpoint, some evidence suggests that energy production from renewable carbon sources may be competitive with production from nonrenewable carbon sources. Industry must play a decisive role in the transition, and the development of cost competitive biomass-based processes is required. Agriculture, a significant global activity, is producing massive amounts of lignocellulosic residues that can be used as feedstock in fractionation/hydrolysis processes to produce fermentable sugars and cellulosic fibers. A large variety of feedstocks such as corn stover, rice barn, wheat straw, corn shell, vegetables and fruits peel, seed, food matrices, etc., can be used as a raw material for extracting valuable resources (phenolic compounds, water-soluble sugars, organic acids, etc.) that could be used in pharmaceutical, cosmetics and biofuel processing industries (Deng et al., 2012). These feedstocks are renewable sources of biofuels and bioproducts which are abundant, inexpensive, and do not compete with food. Industrial food processing residues are another source of biomass that can be transformed into biofuels and biochemicals. Biomass from food waste typically consists of 50–75% carbohydrates (cellulose and hemicellulose), 5–25% lignin (Mohan et al., 2015; Silva et al., 2009), and modest amounts of other substances, including phenolic compounds, which may be processed into higher value-added products (Cardenas-Toro et al., 2014; Yang et al., 2015). Therefore, such biomass feedstocks are ideal sources for obtaining simple sugars (monomers), which are used as fermentation substrates for the production of bioethanol (Prado et al., 2014) and phenolic compounds that possess antioxidant properties or other biological activities in pharmaceutical, food, and health applications (Pourali, Asghari, & Yoshida, 2010; de-Oliviera et al., 2016).

Biofuels represent a class of renewable energy with the potential to contribute significantly to the sustainable energy mix required to meet future energy demands (Awaluddin et al., 2016). Microalgae, which is an aquatic biomass, is heavily researched as feedstock for the production of advanced biofuels as a result of its fast growth rate and the capacity to accumulate high concentrations of biochemical compounds such as lipids and carbohydrates (Chen et al., 2013). Microalgae primary metabolites, such as proteins, fatty acids, and carbohydrates, are produced intracellularly and entrapped within the cells; thus, an effective extraction technology is required to release these biochemical products (De Morais et al., 2015). The primary metabolites are a source of bioactive metabolites, such as vitamins and enzymes, which are commercially beneficial due to their antioxidant, anti-inflammatory, antiangiogenic, anti-obesity, anticancer properties (De Morais et al., 2015). Commonly used extraction technologies via chemical and mechanical methods include expellers, liquid-liquid extraction (organic solvent extraction), supercritical fluid extraction (SFE), and ultrasound techniques (Castejon, Luna, & Senorans, 2017) or subcritical water extraction (SWE) (Ibanez et al., 2003). SWE technology is gaining popularity as a method of valuable material recovery in high yields and high quality of extracted products, is inexpensive, has a short residence time, uses nontoxic solvent (water), and has good selectivity. It can extract different classes of compounds depending on the temperature used for the extraction, with the more polar extracted at lower temperatures and the less polar compounds extracted at higher temperatures which makes it an environmentally friendly technology (Abdelmoez et al., 2014) as compared with other traditional solvent extraction methods (Ravber, Knez, & Skerget, 2015).

Fruits, plants, and vegetable residue contain various bioactive substances, such as ascorbic acid, vitamin E, phenolic compounds also referred to as antioxidants, etc.

and the intake of these substances is an important health-protecting factor (Cheigh et al., 2015). The growing interest in natural food has raised the demand for natural antioxidants, products that have a non-synthetic origin and can prevent or retard oxidation of fats and oils. Antioxidants are important in the food industry not only because of their preservative effect but also for their beneficial effects on human health (Madhavi, Despande, & Salunkhe, 1996). The use of agricultural by-products in food processing is increasing. Natural materials are an emerging field in food science because of their increasing popularity with costumers concerned about health. One application of these materials is the extraction of flavonoids from foodstuffs such as apples, onion skins, and citrus fruit, especially for utilizing the antioxidant effects exhibited by flavones, flavonol, and flavanone. Quercetin is a flavonol that is sometimes added as an antioxidant and nutritional supplement to nutraceuticals. It is a yellow powder with a characteristic smell and a bitter taste. Quercetin is over 77 times more abundant in the inedible part than in the edible part of the white onions (Kang et al., 1998a). Additional functional benefits of quercetin include anti-inflammatory activity, antihistamine effect, allergy medication, and antivirus activities. It has also been claimed that quercetin reduces blood pressure in hypertensive subjects (Boots, Haenen, & Bast, 2008). Subcritical water has the potential to be an environmentally friendly solvent for applications including hydrolysis, liquefaction, extraction, and carbonization (Lachos-Perez et al., 2017).

1.2 SUBCRITICAL WATER TECHNOLOGY

Subcritical water technology gained much popularity being utilized in different applications due to its advantage of being a green, cost-effective, environmentally friendly technology with a higher quality of extracted product and considerably short reaction time as compared with traditional solvent extraction methods (Ravber, Knez, & Skerget, 2015). Subcritical water (also called superheated water or hydrothermal or pressurized hot water) is a term commonly used for water heated under adequate pressure (above its vapor pressure) to between its atmospheric boiling temperature (100°C) and its critical point (374°C); under this condition the water maintains its liquid state and the thermal motion of water molecules increases markedly changing the water properties (Yoshida et al., 2014). Unlike ambient water, the highly hydrogen-bonded structure at subcritical condition slowly starts to disintegrate resulting in a decrease of polarity, the dielectric constant of water decreases which makes it act as solvent for hydrophobic matters, an increase of diffusion rate and a decrease of viscosity and surface tension (Smith, 2002). Extractions performed under subcritical water conditions differ quite significantly from conventional extraction methods. First, it is known to be very fast (Aliakbarian et al., 2012; Carr, Mammucari, & Foster, 2011), due to the already mentioned changes in the physical properties. Also, the decrease in polarity provides a tendency for dissolving less polar compounds (Carr, Mammucari, & Foster, 2011). Second, the hydrolytic nature of subcritical water means that this type of extraction medium when used on natural materials will result not only in water-soluble extracts but also in hydrolyzed products of the extract (Fernandez-Ponce et al., 2012; Ruen-ngam, Quitain, & Tanaka, 2012). Furthermore, the insoluble cell material (numerous complex polymeric structures such as:

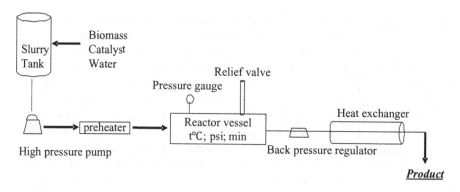

FIGURE 1.1   Basic diagram representation of a continuous SWE system.

proteins, polysaccharides, etc.), can be simultaneously hydrolyzed during extraction which produces various water-soluble products such as amino acids and sugars, thus increasing the overall extraction yield. The destruction of the complex structures would result in formation of less stable emulsions (Sanguansri & Augustin, 2010).

The increased acidity of the medium does not result only in hydrolytic reactions but also in other hydrothermal reactions characteristic to SWE such as dehydration and decarboxylation (Pavlovic, Knez, & Skerget, 2013), which means that the obtained hydrolyzed products can react even further with water molecules resulting in a variety of other products of hydrothermal degradation e.g., furfurals from carbohydrates. A basic diagram of a SWE system is shown in Figure 1.1.

In recent years, SWE technology has been chosen for the extraction of various active compounds from biomass, to remove non-oil biomass leaving an oil-rich biomass for biodiesel production (Levine, Pinnarat, & Savage, 2010) or to release nitrogen containing compounds (amino acids and peptides) to the water phase producing an oil-rich biomass residue (Moscoso-Garcia et al., 2013). SWE was also applied to microalgae and yeasts biomass and the aim was to recover polysaccharides and protein-derived products (Chakraborty et al., 2012; Moscoso-Garcia, Teymouri, & Kumar, 2015), or potentially can be used to produce bio-crude which can be subsequently upgraded to liquid hydrocarbons (Moscoso-Garcia, Teymouri, & Kumar, 2015). Using flash hydrolysis of microalgae under subcritical water conditions, the proteins are solubilized as water-soluble peptides and amino acids and the remaining material which comprises lipids and algal cell can be filtered and used in biofuel production (Kumar et al., 2014).

1.2.1   Physicochemical Properties of Water

Water at normal conditions (25°C and 0.1MPa) is a polar solvent, immiscible with hydrocarbons, with a dielectric constant (ε) of 79.9, and a density of 1000 kg/m³. A high dielectric constant suggests that water has the ability to screen charges (water molecules will surround both anions and cations in a solution and diminish the attraction of the two charges). Under high temperature and pressure water has different properties, its hydrogen bonds break down triggering its dielectric constant to drop (Cheigh et al., 2015; Lu et al., 2014). A low dielectric constant allows subcritical water to dissolve organic compounds.

1.2.2 Dielectric Constant (ε)

Water is extremely unique and has a high dielectric constant that indicates the affinity of water to being a great reaction media (Singh & Saldana, 2011). When water is heated at higher temperature and high pressure is applied (e.g., 250°C; 5MPa) the dielectric constant of water drops from 79.9 to 32.5 and 27, the equivalent of the dielectric constants of methanol and ethanol, respectively (Singh & Saldana, 2011). A lower dielectric constant allows water to dissolve organic compounds acting as an organic solvent.

1.2.3 Ionic Product

An important characteristic of water under subcritical conditions is ions H_3O^+ and OH^- product, therefore water can act as an acid or a base, it gives and takes protons. The amount of these ionic products can be orders of magnitude higher than that of normal water (Toor, Rosendahl, & Rudolf, 2011; Kruse & Dinjus, 2007). Thus, water has a catalytic function and the ionic constant (Kw) of water increases with the increased reaction temperature and is about three times higher than at room temperature

The ion product of water is defined as $K_w = [H^+] [OH^-]$ concentration, under subcritical conditions the high ionization provides an acidic medium, therefore water acts as an acid hydrolysis catalyst (Carr, Mammucari, & Foster, 2011).

1.2.4 SWE Process Parameters and Mechanisms

The essential parameters of SWE that are affecting the efficiency of the extraction are temperature, pressure, and residence time. In addition, solid to water ratio, sample particle size, pH, and catalyst addition, have an impact on the extraction process. In particular, temperature and pressure have a mutual relationship. When increasing the temperature, the pressure inside the reactor will increase accordingly, and the water is maintained in the liquid phase by controlling the pressure. Temperature significantly affects physicochemical properties of water, increasing the temperature will result in a considerable increase of mass transfer rate and high solubility of bioactive compounds (Jintana & Shuji, 2008). At high pressure, the solvent molecules could efficiently infiltrate the sample's matrix and subsequently the characteristics of the extracted product would be different. As the temperature increases, the viscosity and the surface tension of the extraction solvent decreases; thus, the SWE should be done at the maximum allowable temperature, however not exciding the permitted value, in order to avoid degradation of the active compounds (Asl & Kajenoori, 2013). The most important reactions that occur during SWE are hydrolysis, dehydration, decarboxylation, polymerization, and aromatization (Kruse & Dinjus, 2007; Funke & Ziegler, 2010). Hydrolysis, the reaction that degrades the main composition of biomass, is basically breaking down complex

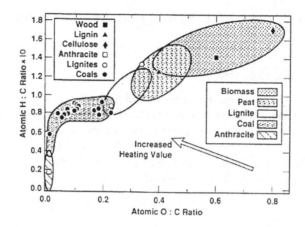

FIGURE 1.2 The Van Krevelen diagram.

compounds to its monomers. As an example, the chain hydrolysis of cellulose could yield oligomers and glucose during SWE, which can be further decomposed to 5-hydroxymethylfurfural, other organic acids, aldehydes, etc. Such reactions occur at temperatures above 220°C. Dehydration (removing water from the sample's matrix) reaction plays an important role during SWE, reducing the atomic ratio of H/C and O/C as the carbonization process progresses, eventually reaching the level of low rank coal, as shown in the Figure 1.2.

A decarboxylation reaction is involved in removing carboxyl and/or carbonyl groups yielding CO_2 as a major gaseous product, and CO. Also, CO_2 could result from condensation reactions and the cleavage of the intramolecular bonds. Polymerization and aromatization reactions occur as succeeding reactions after hydrolysis of cellulose resulting in the formation of other soluble polymers (Asghari & Yoshida, 2006). Aromatization and aldol condensation reactions could occur simultaneously (Sevilla & Fuertes, 2009).

1.3 SUMMARIZATION OF VARIOUS APPLICATIONS OF SWE

The biomass residue from agriculture, forest wood production, and the food and feed processing industry, could be an important feedstock for producing high-quality products at lower production cost and high efficiency. SWE has recently been applied successfully in the advancement of extraction methods for various active compounds. SWE has demonstrated its capability to selectively extract diverse classes of compounds depending on the temperature used, with the more polar compounds extracted at lower temperatures and the less polar compounds extracted at higher temperatures.

1.3.1 PHENOLIC COMPOUNDS AND ANTIOXIDANTS EXTRACTION

Valuable biological active compounds, such as antioxidants (Cheigh et al., 2015) an important group of health-protecting compounds, could be extracted under subcritical water conditions from agricultural residual biomass such as fruits, plants, and vegetable. SWE proved to be very effective in phenolic compounds extraction. Singh & Saldana (2011), show the efficacy of SWE of polyphenols from potato peel; it was found that at 180°C within 30 minutes, 81.83 mg/g of phenolic compounds were recovered as compared to methanol extraction that yielded only 46.59 mg/g after three hours residence time. In another study, (Tunchaiyaphum, Eshtiaghi, & Yoswathana, 2013), successfully applied SWE to separate phenolic compounds from mango peel. At 180°C 50.25 mg/g were obtained after 90 minutes, with a 1:40 solid water ratio, whereas using soxhlet extraction techniques required more than two hours to achieve maximum phenolic yield. Grapes is one of the largest fruit crop worldwide and is a major dietary source of phytochemicals such as polyphenols which include flavonoids and anthocyanins. The winemaking industry produces important amounts of by-product (pomace) including skins and seeds rich in valuable compounds that could be extracted and used in food, cosmetic, or pharmaceutical industries (Aliakbarian et al., 2012). SWE was also used to yield higher polyphenols, flavonoids, and antioxidant activity from grape pomace (Aliakbarian et al., 2012). In this study, raising the extraction temperature from 100°C to 140°C at 8MPa considerably enhanced the yield of total polyphenols from 12.78 mgGAE/g DP to 32.49 mgGAE/g DP (gallic acid equivalents per gram of dried pomace). The highest amount of flavonoids was obtained at 140°C and 11.5 MPa (15.28 mg CE/g DP; catechin equivalents per gram of dried pomace) and the best antioxidant activity was observed when the extraction was performed at 140°C and 15 MPa (13.85 µgDPPH/µl extract), as shown in the Table 1.1.

Two abundant coffee waste residues (powder and defatted cake) were treated using SWE conditions for hydrolysis and extraction of fermentable sugars and total phenolic compounds, in order to investigate the valorization potential of such residue for the production of chemicals and biofuels (Mayanga-Torres et al., 2017). From coffee powder the maximum of reducing sugars was 6.3% at 150°C and 30MPa,

TABLE 1.1

Dried Pomace Extractives Values under Different Subcritical Water Conditions

Temperature (°C)	Pressure (MPa)	Total Polyphelols (mgGAE/gDP)[a]	Total Flavonoids (mgCE/gDP)[b]	Antioxidant Power µgDPPH/µl extract)
100	8	12.78 ± 1.57	8.34 ± 0.43	5.90 ± 0.50
100	11	16.72 ± 1.35	9.99 ± 0.27	7.08 ± 0.54
100	15	16.46 ± 1.23	10.02 ± 0.71	7.11 ± 0.62
140	8	32.49 ± 2.63	15.11 ± 1.34	12.58 ± 1.10
140	11	30.8 ± 3.38	15.28 ± 1.02	12.87 ± 1.25
140	15	28.50 ± 1.50	12.33 ± 1.26	13.85 ± 1.26

Source: Adapted from Aliakbarian, 2012.

[a] (GAE) gallic acid equivalents per gram of (DP) dried pomace.

[b] (CE) catechin equivalents per gram of (DP) dried pomace.

and 9% total reducing sugars at the same conditions. The maximum of total phenolic compounds attained was 26.64 mgGAE/g of powder coffee at 200°C and 22.5MPa. From defatted coffee cake the maximum sugar yields were 8.79% (reducing sugars) and 17.23% (total reducing sugars) both obtained under SWE hydrolysis at 175°C and 22.5MPa. Also, at 175°C the highest total phenolic compounds were extracted, the yield was 55.31 mgGAE/g of defatted coffee cake (Mayanga-Torres et al., 2017). Garcia-Marino et al. (2006) have used subcritical water for the extraction of catechins and proanthocyanidins from grape seeds (27). In their study, sequential extractions were performed under subcritical water conditions at 10.3 MPa and within the temperature range of 50–150°C. The recovery of catechin and epicatechin was enhanced two-fold, compared to conventional methanol: water (75:25) extraction using an ultrasonic bath for 15 min. Rodriguez et al. employed superheated ethanol-water mixture for the extraction of anthocyanins and other phenolic compounds from grape pomace. In this study, neutral ethanol, water and acidified ethanol, and water with 0.8% (v/v) HCl were used for extraction. The yield of extraction increased 7 and 12-fold by increasing pressure and temperature from atmospheric conditions to 8 MPa conditions and 120°C when using neutral and acidified solvents, respectively. Monrad et al. (2010) studied the effect of accelerated solvent extraction on the recovery of anthocyanins from grape pomace at different temperature (40–140°C) and ethanol in water ratios (10–70%v/v). The results of this study showed that the optimum ranges for temperature and ethanol concentration are 80–120°C and 50–70%(v/v). It was concluded that despite lower extraction yield, running cost of the extraction process can be substantially decreased by using a lower concentration of ethanol in hydroethanolic solvent.

As previously mentioned, natural antioxidants, products of non-synthetic origin that are able to prevent or retard oxidation in fats and oils, are in high demand since there is a growing interest in natural foods. They are important in food industry for their preservation effect as well as for their beneficial effects on human health. Ibanez used SWE to isolate antioxidants compounds from rosemary plants. At the lowest temperature, the more polar compounds, such as scutellarein, rosmanol, rosmanol isomer, and genkwanin were extracted, while at higher temperature (200°C), the dielectric constant of water was reduced to values similar to those of methanol or acetonitrile, which increased the solubility of less polar compounds such as carnosol, canosic acid, and methyl carnosate by several orders of magnitude. The antioxidant activity of the obtained fractions at different temperatures was very high with values around 11.3 µg/ml.

Quercetin is the aglycone form of several other flavonoid glycosides such as rutin and quercetrin that are found in citrus fruits, buckwheat, and onions, a polyphenolic compound that exists as glucosides in vegetables or fruits (Makris & Rossiter, 2001). Flavonoids in fruits and vegetables have a low-polarity chemical structure and so dissolve in organic solvent. The quercetin is commonly extracted from vegetables such as onion using ethanol or aqueous-based ethanol or methanol solutions. Quercetin can be efficiently extracted from onion using ethanol at room temperature for two hours or water at 100°C for three hours (Kang et al., 1998b). SFE of quercetin using 7.6% ethanol-modified carbon dioxide at 393 Bar and 40°C takes two and a half hours (Martino & Guyer, 2004).

Though ethanol is classified as a generally-recognized-as-safe solvent, its utilization in this application is restricted by the long extraction time and varying regulation throughout countries. Subcritical water could be an excellent alternative medium for extracting flavonoids such as nonpolar quercetin due to its temperature-dependent selectivity, environmental acceptability, efficiency, and lower cost. In this study, the flavonol quercetin was extracted from onion skin under SWE conditions. The maximum yield of quercetin, about 16.29 mg/g onion skin (accounts for approximately 92.40% of the total amount of quercetin found in onion extract), was obtained at extraction temperature of 165°C and 15 minutes' residence time. In addition, quercetin-4'-glucoside (3.72 mg/g onion skin), kempherol (0.88mg/g onion skin), iso-rahmnetin (0.74 mg/g onion skin), and rutin (0.93 mg/g onion skin) were detected in the onion skin extract. Xiao et.al. (2017) successfully optimized the extraction of ursolic acid from *H. diffuza*, using SWE technology. Ursolic acid, a pentacyclic triterpene acid, and its derivatives have been reported to have several bioactivities including anti-inflammatory, hepatoprotective, antitumor, antiviral, anti-HIV, antimicrobial, antimalarial, antidiabetic, gastroprotective, and antihyperlipidemic effects. The optimum extraction yield was 6.45 mg/g of material and was obtained at an extraction temperature of 157°C, solvent solid ratio of 30 ml/g, and a particle size of 80 mesh. Also, the results indicated that SWE conditions do not require high pressure to obtain the optimum Ursolic acid extraction yield. A summary of optimum reaction temperature and time for extraction of varied valuable products from diverse biomass is shown in the Table 1.2.

1.3.2 OILS AND SUGARS EXTRACTION

The production of microbial lipids is directly correlated with a high conversion of the substrate into intracellular lipids combined with a high extraction efficiency (Bharathiraja et al., 2017). Usually a form of biomass pretreatment, such as cell disruption, it is required to remove the protective cell walls of microorganisms to make the intracellular lipids more accessible to solvent extraction (see Figure 1.3). SWE technology has been chosen for the extraction of various active compounds from biomass, to remove non-oil biomass leaving an oil-rich biomass for biodiesel production (Levine, Pinnarat, & Savage, 2010) or to release nitrogen-containing compounds (amino acids and peptides) to the water phase producing an oil-rich biomass residue (Moscoso-Garcia et al., 2013). Subcritical water extraction was also applied to microalgae and yeasts biomass and the aim was to recover polysaccharides and protein-derived products (Chakraborty et al., 2012; Moscoso-Garcia, Teymouri, & Kumar, 2015), or potentially can be used to produce bio-crude which can be subsequentlyupgraded to liquid hydrocarbons (Moscoso-Garcia et al., 2015). Using flash hydrolysis of microalgae under subcritical water conditions, the proteins are solubilized as water-soluble peptides and amino acids and the remaining material which comprises lipids and algal cell can be filtered and used in biofuel production (Kumar et al., 2014).

The first- and second-generation biodiesel are derived from edible and nonedible plants oil, while the microbial oils are primarily used to produce the third-generation or advanced biodiesel. Currently, the biodiesel is produced from plants oil

TABLE 1.2

Optimum Reaction Temperature and Time for Extraction of Assorted Valuable Products from Diverse Biomass

Raw Material	Extracted Product	Temperature Range	Optimum Temp. (°C)	Optimum Time (min.)	References
Potato peel	Glucose	140–240	240	15	Abdelmoez et al. (2011)
Mango peel	phenolic	160–220	180	90	Tunchaiyaphun et al. (2013)
M. chamonilla	Essential oil	100–175	150	120	Khajenoori (2013)
H. diffusa	Ursolic acid	120–200	157	20	–
Corn stalks	Fermentable hexose	180–392	280	27	Zhao et al. (2009)
Wheat straw	Fermentable hexose		280	54	Zhao et. al. (2009)
Wheat straw	Redusing sugars	170–210	190	30	Abdelmoez et al. (2014)
Sugar cane bagasse	Redusing sugars	200–240	240	2	Zhu et al. (2013)
Defatted rice barn	Sugars & Proteins	200–260	200	5	Hata et al. (2008)
Cotton seed	Oil	180–280	270	30	Abdelmoez et al. (2011)
Jojoba seed	Oil	180–260	240	30	Yoshida et al. (2014)
Sunflower seed	Oil	60–160	130	30	Ravber, Knez, & Skerget (2015)
Chia seed	Omega-3	60–200	90	10	Castqon et al. (2017)
Pomegranate seed	Phenolic	100–220	140	30	He et al. (2012)
Grape pomace	Polyphenols	100–140	140	130	Aliakbarian et al. (2012)
Onion skin	Quercetin	100–190	165	15	Ko et al. (2011)

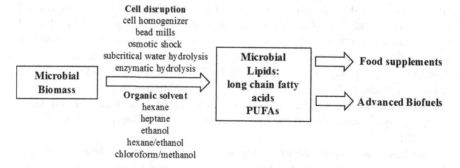

FIGURE 1.3 Summary of microbial lipids extraction methods and utilization potential.

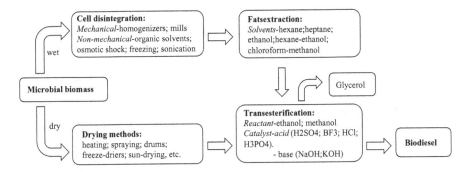

FIGURE 1.4 Summary of biodiesel production from oleaginous microorganisms via conventional transesterification with extracted lipids vs direct transesterification of microbial biomass.

by transesterification with an alcohol (methanol or ethanol), for the reason of cost methanol is used most frequently, in the presence of a base, an acid, or an enzyme as a catalyst. The transesterification is performed with methanol and triacylglycerides extracted from dried microbial biomass. However, the lipids transesterification can be done also via direct alcholysis of dried microbial biomass without previous lipid extraction (see Figure 1.4). Biodiesel production from microbial lipids is currently receiving an increasing amount of attention as a cost-effective, sustainable alternative.

Currently, microorganisms are referred to as sources of long-chain polyunsaturated fatty acids (PUFAs) that are being used as high-grade nutraceuticals for humans and animal consumption. In addition, microorganisms may utilize low-cost substrates like agricultural wastes or industrial by-products. Studies indicate that microbial oils could be used in green-fuels production, pharmaceutical and cosmetic applications, food additives, or biopolymer production (Bharathiraja et al., 2017). In oleaginous microorganisms, the lipid accumulation starts when the carbon source is present in excess and an element in the growth media becomes limiting. Also, studies showed that isocitrate dehydrogenase, ATP-citrate lyase, and malic enzyme are the three key enzymes in lipid accumulation (Christophe et al., 2012). Some microorganisms can produce PUFAs such as omega-3 and omega-6 series, known for their benefits on human health. PUFAs play a substantial role as precursors of eicosanoids and structural components of membrane phospholipids (Sakuradani et al., 2009). PUFAs are components of thrombocytes, neuronal and muscle cells, cerebral cortex, as well as the immunocompetent cells. The potential demand for omega-3 PUFAs (based on 500 mg/day) is approximately 1.2 million tons and the supply by fish is about 0.84 million tons, so the gap is 0.43 million tons that should be obtained from other sources. Conjugated linoleic acid that is a mixture of isomers of linoleic acid, also possess well-known health benefits such as: anticarcinogenic, anti-obesity, antidiabetic, antihypertensive, antiatherogenic, immunomodulatory, and osteosynthetic properties (Vela-Guravic et al., 2014; Beligon et al., 2016). On the other hand, the same kind of oils were investigated concerning their potential to be used as possible alternative to vegetable oils for biodiesel production avoiding competition with human food (Christophe et al., 2012). Nowadays, PUFAs production using various microbial strains is an industrial reality.

Some PUFAs, such as arachidonic acid (ARA) and docosahexaenoic acid (DHA) are now produced by various microorganisms at commercial level. These types of oils are extensively used as dietary supplements in infant formulations (Beligon et al., 2016). Recently, Paik, Sim, & Jeon (2017) showed that *C. reinhardtii* cultivated under special conditions (perfusion microfluidic chip) under a continuous feed system showed a 2.4-fold increase in triacylglyceride production compared to the level obtained under only-ammonia-depleted conditions. Also, the produced triacylglyceride that is a bio-derived neutral lipid, has been regarded as a precursor for biodiesel fuel as well as for ricinoleic acid that has healthcare properties and industrial uses (Kajikawa et al., 2016). Numerous studies investigated the potential of various microorganisms such as microalgae, bacteria, molds, or fungi to produce oils. Some of these microorganisms and their oil content is shown in Table 1.3.

SWE was applied as an alternative greener and faster method for simultaneous removal of oil and water-soluble extract from sunflower seeds (Ravber, Knez, &

TABLE 1.3
Lipid Contents in Various Microorganisms

Microorganism	Total Lipid Content (% as Dry Weight)	Microorganism	Total Lipid Content (% as Dry Weight)
Microalgae		**Bacteria**	
Tetraselmis sueica	15–23	*Bacillus alcalophilus*	18–24
Cylindrotheca sp.	16–37	*Acinetobacter calcoaceticus*	27–38
Chlorella vulgaris	16.1	*Arthrobacter sp.*	>40
Dunaliella tertiolecta	16–71	*Gordonia sp.*	72
Phaeodactylum tricornutum	18–57	*Rhodococcus opacus*	87
Isochrysis galbana	19.5	**Yeast**	
Crypthecodinium cohnii	20	*Candida curvata*	58
Monallanthus salina	>20	*Rhodosporidium toruloides*	58
Nannochloris sp.	20–35	*Cryptococcus curvatus*	58
Dunaliella primolecta	23	*Cryptococcus albidus*	65
Thalassiosira pseudonata	24.5	*Lipomyces tetrasporus*	67
Isochrysis sp.	25–33	*Lipomyces starkei*	68
Chlorella emersonii	25–63	*Rhodotorula glutinis*	72
Botryococcus braunii	25–75	*Saccharomyces cerevisiae*	84
Nannochloropsis oculata	27.2	**Fungi**	
Chlorela sp.	28–32	*Cunninghamella echinulata*	35
Neochloris oleoabundans	29–65	*Aspegillus oryzae*	57
Nannochloropsis sp.	31–68	*Cunninghamella japonica*	60
Parietochloris incisa	>35	*Mucor mucedo*	62
Monodus subterraneus	39.3	*Mortierella vinacea*	66
Nitzschia sp.	45–47	*Humicola lanuginosa*	75
Schizochytrium sp.	50–77	*Mortierella isabelina*	86

Source: Bharathiraja et al., 2017; Shi et al. 2011; Subramaniam et al. 2010.

Skerget, 2015). In this study, the results showed that optimal extraction yield of oil was obtained at 130°C after 30 minutes. However, a minimal degree of oils hydrothermal degradation was observed, the water-soluble fraction was more susceptible toward hydrothermal degradation of which the carbohydrates showed the lowest stability. Hydrolysis of ester and glycoside bonded antioxidants occurred, which produced oil with higher antioxidant capacities than that extracted using the Soxhlet method (Ravber, Knez, & Skerget, 2015). SWE was employed to extract biochemical compounds, mainly protein and total carbohydrates from *C. vulgaris*, showing that 83.5 wt% was high volatile matter and carbon content of 47.11wt%, which qualified the feedstock for biofuel production. The maximum total carbohydrate content and protein yields were of 14.2 g/100 g and 31.2 g/100 g, respectively, at 277°C, 5% of algal biomass loading, and after five minutes reaction time (Awaluddin et al., 2016). The results indicated the potential of SWE of microalgal biomass for large-scale production of biochemical compounds, such as proteins and carbohydrates that have extensive applications in the production of algae-based biofuels and other valuable products. The increasing energy demands and heavy depletion of nonrenewable fossil fuels have accentuated the development and utilization of lignocellulosic biomass as a sustainable source for renewable energy production. SWE was used to hydrolase the rice straw lignocellulose to reducing sugars (e.g., glucose and xylose) prior to fermentation (Lin et al., 2015). The best yield of 0.346 g/g rice straw was obtained at 280°C, 20MPa and a rice straw concentration of 5 wt%.

1.3.3 SWE of Hemicelluloses from Various Feedstocks

SWE extraction and fractionation of valuable components from lignocellulosic biomass has been actively studied in recent decades (Allen et al., 1996; Cantero, 2015b; Lachos-Perez et al., 2016; Mayanga-Torres et al., 2017; Prado et al., 2014). The physicochemical properties of SWE can be tuned by varying the temperature and pressure. As the temperature is increased from 100°C to the critical point, the dielectric constant, viscosity, and surface tension decrease while the diffusion rate increases (Lachos-Perez et al., 2016). As a result of the temperature dependence of the dielectric constant, water extracts polar molecules most efficiently at temperature less than about 150°C, while extraction of less polar molecules benefits from increasing the extraction temperature. Moreover, when extraction is performed at temperature above 150°C, the natural biopolymers present in biomass—namely hemicellulose and cellulose—undergo hydrolysis reactions to yield simple sugars and sugar oligomers. These valuable products can be extracted for use in fermentation and other applications (Carr, Mammucari, & Foster, 2011).

Subcritical water was successfully used for extracting the hemicelluloses from wood prior to pulping. This technology has attracted much attention in recent years, mainly because pure water is the only solvent used, hemicelluloses dissolved in the liquid phase may be recovered and converted into high-value-added products. After the extraction, the wood residue composed mostly of cellulose and lignin can be further subjected to pulping, using lesser chemicals and shorter pulping residence time than in the case of untreated wood (Kubikova et al., 1996; Yoon & van Heiningen, 2008). The pulp generally shows lower yield and reduced strength properties than pulps produced from untreated wood due to the removal of the hemicelluloses (Yoon & van Heiningen, 2008; Helmerius et al., 2010; Al-Dajani & Tschirner, 2010). In the production of high-quality paper-grade pulp, a low hemicellulose content is not desired, while it may well be favorable to produce dissolving pulps. Hot water extractions, also called auto hydrolysis, have been a common treatment often used in the pretreatment of hardwoods, in order to remove some of the lignin and hemicellulose. The higher amount of acetyl groups linked to the hemicelluloses and a good delignification efficiency, with lesser tendency for lignin condensation, make hardwoods more suitable to water extraction than softwoods (Borrega & Sixta, 2011). The particle size of the raw material plays a crucial role; this varies from industrial-size chips to fine wood sawdust. The yield of extracted products is directly correlated with the particle size of the raw material, the mass and heat transfer efficiency (e.g., large particle size, limited mass and heat transfer, result in lower yields of extraction) (Song et al., 2008). The extraction efficiency it is also correlated with the liquid to wood ratio of the process, due to solubility limitation (Chen, Lawoko, & van Heiningen, 2010). The temperatures used in hot water extractions commonly range between 130°C and 240°C. At temperatures above 200°C, hemicelluloses can be quantitatively removed while cellulose is mostly preserved (Mok & Antal, 1992). During hot water treatments, simultaneous lignin depolymerization and condensation reactions may occur. The maximum amount of lignin extracted shifts to shorter reaction times with increasing temperature before condensation reactions overcome (Bobleter 1994; Li & Gellerstedt, 2008). Hot water extractions appear to increase the reactivity of the lignin remaining in the wood residue, improving its removal in a subsequent pulping process (Lora & Wayman, 1978). Lignin condensation hinders its removal during the pulping stage (Lora & Wayman, 1978). A hot water treatment at elevated temperatures may be used to remove the hemicelluloses and some of the lignin from wood, prior to pulping to produce cellulosic pulp but the cellulose should be preserved to maximize the final pulp yield. Excessive condensation of lignin during the hot water extraction should be avoided if a high degree of delignification at the lowest cost is to be accomplished.

Sugarcane bagasse is one of the main residues of sugarcane processing, it is abundant, inexpensive, and readily available in sugarcane mills. For each ton of sugarcane processed, an estimated 140 kg of straw are generated (Oliveria et al., 2013; Vardanega, Prado, & Meireles, 2015). Brazil alone produced over 660 million tons of sugarcane in the 2015/2016 harvest, resulting in the production of 80 million tons of straw (dry basis). The straw is either burned or left on the field after mechanical harvest as soil treatment (da Silva et al., 2010). Mechanical harvesting has increased the production of sugarcane straw that is available and used for electricity generation, chemical production, or production of second-generation ethanol (Candido & Goncalves, 2016). This renewable source represents a great morphological heterogeneity, consists of fiber bundles and other structural elements like vessels, parenchyma, and epithelial cells (Rainey, Covey, & Shore, 2006). According to data published in the literature (Gnansounou, 2010; Godshall, 2005; Sasaki, Adschiri, & Arai, 2003, 2004), about 40–50% of the dry residue is cellulose, much of which is in a crystalline structure. Another 25–35% is hemicelluloses, an amorphous polymer usually composed of xylose, arabinose, galactose, glucose, and mannose. The rest of about 10–14% is mostly lignin, waxes, and other compounds. Sugarcane bagasse is usually used in direct combustion for energy production (Neureiter et al., 2002), production of chemical compounds such as furfural or hydroxymethyl-furfural (HMF) (Almazan, Gonzales, & Galvez, 2001), paper paste (Nagieb et al., 2000), extraction of phenolic compounds (Rodrigues et al., 2010) or fermentable sugars to produce ethanol (Dawson & Boopathy, 2008; Hailing & Simms-Borre, 2008). After cellulose and hemicellulose, lignin is considered to be the most abundant natural polymer present on planet Earth (Argyropolous & Menachem, 1988; Glasser et al., 1984). The global desire to utilize more green materials has generated an increased interest in the chemistry and technology of lignin extraction and derivatives production (Felipe et al. 1997; Nimz & Casten, 1986).

Subcritical water technology, having comparable potential as supercritical water technology but being less expensive (Sasaki, Adschiri, & Arai, 2003, 2004) since it requires less energy input (temperatures and pressures are lower), was successfully applied in the extractions of antioxidants (Ibanez et al., 2003). These extractions included the extraction of whitening agents and azo dyes in paper samples (de los Santos et al., 2005), the extraction of anthocyanins from red grape skin (Yu & Howard, 2005), the extraction of dioxins from contaminated soil (Hashimoto et al., 2004), as well as in biomass conversion. Subcritical water can also be used for liquefying biomass into bio-oil and other liquid fuels (Sealock et al., 1993), such as liquification of corn stalks to bio-oil (Song et al., 2004). Catalysts addition can also enhance the subcritical water reaction (Song et al., 2004).

Hydrolyzing sugarcane straw in a semicontinuous reactor at temperatures ranging from 190–260°C and pressures ranging from 9–16 MPa, glucose, xylose, arabinose, and galactose in addition to 5-hydroxymethylfurfural (HMF) and furfural as minor by-products, were extracted. However, a temperature of 200°C provided the greatest total reducing sugars yield with a minimum char formation, reaching values (32%) that indicate economic potential (Lachos-Perez et al., 2017). Table 1.4 shows

TABLE 1.4

The Composition of Sugarcane Bagasse

Composition (wt%)	Lachos-Perez et al. (2017)	del Rio et al. (2015)	Szczerbowski et al. (2014)	Rueda-Ordonez (2015)
Moisture	6.5 ± 0.2	/	/	8.4 ± 0.3
Ash	6.8 ± 0.8	4.7 ± 0.5	6.2 ± 0.2	3.9 ± 0.2
Protein	2.31 ± 0.04	/	3.7 ± 0.1	/
Acetone extractible	3.5 ± 0.1	1.4 ± 0.1	3.5 ± 0.06[a]	5.28[b]
Klason lignin	21 ± 3	17 ± 0.2	20.6 ± 0.2	21.63[c]
Acid-soluble lignin	5.7 ± 0.1	1.9 ± 0.2	0.71 ± 0.03	/
Holocellulose	73 ± 1	72.9 ± 0.7	77.7 ± 0.3	73.09[d]
a-cellulose	37 ± 3	37.9 ± 0.3	37 ± 1	39.81

[a] Organic Solvent; [b] Water & Ethanol; [c] Total Lignin; [d] Cellulose & Hemicellulose

comparison data regarding the main constituents of the sugarcane straw, the results are in good agreement between various data sets on holocellulose, α-cellulose, and Klason lignin content, although the reported values of acid-soluble lignin content varied more substantially. The subcritical water hydrolysis products were analyzed and quantified; the results are shown in Table 1.4.

The yields of each component followed similar trends with respect to pressure and temperature. Glucose was the dominant monosaccharide present in the hydrolysate originates from cellulose, followed by hemicellulose sugars (xylose; galactose, and arabinose). Decreased sugar yields observed at temperatures above 200°C may be attributable in part to sugar decomposition to secondary products such as furfural and HMF. HMF is produced via dehydration of six-carbon sugars and furfural via dehydration of five-carbon sugars resulting from hemicellulose hydrolysis (Asghari & Yoshida, 2010). In Table 1.5 it can be observed that the yields of HMF and furfural increase with temperature.

Hemicelluloses could be isolated from biomass in molecular weights above 3 kDa and can be used for multiple applications, including replacement of synthetic plastics (Svard, Brannvall, & Edlund, 2015), producing hydrogels used as drug carriers (Ye et al., 2012), and absorbing heavy metals ions from aqueous solutions (Peng et al., 2012); xylans, extracted from hemicelluloses hydrolyzed, have the potential to be used in medicine as cholesterol depressant. The predominant hemicelluloses in hardwoods are acetylated xylans whereas in the softwoods the predominant hemicelluloses are galactoglucomannans. Hot water extraction of hemicelluloses uses water at 160–240°C. Maintaining the water in the liquid phase by applying pressure, depending on the temperature and residence time, hemicelluloses can be more or less degraded into oligomers or monomers. Compared to other methods, hydrothermal pretreatments offer several advantages: no toxic or corrosive solvents are used, no special reactors are required, and no cost associated to recovery or disposal of chemicals is incurred (Gallina et al., 2018). Ten different tree species were subjected to hydrothermal hemicellulose extraction performed at 160°C, 130 psi, wood chips size of 1.25–2mm (average solid load 6.57 g/L) with

TABLE 1.5

Sugars and Their Derivatives after Subcritical Water Hydrolysis of Bagasse (g/100g)

Temperature (°C)	Ln(R0)	Glucose	Xylose	Galactose	Arabinose	HMF	Furfural
190	8	1.5 ± 0.2	1.5 ± 0.4	0.6 ± 0.3	0.78 ± 0.05	0.8 ± 0.2	1.1 ± 0.3
200	9	2.1 ± 0.4	2.3 ± 0.7	0.7 ± 0.1	1.0 ± 0.2	0.93 ± 0.2	1.9 ± 0.3
225	11	1.7 ± 0.5	1.4 ± 0.9	0.42 ± 0.001	0.8 ± 0.5	1.79 ± 0.003	2.0 ± 0.1
250	12	1.1 ± 0.1	0.5 ± 0.1	0.4 ± 0.1	0.4 ± 0.2	2.12 ± 0.01	1.6 ± 0.02
260	13	0.89 ± 0.06	0.5 ± 0.1	0.3 ± 0.2	0.3 ± 0.2	0.97 ± 0.02	1.22 ± 0.001

Source: Adapted from Lachos-Perez, 2017.

TABLE 1.6

The Composition of Various Wood Species (g/gwood Dry-Basis)

	Extractives	Cellulose	Lignin	Hemicellulose[a]
Almond	0.071 ± 0.002	0.353 ± 0.003	0.305 ± 0.004	0.261 ± 0.004
Cedar	0.052 ± 0.003	0.314 ± 0.007	0.398 ± 0.002	0.228 ± 0.006
Cherry	0.021 ± 0.006	0.430 ± 0.001	0.240 ± 0.002	0.303 ± 0.002
Elm	0.022 ± 0.002	0.541 ± 0.003	0.189 ± 0.001	0.248 ± 0.003
Eucalyptus	0.013 ± 0.001	0.461 ± 0.003	0.251 ± 0.006	0.260 ± 0.001
Linden	0.014 ± 0.004	0.420 ± 0.006	0.278 ± 0.002	0.214 ± 0.004
Maple	0.012 ± 0.002	0.299 ± 0.008	0.455 ± 0.004	0.238 ± 0.002
Plane	0.023 ± 0.001	0.340 ± 0.007	0.388 ± 0.002	0.242 ± 0.001
Walnut	0.005 ± 0.004	0.414 ± 0.008	0.330 ± 0.004	0.253 ± 0.003
Catalpa	0.002 ± 0.001	0.495 ± 0.007	0.212 ± 0.003	0.250 ± 0.000

Source: Adapted from Gallina, 2018.

[a] Gas Chromatography Determinations.

various residence times (5–80 minutes) (Gallina et al., 2018). The highest yield was obtained with eucalyptus wood 40.3wt % after 80 minutes. The results are summarized in Table 1.6.

1.3.4 SWE in Cellulose Nanocrystals Production

Environmental concerns over the past decade have strongly increased the research on nanomaterials from renewable sources. In this context, cellulose shows great potential due to its well-known structure and abundance in nature.

The ability of water ho hydrolyze polysaccharides is well known, as seen in hydrothermal processes of hemicellulose removal (Akhlaghi, Berry, & Tam, 2013). The key points for an extensive hydrolysis rate are both the presence of H_3O^+ species and the availability of water molecules (Xiang et al., 2003). Sub- and supercritical water presents lower values of Kw and, consequently, higher concentrations of ionized species (Bandura & Lvov, 2006). Thus, their use is effective for the hydrolysis reactions. Several studies have used water at high temperature and pressures to hydrolyze lignocellulosics (Cantero 2015a), gasify the biomass (Louw et al., 2014), and liquefy cellulose/hemicellulose (Cantero 2015a; Meillisa, Woo, & Chun, 2015). Using supercritical water and controlling temperature (up to 400°C) and pressure (up to 27 MPa), Cantero et al. (Cantero, 2015b) achieved the complete hydrolysis of cellulose avoiding the secondary sugar reactions. Thus, incomplete hydrolysis of cellulose could be achieved by using less severe reaction conditions. The mechanism of the hydrolysis under subcritical water conditions is not fully understood. The increase in the diffusion coefficient and reduction of the dielectric constant (polar solvents under subcritical conditions) could have a positive effect on the disruption of cellulose amorphous domains and thus promote the accessibility of water to cleave the glycosidic bonds (Adschiri et al. 2011).

Several types of nanocellulose can be prepared either by mechanical treatment and are called microfibrillated cellulose or cellulose nanofibers (Lavoine et al., 2012) or they can be prepared by chemical hydrolysis and are identified as cellulose nanocrystals or whiskers (Dufresne, 2013). Various applications of cellulose nanocrystals were investigated from the 1960s, and strongly increased with their industrialization since 2012. Among the applications: nanocomposite materials as a filter to change physiochemical and mechanical properties (Mariano, El Kissi, & Dufresne, 2014; Yang et al., 2014); in drug-delivery systems (Lin & Dufresne, 2013; Akhlaghi, Berry, & Tam, 2013); in smart materials (Dagnon et al., 2013; Chen et al., 2014); and in antimicrobial composites formulations (Azizi et al., 2013). Cellulose nanocrystals are mainly obtained by hydrolyzing the amorphous or semicrystalline regions of the cellulose that are less resistant to hydrolysis (Pirani & Hashaikeh, 2013; Li et al., 2014). Cellulose nanocrystals could be produced via cellulose hydrolysis under strong acidic conditions. The most common process uses concentrated sulfuric acid solutions, or other acids (Mariano, El Kissi, & Dufresne, 2014; Moon et al., 2011). There are several different methods to produce cellulose nanocrystals from various raw materials described in the literature (Jonoobi, 2015). Hydrolysis done using strong inorganic acid, such as sulfuric acid or hydrochloric acid for the production of cellulose nanocrystals (Habibi, Lucia, & Rojas, 2010; Moon et al., 2011) are usually sensitive to temperature due to the presence of acidic moieties at their surface (Lin & Dufresne, 2013), which limits their use in some applications. Because of their highly concentrated media, these reactions are costly and produce large amounts of waste that need treatment. Therefore, acid hydrolysis techniques become inadequate because they produce large amounts of effluents and require large amounts of water for the washing steps. This washing step is usually the main limitation in its industrialization (Chauve & Bras, 2014). Some greener alternatives have been investigated, like the use of enzyme-assisted hydrolysis (Siqueria et al., 2011), the use of ultra-sonication to improve the acid hydrolysis (Guo et al., 2016; Lu et al., 2015, 2013), and the use of ionic liquids to promote the hydrolysis (Mao et al., 2015).

Alternative methods to produce cellulose nanocrystals have also been studied. Different oxidative reagents have been used to hydrolyze cellulose for the production of cellulose nanocrystals, e.g., ammonium persulfate (Leung et al., 2011; Cheng et al., 2014), and sodium metaperiodate (Visanko et al., 2014). However, these methods produce cellulose nanocrystals that already have a surface modification and the reagents used are expensive, reactive, corrosive, and toxic. In this sense, the use of a green medium to hydrolyze cellulose would improve the production of cellulose nanocrystals.

Recently, ionic liquids have been successfully applied to the production of cellulose nanocrystals because of their ability to well solvate cellulose. In fact, Lazko et al. (2014), used 1-butly-3- methylimidazolium chloride as a solvent medium to achieve an increase in the accessibility to the hydrolytic cleavage of specific cellulose sites, decreasing the sulfuric acid consumption, when compared with traditional cellulose nanocrystals extraction. However, hydrolysis was still preformed with aqueous sulfuric acid solution, which increases complications in recovery of the expensive solvent medium, i.e., the ionic liquid (Man et al., 2011). The use of a similar ionic liquid with the hydrogen sulfate anion enabled it to act simultaneously as solvent media and as an acidic reactant. The use of ionic liquids is still limited because of their availability, toxicity, and high recovery costs. Cellulose nanocrystals can also be obtained from natural fibers by enzymatic hydrolysis after a combination of chemical pretreatment and high mechanical shearing forces (De Figueiredo et al., 2012). Different processes of enzymatic hydrolysis were described in the literature to obtain globular (Chen et al., 2012) and rod-like cellulose nanocrystal (Xu et al., 2013; Zhang et al., 2012). The viability of these reactions requires the use of both chemical and mechanical pretreatments to increase accessibility to amorphous regions and extremely long isolation times, in comparison to acid hydrolysis. In addition, the scaling up of these methods toward industrial production would imply the recovery of the enzymes, an important issue that has not yet been developed or even investigated.

Subcritical water could also induce the hydrolysis of amorphous and semicrystalline regions of cellulose. This process showed great potential for cellulose nanocrystals production because it uses water exclusively as the hydrolyzing agent, although it demands more energy due to the use of a high pressurized reactor. However, the overall process would be less expensive since it does not require many washing steps (Novo et al., 2015). Novo et al. (2015) proposed an innovative greener method, that of using pressurized water to produce cellulose nanocrystals. The use of subcritical water under the conditions of 120°C and 20.3 MPa for 60 minutes partially hydrolyzed cellulose and produced cellulose nanocrystals with a 21.9% yield. Various other reaction conditions were also used, and the yield of cellulose nanocrystals increased linearly with increasing pressure of the reaction system. A 100-mL stainless steel reactor of SFT-250 SFE/SFR System (Subcritical Fluid Technologies Inc.) operating in batch mode, was used for the experiments. The parameters were: temperature in the range of 120–200°C, pressure in the range of 8.1–20.3 MPa, for each reaction 1g of raw cellulose (dry basis) was used, and the reaction time was 60 minutes. The classical production of cellulose nanocrystals uses strong acids solutions to promote hydrolysis of amorphous regions and hemicelluloses, requires long duration washing steps, and the produced nanocrystals have low temperature resistance, limiting their larger industrialization. On the other hand, the yields observed when subcritical water was employed were lower than those observed in the classical production of cellulose nanocrystals and crystallinity indexes lower than 80% are undesired (Novo et al., 2015; Novo et al., 2016).

1.4   INDUSTRIAL APPLICATIONS

Sub- and supercritical fluids technologies have been tested in a wide variety of industrial applications such as: biofuels, food, cosmetics, pharmaceuticals, and others, and showed significant progress in recent years, especially in extractions and biopolymers (e.g., cellulose, proteins, and hemicelluloses) hydrolysis. Some of the applications are described in this chapter and generally the extraction techniques are considered to be a cost-effective and environmentally benign process; however, there are lots of applications that need to be investigated further at pilot or larger scales. A few small companies, such as Tyton Biosciences located in Virginia, developed new methods based on subcritical water technologies and built one of the first pilot-scale (4 MT/day biomass throughput) plants in the United States that was successfully tested on lignocellulosic sugars extraction or various other applications that requires partial hydrolysis. A massive level of interest was directed toward algae conversion to transportation fuels and bioproducts production. Nevertheless, the road to commercialization will be associated with energy demands to avoid the production of biofuel with higher greenhouse gas emissions than conventional fuel. Beyond technology challenges, prospective algae cultivation systems will play a key role in improving this technology.

Using subcritical water technology to produce cellulose nanocrystals could be done at a price of 0.02 $/kg. That's a theoretical value 77-fold lower than that when the classical sulfuric acid hydrolysis is used, with a price of 1.54 $/kg (Novo et al., 2015). Using subcritical water technology to produce nanocrystals showed that the yield increased linearly with increasing the pressure, therefore, the production process requires a reactor that supports high pressures but results in a greener method and lower effluent generation.

1.5 CONCLUSIONS

SWE technology is undoubtedly very promising, and this chapter summarizes the recovery efficiency of a large assortment of biological active compounds and other valuable products using this technology, especially at an industrial scale. SWE is considered to be an environmentally friendly process having the advantage of short time reaction, low cost, and no organic solvent is used, therefore, the product is free of residual solvent, which make it suitable for the production of organic products of functional foods and nutraceuticals. SWE could completely decompose the cellulose and hemicellulose parts of biomass samples in a very short reaction time compared to the organic solvent extraction methods (maceration extraction; heat reflux extraction; ultrasonic extraction; microwave-assisted extraction; etc.). The use of SWE for the generation of bioproducts from microalgal biomass is expected to be representative for future bio-extraction technologies.

REFERENCES

Abdelmoez W., Abdelfatah R. & Tayeb A. (2011) Extraction of cottonseed oil using subcritical water technology. *AIChE J.* 57(9): 2353–2359.

Abdelmoez W., Nage S.M. & Bastawess A. et al. (2014) Subcritical water technology for wheat straw hydrolysis to produce value added products. *J. Cleaner Prod.* 70: 68–77.

Adschiri T., Lee Y.W., Goto M. & Takami S. (2011) Green materials synthesis with supercritical water. *Green Chem.* 13: 1380.

Akhlaghi S.P., Berry R.C. & Tam K.C. (2013) Surface modification of cellulose nano-crystals with chitosan oligosaccharide for drug delivery applications. *Cellulose* 20: 1747–1764.

Al-Dajani W.W. & Tschirner U.W. (2010) Pre-extraction of hemicelluloses and subsequent ASA and ASAM pulping: Comparison of autohydrolysis and alkaline extraction. *Holzforschung* 64: 411.

Aliakbarian B., Fathi A., Perego P. & Dehghani F. (2012) Extraction of antioxidants from winery wastes using subcritical water. *J. Subcrit. Fluids* 65: 18–24.

Allen S.G., Kam L.C., Zeamann A.J. et al. (1996) Fractionation of sugar cane with hot compressed liquid water. *Ind. Eng. Chem. Res.* 35: 2709–2715.

Almazan O., Gonzales L. & Galvez L. (2001) The sugarcane, its by-product and co-products. *Sugar Cane Int.* 7: 3–8.

Alvarez V.H. & Saldana M.D.A. (2013) Hot pressurized fluid extraction optimization of potato peel using response surface and the taguchi method. In *III Iberoamerican Conference on Supercritical Fluids Cartagena de Indias* (pp. 1–8), Columbia.

Asghari F.S. & Yoshida H. (2006) Acid-catalyzed production of 5-hydroxymethyl furfural from d-fructose in subcritical water. *Ind. Eng. Chem. Res.* 45: 2163–2173.

Asghari F.S. & Yoshida H. (2010) Conversion of Japanese red pine wood (*Pinus densiflora*) into valuable chemicals under subcritical water conditions. *Carbohydr. Res.* 345: 124–131.

Asl A.H. & Khajenoori M. (2013) Subcritical water extraction. In *Mass Transfer – Advances in Sustainable Energy and Environment Oriented Numerical Modeling* (pp. 457–487). IntechOpen, doi:10.5772/54993.

Argyropolous D.S. & Menachem S.B. (1988) Lignin. In D.L. Kaplan (Ed.). *Biopolymers from Renewable Resources* (pp. 292–322). Berlin: Springer.

Awaluddin S.A., Thiruvenkadam S., Izhar S. et al. (2016) Subcritical water technology for enhanced extraction of biochemical compounds from *Chlorella vulgaris*. *BioMed Res. Int.* ID: 10.

Azizi S., Ahmad M., Mahdavi M. et al. (2013) Preparation, characterization and anti-microbial activities of ZnO nanoparticles/cellulose nanocrystal nanocomposites. *BioResources* 8: 1841–1851.

Bandura A.V. & Lvov S.N. (2006) The ionization constant of water over wide ranges of temperature and density. *J. Phys. Chem. Ref. Data* 35(1): 15.

Beligon V., Christophe G., Fontanille P. et al. (2016) Microbial lipids as potential source to food supplements. *Curr. Opin. Food Sci.* 7: 35–42.

Bharathiraja B., Sridharan S., Sowmya V. et al. (2017) Microbial oil – A plausible alternate resource for food and fuel. *Bioresour. Technol.* 233: 423–432.

Bobleter O. (1994) Hydrothermal degradation of polymers derived from plants. *Prog. Polym. Sci.* 19: 797.

Boots A., Haenen G. & Bast A. (2008) Health effects of quercetin: From antioxidant to nutraceutical. *Eur. J. Pharmacol.* 585: 325–337.

Borrega M. & Sixta H. (2011) Production of cellulosic pulp by subcritical water extraction followed by mild alkaline pulping. *Conference: 16th International Symposium on Wood, Fiber and Pulping Chemistry – Proceedings*, ISWFPC.

Candido R.G. & Goncalves A.R. (2016) Synthesis of cellulose acetate and carboxymethylcellulose from sugarcane straw. *Carbohydr. Polym.* 152: 679–686.

Cantero D.A., Martinez C., Bermejo M.D. et al. (2015a) Simultaneous and selective recovery of cellulose and hemicellulose fractions wheat bran by supercritical water hydrolysis. *Green Chem.* 17: 610–618.

Cantero D.A., Vaquerizo L., Mato F. et al. (2015b) Energetic approach of biomass hydrolysis in supercritical water. *Bioresour. Technol.* 179: 136–143.

Cardenas-Toro F.P., Alcatraz-Alay S.C., Forster-Carniero T. et al. (2014) Obtaining oligo- and monosaccharides from agroindustrial and agricultural residues using hydrothermal treatments. *Food Public Health* 4: 123–139.

Carr A.G., Mammucari R. & Foster N.R. (2011) A review of subcritical water as a solvent and its utilization for the processing of hydrophobic organic compounds. *Chem. Eng. J.* 172: 1–17.

Castejon N., Luna P. & Senorans J. (2017) Ultrasonic removal of mucilage for pressurized liquid extraction of omega-3 rich oil from Chia seed (*Salvia hispanica* L.). *J. Agric. Food Chem.* 65: 2572–2579.

Chakraborty M., Miao C., McDonald A. et al. (2012) Concomitant extraction of bio-oil and value added polysaccharides from *Chlorella sorokiniana* using a unique sequential hydrothermal technology. *Fuel* 95: 63–70.

Chauve G. & Bras J. (2014) Industrial point of view of nanocellulose materials and their possible applications. In *Handbook of Green Materials* (pp. 233–252). Intech: Rijeka, Croatia.

Cheigh C.I., Yoo S.Y., Ko M.J. et al. (2015) Extraction characteristics of subcritical water depending on the number of hydroxyl group in flavonols. *Food Chem.* 168: 21–26.

Chen C.Y., Zhao X.Q., Yen H.W. et al. (2013) Microalgae-based carbohydrates fir biofuel production. *Biochem. Eng. J.* 78: 1–10.

Chen Q., Liu P., Nan F. et al. (2014) Tuning the iridescence of chiral-nematic cellulose nano-crystal films with a vacuum assisted self-assembly technique. *Biomacromolecules* 15: 4343–4350.

Chen X., Deng X., Shen W. & Jiang L. (2012) Controlled enzymolysis preparation of nano-crystalline cellulose from pretreated cotton fibers. *BioResources* 7: 4237–4248.

Chen X., Lawoko M. & van Heiningen A. (2010) Kinetics and mechanism of autohydrolysis of hardwoods. *Bioresour. Technol.* 101: 7812.

Cheng M., Qin Z., Liu Y. et al. (2014) Efficient extraction of carboxylated spherical cellu-lose nanocrystals with narrow distribution through hydrolysis of lyocell fibers by using ammonium persulfate as an oxidant. *J. Mater. Chem.* 2: 251.

Christophe G., Kumar V., Nouaille R. et al. (2012) Recent developments in microbial oils production: A possible alternative to vegetable oils for biodiesel without competition with human food? *Braz. Arch. Biol. Technol.* 55(1): 29–46.

da Silva A.S.A., Inoue H., Endo T. et al. (2010) Milling pretreatment of sugarcane bagasse and straw for enzymatic hydrolysis and ethanol fermentation. *Bioresour. Technol.* 101: 7402–7409.

Dawson L. & Boopathy R. (2008) Cellulosic ethanol production from sugarcane bagasse without enzymatic saccarification. *BioResources* 3: 452–460.

De Figueiredo M.C.B., De Freitas Rosa M., Lie Ugaya C.M. et al. (2012) Life cycle assess-ment of cellulose nanowhiskers. *J. Cleaner Prod.* 35: 130–139.

de los Santos M., Batlle R., Salafranca J. & Nerin C. (2005) Subcritical water and dynamic sonication-assisted solvent extraction of fluorescent whitening agents and azo dyes in paper samples. *J. Chromatogr.* 1064: 135–141.

del Rio J.C., Lino A.G., Colodette J.L. et al. (2015) Differences in the chemical structure of the lignins from sugarcane bagasse and straw. *Biomass Bioenergy* 81: 322–338.

De-Morais M.G., Vaz B.D.S., De-Morais E.G. et al. (2015) Biologically active metabolites synthesized by microalgae. *BioMed Res. Int.* ID: 15.

de-Oliviera L.A.B., Pacheco H.P. & Scherer R. (2016) Flutriafol and pyraclostrobin residues in Brazilian green coffees. *Food Chem.* 190: 60–63.

Deng G.F., Shen C., Xu X.R. et al. (2012) Potential of fruit wastes as natural resources of bioactive compounds. *Int. J. Mol. Sci.* 13: 8308–8323.

Dragon K.L., Way A.E., Carson S.O. et al. (2013) Controlling the rate of water-induced switching in mechanically dynamic cellulose nanocrystal composites. *Macromolecules* 46: 8203–8212.

Dufresne A. (2013) Nanocellulose: A new ageless bionanomaterial. *Mater. Today* 16: 220–227.

Felipe M., Vitolo M., Mancillia I.M. et al. (1997) Environmental parameters affecting xylitol production from sugarcane bagasse hemicellulosic hydrolysate by *Candida guilliermondii. J. Basic Microbiol.* 18: 251–254.

Felipe M.G.A., Alvez L.A., Silva S.S. et al. (1996) Fermentation of eucalyptus hemicellulosic hydrolysate to xylitol by *Candida guilliermondii. Bioresour. Technol.* 56: 281–283.

Fernandez-Ponce M.T., Casas L., Mantell C. et al. (2012) Extraction of antioxidant compounds from different varieties of *Mangigfera indica* leaves using green technologies. *J. Supercrit. Fluids* 72: 168–175.

Funke A. & Ziegler F. (2010) Hydrothermal carbonization of biomass: A summary and discussion of chemical mechanisms for process engineering. *Biofuel. Bioprod. Bior.* 4: 160–177.

Gallina G., Cabeza A., Grenman H. et al. (2018) Hemicellulose extraction by hot pressurized water pretreatment at 160°C for ten different woods: Yield and molecular weight. *J. Supercrit. Fluids* 133: 716–725.

García-Marino M., Rivas-Gonzalo J.C., Ibáñez E. & García-Moreno C. (2006) Recovery of catechins and proanthocyanidins from winery by-products using subcritical water extraction. *Anal Chim Acta.* 563: 44–50.

Glasser W.G., Barnett C.A., Rials T.G. & Saraf V.P. (1984) Engineering plastics from lignin. Characterization of hydrohyalkyl lignin derivatives. *J. Appl. Polym. Sci.* 29: 1815–1930.

Gnansounou E. (2010) Production and use of lignocellulosic bioethanol in Europe: Current situation and perspectives. *Bioresour. Technol.* 101: 4842–4850.

Godshall M.A. (2005) Enhancing the agro-industrial value of the cellulosic residues of sugarcane. *Int. Sugar J.* 107: 53–60.

Guo J., Guo X., Wang S. & Yin Y. (2016) Effects of ultrasonic treatment during acid hydrolysis on the yield, particle size and structure of cellulose nanocrystals. *Carbohydr. Polym.* 135: 248–255.

Habibi Y., Lucia L.A. & Rojas O.J. (2010) Cellulose nanocrystals: Chemistry, self-assembly, and applications. *Chem. Rev.* 110: 3479–3500.

Hailing P. & Simms-Borre P. (2008) Overview of lignocellulosic feedstock conversion into ethanol – Focus on sugarcane bagasse. *Int. Sugar J.* 110: 191–194.

Hashimoto S., Watanabe K., Nose K. & Morita M. (2004) Remediation of soil contaminated with dioxins by supercritical water extraction. *Chemosphere* 54: 89–96.

Hata S., Jintana W., Maeda A. et al. (2008) Extraction of defatted rice bran by subcriticatl water treatment. *Biochem. Eng. J.* 40: 44–53.

He L., Zhang X., Xu H. et al. (2012) Subcritical water extraction of phenolic compounds from pomegranate (*Punica granatum* L.) seed residue and investigation into their antioxidant activities with HPLC-ABTS+ assay. *Food Bioprod. Process.* 90: 215–223.

Helmerius J., von Walter J.V., Rova U. et al. (2010) Impact of hemicellulose pre-extraction for bioconversion on birch Kraft pulp properties. *Bioresour. Technol.* 101: 5996.

Ibanez E., Kubatova A., Señoráns F.J. et al. (2003) Subcritical water extraction of antioxidant compounds from rosemary plants. *J. Agric. Food Chem.* 51: 375–382.

Jintana W. & Shuji A. (2008) Extraction of functional substances from agricultural products or by-products by subcritical water treatment. *Food Sci. Technol. Res.* 14: 319–328.

Jonoobi M., Oladi R., Davoudpour Y. et al. (2015) Different preparation methods and properties of nanostructured cellulose from various natural resources and residues: a review. *Cellulose* 22: 935–969.

Kajikawa M., Tatsuki A., Kentaro I. et al. (2016) Production of ricinoleic acid containing monoestolide triacylglycerides in an oleaginous diatom *Chaetoceros gracillis. Sci. Rep.* 6: 36809.

Kang S., Kim Y., Hyun K. et al. (1998a) Development of separating techniques on quercetin-related substances in onion (*Allium cepa* L.). Contents and stability of quercetin-related substances in onion. *J. Korean Soc. Food Sci. Nutr.* 27: 682–686.

Kang S., Kim Y., Hyun K. et al. (1998b) Development of separating techniques on quercetin-related substances in onion (*Allium cepa* L.). Optimal extracting conditions of quercetin-related substances in onion. *J. Korean Soc. Food Sci. Nutr.* 27: 687–692.

Khajenoori M. (2013) Subcritical water extraction of essential oils from *Matricaria chamomilla* L. *Int. J. Eng.* 26: 489–494.

Ko M.J., Cheigh C.I., Cho S.W. et al. (2011) Subcritical water extraction of flavonol quercetin from onion skin. *J. Food Eng.* 102: 327–333.

Kruse A. & Dinjus E. (2007) Hot compressed water as reaction medium and reactant: Properties and synthesis reactions. *J. Supercrit. Fluids* 39: 362–380.

Kubicova J., Zeamann A., Krkoska P. et al. (1996) Hydrothermal pretreatment of wheat straw for the production of pulp and paper. *Tappi J.* 79: 163.

Kumar S., Hablot E., Moscoso-Garcia J.L. et al. (2014) Polyurethanes preparation using proteins obtained from microalgae. *J. Mater. Sci.* 49(22): 7824–7833.

Lachos-Perez D., Martinez-Jimenez F., Rezende C.A. et al. (2016) Subcritical water hydrolysis of sugarcane bagasse: An approach on solid residues characterization. *J. Supercrit. Fluids* 108: 69–78.

Lachos-Perez D., Tompsett G.A., Guerra P. et al. (2017) Sugars and char formation on subcritical water hydrolysis of sugarcane straw. *Bioresour. Technol.* 243: 1069–1077.

Lavoine N., Desloges I., Dufresne A. et al. (2012) Microfibrillated cellulose – Its barrier properties and applications in cellulose materials: A review. *Charbohydr. Polym.* 90: 735–764.

Lazko J., Senechal T., Landercy N. et al. (2014) Well defined thermostable cellulose nanocrystals via two-step ionic liquid swelling-hydrolysis extraction. *Cellulose* 21: 4195–4207.

Leung A.C.W., Hrapovic S., Lam S. et al. (2011) Characteristics and properties of carboxylated cellulose nanocrystals prepared from a novel one-step procedure. *Small* 7: 302–305.

Levine R.B., Pinnarat T. & Savage P.E. (2010) Biodiesel production from wet algal biomass through in situ lipid hydrolysis and supercritical transesterification. *Energy Fuels* 24: 5235–5243.

Li J. & Gellerstedt G. (2008) Improved lignin properties and reactivity by modifications in the autohydolysis process of aspen wood. *Ind. Crop. Prod.* 27: 175.

Li Y., Li G., Zou Y. et al. (2014) Preparation and characterization of cellulose nanofibers from partly mercerized cotton by mixed acid hydrolysis. *Cellulose* 21: 301–309.

Lin N. & Dufresne A. (2013) Supramolecular hydrogels from in situ host – Guest inclusion between chemically modified cellulose nanocrystals and cyclodextrin. *Biomacromolecules* 14: 871–880.

Lin R., Cheng J., Lingkan D. et al. (2015) Subcritical water hydrolysis of rice straw for reducing sugar production with focus on degradation by-products and kinetic analysis. *Bioresour. Tech.* 186: 8–14.

Lora J.H. & Wayman M. (1978) Delignification of hardwoods by autohydrolysis and extraction. *Tappi J.* 61: 47.

Louw J., Schwarz C.E., Knoetze J.H. et al. (2014) Thermodynamic modeling of supercritical water gasification: Investigating the effort of biomass composition to aid in the selection of appropriate feedstock material. *Bioresour. Technol.* 174: 12–23.

Lu J., Feng X., Han Y. et al. (2014) Optimization of subcritical fluid extraction of carotenoids and chlorophyll a from (luminaria japonica aresch) by response surface methodology. *J. Sci. Food Agric.* 94: 139–145.

Lu Q., Lin W., Tang L. et al. (2015) A mechanochemical approach to manufacturing bamboo cellulose nanocrystals. *J. Mater. Sci.* 50: 611–619.

Lu Z., Fan L., Zheng H. et al. (2013) Preparation characterization and optimization of nanocellulose whiskers by simultaneously ultrasonic wave and microwave assisted. *Bioresour. Technol.* 146: 82–88.

Madhavi D.L., Despande S.S. & Salunkhe D.K. (1996) *Food Antioxidants* (p. 1). New York: Dekker.

Makris D.P. & Rossiter J.T. (2001) Domestic processing of onion bulbs (*Allium cepa* L.) and Asparagus Spears (*Asparagus officinalis*): Effect on flavonol content and antioxidant status. *J. Agric. Food Chem.* 49: 3216–3222.

Man Z., Muhammad N., Sarwono A. et al. (2011) Preparation of cellulose nanocrystals using an ionic liquid. *J. Polym. Environ.* 19: 726–731.

Mao J., Heck B., Reiter G. et al. (2015) Cellulose nanocrystals production in near theoretical yields by 1-butyl-3-methylimidazolium hydrogen sulfate – Mediated hydrolysis. *Carbohydr. Polym.* 117: 443–451.

Mariano M., El Kissi N. & Dufresne A. (2014) Cellulose nanocrystals and related nanocomposites: Review of some properties and challenges. *J. Polym. Sci. Part B Polym. Phys.* 52: 791–806.

Martino K. & Guyer D. (2004) Supercritical fluid extraction of quercetin from onion skin. *J. Food Process Eng.* 27: 17–28.

Mayanga-Torres P.C., Lachos-Perez D., Rezende C.A. et al. (2017) Valorization of coffee industry residues by subcritical water hydrolysis: Recovery of sugars and phenolic compounds. *J. Supercrit. Fluids* 120: 75–85.

Meillisa A., Woo H.C. & Chun B.S. (2015) Production of monosaccharides and bio-active compounds derived from marine polysaccharides using subcritical water hydrolysis. *Food Chem.* 171: 70–77.

Mohan M., Banerjee T. & Goud V.V. (2015) Hydrolysis of bamboo biomass by subcritical water treatment. *Bioresour. Technol.* 191: 244–252.

Mok W.S.L. & Antal Jr. M.J. (1992) Uncatalyzed solvolysis of whole biomass hemicellulose by hot compressed liquid water. *Ind. Eng. Chem. Res.* 31: 1157.

Monrad J.K., Howard L.R., King J.W. & Mauromoustakos A. (2010) Subcritical solvent extraction of anthocyanins from dried red grape pomace. *J. Agric. Food Chem.* 58: 2862–2868.

Moon R.J., Martini A., Nairn J. et al. (2011) Cellulose nanomaterials review: Structure, properties and nanocomposites. *Chem. Soc. Rev.* 40: 3941–3994.

Moscoso-Garcia J.L., Obeid W., Kumar S. et al. (2013) Flash hydrolysis of microalgae (*Scenedesmus* sp.) for protein extraction and production of biofuels intermediates. *J. Supercrit. Fluids* 82: 183–190.

Moscoso-Garcia J.L., Teymouri A. & Kumar S. (2015) Kinetics of peptides and arginine production from microalgae (*Scenedesmus* sp.) by flash hydrolysis. *Ind. Eng. Chem. Res.* 54(7): 2048–2058.

Nagieb Z.A., Abd-El-Sayed E.S., E-I-Sakhawy M. & Khalil E.M. (2000) Hydrogen peroxide alkaline pulping of bagasse. *IPPTA* 12: 23–34.

Neureiter M., Danner H., Thomasseer C. et al. (2002) Dilute-acid hydrolysis of sugarcane bagasse at varying conditions. *Appl. Biochem. Biotechnol.* 98: 49–58.

Nimz H.H. & Casten R. (1986) Chemical processing of lignocellulosics. *Hols Roh Werkst.* 44: 207–212.

Novo L.P., Bras J., Garcia A. et al. (2015) Subcritical water: A method for green production of cellulose nanocrystals. *ACS Sustain. Chem. Eng.* 3: 2839–2846.

Novo L.P., Bras J., Garcia A. et al. (2016) A study of the production of cellulose nanocrystals through subcritical water hydrolysis. *Ind. Crop. Prod.* 93: 88–95.

Oliviera F.M.V., Pinheiro I.O., Souto-Maior A.M. et al. (2013) Industrial-scale steam explosion pretreatment of sugarcane straw for enzymatic hydrolysis of cellulose for production of second generation ethanol and value-added products. *Bioresour. Technol.* 130: 168–173.

Paik S.M., Sim S.J. & Jeon N.L. (2017) Microfluidic perfusion bioreactor for optimization of microalgal lipid productivity. *Bioresour. Technol.* 233: 433–437.

Pavlovic I., Knez Z. & Skerget M. (2013) Hydrothermal reactions of agricultural and food processing wastes in sub- and supercritical water. *J. Agric. Food Chem.* 61: 8003–8025.

Peng X.W., Zhong L.X., Ren J.L. & Sun R.C. (2012) Highly effective adsorption of heavy metal ions from aqueous solutions by macroporous xylan-rich hemicelluloses-based hydrogel. *J. Agri. Food Chem.* 60: 3909–3916.

Pirani S. & Hashaikeh R. (2013) Nonocrystalline cellulose extraction process and utilization of the byproduct for biofuels production. *Carbohydr. Polym.* 93: 357–363.

Pourali O., Asghari F.S. & Yoshida H. (2010) Production of phenolic compounds from rice bran biomass under subcritical water conditions. *Chem. Eng. J.* 160: 259–266.

Prado J.M., Follegatti-Romero L.A., Forster-Carneiro T. et al. (2014) Hydrolysis of sugarcane bagasse in subcritical water. *J. Supercrit. Fluids* 86: 15–22.

Rainey T.J., Covey G. & Shore D. (2006) An analysis of Australian sugarcane regions from bagasse paper manufacture. *Int. Sugar J.* 108: 640–644.

Ravber M., Knez Z. & Skerget M. (2015) Simultaneous extraction of oil and water-soluble phase from sunflower seeds with subcritical water. *Food Chem.* 166: 316–323.

Rueda-Ordonez Y.J. & Tannous K. (2015) Isoconversional kinetic study of the thermal decomposition of sugarcane straw for thermal conversion processes. *Bioresour. Technol.* 196: 136–144.

Rodriguez R.C.L.B., Rocha G.J.M., Rodriguez D. et al. (2010) Scale-up of diluted sulfuric acid hydrolysis for producing sugarcane bagasse hemicellulosic hydrolysate (SBHH). *Bioresour. Technol.* 101: 1247–1253.

Ruen-ngam D., Quitain A.T. & Tanaka M. (2012) Reaction kinetics of hydrothermal hydrolysis of hesperidin into more valuable compounds under supercritical carbon dioxide conditions. *J. Supercrit. Fluids* 66: 215–220.

Sakuradani E., Ando E., Ogawa A. et al. (2009) Improved production of various polyunsaturated fatty acids through filamentous fungus *Mortierellq alpine* breeding. *Appl. Microbiol. Biotechnol.* 84(1): 1–10.

Sanguansri L. & Augustin A.M. (2010) Microencapsulation in functional food product development. In J. Smith & E. Charter (Eds.). *Functional Food Product Development* (Vol. 1, pp. 1–23). Oxford: Wiley-Blackwell.

Sasaki M., Adschiri T. & Arai K. (2003) Fractionation of sugarcane bagasse by hydrothermal treatment. *Bioresour. Technol.* 86: 301–304.

Sasaki M., Adschiri T. & Arai K. (2004) Kinetics of cellulose conversion at 25MPa in sub- and supercritical water. *AIChE J.* 50: 192–202.

Sealock L.J., Eliot D.C., Baker E.G. et al. (1993) Chemical processing in high pressure aqueous environments. Historical perspective and continuing developments. *Ind. Eng. Chem. Res.* 32: 1535–1541.

Sevilla M. & Fuertes A.B. (2009) The production of carbon materials by hydrothermal carbonization of cellulose. *Carbon* 47: 2281–2289.

Shi S., Valle-Rodriguez J.O., Siewers V. et al. (2011) Prospects for microbial biodiesel production. *Biotechnol. J.* 6: 277–285.

Silva R., Haraguchi S., Muniz E. et al. (2009) Aplicacoes de fibras lignocelulosicas na quimica de polimeros e em compositos. *Quim. Nova* 32: 661–671.

Singh P.P. & Saldana M.D.A. (2011) Subcritical water extraction of phenolic compounds from potato peel. *Food Res. Int.* 44: 2452–2458.

Siqueira G., Tapin-Lingua S., Bras J. et al. (2011) Mechanical properties of natural rubber nanocomposites reinforced with cellulosic nanoparticles obtained from combined mechanical shearing and enzymatic and acid hydrolysis of sisal fibers. *Cellulose* 18: 57–65.

Smith R.M. (2002) Extractions with superheated water. *J. Chromatogr.* 975: 31–46.

Song C., Hu H., Zhu S. et al. (2004) Nonisothermal catalytic liquefaction of corn stalk in subcritical and supercritical water. *Energy Fuels* 18: 90–96.

Song T., Pranovich A., Sumerskiy I. et al. (2008) Extraction of galactoglucomannan from spruce wood with pressurized hot water. *Holzforschung* 62: 659.

Svard A., Brannvall E. & Edlund U. (2015) Rapeseed straw as a renewable source of hemicelluloses: Extraction, characterization and film formation. *Carbohydr. Polym.* 133: 179–186.

Subramaniam R., Dufreche S., Zappi M. et al. (2010) Microbial lipids from renewable resources: production and characterization. *J. Ind. Microbiol. Biotechnol.* 37: 1271–1287.

Szczerbowski D., Pitarelo A.P., Zandona F.A. et al. (2014) Sugarcane biomass for biorefineries: comparative composition of carbohydrate and non-carbohydrate components of bagasse and straw. *Carbohydr. Polym.* 114: 95–101.

Toor S.S., Rosendahl L. & Rudolf A. (2011) Hydrothermal liquefaction of biomass: A review of subcritical water technologies. *Energy* 36: 2328–2342.

Tunchaiyaphum S., Eshtiaghi M.N. & Yoswathana N. (2013) Extraction of bioactive compounds from mango peels using green technology. *Int. J. Chem. Eng. Appl.* 4: 194–198.

Vardanega R., Prado J.M. & Meireles M.A.A. (2015) Adding value to agri-food residues by means of supercritical technology. *J. Supercrit. Fluids* 96: 217–227.

Vela Gurovic M.S., Gentili A.R., Oliviera N.L. et al. (2014) Lactic acid bacteria isolated from fish gut produce conjugated linoleic acid without the addition of exogenous substrate. *Process Biochem.* 49: 1071–1077.

Venkata M.S., Nikhil G.N., Chiranjeevi P. et al. (2016) Waste biorefinery models towards sustainable circular bioeconomy: Critical review and future perspectives. *Bioresour. Technol.* 215: 2–12.

Visanko M., Liimatainen H., Sirvio J.A. et al. (2014) Amphiphilic cellulose nanocrystals from acid-free oxidative treatment: Physicochemical characteristics and use as an oil-water stabilizer. *Biomacromolecules* 15: 2769–2775.

Xiang Q., Lee Y.Y., Peterson P.O. et al. (2003) Heterogeneous aspects of acid hydrolysis of alpha-cellulose. *Appl. Biotechnol.* 107: 505–514.

Xiao S., Xi X., Tang F. et al. (2017) Subcritical water extraction of ursolic acid from *Hedyotis diffusa*. *Appl. Sci.* 7: 187–200.

Yamamoto M., Iakovlev M., Bankar S. et al. (2014) Enzymatic hydrolysis of hardwood and softwood harvest residue fibers released by sulfur dioxide-ethanol-water fractionation. *Bioresour. Technol.* 167: 530–538.

Yang J., Han C.R., Zhang X.M. et al. (2014) Cellulose nanocrystals mechanical reinforcement in composite hydrogels with multiple cross-links: Correlations between dissipation properties and deformation mechanisms. *Macromolecules* 47: 4077–4086.

Yang L., Xu F., Ge X. et al. (2015) Challenges and strategies for solid-state anaerobic digestion of lignocellulosic biomass. *Renew. Sustain. Energy Rev.* 44: 824–834.

Ye Q., Sun X.F., Jing Z.X. et al. (2012) Preparation of pH-sensitive hydrogels based on hemicellulose and its drug release property. *Xiandai Huagong/Mod. Chem. Ind.* 32: 62–66.

Yoon S.H. & van Heiningen A. (2008) Kraft pulping and papermaking properties of hot-water pre-extracted loblolly pine in an integrated forest products biorefinery. *Tappi J.* 7: 22.

Yoshida H., Izhar S., Nishio E. et al. (2014) Recovery of indium from TFT and CF glasses in LCD panel wastes using subcritical water. *Sol. Energy Mater. Sol. Cells* 125: 14–19.

Yu Z.Y. & Howard L.R. (2005) Subcritical water and sulfured water extraction of anthocyanins and other phenolics from dried red grape skin. *J. Food Sci.* 70: 270–276.

Zhang Y., Lu X.B., Gao C. et al. (2012) Preparation and characterization of nano crystalline cellulose from Bamboo fibers by controlled cellulose hydrolysis. *J. Fiber Bioeng. Informat.* 5: 263–271.

Zhao Y., Lu W.J., Wang H.T. et al. (2009) Fermentable hexose production from corn stalks and wheat straw with combined supercritical and subcritical hydrothermal technology. *Bioresour. Technol.* 100(23): 5884–5889.

Zhu G., Xiao Z., Zhu X. et al. (2013) Reducing sugars production from sugarcane bagasse wastes by hydrolysis in sub-critical water. *Clean Technol. Environ. Policy* 15: 55–61.

2 Hydrothermal Pretreatment of Lignocellulosic Biomass

Eleazer P. Resurreccion and Sandeep Kumar

CONTENTS

2.1 INTRODUCTION

Lignocellulosic biomass consists of cellulose, hemicellulose, lignin, and other minor components. Cellulose (e.g., glucose) and hemicellulose (e.g., arabinose, xylose) are polymers of sugars that can be converted into sugar products. Cellulose is more rigid than hemicellulose and thus requires rigorous pretreatment. Hemicellulose is easily hydrolyzable using mild acid or base. Lignin is a polymer of aromatic alcohol (e.g., sinapyl alcohol, coumaryl alcohol). It is recalcitrant and is located in the exterior of the biomass. Initial biomass processing consists of size and moisture reduction. Thereafter, processing of lignocellulose is achieved via two subsequent routes.

i. Pretreatment of lignocellulose to expose cellulose fraction for further processing. This route may be accompanied by hemicellulose hydrolysis and separation of the biomass into two fractions, essentially increasing the

biomass' surface area and penetrability to chemicals and/or enzymes: liquid hydrolyzate (hemicellulose) and solid residue (pretreated biomass).
ii. Enzymatic hydrolysis of pretreated biomass to recover monomeric sugars.

Pretreatment of lignocellulosic natural fibers is a critical step in the cellulosic biomass-to-biofuel conversion as it has the ability to increase yield and decrease process costs. Pretreatment is dependent upon chemical, structural, or physiochemical factors (Sasmal and Mohanty, 2018). The objectives of pretreatment in lignocellulosic energy production are as follows (Sasmal and Mohanty, 2018; De Jong and van Ommen, 2015).

    i. Minimization of energy use while increasing fuel energy density
    ii. Generation of reactive cellulose fibers suitable for producing simple sugars (glucose, xylose, etc.)
    iii. Maintenance of biomass' hemicellulose content
    iv. Reduction in byproducts formation which lowers sugar yield and enhances the creation of inhibitory chemicals (furfural)
    v. Enhancement in the use of low cost, less hazardous chemicals in downstream processing
    vi. Decrease in size reduction
    vii. Reduction in moisture content and ash formation
    viii. Improvement in feedstock storage ability

Challenges associated with pretreatment technologies include high cost from high energy and/or chemicals use (Harmsen et al., 2010), byproducts formation (e.g., phenols from lignin, furans from cellulose, and acetic acid from hemicellulose) that inhibit ethanol formation, low effectiveness of enzymatic cellulosic hydrolysis due to incomplete lignin-cellulose separation, and significant waste formation. The search for an economical and environmentally sustainable pretreatment approach is crucial in the widespread lignocellulose-based biorefinery. This chapter focuses in detail on the four types of pretreatment routes: physical (mechanical), chemical, physicochemical, and biological. It also includes initial biomass processing consisting of harvesting, transport, and storage. Several recent review articles have provided significant knowledge on pretreatment technologies (Alvira et al., 2010; Carvalheiro et al., 2008; Hendriks and Zeeman, 2009; Karimi and Taherzadeh, 2016).

2.2 HARVESTING AND TRANSPORT

There are two linear polymers (cellulose and hemicellulose) and one nonlinear polymer (lignin) in a lignocellulose making it the most abundant feedstock for biofuels production on Earth (Watanabe and Tokuda, 2001). It also contains other components such as pectin, protein, and ash. Lignocellulose is very hard and resistant to degradation. Lignocellulose is ideal for carbon sequestration because it incorporates biogenic carbon in its chemical composition. There is a wide variety in the composition of lignocellulose depending upon the origin and type of biomass involved, but generally, the weight average composition are as follows: 35–40% cellulose,

20–40% hemicellulose, and 5–30% lignin (Lynd et al., 2002; Kumar and Gupta, 2009). Table 2.1 shows typical variations of cellulose, hemicellulose, and lignin compositions among various lignocellulosic biomass.

The harvesting method for lignocellulosic feedstock will depend on biomass type. Each biomass has different lower heating value and will be harvested at unique energy input. Harvesting parameters such as time, transportation mode to biorefineries, and travel distance will also vary. The following section discusses the various harvesting techniques for each biomass identified in Table 2.1.

Wood: Wood is harvested by loggers using chainsaw and other special equipment. Delimbing is a special process of cutting the branches once a tree is felled. Branches left in the woods are called *slash* and they protect new tree seedlings from deer that are browsing for food. As slash breaks down through decay over a few years, nutrients return to the soil. The trees are then pulled (*skidded*) through the harvest area to an open place called the landing done by *forwarders*. At times, *cable haulers* are used to haul extracted wood from mountainous or hilly landscapes. Rubber-tired skidders, bulldozers, tractors, or horses are used to move trees. Trees are then cut (*bucked*) into smaller pieces (*section*) at the landing. Effort must be made to protect forest soils and to control the direction and amount of water flowing into the roads, landings, and skid trails as excessive water can cause topsoil wash. The logs are loaded onto trucks and delivered to biorefineries for further processing. If the logs are to be sent out to pulp and paper milling, removal of the bark is done (*debarking*). The harvested wood is then compressed, bundled, and tied together (*baled*) to enable energy dense wood transportation. In perspective, one ton of green wood chips equals 8.6 million BTU, which is roughly equal to 661 kWh electricity, 40 gallons gasoline/diesel, or 8,000 ft^3 of natural gas (Bergman et al., 2010).

Corn: Harvest of corn generally consists of cutting, gathering, densifying, and transporting from the field to the field-side storage (Hess et al., 2009). A typical corn harvesting method is the Conventional Bale Stover system which includes a two-step process of harvesting the grain and collecting the residue (stover). The grain stover is cut, conditioned, and windrowed for baling. A flail shredder is used to condition

TABLE 2.1

Overview of the Cellulose, Hemicellulose, and Lignin Components of Different Lignocellulosic Biomass

Biomass	Cellulose	Hemicellulose	Lignin	Reference
Hardwoods	40–55	24–40	18–25	Bajpai, 2016, Pettersen, 1984
Softwoods	45–50	25–35	25–35	Bajpai, 2016, Pettersen, 1984
Corn cobs	45	35	15	Bajpai, 2016
Bagasse	42	25	20	Kim and Day, 2011
Wheat straws	29–35	26–32	16–21	McKendry, 2002
Rice straws	32.1	24	18	Prassad et al., 2007
Grasses	25–40	35–50	10–30	Swart et al., 2008
Switchgrass	45	31.4	12	Bajpai, 2016

the stover, at which point the moisture is reduced from approximately 50 wt% to ~12 wt% in the windrow (Hoskinson et al., 2007). The bulk DM density prior to bailing is approximately 0.8–0.9 lb/ft³ (1.04 ton/1,000 windrow-ft), where windrow size is based on a 15-ft swath 9 yield/acre (windrow bulk density is estimated at 10% of bale bulk density; however, biomass material size and weathering can greatly influence windrow volume). After bailing, the bulk DM density becomes 9 lb/ft³ (Shinners et al., 2012). The bales are then collected and distributed to bale collection points at the sides of the road, often referred to as "roadsiding." They are then loaded to trucks for transportation to biorefineries.

Bagasse: Bagasse is the fibrous material remaining after removing the juice from the sugarcane delivered to the mill. The largest sugarcane producing countries are, measured in millions of tons: Brazil (721), India (347), China (123), and Thailand (Statista, 2019). The sugarcane industry typically produces 140 kg of bagasse for every ton of sugarcane processed (Bezerra and Ragausks, 2016). The sugarcane bagasse consists of cellulose/glucan, a linear polysaccharide made up by ß-1,4-D-glucose unities; hemicellulose, a heterogeneous polysaccharide composed of differentiated amounts of hexose and pentose sugars; and lignin, an aromatic macromolecule, which gives a high complexity and recalcitrance to the lignocellulosic structure (Kim and Day, 2011). Sugar cane grows for 12 to 16 months and are harvested between June and December each year. Upon harvest, the cane stands 2–4 meters high. There are two methods used to harvest cane. First, the leftover cuttings form a mulch which keeps in moisture, stops the growth of weeds, and helps prevent soil erosion. Second, the sugar cane is burnt to remove leaves, weeds, and other matter which can make harvesting and milling operations difficult. In both processes, the harvester moves along the rows of sugar cane removing the leafy tops of the cane stalks, cutting the stalks off at ground level and chopping the cane into small lengths called *billets*. These are loaded into a haul-vehicle traveling alongside the harvester. The cane is then delivered to the appropriate biorefineries. The stubble left behind grows new shoots, producing a "ratoon" crop. Two or three ratoon crops can be grown before the land is rested or planted with an alternative crop such as legumes, ploughed, and replanted for the cycle to start again.

Straw: The harvest of wheat, rice, and other straw biomass uses a machine called a combine. A combine performs reaping, threshing, and winnowing the grains in one single machine. The grain is harvested in a three-step process (McKendry, 2002; Prassad et al., 2007). The first step is *reaping,* which means cutting the crops down. The second step is *threshing.* This process separates the edible part of the grain from the inedible chaff that surrounds it. This was originally done by beating the grain on the threshing floor, though later it was done by shaking the crops in a large bin so the chaff could fall through to a lower container. The threshing process was originally the most time-consuming part of the harvest—it took one hour to thresh just a bushel of wheat. The last step in the harvest process is *winnowing.* While threshing caused the grain to separate from its surrounding chaff, the two were still all mixed together on the threshing floor. The grain is winnowed up in the air so the lighter chaff floats away while the heavier grain falls to the ground. Similar to corn, the straws are picked up using mechanical balers. Bales are transported because its transportation is much cheaper than raw biomass (Van Loo and Koppejan, 2004).

Grass: Grasses are harvested with traditional hay swathers and balers, producing either round or square bales. Large square bales are preferred over round bales because they are easier to manage for transportation and long-term storage. Baled grass can be stored outside in dry areas without any protection. Significant loss in dry matter is to be expected if the baled grass is left in areas with high rainfall (> 30 inches). The ideal storage mechanism for harvested grass is through barns because these structures reduce biomass losses but increase overall production costs. Prior to storage in barns, the harvested grass is subjected to quick drying by exposing the grass to airflow and solar irradiation in a process known as *tedding*. It is important to reduce moisture from grasses as water leads to auto-combustion during storage. Moisture also induces grass decay that produces a foul smell. Ideally, grass moisture must be maintained between 30 and 50 wt%. The quality and yield of the harvested grass is critical for efficient biofuel production. If grass is to be used as feedstock for fermentation and gasification processes, it is beneficial to harvest during the fall season due to high biomass yields. In the event grass will be fed as pellets via direct combustion, grass harvested in spring is favorable due to its lower mineral concentration caused by winter leaching. For both fall and spring grasses, a 15 wt% moisture is ideal to: (1) ensure high-quality feedstock for processing in a biorefinery, and (2) facilitate easy baling and transportation. For switchgrass co-fired in coal-fired power plants, typical moisture content is 12–13 wt% (Wu, 2019). In the case of hay, moisture a little above 13 wt% minimizes handling losses during swathing and baling.

2.3 STORAGE

Storage allows for continuity in feedstock supply to the biorefinery without putting strain in the cultivation and harvesting process. Storage is not considered as pretreatment, but it maintains the quality of feedstock prior to processing. Storage conditions directly affect biomass' moisture content, dry matter content, and heating value. These physical properties dictate the type of pretreatment technology to be applied to the biomass (Maciejewska et al., 2008). Among these three characteristics, the most consequential is moisture content because it is associated with safety and environmental aspects such as loss of amount of dry matter in the biomass (Hunder 2005), increase in greenhouse gas emissions, particularly CH_4 and N_2O (Wihersaari, 2005), and changes in energy content as a result of microbiological action which stimulates self-heating. It is advisable to store biomass at an average moisture of 20 wt% in an airtight vessel/container to avoid further loss due to degradation. If possible, unreduced-sized biomass is recommended for storage because it promotes airflow and ventilation (Maciejewska et al., 2008). In a previous study, wood chips stored at 40 wt% moisture generates 58 kg CO_2-eq/MWh fuel while the same wood chips stored at 60 wt% moisture increases CH_4 and N_2O emissions by 2.5 times (144 kg CO_2-eq/MWh fuel) (Wihersaari, 2005). There are two types of storage mechanism: outdoor and indoor. Outdoor storage entails placing the biomass outside, below some sort of shed, with or without a plastic cover. Indoor storage uses silos, barns, or bunkers depending on packing biomass morphology. Typically, pre-compaction (baling or pelletizing) is applied prior to storage to cut costs. According to Kaltschmitt et al. (2001), the risks associated with biomass storage are: (1) self-heating and ignition,

(2) bad odor, (3) fungi growth, (4) moisture reabsorption, (5) material loss due to bacterial activity, (6) particle coagulation, and (7) explosion potential (Kaltschmitt et al., 2001).

2.4 PHYSICAL PRETREATMENT

2.4.1 MILLING

Milling reduces the size of the biomass (comminution) which makes it susceptible to enzymatic hydrolysis. Previously, manual milling (mortar and pestle) was used but now, different milling methods are used: ball milling, vibratory ball milling, knife milling, hammer milling, two-roll milling, wet-disk milling, jet milling, and centrifugal milling (Taherzadeh and Karimi, 2008). A parameter called the Hardgrove grindability index (HGI) is an empirically determined test used to characterize the pulverization property of coal. HGI is also applied to lignocellulosic biomass. A 50 cm^3 sample biomass is grinded in a milling machine to 0.6–1.18 mm at 60 revolutions (Bridgeman et al., 2010). HGI is determined as the mass of the biomass that passes through a 75 μm sieve, as calibrated using coal results of known HGI. Decreasing biomass particle size enhances chemical and biochemical depolymerization and increases the bulk density for storage and transportation. Milling energy decreases with increasing particle size (Spliethoff, 2010) and can be expressed using Equation 2.1.

$$E = C \int_{1}^{2} \frac{dL}{L^n}$$
(2.1)

where E is the specific energy consumption (kJ·kg^{-1}), C is a constant, dL is the differential size (dimensionless), and L is the screen opening size (mm).

2.4.2 IRRADIATION

Irradiation applies a uniform heat to lignocellulosic biomass at a very short time (usually in minutes) (Li et al., 2016). The method can be applied to a wide range of mediums such as glycerol, wherein 11.3% and 5.4% degradation of xylan and lignin, respectively is achieved in a five-minute treatment time (Motevali et al., 2014). In another research, 1-butyl-3-methylimidazolium cations and various anions containing ionic liquid were subjected to microwave irradiation for two–five seconds (Swatloski et al., 2002) resulting in a 25% cellulose degradation. Irradiation results in higher sugar yield with no inhibitory compounds but needs special design of equipment and process.

2.5 HYDROTHERMAL PRETREATMENT

Lignocellulose biomass, which mainly contains cellulose, hemicelluloses, and lignin biopolymers, can be used as fuel by direct combustion or by first gasifying and then burning the gas. There is a great deal of interest in converting these resources to bioethanol and other chemicals. The biochemical pathways which can be realized at very moderate process conditions using cellulase enzyme to convert holocelluloses to fermentable sugars are the most promising ones for large-scale bioethanol production. But the efficiency of this technology is limited due to the complex chemical structure of lignocellulose biomass and the inaccessibility of β-glycosidic linkages to cellulase enzymes because of the low surface area and small size of pores in multicomponent structure. Hence, pretreatment is nowadays viewed as a critical step in lignocellulose processing. Pretreatment alters both the structural barrier (removal of lignin and hemicelluloses) and physical barrier (surface area, crystallinity, pore size distribution, degree of polymerization) which help in improving the accessibility of enzyme for hydrolysis (Mosier et al., 2005; Laxman and Lachke, 2008; Gupta and Lee, 2008). The pretreatment enhances the rate of production and the yield of monomeric sugars from biomass. But the pretreatment is among the costliest step in the bioethanol conversion process as it may account for up to 40% of the processing cost. Moreover, it also affects the cost of upstream and downstream processes (Zhang et al., 2009; Wyman et al., 2005; Lynd, 1996; Pérez et al., 2008). Hence, an efficient, less energy intensive and cost-effective pretreatment method is a necessity for producing ethanol at an economically viable cost. Different pretreatment methods are broadly classified into physical, chemical, physiochemical, and biological processes (i.e., conventional pretreatment). Hydrothermal pretreatment is unlike conventional pretreatment methods. The latter utilize acids or alkalis which have serious economic and environmental constraints due to the heavy use of chemicals and chemical resistant materials (Hsu, 1996; Sun and Cheng, 2002; Diaz et al., 2010).

Hydrothermal pretreatment employing subcritical water has attracted much attention because of its suitability as a non-toxic, environmentally benign and inexpensive media for chemical reactions (Kumar et al., 2009b; Matsumura et al., 2006). One of the most important benefits of using water instead of acid as pretreatment media is that there is no need for an acid recovery process and related solid disposal and handling cost (Diaz et al., 2010; Kadam et al., 2009). Liquid water, below the critical point is conducive for conducting ionic reactions. Ambient water is polar, has extensive network of H-bonding, and does not solubilize most organics. As water is heated, the H-bonds start weakening, allowing dissociation of water into acidic hydronium ions (H_3O^+) and basic hydroxide ions (OH^-). Below the critical point, the ionization constant of water increases with temperature and is about three orders of magnitude higher than that of ambient water. Also, the dielectric constant of water decreases with temperature. A low dielectric constant allows liquid water to dissolve organic compounds, while a high ionization constant provides an acidic medium for the hydrolysis of biomass components via the cleavage of ether and ester bonds and favor the hydrolysis of hemicelluloses (Franck, 1987; Miyoshia et al., 2004; Savage, 1999; Kumar and Gupta, 2009). The structural alterations due to the removal of hemicelluloses increase the accessibility and enzymatic hydrolysis of cellulose. Enzyme accessibility is increased as a result of the increase in mean pore size of the substrate which enhances the probability of the hydrolysis of glycosidic linkage (Hendriks and Zeeman, 2009).

Hydrothermal pretreatment is typically conducted in the range of 150–220°C. The temperature range, aiming for the fractionation of hemicelluloses, are decided since at temperature below 100°C, less/small extent of hydrolytic reaction is observed; whereas cellulose hydrolysis and degradation become significant above 210°C (Heitz et al., 1986; Garrote et al., 1999). Ether bonds of the hemicelluloses are most susceptible to breakage by the hydronium ions. Depending on the operational conditions, hemicelluloses are depolymerized to oligosaccharides and monomers, and the xylose recovery from biomass can be as high as 88–98%. For example, Suryawati et al., has reported 90% removal of hemicelluloses from Kanlow switchgrass at 200°C (Suryawati et al., 2008). Acetic acid is also generated from the splitting of thermally labile acetyl groups of hemicelluloses. In further reactions, the hydronium ions generated from the autoionization of acetic acid also acts as catalyst and promotes the degradation of solubilized sugars. In fact, the formations of hydronium ions from acetic acid is much more than from water (Heitz et al., 1986; Garrote et al., 1999).

The low pH (< 3) of the medium causes the precipitation of solubilized lignin and catalyzes the degradation of hemicelluloses. To avoid the formation of inhibitors, the pH should be kept between 4 and 7 during the pretreatment. This pH range minimizes the formation of monosaccharides, and therefore the formation of degradation products that can further catalyze hydrolysis of the cellulosic material during pretreatment (Laxman and Lachke, 2008; Hendriks and Zeeman, 2009; Bobleter, 1994; Sierra et al., 2008; Kohlmann et al., 1995; Mosier et al., 2005a; Weil et al., 1997). Maintaining the pH near neutral (5–7) helps in avoiding the formation of fermentation inhibitors during the pretreatment.

In general, the concentrations of solubilized products are lower in hydrothermal pretreatment compared to the steam pretreatment (Bobleter, 1994). Since the hot compressed water is used instead of steam, the latent heat of evaporation is saved which makes it easier to apply for a continuous process (Kobayashi et al., 2009). Earlier, Yang and Wyman have reported that flow through process fractionated more hemicelluloses and lignin from corn stover as compared to batch system under the conditions of similar severity (Yang and Wyman, 2004). In a flow through system, the product is continuously removed from the reactor which reduces the risk on condensation and precipitation of lignin components, making the biomass less digestible. The soluble lignin is very reactive at the pretreatment temperature and if not removed rapidly part of these compounds recondense and precipitate on the biomass (Liu and Wyman, 2003).

2.5.1 CELLULOSE PRETREATMENT IN SUBCRITICAL WATER

Biomass pretreatment with acid or alkali followed by enzymatic saccharification is a conventional method used to produce fermentable sugars for ethanol production. However, this treatment is associated with the serious economic and environmental constraints due to the heavy use of chemicals (Hsu, 1996; Sun and Cheng, 2002). Supercritical (>374°C, >22.1 MPa) and subcritical water have attracted much attention because of their suitability as a nontoxic, environmentally benign and inexpensive media for chemical reactions. Supercritical water technology provides a novel method to quickly convert cellulose to sugar and to conduct tunable reactions for the synthesis of specialty chemicals from biomass (Matsumura et al., 2006). In the subcritical region, the ionization constant (K_w) of water increases with temperature and is about three orders of magnitude higher than that of ambient water. On the other hand, the dielectric constant (ε) of water drops from 80 to 20, which help solubilize the organic compounds in subcritical water. Therefore, subcritical water potentially provides an acidic medium for the hydrolysis of cellulose (Franck, 1987; Miyoshia et al., 2004; Savage, 1999).

Experimental studies conducted on cellulose hydrolysis in subcritical and supercritical (320–400°C; 0.05–10.0 seconds) showed that cellulose can be effectively converted to hydrolysis products (i.e., oligosaccharides, cellobiose, monosaccharides) and aqueous decomposition products of glucose (i.e., levoglucosan, 5-hydroxymethyl furfural, erythrose, glycolaldehyde, dihydroxyacetone) in a fraction of a second within a flow type reactor (Ehara and Saka, 2009; Ehara and Saka, 2005; Matsumura et al., 2005; Sasaki et al., 1996; Sasaki et al., 1998; Sasaki et al., 2000). The hydrolysate (water-soluble portion) and precipitate obtained by supercritical water treatment were treated with alkali or wood charcoal for reducing inhibitory effects of the various decomposition compounds of cellulose on enzyme activity and fermentation. High efficiency of ethanol production was achieved, and the studies concluded that supercritical water treatment could be a promising pretreatment for ethanol production from lignocellulosic biomass (Hisashi et al., 2005; Shiro and Hisashi, 2007; Nakata et al., 2006).

Hydrolysis of cellulose in supercritical water is very sensitive to residence time. A high residence time enhances the formation of organic acids such as acetic acid, formic acid, and lactic acid. Formation of acids makes the reaction medium more acidic, which is conducive for further degradation of hydrolysis products. After hydrolysis of cellulose in water at 320°C and 25 MPa for 9.9 s, more than half of the cellulose was converted to organic acids (Sasaki et al., 2000). The solid cellulose-like residues have been inevitably observed in all the studies because of the rapid change in the polarity of water in going from reaction condition to room temperature. These residues have been reported to have less viscosity-average degree of polymerization (DP_v) with no significant change in crystallinity as compared to untreated cellulose.

Supercritical water treatment requires an additional process of alkaline or wood charcoal treatment for reducing the inhibitory effect of degradation products present in aqueous solution. To avoid the degradation products, the use of hot compressed (subcritical) water as an effective pretreatment medium for biomass is used. However, a clear fundamental understanding is lacking. The early fundamental work in this area was performed by Bobleter and coworkers in which they established that hemicelluloses could be completely solubilized from the lignocellulosic biomass, together with lignin (Bobleter, 1994).

2.5.2 State of Research on Hydrothermal Pretreatment

In a study by Kumar et al., switchgrass and corn stover were hydrothermally pretreated in a flow through reactor to enhance and optimize the enzymatic digestibility (Kumar et al., 2011). Figure 2.1 shows the percentage removal (oven dry basis) of hemicelluloses and lignin from switchgrass upon pretreatment 3.4 MPa and 20 minutes of steady operation time (run 1, 4, and 6) and varying temperatures. In Figure 2.2, the 72-h glucan digestibility for pretreated switchgrass is presented.

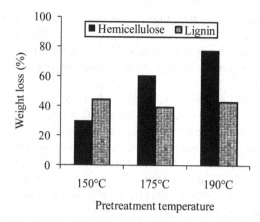

FIGURE 2.1 Percentage removal of hemicelluloses and lignin with temperature for the switchgrass pretreated at 3.4 MPa and 20-minute steady operation time. (From Kumar et al., 2011.)

The untreated switchgrass has almost negligible enzymatic reactivity. Results from this study indicate that more than 80% of glucan digestibility was achieved by pretreatment at 190°C. Lower operating temperature and enhanced pretreatment rate are achieved using a small amount of K_2CO_3 (0.45–0.9 wt%). Switchgrass pretreated in both water at 190°C and K_2CO_3 showed similar digestibility, although switchgrass pretreated at 190°C only with water had higher internal surface area than that pretreated in the presence of K_2CO_3. Comparatively, corn stover required milder pretreatment conditions than switchgrass. A high value-added product known as carbon microspheres are generated via the hydrothermal carbonization of the liquid hydrolysate.

To expand research on cellulose pretreatment in subcritical water, Kumar et al. carried out a fundamental study on how subcritical water conditions affect cellulose structure and digestibility (Kumar et al., 2009b). Microcrystalline cellulose (MCC) was pretreated with subcritical water in a continuous flow reactor for enhancing its

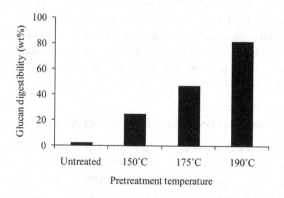

FIGURE 2.2 Glucan digestibility (72 hours) of hydrothermally pretreated switchgrass. (From Kumar et al., 2011.)

enzymatic reactivity with cellulase enzyme. Cellulose/water suspension was mixed with subcritical (i.e., pressurized and heated) water and then fed into the reactor maintained at a constant temperature and pressure. Results revealed that the DP_v of cellulose steadily decreased with increase in the pretreatment temperature, with a rapid drop occurring above 300°C. XRD analysis did not show any decrease in crystallinity upon pretreatment but, partial transformation of cellulose I to cellulose II structure was noticed in the MCC treated at ≥300 °C. Enzymatic reactivity was increased after the treatment at ≥300 °C.

In the hydrothermal pretreatment approach, lignocellulosic biomass is exposed to different temperatures under high pressures resulting in the degradation of biomass. In subcritical water-based processes, water is kept in a liquid phase by applying pressure greater than the vapor pressure of water. In this way, the latent heat required for phase change of water is avoided, which requires less energy than steam generation (Kumar, 2013). In a study by Popov et al., few types of oilseeds (cotton-, flax-, mustard-, canola-, and jatropha seeds) with different morphologies were subjected to hydrothermal pretreatment and oil extraction followed by hydrothermal carbonization of the extracted seedcake to hydrochar at different temperatures (Popov et al., 2016). This was done to maximize the extractability of oils without changing its quality. The seeds were subjected to hydrothermal pretreatment (120–210°C) for 30 minutes. Oils were extracted from the pretreated seeds using n-hexane in a Soxhlet apparatus for 120 minutes. Figure 2.3 depicts the improvement in oil yields after hydrothermal pretreatment for different oilseeds. The crude oil yields from the pretreated seeds at 180°C and 210°C were significantly higher (up to 30 wt%) than those from the respective untreated ground seeds. The produced hydrochar had higher heating value of 26.5 kJ/g comparable to that of bituminous coal. The BET surface area, pore volume, and pore size of the hydrothermally pretreated oilseeds are greater than those of the raw seeds, which attributes to their

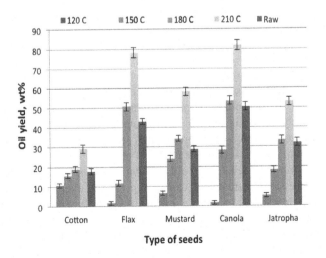

FIGURE 2.3 Oil yields from raw and hydrothermally pretreated oilseeds at different temperatures. (From Popov et al., 2016.)

better oil extractability and shorter extraction time. The partial hydrolysis of the oilseeds with degradation of the cellulose hulls, swelling of the kernel matrix, and removal of carbohydrates and proteins from the kernels to the aqueous phase in the pretreatment step helps oil extraction. This retains lipids in the solid phase and thus increases their concentration in the pretreated seeds. Hydrothermal pretreatment changes the oilseeds morphology and makes their structure more porous with a greater surface area, which reduces the mass transfer resistance and makes the oils more accessible to organic solvents. Hydrothermal pretreatment has the following advantages: improved oil extractability, shorter extraction time, tolerance to high moisture content feedstock, avoidance of preparation stages, and utilization of aqueous phase obtained after the pretreatment step for heating plant residues. It can also be adapted to lignocellulosic biomass.

Inbicon, a Danish company involved in commercial bioethanol production, utilizes hydrothermal pretreatment method for conversion of cellulosic material, such as chopped straw and corn stover, and household waste, to bioethanol and other products. The cellulosic material is subjected to continuous hydrothermal pretreatment and subsequently the fiber fraction is subjected to enzymatic liquefaction and saccharification. Their patented method focuses on creating a fiber fraction and a liquid fraction by selecting the temperature and residence time for the hydrothermal pretreatment, so that the fibrous structure of the feedstock is maintained and at least 80% of the lignin is maintained in the fiber fraction. A pressurized water at temperatures on the order of 160–230°C is used to gently melt hydrophobic lignin that is intricately associated with hemicellulose and cellulose strands and to solubilize a major component of hemicellulose and to disrupt cellulose strands so as to improve accessibility to productive enzyme binding.

Inbicon utilizes a hydrothermal pretreatment process after the mechanical conditioning of the feedstock. The premise of this process is to produce a fiber fraction and a liquid fraction through an extraction using hot water. In the fiber fraction this process achieves greater than 80% of the lignin present in the initial feedstock. Whereas the liquid fraction contains: C5 sugars, alkali chlorides, and fermentation inhibitors. The main fermentation inhibitor present is acetic acid. However, the fermentation inhibitors can later be removed through detoxification with NH_3 at a feasible cost. Since this pretreatment process does not use acids or bases and only water it eliminates the need to extract pretreatment chemicals once completed. During the initial step of the hydrothermal pretreatment process the feedstock is soaked and simultaneously placed in temperatures of up to 100°C and ambient pressure. This allows for the extraction of air present and saturates the feedstock with water. The next step entails a pressurized treatment at elevated temperatures in the range of 170 to 230°C through the addition of hot water or steam for approximately 5–15 minutes. This pressurized treatment can be repeated at altered temperatures and pressures in different zones increasing each subsequent time, therefore, labeled as a countercurrent process. The steam released during the hydrothermal pretreatment process is collected and reused in downstream evaporation processes. Throughout the hydrothermal pretreatment, there are acids formed that causes the fiber fraction to be nearly neutral in pH. The main benefit of this is that the pH will have to barely be adjusted for enzymatic liquefaction.

2.6 CHEMICAL PRETREATMENT

2.6.1 ACID AND ALKALI PRETREATMENT

Acid pretreatment entails the application of dilute or concentrated acid (typically H_2SO_4) to the lignocellulosic biomass which ruptures the rigid lignin matrix. Currently, dilute sulfuric acid pretreatment has been researched extensively using various lignocellulosic biomass: switchgrass (Digman et al., 2010), corn stover (Xu et al., 2009), and poplar (Wyman et al., 2009). The research by Cara et al. achieved a 76.5% hydrolysis yield when olive tree biomass is pretreated with 1.4% H_2SO_4 at 210°C (Cara et al., 2008). Dilute acid pretreatment (high temperature/short exposure time; low temperature/long exposure time) is preferable over concentrated acids due to higher reaction rates, shorter reaction times, and less damage to the carbohydrate component of biomass (Huang et al., 2009). Other acids used in pretreatment include H_3PO_4, H_2SO_4, and HCl, with the latter two being highly effective. Bases such as NaOH, NH_4OH, KOH, $Ca(OH)_2$, NH_3, and $(NH_4)_2SO_3$ are also used for lignocellulosic pretreatment. However, NaOH is preferred because strong bases cleave the carbon-carbon bond in lignin which facilitates enzymatic hydrolysis (Kim et al., 2016). There are two types of chemistry in a high pH (alkali pretreatment) (1) degradation: disentanglement of lignin fraction and (2) condensation: increase in lignin fraction size to cause its precipitation. Examples of alkali pretreatment include the ammonia fiber explosion (AFEX) method, ammonia recycle percolation (ARP) method, and soaking in aqueous ammonia (SAA) (Clifford, 2017). The AFEX method (see Figure 2.4a) utilizes liquid ammonia at around 60–160°C that causes lignocellulosic *swelling* followed by release of pressure that causes *expansion*. Composition is not changed but lignin is relocated, cellulose is decrystallized, and hemicellulose is depolymerized. In the ARP method, the biomass is pretreated with flowing aqueous ammonia at elevated pressure to prevent flash evaporation. The solid fraction (cellulose and hemicellulose rich) is separated from the liquid. The liquid is then evaporated via steam to extract lignin, sugar, and recover ammonia. The ammonia is recycled back to the process while the liquid fraction is crystallized and washed. The SAA (see Figure 2.4b) method is carried out at low temperatures to separate the lignin from the cellulose/hemicellulose fraction which can then be hydrolyzed to fermentable sugars.

2.6.2 ORGANOSOLVOSIS

Organic solvents are ideal in pretreating lignocellulose as they are selective in dissolving both lignin and hemicellulose while preserving the cellulosic component. A previous study showed inputs to include water, biomass, and H_2SO_4 at 180°C, 60 min, 1.25 wt% H_2SO_4, and 60 wt% ethanol (Pan et al., 2006). Typical solvation conditions are: 35–70% (w/w) alcohol solvent at cooking temperature ranges from 180–195°C in 2.0–3.8 pH range for 30–90 minutes. The liquid phase (lignin-rich) is precipitated to recover the solvent, extract soluble sugars, and separate lignin. The solid phase (cellulose-rich) is washed, saccharified, and fermented. Ethanol is the most commonly used alcohol solvent because it is cheap, easy to recover, and

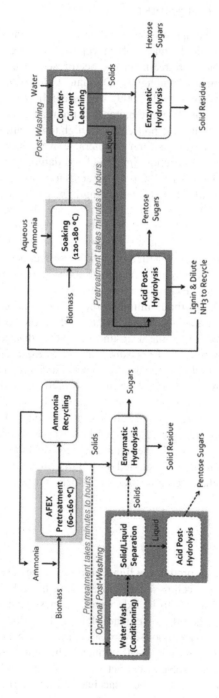

FIGURE 2.4 Process flows for (a) the AFEX method and (b) the SAA method. (From Clifford, 2017.)

has a low boiling point. Organosolvosis is beneficial because: (1) it produces high-quality lignin for high-value chemicals production, and (2) it lowers enzyme cost because the process removes lignin prior to cellulose hydrolysis thereby minimizing the absorption of cellulase enzymes to lignin.

An example of organosolvosis is aqueous organosolvation extraction (Imai et al., 2004). Solvent is prepared using 50:50 (w/w) ethanol water at ~200°C and 400 psi to extract lignin from a pulp with 6.4–27.4% (w/w) lignin. Lignin is precipitated by flashing pulping liquor at atmospheric pressure, then quickly diluting it with water. The supernatant water contains hemicellulose sugar and furfural, which are also recovered. The work by Arato et al. utilized the lignol process on woody biomass and found out that various chemical hydrolysis reactions take place when the biomass is "cooked" in water and ethanol under selected temperatures and reaction times (Arato et al., 2005). More specifically, Pan et al. found out that at 48-h reaction time, all pulps were hydrolyzed in a low-lignin pulp (<18.4% residual lignin) and more than 90% of the cellulose is hydrolyzed to glucose (Pan et al., 2005).

2.6.3   Ionic Liquids and Surfactants

Ionic liquids (ILs) are excellent solvation agents and have a melting point below 100°C. Additionally, ILs are less volatile, nonabrasive, effective, and are able to solubilize the entire lignocellulosic biomass. Because of their inorganic anion (e.g., Cl^-) and organic cation (Na^+) structure, ILs can selectively dissolve lignin or cellulose component individually at ambient conditions without the use of acids or bases. Once dissolved, the components are easily separated from the organic salt using polar liquids: ethanol, methanol, or water. Interestingly, ILs are used to separate compounds for the generation of high-value product. Despite these structural advantages in delignification, there are no ILs-based lignocellulosic applications in industrial biorefinery to date. Factors include limited information of ILs action on lignocellulose (Harmsen et al., 2010), IL recoverability, toxicity, and IL combination with water. An example of IL is the EmimAc (1-ethyl-3-methylimidazolium acetate) which is capable of completely solubilizing both lignin and cellulose in switchgrass and a significant change in cellulose from crystalline to amorphous structure was observed after reprecipitation at 120°C (Singh et al., 2009).

2.6.4   Oxidative Delignification

Oxidative delignification is a collective term used in degrading lignocellulose using oxidizing agents such as hydrogen peroxide, ozone, oxygen, or air. The aromatic ring in lignin reacts with the oxidizing chemicals and is converted to carboxylic acids. Such acids are neutralized or removed because they inhibit subsequent fermentation to sugar. In contrast, the hemicellulose portion of the biomass can no longer be utilized for sugar production.

Hydrogen peroxide (H_2O_2) can solubilize around 50% of the lignin and almost all of the hemicellulose at only 2% H_2O_2 at 30°C within eight hours (Azzam, 1989). The cellulose, on the other hand, is 95% hydrolyzed to glucose using cellulase at 45°C for 24 hours (Azzam, 1989). Ozone, one of the strongest oxidizing agents, is also used in lignocellulosic pretreatment. Ozone acts by attacking the electron-rich lignin, cleaving its aromatic ring structures, but maintains the carbohydrate part (hemicellulose and cellulose). Ozonolysis of lignocellulosic biomass aids in enzymatic hydrolysis as it effectively destroys the lignin without generating inhibitory compounds. It is also applicable to a wide array of lignocellulose such as bagasse, wheat straw, cotton straw, peanut, and pine. However, ozone is highly reactive and inflammable. Several studies have shown that different lignocellulose produce different yields upon ozonolysis. In other studies, wheat pretreated with ozone had a yield between 158–168% hydrogen relative to their non-pretreated counterparts (Wu et al., 2013a; Wu et al., 2013b; Schultz-Jensen et al., 2011). Wheat was also found to have 60% lignin reduction when pretreated with ozone for three hours, subsequently producing 52% alcohol upon saccharification and fermentation (SSF) (Schultz-Jensen, 2011).

Water and air (with oxygen) can also act as oxidant to pretreating lignocellulose at elevated temperature and pressure (McGinnis et al., 1983). It operates by treating high organic matter wastes at temperatures between 150–350°C and pressures between 5–20 MPa, sufficiently oxidizing soluble or suspended materials (Jorgensen et al., 2007). The main degradation products from wet oxidation of lignin and hemicellulose are carboxylic acids, CO_2, and H_2O. Wet oxidation opens the crystalline structure of the lignocellulose (Panagiotou and Olsson, 2007) and increases the rate of lignin and hemicellulose dissolution suitable for enzymatic hydrolysis (Martin et al., 2007). Unlike other pretreatment processes, wet oxidation prevents the proliferation of microbial growth inhibitors such as furfural and hydroxymethylfurfural. The Schimdt group has applied wet oxidation to hardwood and wheat (Schmidt et al., 1996) while Klinke et al. determined a 65% delignification in wheat following wet oxidation (Klinke et al., 2002). Wet oxidation pretreatment can also be combined with alkali hydrolysis, such as in the case of wheat straw achieving 85% cellulose-to-glucose conversion at 20 g/L straw concentration, 170°C temperature, and five-1ten-minute retention time (Bjerre et al., 1996).

2.7 BIOLOGICAL PRETREATMENT

Except for dilute acid and alkali pretreatment, most pretreatment technologies are costly requiring specialized equipment, are energy intensive, and produce waste products and inhibitors that are challenging to treat. Significant research into environmentally friendly and cost-effective alternatives has led to biological pretreatment. In biological processes, microorganisms such as white-rot, brown-rot and soft-rot fungi are used to efficiently and effectively degrade lignin and hemicellulose. White-rot and soft-rot fungi attack both cellulose and lignin while brown-rot fungi attack only cellulose. White-rot fungi such as *Phenerochaete crysosporium*, *Pycnoporus cinnabarinus*, and *Ceriporia lacerate* have been used to pretreat biomass for decades and are more effective for biological pretreatment (Fan et al., 1987). More recently, soft-rot fungi such as *Daldinia concentrica* and brown-rot fungi such as *Serpula lacrymans* and *Coniophora puteana* have been explored for the same purpose. Nineteen white-rot fungi (*Pleurotus ostreatus*) were applied by Hatakka et al. to wheat straw and who found that 35% of the straw was converted to glucose within five weeks (Hatakka, 1983). Similar wheat-to-sugar conversions were realized when *Phanerochaete sordida* (Ballesteros et al., 2006) and *Pycnoporus cinnabarinus* (Okano et al., 2005) were utilized for a retention time of four weeks. The white-rot fungus *P. chrysosporium* generates enzyme found in their extracellular filtrates which are responsible for wood cell wall degradation (Boominathan and Reddy, 1992; Kirk and Farrell, 1987; Waldner et al., 1988). One study determined that 51 isolates of the white-rot fungus *Punctularia* sp. mediated a 50% bamboo lignin removal (Suhara et al., 2012) while another white-rot fungus, *Irpex lacteus*, has hydrolyzed 82% of corn stalk (Du et al., 2011). In addition to fungi, bacteria such as *Bacillus* spp. can be cocultured to provide enhanced saccharification (Akhtar et al., 2015; Akhtar et al., 2013; Singh et al., 2008). The advantages of biological pretreatment over physical, thermochemical, or chemical pretreatment makes it worthy of further research. These advantages include low energy requirement, mild environmental conditions, production of less inhibitors, and low probability for corrosion. Despite these advantages, focus must be made on enhancing the rate of hydrolysis in most biological pretreatment processes.

REFERENCES

Akhtar N., Goyal, D., Goyal A. (2015) Biodegradation of leaf litter biomass by combination of Bacillus sp. and Trichoderma reesei MTCC 144. *Minerva Biotecnol* 27, (4), 191–200.

Akhtar N., Sharma A., Deka D. (2013) Characterization of cellulase producing *Bacillus sp.* for effective degradation of leaf litter biomass. *Environ Prog Sustain Energy* 32, (4), 1195–1201.

Alvira P., Tomás-Pejó E., Ballesteros M., Negro M. J. (2010) Pretreatment technologies for an efficient bioethanol production process based on enzymatic hydrolysis: A review. *Bioresour Technol* 101, (13), 4851–4861.

Arato C., Pye E. K., Gjennestad G. (2005) The lignol approach to biorefining of woody biomass to produce ethanol and chemicals. *Appl Biochem Biotechnol* 123, (1–3), 871–882.

Azzam M. (1989) Pretreatment of cane bagasse with alkaline hydrogen peroxide for enzymatic hydrolysis of cellulose and ethanol fermentation. *J Environ Sci Health B* 24, (4), 421–433.

Bajpai P. (2016) Structure of lignocellulosic biomass. In: Sharma S. K., Rajasthan J. (Eds.) *Pretreatment of Lignocellulosic Biomass for Biofuel Production.* Springer Science + Business Media Singapore PTE Ltd., Singapore, Singapore. 9789811006869.

Ballesteros I., Negro M. J., Oliva J. M., Cabañas A., Manzanares P., Ballesteros M. (2006) Ethanol production from steam explosion pretreated wheat straw. *Appl Biochem Biotechnol* 129–132, 496–508.

Bergman R., Cai Z., Carll C. G., Clausen C. A., Dietenberger M. A., Falk R. H., Frihart C. R., Glass S. V., Hunt C. G., Ibach R. E., Kretschmann D. E., Lebow S. T., Rammer D. R., Ross R. J., Stark N. M., Wacker J. P., Wang X., Wiedenhoeft A. C., Wiemann M. C., Zelinka S. L. (2010) *Wood Handbook: Wood as an Engineering Material.* A general technical report by the U.S. Department of Agriculture, Forest Service, Forest Products Laboratory, Madison, WI.

Bezerra T., Ragausks A. J. (2016) A review of sugarcane bagasse for second-generation bioethanol and biopower production. *Biofuel Bioprod Bior* 10, (5), 634–647.

Bjerre A. B., Olesen A. B., Fernqvist T. (1996) Pretreatment of wheat straw using combined wet oxidation and alkaline hydrolysis resulting in convertible cellulose and hemicellulose. *Biotechnol Bioeng* 49, (5), 568–577.

Bobleter O. (1994) Hydrothermal degradation of polymers derived from plants. *Prog Polym Sci* 19, (5), 797–841.

Boominathan K., Reddy C. A. (1992) cAMP-mediated differential regulation of lignin perox-idase and manganese-dependent peroxidases production in the white-rot basidiomycete *Phanerochaete chrysosporium. Proc Natl Acad Sci USA* 89, (12), 5586–5590.

Bridgeman T. G., Jones J. M., Williams A., Waldron D. J. (2010) An investigation of the grindability of two torrefied energy crops. *Fuel* 89, (12), 3911–3918.

Cara C., Ruiz E., Oliva J. M., Sáez F., Castro E. (2008) Conversion of olive tree biomass into fermentable sugars by dilute acid pretreatment and enzymatic saccharification. *Bioresour Technol* 99, (4), 1869–1876.

Carvalheiro F., Luís C. H., Gírio F. M. (2008) Hemicellulose biorefineries: A review on bio-mass pretreatments. *J Sci Ind Res* 67, (11), 849–864.

Clifford, C. B. (2017) *EGEE 439: Alternative Fuels from Biomass Sources: High pH (Alkaline) Pretreatment.* College of Earth and Mineral Sciences, The Pennsylvania State University, State College, PA. https://www.e-education.psu.edu/egee439/node/657 (Accessed 01/11/2019).

De Jong W., van Ommen JR. (Eds.) (2015) *Biomass as a Sustainable Energy Source for the Future: Fundamentals of Conversion Processes.* John Wiley & Sons, Inc., Hoboken, NJ. 9781118916636.

Diaz M. J., Cara C., Ruiz E., Romero I., Moya M., Castro E. (2010) Hydrothermal pre-treat-ment of rapeseed straw. *Bioresour Technol* 101, (2010), 2428–2435.

Digman M. F., Shinners K. J., Casler M. D., Dien B. S., Hatfield R. D., Jung H.-J., Muck R. E., Weimer P. J. (2010) Optimizing on-farm pretreatment of perennial grasses for fuel ethanol production. *Bioresour Technol* 101, (14), 5305–5314.

Du W., Yu H., Song L., Zhang J., Weng C., Ma F., Zhang X. (2011) The promising effects of by-products from *Irpex lacteus* on subsequent enzymatic hydrolysis of bio-pretreated corn stalks. *Biotechnol Biofuel* 4, 37.

Ehara K., Saka S. (2005) Decomposition behavior of cellulose in supercritical water, subcriti-cal water, and their combined treatments. *J Wood Sci* 51, (2), 148–153.

Ehara K., Saka S. (2009) A comparative study on chemical conversion of cellulose between the batch-type and flow-type systems in supercritical water. *Cellulose* 9, (3–4), 301–311.

Fan L.-T., Gharpuray M. M., Lee Y.-H. (Eds.) (1987) *Biotechnology Monographs, Volume 3: Cellulase Hydrolysis.* Springer-Verlag, Berlin, Germany 57. 9783540176713.

Franck E. U. (1987) Fluids at high pressures and temperatures. *Pure Appl Chem* 59, (1), 25–34.

Garrote G., Dominguez H., Parajo J. C. (1999) Hydrothermal processing of lignocellulosic materials. *Eur J Wood Wood Prod (Holz Roh Werkst)* 57, (3), 191–202.

Gupta R., Lee Y. Y. (2008) Mechanism of cellulase reaction on pure cellulosic substrates. *Biotechnol Bioeng* 102, (6), 1570–1581.

Harmsen P. F. H., Huijgen W. J. J., Bermúdez López L. M., Bakker R. R. C. (2010) *Literature Review of Physical and Chemical Pretreatment Processes for Lignocellulosic Biomass.* A joint report by the Food & Biobased Research of the Wageningen University & Research Centre, Energy Research Centre of the Netherlands, and Abengoa Bioenergia Nuevas Tecnologias. https://www.ecn.nl/docs/library/report/2010/e10013.pdf (Accessed 01/15/2019).

Hatakka A. I. (1983) Pretreatment of wheat straw by white-rot fungi for enzymatic sacchari-fication of cellulose. *Eur J Appl Biochem Biotechnol* 18, 350–357.

Heitz M., Carrasco F., Rubio M., Chauvette G., Chornet E., Jaulin L., Overend R. P. (1986) Generalized correlations for aqueous liquefaction of lignocellulosics. *Can J Chem Eng* 64, 647–650.

Hendriks A. T. W. M., Zeeman G. (2009) Pretreatments to enhance the digestibility of ligno-cellulosic biomass. *Bioresour Technol* 100, (1), 10–18.

Hess J. R., Kenney K. L., Wright C. T., Perlack R., Tuhhollow A. (2009) Corn stover analysis for biomass conversion: Situation analysis. *Cellulose* 16, (4), 599–619.

Hisashi M., Toshiki N., Katsunobu E., Shiro S. (2005) Fermentability of water-soluble portion to ethanol obtained by supercritical water treatment of lignocellulosics. *Appl Biochem Biotechnol* 124, (1–3), 963–972.

Hoskinson R. L., Kalen D. L., Birrell S. J., Radtke C. W., Wilhelm W. W. (2007) Engineering, nutrient removal, and feedstock conversion evaluations of four corn stover harvest scenarios. *Biomass Bioenergy* 31, (2–3), 126–136.

Hsu T.-A. (1996) Pretreatment of biomass. In: *Handbook on Bioethanol: Production and Utilization*. Taylor and Francis, Washington, DC.

Huang H. J., Lin W. L., Ramaswamy S., Tschirner U. (2009) Process modeling of comprehensive integrated forest biorefinery: An integrated approach. *Appl Biochem Biotechnol* 154, (1–3), 205–216.

Hunder M. (2005) Some aspects of wood chips' storage and drying for energy use. *BioEnergy* 2005, 257–260, *International Bioenergy in Wood Industry Conference and Exhibition*, Jyväskylä, Finland.

Imai M., Ikari K., Suzuki I. (2004) High-performance hydrolysis of cellulose using mixed cellulase species and ultrasonication pretreatment. *Biochem Eng J* 17, (2), 79–83.

Jorgensen H., Kristensen J. B., Felby C. (2007) Enzymatic conversion of lignocellulose into fermentable sugars: Challenges and opportunities. *Biofuel Bioprod Bior* 1, (2), 119–134.

Kadam K. L., Chin C. Y., Brown L. W. (2009) Continuous biomass fractionation process for producing ethanol and low-molecular-weight lignin. *Environ Prog Sustain Energy* 28, (1), 89–99.

Kaltschmitt M., Hartmann H., Hofbauer H. (2001) *Energie aus biomasse: Grundlagen, techniken und verfahren*. Springer-Verlag, Berlin, Germany.

Karimi K., Taherzadeh M. J. (2016) A critical review of analytical methods in pretreatment of lignocelluloses: Composition, imaging, and crystallinity. *Bioresour Technol* 200, 1008–1018.

Kim J. S., Lee Y. Y., Kim T. H. (2016) A review on alkaline pretreatment technology for bioconversion of lignocellulosic biomass. *Bioresour Technol* 199, 42–48.

Kim M., Day D. F. (2011) Composition of sugar cane, energy cane, and sweet sorghum suitable for ethanol production at Louisiana sugar mills. *J Ind Microbiol Biotechnol* 38, (7), 803–807.

Kirk T. K., Farrell R. L. (1987) Enzymatic combustion: The microbial degradation of lignin. *Annu Rev Microbiol* 41, 465–505.

Klinke H. B., Ahring B. K., Schmidt A. S., Thomsen A. B. (2002) Characterization of degradation products from alkaline wet oxidation of wheat straw. *Bioresour Technol* 82, (1), 15–26.

Kobayashi N., Okada N., Hirakawa A., Sato T., Kobayashi J., Hatano S., Itaya Y., Mori S. (2009) Characteristics of solid residues obtained from hot-compressed-water treatment of woody biomass. *Ind Eng Chem Res* 48, (1), 373–379.

Kohlmann K. L., Westgate P. J., Sarikaya A., Velayudhan A., Weil J., Hendricson R., Ladisch M. R. (1995) Enhanced enzyme activities on hydrated lignocellulosic substrates. In: *207th American Chemical Society National Meeting, ACS Symposium Series No. 618*, 237–255.

Kumar P., Barrett D. M., Delwiche M. J., Stroeve P. (2009a) Methods for pretreatment of lignocellulosic biomass for efficient hydrolysis and biofuel production. *Ind Eng Chem Res* 48, (8), 3713–3729.

Kumar S. (2013) Sub- and supercritical water technology for biofuels. In: Lee J. W. (Ed.) Advanced Biofuels and Bioproducts. Springer, New York, NY, 147–183.

Kumar S., Gupta R. B. (2009) Biocrude production from switchgrass using subcritical water. *Energy Fuels* 23, (10), 5151–5159.

Kumar S., Gupta R., Lee Y. Y., Gupta R. B. (2009b) Cellulose pretreatment in subcritical water: Effect of temperature on molecular structure and enzymatic reactivity. *Bioresour Technol* 101, (2010), 1337–1347.

Kumar S., Kothari U., Kong L., Lee Y. Y., Gupta R. B. (2011). Hydrothermal pretreatment of switchgrass and corn stover for production of ethanol and carbon microspheres. *Biomass Bioenergy* 35, (2), 956–968.

Laxman R. S., Lachke A. H. (2008) Bioethanol from lignocellulosic biomass, Part 1: Pretreatment of the substrates. In: Pandey A. (Ed.) *Handbook of Plant-based Biofuels*. CRC Press, Boca Raton, FL, 121–139.

Li H., Qu Y., Yang Y., Chang S., Xu J. (2016) Microwave irradiation – A green and efficient way to pretreat biomass. *Bioresour Technol* 199, 34–41.

Liu C., Wyman C. E. (2003) The effect of flow rate of compressed hot water on xylan, lignin and total mass removal from corn stover. *Ind Eng Chem Res* 42, (21), 5409–5416.

Lynd L. R. (1996) Overview and evaluation of fuel ethanol from cellulosic biomass: Technology, economics, the environment, and policy. *Ann Rev Energy Environ* 21, 403–465.

Lynd L. R., Weimer P. J., van Zyl W. H., Pretorius I. S. (2002) Microbial cellulose utilization: Fundamentals and biotechnology. *Microbiol Mol Biol Rev* 66, (3), 506–577.

Maciejewska A., Veringa H. J., Sanders J., Peteves S. D. (2008) *Co-firing of Biomass with Coal: Constraints and Role of Biomass Pretreatment*. Europe Commissie – Joint Research Centre, Petten, Netherlands, Report No. EUR 22461 EN.

Martín C., Klinke H., Marcet M., García L., Hernández E., Thomsen A. B. (2007) Study of the phenolic compounds formed during pretreatment of sugarcane bagasse by wet oxidation and steam explosion. *Holzforschung* 2007, 61, 483–487.

Matsumura Y., Minowa T., Photic B., Kersten S. R. A., Prins W., Swaaij W. P. M. V., Beld B. V. D., Elliott D. C., Neuenschwander G. G., Kruse A., Antal Jr. M. J. (2005) Biomass gasification in near- and super-critical water: Status and prospects. *Biomass Bioenergy* 29, (4), 269–292

Matsumura Y., Sasaki M., Okuda K., Takami S., Ohara S., Umetsu M., Adschiri T. (2006) Supercritical water treatment of biomass for energy and material recovery. *Combust Sci Technol* 178, (1–3), 509–536.

McGinnis G. D., Wilson W. W., Mullen C. E. (1983) Biomass pretreatment with water and high-pressure oxygen. The wet-oxidation process. *Ind Eng Chem Prod Res Dev* 22, (2), 352–357.

McKendry P. (2002) Energy production from biomass (part 1): Overview of biomass. *Bioresour Technol* 83, (1), 37–46.

Miyoshia H., Chena D., Akai T. (2004) A novel process utilizing subcritical water to recycle soda-lime-silicate glass. *J Non-Cryst Solids* 337, (3), 280–282.

Mosier N., Hendrickson R., Ho N., Sedlak M., Ladisch M. R. (2005a) Optimization of pH controlled liquid hot water pretreatment of corn stover. *Bioresour Technol* 96, (18), 1986–1993.

Mosier N., Wyman C., Dale B., Elander R., Lee Y. Y., Holtzapple M., Ladisch M. (2005) Features of promising technologies for pretreatment of lignocellulosic biomass. *Bioresour Technol* 96, (6), 673–686.

Motevali A., Minaei S., Banakar A., Ghobadian B., Khoshtaghaza M. H. (2014) Comparison of energy parameters in various dryers. *Energy Convers Manage* 87, 711–725.

Nakata T., Miyafuji H., Saka S. (2006) Bioethanol from cellulose with supercritical water treatment followed by enzymatic hydrolysis. *Appl Biochem Biotechnol* 130, (1–3), 476–485.

Okano K., Kitagaw M., Sasaki Y., Watanabe T. (2005) Conversion of Japanese red cedar (*Cryptomeria japonica*) into a feed for ruminants by white-rot basidiomycetes. *Anim Feed Sci Technol* 120, (3–4), 235–243.

Pan X., Arato C., Gilkes N., Gregg D., Mabee W., Pye K., Xiao Z., Zhang X., Saddler J. (2005) Biorefining of softwoods using ethanol organosolv pulping: Preliminary evaluation of process streams for manufacture of fuel-grade ethanol and co-products. *Biotechnol Bioeng* 90, (4), 473–481.

Pan X., Gilkes N., Kadla J., Pye K., Saka S., Gregg D., Ehara K., Xie D., Lam D., Saddler J. (2006) Bioconversion of hybrid poplar to ethanol and co-products using an organosolv fractionation process: Optimization of process yields. *Biotechnol Bioeng* 94, 851–861.

Panagiotou G., Olsson L. (2007) Effect of compounds released during pretreatment of wheat straw on microbial growth and enzymatic hydrolysis rates. *Biotechnol Bioeng* 96, (2), 250–258.

Pérez, J. A., Ballesteros I., Ballesteros M., Sáez F., Negro M. J., Manzanares P. (2008) Optimizing liquid hot water pretreatment conditions to enhance sugar recovery from wheat straw for fuel-ethanol production. *Fuel* 87, (17–18), 3640–3647.

Pettersen R. C. (1984) The chemical composition of wood. In: Rowell R. M. (Ed.) *The Chemistry of Solid Wood*, Volume 207, 115–116, *Advances in Chemistry Series*. American Chemical Society, Washington, DC.

Popov S. T., Abdel-Fattah T., Kumar S. (2016) Hydrothermal treatment for enhancing oil extraction and hydrochar production from oilseeds. *Renew Energy* 85, 844–853.

Prassad S., Singh A., Joshi H. C. (2007) Ethanol as an alternative fuel from agricultural, industrial and urban residues. *Resour Conserv Recy* 50, (1), 1–39.

Sakaki T., Shibata M., Miki T., Hirosue H., Hayashi N. (1996) Reaction model of cellulose decomposition in near-critical water and fermentation of products. *Bioresour Technol* 58, (2), 197–202.

Sasaki M., Fang Z., Fukushima Y. N., Adschiri T., Arai K. (2000) Dissolution and hydrolysis of cellulose in subcritical and supercritical water. *Ind Eng Chem Res* 39, (8), 2883–2890.

Sasaki M., Kabyemela B., Malaluan R., Hirose S., Takeda N., Adschiri T., Arai K. (1998) Cellulose hydrolysis in subcritical and supercritical water. *J Supercrit Fluids* 13, (1–3), 261–268.

Sasmal S., Mohanty K. (2018) Pretreatment of lignocellulosic biomass toward biofuel production. In: Kumar S., Sani R. (Eds.) *Biorefining of Biomass to Biofuels: Biofuel and Biorefinery Technologies*, Volume 4. Springer Link, New York, NY, 203–221. 9783319676784.

Savage P. E. (1999) Organic chemical reactions in supercritical water. *Chem Rev* 99, (2), 603–622.

Schmidt A. S., Puls J., Bjerre A. B. (1996) Comparison of wet oxidation and steaming for solubilization of the hemicellulose fraction in wheat straw and birch wood. In: Chartier P., Ferrero G. L., Henius U. M., Hultberg S., Sachau J., Winblad M. (Eds.) *Biomass for Energy and the Environment*, Volume 3, 1510–1515, *Proceedings of 9th European Bioenergy Conference*, Copenhagen, Denmark.

Schultz-Jensen N., Kádár Z., Thomsen A., Bindslev H., Leipold F. (2011) Plasma-assisted pretreatment of wheat straw for ethanol production. *Appl Biochem Biotechnol* 165, (3–4), 1010–1023.

Shinners K. J., Bennett R. G., Hoffman D. S. (2012) Single- and two-pass corn grain and stover harvesting. *Trans ASABE* 55, (2), 341–350.

Shiro S., Hisashi M. (2007) Bioethanol production from lignocellulosics using supercritical water. In: *American Chemical Society National Meeting, ACS Symposium Series No. 954*, 422–433.

Sierra R., Smith A., Granda C., Holtzapple M. T. (2008) Producing fuels and chemicals from lignocellulosic biomass. *Chem Eng Prog* 104, (8), S10–S18.

Singh P., Suman A., Tiwari P., Arya N., Gaur A., Shrivastava A. K. (2008) Biological pretreatment of sugarcane trash for its conversion to fermentable sugars. *World J Microbiol Biotechnol* 24, (5), 667–673.

Singh S., Simmons B. A., Vogel K. P. (2009) Visualization of biomass solubilization and cellulose regeneration during ionic liquid pretreatment of switchgrass. *Biotechnol Bioeng* 104, (1), 68–75.

Spliethoff H. (2010) *Power Generation from Solid Fuels.* Springer Verlag, Berlin/Heidelberg, Germany.

Statista (2019) *U.S. Sugar Cane Production by State, 2010–2018.* https://www.statista.com/statistics/191975/sugarcane-production-in-the-us-by-state/ (Accessed 02/14/2019).

Suhara H., Kodama S., Kamei I., Maekawa N., Meguro S. (2012) Screening of selective lignin-degrading basidiomycetes and biological pretreatment for enzymatic hydrolysis of bamboo culms. *Int Biodeter Biodegr* 75, 176–180.

Sun Y., Cheng J. J. (2002) Hydrolysis of lignocellulosic material for ethanol production: A review. *Bioresour Technol* 83, (1), 1–11.

Suryawati L., Wilkins M. R., Bellmer D. D., Huhnke R. A., Maness N. O., Banat I. M. (2008) Simultaneous saccharification and fermentation of Kanlow switchgrass pretreated by hydrothermolysis using *Kluyveromyces marxianus* IMB4. *Biotechnol Bioeng* 101, (5), 894–902.

Swart J. A. A., Jiang J., Ho P. (2008) Risk perceptions and GM crops: The case of China. *Tailoring Biotechnol Soc Sci Technol* 33, 11–28.

Swatloski R. P., Spear S. K., Holbrey J. D., Rogers R. D. (2002) Dissolution of cellulose with ionic liquids. *J Am Chem Soc* 124, 4974–4975.

Taherzadeh M. J., Karimi K. (2008) Pretreatment of lignocellulosic wastes to improve ethanol and biogas production: A review. *Int J Mol Sci* 9, (9), 1621–1651.

Van Loo S., Koppejan J. (Eds.) (2004) *Handbook of Biomass Combustion and Co-firing.* Prepared by task 32 of the implementing agreement on bioenergy under the auspices of the International Energy Agency. Twente University Press, Enschede, Netherlands.

Waldner R., Leisola M. S. A., Fiechter A. (1988) Comparison of ligninolytic activities of selected fungi. *Appl Microbiol Biotechnol* 29, (4), 400–407.

Watanabe H., Tokuda H. (2001) Animal cellulases. *Cell Mol Life Sci* 58, (9), 1167–1178.

Weil J. R., Brewer M., Hendrickson R., Sarikaya A., Ladisch M. R. (1997) Continuous pH monitoring during pretreatment of yellow poplar wood sawdust by pressure cooking in water. *Appl Biochem Biotechnol* 70–72, 99–111.

Wihersaari M. (2005) VTT processes, evaluation of greenhouse gas emission risks from storage of wood residue. *Biomass Bioenergy* 28, (5), 444–453.

Wu J., Ein-Mozaffari F., Upreti S. (2013a) Effect of ozone pretreatment on hydrogen production from barley straw. *Bioresour Technol* 144, 344–349.

Wu J., Upreti S., Ein-Mozaffari F. (2013b) Ozone pretreatment of wheat straw for enhanced biohydrogen production. *Int J Hydrogen Energy* 38, (25), 10270–10276.

Wu Y. (2019) *Switchgrass: Harvesting.* United States Department of Plant and Soil Sciences, Oklahoma State University, Stillwater, OK. http://switchgrass.okstate.edu/harvesting (Accessed 02/14/2019).

Wyman C. E., Dale B. E., Elander R. T., Holtzapple M., Ladisch M. R., Lee Y. Y. (2005) Coordinated development of leading biomass pretreatment technologies. *Bioresour Technol* 96, (18), 1959–1966.

Wyman C. E., Dale B. E., Elander R. T., Holtzapple M., Ladisch M. R., Lee Y. Y., Mitchinson C., Saddler J. N. (2009) Comparative sugar recovery and fermentation data following pretreatment of poplar wood by leading technologies. *Biotechnol Progr* 25, (2), 333–339.

Xu J., Thomsen M. H., Thomsen A. B. (2009) Pretreatment of corn stover with low concentration of formic acid. *J Microbiol Biotechnol* 19, (8), 845–850.

Yang B., Wyman, C. E. (2004) Effect of xylan and lignin removal by batch and flow through pretreatment on the enzymatic digestibility of corn stiver cellulose. *Biotechnol Bioeng* 86, (1), 88–95.

Zhang, Y.-H. P., Berson E., Sarkanen S., Dale B. E. (2009) Sessions 3 and 8: Pretreatment and biomass recalcitrance: Fundamentals and progress. *Appl Biochem Biotechnol* 153, (1–3), 80–83.

3 Hydrothermal Liquefaction of Terrestrial and Aquatic Biomass

Florin Barla and Sandeep Kumar

CONTENTS

3.1 INTRODUCTION

Biomass is one of the most abundant sources of renewable energy and rather than being subjected to direct combustion the attention was recently focused on biomass conversion into liquid energy carriers. Biomass meets all the requirements of renewable and CO_2 neutral materials, it is one of the largest sources of energy in the world. Biomass sources such as wood wastes, energy crops, aquatic plants, agricultural residues, municipal and animal wastes are considered potential sources of fuels and chemical feedstocks. There is growing attention to the development of technologies to convert biomass into more valuable fuels (bio-oil; bio-syngas) and chemicals. Thermochemical conversion of biomass into liquid via hydrothermal liquefaction (HTL) takes place in a hot (250–370°C), pressurized environment (4–22 MPa) for a sufficient amount of time (5 to 60 min) to decompose the biopolymeric structure to mainly liquid components. The temperature initiates the pyrolytic mechanism and the pressure is sufficient to maintain the water into liquid phase. HTL it is also known as direct liquefaction and is essentially pyrolysis in hot and pressurized liquid water. Recently, biomass and its derivatives have generated high expectations for the production of fuels and fine chemicals.

The biofuels industry has focused on using lignocellulosic biomass mainly from agricultural/forestry waste such as corn stover, straw, wood, and other by-products, as well as dedicated crops like switchgrass, miscanthus, hybrid poplar, and energy tobacco to produce second-generation or advanced biofuels (Adrianov et al., 2010). In addition to these terrestrial crops, aquatic biomass such as microalgae has attracted much attention because it has shown great productivity compared to terrestrial plants. Nevertheless, the main challenge with microalgae is the high-water content, with usually only 1 g of dry algal biomass recovered per liter of water (Kumar, 2012).

3.2 HTL OF BIOMASS—REACTION PATHWAYS

Many biomass feedstocks, such as agricultural residuals, food processing wastes, municipal and agricultural sludge contains significant amounts of water and vary in chemical composition. The main components of various biomasses feedstocks are lignocellulose, fatty acids, and proteins. These products can be hydrothermally transformed to produce a range of derivative products and the basic reaction mechanism is described as: depolymeration of biomass; decomposition of biomass by cleavage, dehydration, decarboxylization, and deamination and recombination of highly reactive products (Peterson et al., 2008; Toor et al., 2011). Variations of physical properties of water under hydrothermal conditions can facilitate efficient separations of product and by-product. Various complex reactions take place during the HTL process and the biomass is transformed into crude oil like products. The process mechanism is classified for two types of feedstock: lignocellulosic biomass (dry feedstock) and algal biomass (wet feedstock). (Dimitriadis and Bezergianni, 2017).

3.3 CONVERSIONS OF CARBOHYDRATES

Any biomass contains a considerable amount of carbohydrates, polysaccharides, cellulose, hemicellulose, and starch. Subjected to subcritical water conditions these components are rapidly hydrolized to their monomers (glucose, xylose, and other saccharides) which can be further degraded. Hemicelluloses and starch are hydrolyzed much faster than cellulose. Table 3.1 summarizes the HTL of different carbohydrates.

3.3.1 CELLULOSE

Cellulose $(C_6H_{10}O_5)_n$, a nonpolar compound like starch, is a plysaccharide and consists of glucose monomers linked via β-(1$\longrightarrow$4)-glycosidic bonds, which allows strong intra- and inter-molecular hydrogen bonds to form that makes them crystalline, resistant to swelling in water, and to the enzymatic hydrolysis. Cellulose is the most abundant organic polymer; it is found in all terrestrial plants and even in many aquatic species such as algae; it is distributed in all plants from higher plants

TABLE 3.1
HTL Products from Saccharides and Furans

Substrate	Reaction Conditions	Products
Glucose	300–400°C 25–40 Mpa 0.02–2 $ Ph - neutral	Fructose Dihydroxyacetone Glyceraldehyde Erythrose Glycoialdehyde Pyruvaldehyde 1,6-Anhydroglucose Acetic acid Formic acid 5-HMF
D-glucose and other monosaccharides	340°C 27.5 MPa 25–204 s (Acid & base catalyst)	5-HMF Glycoialdehyde Glyceraldehyde Formic acid Acetic acid Acrylic acid 2-furaldehyde 1,2,4-Benzenetriol
D-Fructoze	200–320°C 120 s pH range 1.5-5	5-HMF Formic acid Lavulinic acid
Micricrystalline cellulose	320400°C 25 MPa 0.05-10 5 (pH - neutral)	1,6-Anhydroglucose Erythrose Glyceraldehyde Glycoialdehyde Pyruvaldehyde Dihydroxyacetone Furfural 5-HMF
5-HMF	290–400°C 27.5 MPa 0.1–0.308 min	1,2,4-Benzenetriol 4-Oxopentanoic acid
Cellulose	250407°C Alkaline media	o-,m- or p-Xylene Ethylbenzene n-Propyl benzene 1-methyl-2-ethylbenzene 1-methyl-3-othylbenzene Phenol 0-, m-, p-Cresol 2-Phenoxyethanol

Source: Toor, Rosendahl, and Rudolf, 2011.

to primitive organisms such as seaweeds, flagellates, and bacteria. At the molecular level, the crystalline cellulose core of microfibrils possess a strong resistance to chemical and biological hydrolysis due to chains of cellodextrins that are very precisely arranged. Under subcritical conditions (elevated temperature and pressure) cellulose is solubilized and hydrolyzed to its constituents. Some kinetic studies showed that cellulose hydrolysis rate at 25 MPa increased tenfold between 240–310°C and was considerably slower than starch hydrolysis (Rogalinski et al., 2008). On the other hand, cellulose from different biological sources has different properties and both physical and chemical structure can affect its behavior. Due to these differences, the conditions under which the cellulose is degraded vary according with studies published in the literature (Schwald and Bobleter, 1989; Adschiri et al., 1993; Sasaki et al., 2000; Kamio et al., 2008). Also, at temperatures below 260°C, when carbon dioxide was added to the reaction, the hydrolysis rate increased significantly due to carbonic acid formation that acted as a catalyst (Roglinski et al., 2008).

3.3.2 HEMICELLULOSES

Hemicelluloses are heteropolymer with amorphous structure, composed of sugars monomers such as xylose, mannose, glucose, galactose, and others in different ratio depending on feedstocks sources, and make up 20–40% of plant biomass. Hemicelluloses are alkali soluble after the removal of pectic substances. Hemicellulose does not form crystalline structure and therefore it is much more susceptible to hydrothermal extraction and hydrolysis. It is the most complex component of the cell wall that forms hydrogen bonds with cellulose, covalent bonds with lignin (a-benzyl ether linkages), and ester linkages with acetyl units and hydroxycinnamic acids. Hemicellulose easily dissolved in water at temperatures above 180°C, and more than 95% of hemicellulose as monomers sugars where extracted at 200–230°C, 34.5 MPa in just a few minutes (Mok and Antal, 1992; Bobleter, 1994). Hemicellulose has a weaker structure and is less resistant to intramolecular hydrogen bonding that is easier to disintegrate via hydrothermal treatment.

3.3.3 STARCH

Starch $(C_6H_{10}O_5)_n - (H_2O)$, is another main biomass component, a polysaccharide consisting of glucose monomers bound with α-(1$\longrightarrow$4) and α-(1$\longrightarrow$6) bonds, forming amylose a linear structure and amylopectin a more branched structure. Unlike cellulose, starch is easily hydrolyzed without the addition of acids or enzymes and the reported yields of glucose are lower in hydrothermal treatment compared with conventional enzymatic methods due to further decomposition of glucose. Some authors indicate yields of glucose over 60% from sweet potato starch under subcritical conditions at 200°C for 30 minutes, with no pressure specified. Under more severe conditions (240°C; ten minutes residence time) the glucose yield decreased dramatically due to glucose degradation to 5-hydroxymethylfurfural (HMF) (Nagamori and Funazukuri, 2004). Miyazawa and Funazukui (2005) in a similar study, reported lower glucose yields of 3.7% after 15 minutes residence time at 200°C but the key of this study was that CO2 addition at the ratio of 0.1g CO_2 per g of water, increased the glucose yield significantly at 53%. The CO2 addition increased the acidity of hydrothermal media.

3.3.4 SACCHARIDES

As mentioned above, the carbohydrates in biomass are fundamentally polymers of monosaccharides that are rapidly hydrolyzed to their monomers under subcritical conditions. Under subcritical conditions, cellulose breaks down to glucose and other products, hemicellulose breaks down to monosaccharides among which the most prevalent is 5-carbon sugar xylose. Industrially, most of the global production of furfural is produced from hemicellulose-derived xylose (Peterson et al., 2008). When glucose or fructose dissolves in water, it exists in three forms: as an open chain, pyranose ring, and furanose ring, therefore when glucose and fructose is present in water their forms exist too. Also, glucose reversibly isomerizes into fructose via Lobry de Bruyn, Alberda van Ekenstein (LBAE) transformation, and studies indicate that fructose is more reactive than glucose, at least in the presence of phosphoric acid, when after two minutes at 340°C, 98% of the fructose was degraded whereas only 52% of the glucose was destroyed (Salak and Yoshida, 2006; Toor et al., 2011).

3.4 CONVERSION OF LIGNIN

Lignin is another fraction of lignocellulosic biomass, a heteropolymer with a high molecular weight and a more random structure than hemicellulose, consisting of p-hydroxyphenylpropanoid units held togheter by C-C or C-O-C bonds. The prevalent building blocks in lignin are all phenylpropane derivatives: p-coumaryl alcohol, coniferyl alcohol, and sinapyl alcohol. As hemicellulose, also the lignin possesses similar morphological charachteristics of amorphous form, less solubility similar to cellulose, and the peculiar behavior of its hydrophobic nature. It is relatively resistant to chemical or enzymatic degradation but under subcritical conditions various phenols and methoxy phenols are formed via hydrolysis of ether-bonds. Also, lignin degradation is catalyzed by alkaline pH and hydrothermal liquefaction produces significant amounts of solid residue.

3.5 CONVERSION OF LIPIDS

Fats and oils are nonpolar compounds with mainly aliphatic character also referred to as triacylglycerides triesters of fatty acids and glycerol, their chemical structures are similar to hydrocarbon fuels. In subcritical water processing, the dielectric constant of water is significantly lower allowing greater miscibility (Torr et al., 2011; Khuwijitjaru et al., 2002). Under subcritical conditions the lipids are hydrolyzed to fatty acids that are stable in subcritical water, 90–100% of fatty acids were extracted from soybean oil in 10–15 minutes at 330–340°C and 13.1 MPa (King et al., 1999).

3.5.1 GLYCEROL

Triglyceride hydrolysis generates glycerol that is also a major coproduct of bio-diesel production. It can be used to synthetize specialty chemicals, an important source of energy or fuels. During hydrothermal liquefaction, glycerol is converted into some

TABLE 3.2

Hydrothermal Liquefaction Products

Compound	Reaction Conditions	Products
Glycerol	360°; 34MPa	Acrolein
	$ZnSO_4$—catalyst	
	0–180 seconds	
Glycerol	349–475°	Methanol
	25-45MPa	acetaldehyde
	32–165 seconds	Ethanol; CO
		Formaldehyde
		CO_2; H_2
Stearic acid	400°; 25MPa	C-17 Alkane
	NaOH; KOH—catalyst	CO_2
	30 min.	

Source: Toor, Rosendahl, and Rudolf (2011).

water-soluble compounds. The reaction conditions and the resulting products of some studies are shown in Table 3.2.

3.5.2 Fatty Acids

Under subcritical conditions, fatty acids are relatively stable, but they can be partially degraded to long-chain hydrocarbons that are excellent precursors of fuels. Two major products, heptadecane (C17H36) and cetene (C16H32), were found after processing stearic acid at 400°C and 25 MPa for 30 minutes in a batch reactor (see Table 3.2). Also, alkaline addition (KOH) accelerated the decomposition and the yield was 32%, and the hot compressed water stabilized the fatty acids and suppressed the degradation (Watanabe et al., 2006).

3.6 CONVERSION OF PROTEINS

Proteins are biomass components and their building blocks, the amino acids, have high commercial value (use in feed, food, pharmaceuticals, and cosmetics). The amino acids are linked together by peptide bonds. A considerable fraction of the nitrogen from proteins will be incorporated into bio-oil during hydrothermal liquefaction. Understanding the degradation of proteins via hydrothermal treatment is important, the protein degradation will affect the smell, combustion, and other properties of the product. The C-N bond between carboxyl and amine groups present in all amino-acids, will rapidly hydrolyze in hydrothermal conditions. The peptide bonds are more stable than the glyosidic bonds found in cellulose and starch and reports indicate that slow protein hydrolysis occurs only at temperatures below 230°C (Rogalinski et al., 2008; Brunner, 2009). Some reports showed that with CO_2 addition at 250°C and 25 MPa for about five minutes, the total amino acids yield quadrupled from 3.7 to 15% (Rogalinski et al., 2008). In general, the total hydrolysis yields are low due to degradation.

3.6.1 AMINO ACIDS

Under hydrothermal conditions the amino acids are rapidly degraded compared with other biomass monomers. Amino acids have the same peptide backbone, and, due to their heterogeneity, to describe their degradation is challenging; however, they undergo similar decarboxylation and deamination reactions. Klinger et al. (2007) showed, in a study focused on hydrothermal degradation of glycine and alanine, that the primary mechanism of degradation is decarboxylation and deamination and 50% of the starting material degraded at 350°C and 34 MPa in 5–15 seconds. Also, there was no additional effect observed when the temperature and pressure varied between 300–350°C and 24–34 MPa, respectively. Among the major decomposition products were acetaldehyde, acetaldehyde-hydrate, diketopiperazine, ethylamine, methylamine, formaldehyde, lactic acid, and propionic acid; a summary of the studies on hydrothermal degradation of amino acids is shown in Table 3.3.

3.7 HTL OF LIGNOCELLULOSIC BIOMASS

Hydrothermal liquefaction is an effective method of converting biomass into liquid products which are potential intermediates to produce biofuels and chemicals, although the mechanism of the hydrothermal liquefaction has not been elucidated. The pathway of this process comprises: depolymerization, decomposition, and recombination (Sohail et al., 2011). Biomass is decomposed and depolymerized into

TABLE 3.3
Products from Amino Acids Degradation in Subcritical Water

Substrate (Amino Acid)	Reaction Conditions	Hydrothermal Products
Valine	220–270°C	NH_3, CO_2, CO, Propane, Butane,
Leucine	1hour	Isobutane, Sopentene, 3-methyl-1-butene
Isoleucine	No water	2-methyl-1-butene, Propene, Butene,
		Isobutylene, Acetone, So-butylamine
Glycine	350°C	Acetaldehyde, Acetaldehyde-hydrate,
Alanine	34MPa	Diketopiperazine, Ethylamine, Methylamine,
	5-15s	Formaldehyde, Actic and Formic Acids
Alanine	300°C	NH3, Carbonic Acid, Lactic Acid, Pyruvic Acid
	20MPa	Acrylic Acid, Acetic Acid, pPopionic Acid,
		Formic Acid
Bovine serum	310°C	CO_2, CO, H_2, and CH_4, Acetic Acid,
Albumin (BSA)	25MPa	Propanoic Acid, n-butyric Acid
	30s	Iso-butyric Acid, Iso-valeric Acid

Source: Toor, Rosendahl and Rudolf, 2011.

small components that could be very reactive and they could re-polymerize. The temperature and residence time are critical process parameters of HTL and since biomass is a complex mixture of carbohydrates, lignin, proteins, and lipids, the reaction chemistry and mechanism of biomass liquefaction are therefore also very complex. Depolymerization of biomass is a sequential dissolving of macromolecules that mimics the natural geological processes of producing fossil fuels. Basically, the long chain polymers consisting of hydrogen, oxygen, and carbon are hydrolyzed to shorter chain hydrocarbons under high temperatures and pressure (Zein and Winter, 2000). Further, the decomposition of biomass monomers consists in cleavage, dehydration (loss of water), decarboxylation (loss of CO_2), and deamination (removal of amino acid content). Dehydration and decarboxylation induces oxygen removal from biomass in the form of H_2O and CO_2. The hydrolysis of macromolecules generates polar compounds such as oligomers and monomers, in the case of cellulose the hydrogen bonds are broken down to its glucose monomers. The fructose is more reactive then glucose, therefore it is rapidly degraded via isomerization, hydrolysis, dehydration, reverse-aldol defragmentation, and rearrangement and recombination reactions to a variety of products (Zhang et al., 2016). Most of the degradation products such as polar molecules, furfurals, glycoaldehydes, phenols, and organic acids are highly soluble in water. Finally, the recombination and repolymerization of the reactive fragments, which is the reverse of the first steps (depolymerization and decomposition) occurs due to the unavailability of hydrogen compound (Sohail et al., 2011). When hydrogen is available in the organic media for the HTL process, the free radicals will be capped yielding the stable molecular weight species. On the other hand, under conditions of hydrogen unavailability or high concentration of free radicals, the fragments recombine or depolymerize to form high molecular weight char compounds (Shah, 2015). Under HTL conditions, the biomass is converted to bio-liquid at moderate temperature (280–370°C) and high pressure (10–25 MPa). Bio-crude (bio-oil) is the main product but along with it also gases and aqueous and solids by-products are produced, high-energy content products. The reaction can be conducted in a continuous or batch reactor. Continuous reactors require a feeding system that operates under pressure, pumps that could pump lignocellulosic slurries, pretreatment of feedstocks such as homogenizing the particle size, removing contaminants, and alkaline treatment. Alkaline pretreatment increases the internal surface area of cellulose, decreases crystallinity, cleaves lignin carbohydrate complexes and solubilizes lignin. Crystalline cellulose undergoes a transformation to an amorphous state followed by complete dissolution at temperature between 330–340°C at 25 MPa (Kumar and Gupta, 2008). Also, alkaline pretreatment produces a lower amount of sugar degradation products when compared with acidic pretreatment processes, and the most used catalysts are: NaOH, KOH, ammonia, or lime (Balan, 2014). In some cases, oxidants like oxygen or hydrogen peroxide are used to improve the performance of the pretreatment by efficiently removing lignin (Carvalheiro et al., 2009). The biomass is processed via HTL at 350°C and 15 MPa for approximately 15 minutes. The separation of gaseous stream of CO2, solid residue, bio-crude, and small traces of aqueous phase, occurs almost spontaneously (Rowbotham et al., 2012). The water /aqueous phase can be recirculated to a HTL unit which reduces the water requirement and enhances the bio-oil yield. The solid phase material can be used as biochar/fertilizer.

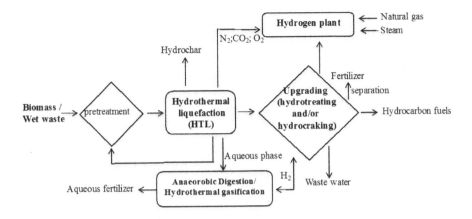

FIGURE 3.1   The simplified diagram of biomass to liquid hydrocarbons via HTL.

However, the bio-crude may not be directly used and needs to be further subjected to a hydrotreatment for commercial utilization; a schematic summarization of these steps, is shown in Figure 3.1. The ultimate target of HTL of biomass is to efficiently convert it to bio-crude that can be utilized as an alternative to the commercial fossil fuels, and the advantage is that the bio-crude obtained via HTL doesn't require much treatment/upgrading procedures for the commercial utilization.

3.8   HTL OF WET/ALGAL BIOMASS

Algal biomass is considered an essential bioenergy feedstock because of their rapid growth rate and for their capacity to harbor significant quantities of biochemical via CO_2 bio-sequestration for biofuel production. Many studies focused on selecting the microalgae strains that can produce large amounts of lipids and optimizing the cultivation conditions (Taleb et al., 2016; Hu et al., 2016; Yee, 2015). There is an interest in producing biofuels from microalgae determined by several microalgae characteristics: able to convert effectively solar energy into biomass, small size, aquatic habitat, high growth rate, can be cultivated on non-arable lands using saline and waste waters as nutrients, cultivation does not require plant protection means (Wijffels et al., 2010). Also, high photosynthetic efficiency, simple life cycle and resource availability for large-scale production, less water intake, and short harvesting periods (Thiruvenkadam et al., 2015). Processing algae (wet biomass) has the potential for recycling nutrients back to cultivation. The aqueous phase generated during the HTL process is recirculated back to the algae cultivation and the CO2 released could be utilized by algae in the photosynthesis process of the next batch. The schematic of the wet biomass hydrothermal processing it is shown in Figure 3.2. Generally, the algae processing has less complexity compared with lignocellulosic feedstock due to the particle size of microalgae and the pumpability of slurries to the reactor. A dewatering step tends to produce a slurry with about 20% solids that is further passed through the HTL reactor to produce bio-crude and then the bio-crude is hydro-treated to produce hydrocarbon fuels (Gollakota et al., 2018).

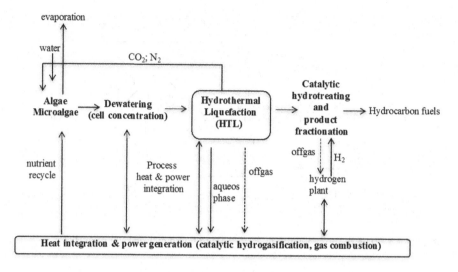

FIGURE 3.2 A general diagram of HTL of microalgae for biofuels production.

Apart from bio-oil, other products such as carbohydrates, polyunsaturated fatty acids, vitamins, minerals, and dietary fibers could be extracted from algae (Thiruvenkadam et al., 2015). The subcritical water technology process uses water as a solvent, that plays dual role as solvent and catalyst, making it more favorable in terms of environmental concerns and lack of solvent recovery operations (Ruiz et al., 2013). The main advantages of this technology are that feedstock pre-drying is not required and a relatively high product yield of all microalgae components such as lipids, proteins, and carbohydrates is attained. The process of biofuel production from microalgae consists in transesterification of fatty carboxylic acids triacylglycerides. The transesterification reaction occurs between low –molecular alcohols (methanol, ethanol) and fatty acids in the presence of alkaline or acid catalyst, as shown in Figure 3.3.

However, the production of bio-diesel involves high energy consumption and uses hazardous organic solvents (methanol, ethanol). In addition, only the lipids are used in bio-diesel production and a greater part of microalgae biomass (carbohydrates, proteins, and other valuable compounds) is not utilized. When algal biomass is subjected to hydrothermal processes, various types of fuels could be obtained at different temperatures such as solid fuel by hydrothermal carbonization (180–250°C), liquid fuel by HTL (250–370°C), or gaseous fuel via hydrothermal gasification at temperatures higher than 400°C. The main product targeted from HTL of algal

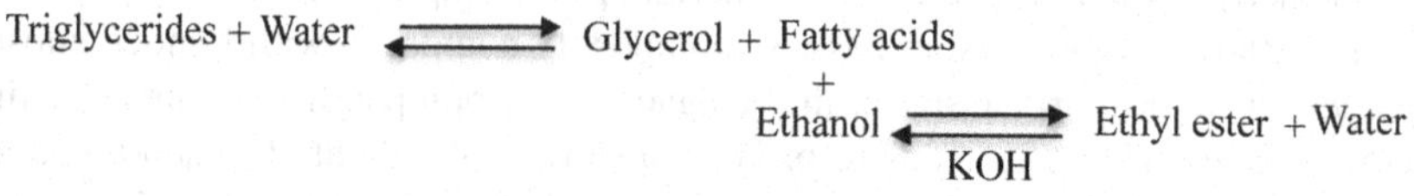

Triglycerides + Water $\rightleftharpoons$ Glycerol + Fatty acids

+

Ethanol $\rightleftharpoons$ Ethyl ester + Water

KOH

FIGURE 3.3 The general reaction of producing biodiesel from lipids.

TABLE 3.4
Summary of Bio-Crude Produced from Various Species of Algae via HTL

Algae Strain	HTL Temperature (t°C)	Residence Time (min)	Catalyst	Bio-Crude (Yield %)
Aurantiochytrium sp.	400	10	/	51.22
Chorellal pyrenoidosa	280	120	/	39.40
Chorella vulgaris	350	60	Pt/Al	38.90
Chorella vulgaris	350	60	$Na_2 CO_3$	30.00
Chorella vulgaris	350	60	HCOOH	30.00
Chorella vulgaris	350	60	$Co/Al_2/O_3$	38.70
Porphyridium cruentum	350	60	$Na_2 CO_3$	30.00
Nannochloropsis sp.	600	1	/	67.00
Nannochloropsis salina	340	30	$Ni-Mo/Al_2O_3$	78.50
Desmodesmus sp.	350	60	/	46.50
Dunaliella tertiolecta	250	5	/	44.80
Nannochloropsisl oculata	350	60	HCOOH	28.00
Spirulina	350	60	$CH_3 COOH$	25.30
Tetraselmis sp.	350	5	/	65.00

Source: Vlaskin, Chernova, Kiseleva, Popel, and Zhuk (2017).

biomass is bio-crude, which is a biodiesel precursor that normally consists of organic acids, various ketones and phenols, and aqueous products of dissolved organics, gas, and solids residue that are coproducts. Bio-crude from HTL of algal biomass does not contain only the lipid fraction but also the proteins and carbohydrate fractions and could be used as a feedstock to produce other bio-products such as resins and polyurethane foam (Xu et al., 2014). Under subcritical conditions, the dielectric constant of water increases and the density decreases from $1g/cm^2$ at 25°C to $0.75g/cm^2$ at 300°C and the polarity of water changes from complete polarity to being moderately nonpolar. This decrease of water polarity at high temperatures favors complete miscibility between lipids and water molecules therefore increasing aqueous lipid extraction yield from algal biomass (Reddy et al., 2013). Bio-crude yield is correlated with the algae strain and it is affected by various factors such as temperature, residence time, solvent/cosolvents used, and the presence of a catalyst. A summary of bio-crude yield obtained via HTL process from various strains of microalgae is shown in Table 3.4.

3.9   INDUSTRIAL APPLICATIONS

About 40 years of research and development were necessary for a few liquefaction technologies to scale and now are capable of processing about 200–3000 tons per year and convert lignocellulosic biomass to biocrudes for use as heavy fuel or for upgrading to biofuels. Biomass liquefaction belongs to the larger field of biomass valorization technologies. The valorization of lignocellulose as residue from agriculture and forestry did have slower progress; however, this feedstock is more preferable as it is available in much larger quantities, at a lower price, and with a lower environmental footprint (Lange, 2018). Technological developments have been slower because lignocellulose is chemically heterogeneous and recalcitrant and, therefore, requires extensive chemistry and processing for upgrading. Various approaches have been proposed, generally based on two main steps, namely "depolymerization" followed by "deoxygenation." The depolymerization step proceeds through gasification (>700°C), pyrolysis (400–500°C), hydrolysis (< 200°C), or hybrid forms of these. Globally, several companies are currently operating at ton level. BioCrak is processing lignocellulose via liquefaction in vacuum gas oil (VGO) at around 375°C and atmospheric pressure, the process is claimed to deliver a yield of 40% of light oil and 39C% of char. The liquefaction process is currently operating at a scale 500–800 tons a^{-1} in Schwechat, Austria. Conoco Phillips also reported the liquefaction of lignocellulose in VGO at 320–400^{0}C to produce bio-crude with 60 wt % yield. Acid-catalyzed liquefaction concept, a multistep process to digest lignocellulose in a g-valerolactone/water mixture to produce furfural, levulinic acid, and lignin, precipitate the lignin by water addition, and upgrade the levulinic acid to g-valerolac- tone for partial recycling as reaction solvent is presently being developed by GlucanBio. Ignite from Canada has developed the catalytic hydrothermal reactor (Cat-HTR) to convert lignocellulose, waste plastic, or lignite to diesel. Its subsidiary, Licella, has partnered with the Canadian pulp company Canfor to deploy the technology for lignocellulose upgrading at 3 kt -^{1} scale. The process is claimed to operate in subcritical water in the presence of base (e.g., NaOH) and hydrogen-donor agent (Na-formate). Steeper Energy is also using supercritical water and reporting to produce a bio-crude with a yield of 45 wt% (or 80% energy efficiency) with an oxygen content of 8–10wt%. The hydrofraction process is being demonstrated at the scale of 200 t a^{-1} in a demonstration unit that is colocated with the Daishowa Marubeni International Alberta Peace River pulp mill in Canada. The technology is claimed to deliver renewable diesel at $140 per BOE or $3.3 per gallon. The Catchlight development was wined down in 2013, but eventually resumed in a collaboration between Chevron and the Iowa State's BioCentury Research Farm. The research is carried out at a scale of 0.5–1 kg h^1 biomass and 2–4 kg h^1 solvent. Bio-oil is produced at 55 wt % yield with hydrocarbon solvents and 67 wt % with phenolic solvents (Lange, 2018). There has been a lot of research and pilot scale demonstration by the U.S. Department of Energy (DOE) labs, biofuels industries, and universities on HTL. In fact, the DOE has identified subcritical water/hydrothermal processes as a viable technology which can process wet biomass for bio-crude production. The most of the DOE supported projects focused on demonstrating high process and carbon efficiencies for HTL with a broad range of feedstocks (e.g., lignocellulosic biomass, municipal solid waste, algae, and lignin) and assess value of bio-crude for fuels and chemicals using the continuous flow scalable reactors design. Researchers from the DOE National Renewable Energy Laboratory (NREL) and Pacific Northwest National Laboratory (PNNL) have released a report on the potential production of bio-crude from organic wastes such as wastewater sludge, animal manure, food waste, and fats, oils, and greases throughout the USA.

3.10 CONCLUSIONS

HTL is a medium-temperature, high-pressure thermochemical process which can convert biomass into a liquid product called bio-oil or bio-crude. During the HTL process, the biomass macromolecules are hydrolyzed and/or degraded into smaller molecules, most of which are unstable and highly reactive and therefore can recombine. Most of the oxygen from biomass is removed via dehydration or decarboxylation. The quality of bio-oil is directly correlated with the biomass substrate composition. HTL technology is attractive from the point of view of energy consumption and process integration making it a promising method for biomass conversion and production of high quality bio-oil, the energy recovery from biomass to fuel could be as high as 73–80% (Toor et al., 2011). Also, reports showed that the bio-oil yield is highly dependent on the chemical composition of the biomass; in the case of walnut shell, the maximum bio-oil obtained was 25% by weight at a temperature of 320°C, under subcritical water conditions. In addition, long residence times lead to high bio-oil yields (de Caprariis et al., 2017). On the other hand, there is an increasing interest in applying microalgae conversion technology to bio-oil. The literature suggests that it is important to solve some challenges such as: identifying microalgae nitrogen-sulfur-low strains that possess high productivity, developing a cultivation system that uses wastewater or seawater, optimization of the HTL process and the solvent system used for bio-oil separation, finding solutions for maximizing the yields and enhancing the quality of bio-oil using various catalysts and/or feedstock activation technology (Vlaskin et al., 2017). The road to commercialization is associated with energy demands to avoid the biofuel production with higher greenhouse gas emissions than conventional fuel. Pilot studies should focus on addressing the current challenges associated with the yield and biomass composition.

REFERENCES

Adschiri T., Hirose S., Malalulan R. and Arai K., (1993) Noncatalytic conversion of cellulose in supercritical and subcritical water. *J. Chem. Eng. Jpn.* 26(6), 676–680.

Andrianov V., et al., (2010) Tobacco as a production platform for biofuel: Overexpression of Arabidopsis DGAT and LEC2 genes increases accumulation and shifts the composition of lipids in green biomass. *Plant Biotechnol. J.* 8(3), 277–287.

Balan V., (2014) Current challenges in commercially producing biofuels from lignocellulosic biomass. *ISRN Biotechnol.*

Bobleter O., (1994) Hydrothermal degradation of polymers derived from plants. *Polym. Sci.* 19, 797–841.

Brunner G., (2009) Near critical and subcritical water. Part I. Hydrolytic and hydrothermal processes. *J. Supercrit. Fluids* 47, 373–381.

Carvalheiro F., et al., (2009) Wheat straw auto-hydrolysis process optimization and products characterization. *Appl. Biochem. Biotechnol.* 153(1–3), 84–93.

de Caprariis B., et al., (2017) Hydrothermal liquefaction of biomass: Influence of temperature and biomass composition on the bio-oil production. *Fuel* 208, 618–625.

Dimitriadis A. and Bezergianni S., (2017) Hydrothermal liquefaction of various biomass and waste feedstocks for bio-crude production: A state of art review. *Renew. Sustain. Energy Rev.* 68, 113–125.

Gollakota A., Kishore N. and Gu S., (2018) A review on hydrothermal liquefaction of bio-mass. *Renew. Sustain. Energy Rev.* 81, 1378–1392.

Hu X., Zhou J., Liu G. and Gui B., (2016) Selection of microalgae for high CO2 fixation efficiency and lipid accumulation from ten Chlorella strains using municipal wastewater. *J. Environ. Sci.* 46, 83–91.

Kamio E., Sato H., Takahashi S., Noda H., Fukuhara C. and Okamura T., (2008) Liquefaction kinetics of cellulose treated by hot compressed water under variable temperature conditions. *J. Mater. Sci.* 43, 2179–2188.

Khuwijitjaru P., Adachi S. and Matsuno R., (2002) Solubility of saturated fatty acids in water at elevated temperatures. *Biosci. Biotechnol. Biochem.* 66, 1723–1726.

King J.W., Holliday R.I. and List G.R., (1999) Hydrolysis of soybean oil in a subcritical water flow reactor. *Green Chem.* 1, 261–264.

Klinger D., Berg J. and Vogel H., (2007) Hydrothermal reactions of alanine and glycine in sub- and supercritical water. *J. Supercrit. Fluids* 43(1), 112–119.

Kumar S., (2012) Sub- and supercritical water-based processes for microalgae to biofuels, in *The Science of Algal Fuels*, R. Gordon and J. Seckbach, Editors. Netherlands: Springer. p. 467–493.

Kumar S. and Gupta R.B., (2008) Hydrolysis of microcrystalline cellulose in subcritical and supercritical water in a continuous flow reactor. *Ind. Eng. Chem. Res.* 47, 9321–9329.

Lange J.P., (2018) Lignocellulose liquefaction to biocrude: A tutorial review. *ChemSusChem* 11, 997–1014.

Miyazawa T. and Funazukuri T., (2005) Polysaccharide hydrolysis accelerated by adding carbon dioxide under hydrothermal conditions. *Biotechnol. Progr.* 21, 1782–1786.

Mok W.S.L. and Antal M.J., (1992) Uncatalyzed solvolysis of whole biomass hemicellulose by hot compressed liquid water. *Ind. Eng. Chem. Res.* 31, 1157–1161.

Nagamori M. and Funazukuri T., (2004) Glucose production by hydrolysis of starch under hydrothermal conditions. *J. Chem. Technol. Biotechnol.* 79, 229–233.

Peterson A.A., et al., (2008) Thermochemical biofuel production in hydrothermal media: A review of sub- and supercritical water technologies. *Energy Environ. Sci.* 1, 32–65.

Reddy H.K., et al., (2013) ASI: Hyrothermal extraction and characterization of bio-crude-oils from wet *Chlorella sorokiniana* and *Dunaleiella tertiolecta*. *Environ. Progr. Sustain. Energy* 32, 910–915.

Rogalinski T., Liu K., Albrecht T. and Brunner G., (2008) Hydrolysis kinetics of biopolymers in subcritical water. *J. Supercrit. Fluids* 46, 335–341.

Rowbotham J., Dyer P., Greenwell H. and Theodorou M., (2012) Thermochemical processing of macroalgae: A late bloomer in the development of third-generation biofuels? *Biofuels* 3, 441–461.

Ruiz H.A., et al., (2013) Hydrothermal processing as an alternative for upgrading agriculture residues and marine biomass according to the biorefinery concept: A review. *Renew. Sustain. Energy Rev.* 21, 35–51.

Salak A. and Yoshida H., (2006) Acid-catalyzed production of 5-hydroxymethyl furfural from D-fructose in subcritical water. *Ind. Eng. Chem. Res.* 45(7), 2163–2173.

Sasaki M., Fang Z., Fukushima Y., Adschiri T. and Arai K., (2000) Dissolution and hydrolysis of cellulose in subcritical and supercritical water. *Ind. Eng. Chem. Res.* 39, 2883–2890.

Schwald W. and Bobleter O., (1989) Hydrothermolysis of cellulose under static and dynamic conditions at high temperatures. J. Carbohydr. Chem. 8(4), 565–578.

Shah Y.T., (2015) *Energy and Fuel Systems Integration*. Boca Raton, FL: Taylor and Francis Group.

Sohail S., Rosendahl L. and Rudolf A., (2011) Hydrothermal liquefaction of biomass: A review of subcritical water technologies. *Energy* 36, 2328–2342.

Taleb A., et al., (2016) Screening of freshwater and sea-water microalgae strains in fully controlled photobioreactors for biodiesel production. *Bioresour. Technol.* 218, 480–490.

Thiruvenkadam S., Izhar S., Yoshida H., Danquah M. and Harun R., (2015) Process application of subcritical water extraction (SWE) for algal bio-products and biofuels production. *Appl. Energy* 154, 815–828.

Toor S.S., Rosendahl L. and Rudolf A., (2011) Hydrothermal liquefaction of biomass: A review of subcritical water technologies. *Energy* 36, 2328–2342.

Vlaskin M.S., Chernova N.I., Kiseleva S.V., Popel O.S. and Zhuk A.Z., (2017) Hydrothermal liquefaction of microalgae to produce biofuels: State of the art and future prospects. *Thermal Eng.* 64(9), 627–636.

Watanabe M., Iida T. and Inomata H., (2006) Decomposition of a long chain saturated fatty acid with some additives in hot compressed water. *Energy Convers. Manage.* 47, 3344–3350.

Wijffels R.H., Barbosa M.J. and Eppink M.H., (2010) Microalgae for the production of bulk chemicals and biofuels. *Biofuel. Bioprod. Bior.* 4, 287–295.

Xu C.C., et al., (2014) Hydrothermal liquefaction of biomass in hot-compressed water, alcohols, and alcohols-water co-solvents for bio-crude production, in *Application of Hydrothermal Reactions to Biomass Conversion*. Springer, New York, NY. p. 171–187.

Yee W., (2015) Feasibility of various carbon sources and plant materials in enhancing the growth and biomass productivity of the fresh-water microalgae *Monoraphidium griffithii* NS16. *Bioresour. Technol.* 196, 1–8.

Zein M. and Winter R., (2000) Effect of temperature, pressure and lipid acyl chain length on the structure and phase behavior of phospholipid-gramicidin bilayers. *Phys. Chem.* 2, 45–51.

Zhang X., Wilson K. and Lee A.F., (2016) Heterogeneously catalyzed hydrothermal processing of C5-C6 sugars. *Chem. Rev.* 116, 12328–12368.

4 Hydrothermal Carbonization for Producing Carbon Materials

Anuj Thakkar and Sandeep Kumar

CONTENTS

4.1 INTRODUCTION

Depletion of fossil fuels and environmental concerns are prompting a search for alternative and renewable energy resources. Various options include wind, hydro/oceanic, geothermal, solar, and biomass energies, among which biomass is the only renewable source capable of directly producing solid, liquid, and gaseous fuels. The additional advantage of biomass utilization is in its wider distribution in the world which may eliminate the regional dependence for the energy sources. Biofuels are CO_2 neutral, as during the growth and utilization process of biomass, the same amount of carbon dioxide and other forms of carbon is absorbed and released, respectively. Hence, the concentration of carbon dioxide in the atmosphere theoretically remains constant in this cycle (Matsumura et al., 2006). The conversion of biomass to biofuels can be realized in several pathways such as pyrolysis, gasification, fermentation, and hydrothermal reaction (Kumar & Gupta, 2009). Hydrogen (Lee, Kim, & Ihm, 2002), ethanol (Motonobu, Ryusaku, & Tsutomu, 2004), organic acid (Kishida, Jin, & Zhou, 2005; Kong, Li, & Wang, 2008; Lourdes & David, 2002), biofuel (Selhan, Thallada, & Akinori, 2005), and hydrochar (Abdullah & Wu, 2009) are produced in different processes. Recently, the research attention was focused on the conversion of biomass to hydrochar which has a higher heating valve (HHV) typically over 28 kJ/g and is richer in carbon content as compared to raw biomass. Hydrochar is defined here as the non-liquefied carbonaceous solid product of a high energy density value containing mainly C, H, O, and ash after the thermal treatment of biomass. Table 4.1 compares some of the advantages of hydrochar as solid fuels as compared to raw biomass.

In nature, coal is formed from plant material undergoing heat and pressure treatment over millions of years. The acceleration of coalification of biomass by a factor of 10^6–10^9 in hydrothermal medium under milder process conditions can be a considerable and technically attractive alternative for hydrochar production (Titirici, Thomas, & Antonietti, 2007). Essentially, all forms of biomass can be converted to hydrochar. Forest thinning, herbaceous grasses, crop residues, manure, and paper sludge are some of the potentially attractive feedstocks.

The enhanced transportation and solubilization properties of sub- and supercritical water (hydrothermal medium) play an important role in the transformation of biomass to high energy density fuels and functional materials (Hu, Yu, Wang, Liu, & Xu, 2008). Here, water acts as reactant as well as reaction medium which helps in performing hydrolysis, depolymerization, dehydration, and decarboxylation reactions. The proton-catalyzed mechanism, direct nucleophilic attack mechanism, hydroxide ion catalyzed mechanism, and the radical mechanism play important roles in the conversion of biomass in hydrothermal medium (Masaru, Takafumi, & Hiroshi, 2004; Savage, 1999). Under the umbrella of hydrothermal process, the conversion of a variety of biomass to chemicals, including organic monomer, biofuel, hydrogen, and hydrochar have been studied widely. Hydrothermal carbonization (HTC) is defined as combined dehydration and decarboxylation of a fuel to increase its carbon content with an objective of achieving a higher heating value (Funke & Ziegler, 2010). Through hydrothermal carbonization, a carbon-rich black

TABLE 4.1

Comparison of Biomass and Biochar as Solid Fuel

Biomass	Biochar
High moisture retention	Low moisture retention and easily dried
Low heating value, so high transportation cost	High heating value, so less ($)/MJ cost during transportation
Perishable on storage	Not perishable, and can be stored longer
Fibrous and so, difficult material handling	Friable, easier to compact and handle
Poor compatibility with coal for co-firing	Better compatibility with coal for co-firing

solid as insoluble product is obtained from biomass in 180–350°C range (Savovaa et al., 2001; Titirici et al., 2007).

The earliest research focused on analyzing the changes in O/C and H/C atomic ratios was to understand the chemical transformations taking place during HTC (Bergius & Specht, 1913). For example, Marta et al. studied the hydrothermal carbonization of three different saccharides (glucose, sucrose, and starch) at temperatures ranging from 170 to 240°C. The result showed that a carbon-rich solid product made up of uniform spherical micrometer-sized particles of diameter 0.4–6 mm range could be synthesized by modifying the reaction conditions. The formation of the carbon-rich solid through the HTC of saccharides was the consequence of dehydration, condensation, polymerization, and aromatization reactions. In another study, the same group used cellulose as starting material and successfully established that highly functionalized carbonaceous materials can be produced by HTC from cellulose in the range of 220 to 250°C (Marta & Antonio, 2009). The formation of this material follows essentially the path of a dehydration process, similar to that previously observed for the hydrothermal transformation of saccharides such as glucose, sucrose, or starch. Titirici et al. compared hydrothermal carbons synthesized from diverse monomeric sugars and their derivatives (5-hydroxymethyl-furfural-1-aldehyde (HMF) and furfural) under hydrothermal conditions at 180°C with respect to their chemical structures. The results showed that type of sugars has an effect on the structure of carbon-rich solids (Titirici, Antonietti, & Baccile, 2008). The majority of the research studies have utilized monomeric sugars instead of real biomass as model compounds for understanding the reaction mechanism and the structural differences in solids formed from the diverse sugar sources in the HTC process.

The traditional method for hydrochar production from biomass sources is slow pyrolysis, where dry biomass is used for the purpose in the range of 500 to 800°C. Antal Jr. and coworkers developed another method for charcoal production which is named flash carbonization. The process is conducted at elevated pressure by the ignition and control of a flash fire within a packed bed of biomass (Varhegyi et al., 1998). Considering the relatively high energy consumption needed in the pyrolysis process, HTC which is typically conducted in subcritical water below 350°C may potentially be an economical and efficient option for hydrochar production. The process is particularly attractive since it can utilize the wet biomass.

4.2 PROCESS PARAMETERS

The purpose of optimization of process parameters is to maximize the production of hydrochar and/or tune its properties. The distribution of biomass into various phases vary with biomass and HTC reaction parameters. Figure 4.1 depicts product distribution in the HTC process. The four most critical process parameters of temperature, pressure, residence time, and pH are discussed here.

4.2.1 TEMPERATURE

Temperature is the most important parameter in HTC as it determines the property of water. Various HTC experiments carried out utilize temperature in the range

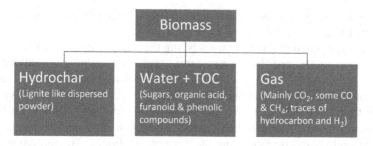

FIGURE 4.1   Product distribution in HTC. (From Funke & Ziegler, 2010.)

of 180°C to 350°C. The concentration of ionic products H⁺ and OH⁻ increases as the temperature in SCWG rises to 300°C, decreasing thereafter (Savage, 2009). At higher temperature, free radical reactions are favored causing dissolution of complex organic compounds and their degradation products to gaseous products (Chuntanapum & Matsumura, 2010). In subcritical conditions where ionic reactions dominate, increase in temperature affects the viscosity of water and thus causes higher water penetration in porous biomass. Higher water penetration further helps in the breakdown of the biomass (Möller, Nilges, Harnisch, & Schröder, 2011). If temperature is not sufficient to break down biomass, pyrolysis like reaction occurs within the biomass (Wang, Zhai, Zhu, Li, & Zeng, 2018). It is observed that H/C and O/C ratio of hydrochar decreases with rise in temperature in various biomass sources (Hwang, Aoyama, Matsuto, Nakagishi, & Matsuo, 2012; Kim, Lee, & Park, 2014; Parshetti, Hoekman, & Balasubramanian, 2013; Sevilla & Fuertes, 2009a, 2009b).

Generally, it is found that a higher heating rate does not favor hydrochar formation. Zhang et al. investigated heating rates in the range of 5–140°C min⁻¹ on the hydrothermal process of grassland perennials and found that hydrochar yield decreased from 22%–23% to 8–9% when the heating rate increased (Zhang, von Keitz, & Valentas, 2009). After the dissolution and stabilization of the fragmented compounds in the hydrothermal process, secondary reactions occurred for the bulk fragmentation of biomass and decreasing the heating rate could result in more solid residue formation (Akhtar & Amin, 2011). A higher heating rate is beneficial for liquefaction, not for carbonization. Hydrochar produced at a lower heating rate has higher heating value than hydrochar produced at a higher heating rate (Wang et al., 2018).

4.2.2 PRESSURE

The reaction pressure influences the reaction network according to the principles of the LeChatelier principle. Reaction equilibrium shifts toward the formation of solid and liquid phases with rise in pressure. Both dehydration and decarboxylation are being depressed at elevated reaction pressure. This effect has been verified experimentally but proved to have a low impact on HTC. The removal of extractables from biomass is generally believed to be facilitated at higher reaction pressure. It is assumed that at higher pressure, encapsulated gases are compacted and dissolved more easily in water which allows biomass higher accessibility to its liquid phase (Funke & Ziegler, 2010).

4.2.3 RESIDENCE TIME

HTC of biomass is relatively a slow reaction. Various residence times ranging from a few minutes to several hours have been reported. Contrary to the expected lower hydrochar yield at higher residence time due to higher solvation, longer residence time significantly improved hydrochar yield (Schuhmacher, Huntjens, & Vankrevelen, 1960; Sevilla & Fuertes, 2009b). This could be due to polymerization and precipitation of dissolved organic compounds in the aqueous phase. The improvement in the heating value of hydrochar with increase in residence time could be because of elimination of oxygen-rich molecules and hydrolysis of hemicellulose which has a heating value lower than the average heating value of biomass. Compared to the long residence time required for carbonization, hydrolysis and extraction happens in a few minutes (Bart, 2006; Hashaikeh, Fang, Butler, Hawari, & Kozinski, 2007; Karagöz, Bhaskar, Muto, & Sakata, 2005b; Mok & Antal Jr, 1992; Peterson et al., 2008). Also, residence time of 30 minutes resulted in formation of hydrochar with a heating value equal to or more than the heating value of lignin which is the component with the highest heating value in lignocellulosic biomass (Inoue, Hanaoka, & Minowa, 2002; Inoue, Uno, & Minowa, 2008). Thus, indicating that elimination of lower heating value product is not the only reason for higher heating value of hydrochar. It is not clear which hydrothermal carbonization reaction mechanism is determining the rate, but biomass decomposition and condensation polymerization are likely to govern the overall rate of reaction (Funke & Ziegler, 2010).

4.2.4 pH

It has been found that the pH of the reaction mixture drops due to formation of organic acids like formic acid, acetic acid, levulinic acid, etc. during HTC (Antal, Mok, & Richards, 1990; Orem, Neuzil, Lerch, & Cecil, 1996; Takeda, Sato, & Machihara, 1990; Tropsch & Phillipovich, 1922). Acids as well as bases have been tested for their effect on HTC. Different pH conditions have a significant effect on reaction rate, product distribution and product properties (Schuhmacher et al., 1960; Titirici et al., 2007). Higher pH leads to higher H/C ratio of the product and bitumen content in coal (Blazsó et al., 1986; Khemchandani, Ray, & Sarkar, 1974). This phenomenon is generally used in hydrothermal liquefaction. It has been found that acidic pH enhances cellulose hydrolysis compared to neutral pH (Funke & Ziegler, 2010). It has been also reported that weakly acidic conditions enhances the overall rate of reaction during HTC (Titirici et al., 2007).

4.2.5 CATALYST

HTC can be classified as direct HTC and catalytic HTC. High temperature and pressure conditions using only water and biomass is direct HTC while addition of any catalysts to it makes it catalytic HTC. Catalysts change properties of water and biomass mixture and thus ultimately has an effect on properties of the final product. The selection of catalyst depends on the properties desired in hydrochar. The formation of hydrochar can be increased by using acid catalysts which facilitate hydrolysis. While the hydrochar formation can be decreased by applying basic catalysts to enhance the formation of liquid products (Nizamuddin et al., 2017). NO_x emission during combustion of hydrochar can be due to the fixation of nitrogen in the air and oxidation of N in the fuel. NO_x emission can be controlled by controlling N content of the hydrochar. It has been shown that replacing the pure water with more acidic and basic aqueous solution during HTC by using catalysts can increase the removal of N (Zhao, Shen, Ge, Chen, & Yoshikawa, 2014). Mumme et al. studied the impacts of natural Zeolite in HTC of an agricultural digestate and cellulose and saw an increase in energy and carbon content of the hydrochar produced from digestate considerably and from cellulose slightly. Also, the surface area and pore volume of the products from catalytic HTC were higher (Mumme et al., 2015). Hamid et al. found that using Lewis acid catalysts $FeCl_2$ and $FeCl_3$ in HTC can lead to complete carbonization at temperatures as low as 200°C (Hamid, Teh, & Lim, 2015). Karagöz et al. studied the effect of basic catalysts RbOH and CsOH on pine wood and observed an increase in liquid product and decrease in solid product (Karagöz, Bhaskar, Muto, & Sakata, 2005a). A similar effect was observed by using K_2CO_3 (base) on sawdust (Karagöz, Bhaskar, Muto, & Sakata, 2006). Good catalysts for HTC are the ones that are thermally stable, effective, cost-effective, and have high selectivity toward the required yield. Figure 4.2 shows a variety of catalysts used in biomass conversion technologies, the favored reactions in the dark-gray box and the advantages and disadvantages in the light-gray box.

4.3   HTC CHEMISTRY

The HTC process overall is a governed decarboxylation and dehydration reaction (Peterson et al., 2008); hence it is an exothermic reaction. HTC is a result of many complex reactions and thus, the reaction network is not very well known. The reactions that have been identified and can provide reasonable information about the chemistry behind HTC are hydrolysis, dehydration, decarboxylation, condensation polymerization, and aromatization.

Hydrolysis of biomass is the cleavage of ether and ester bond by the addition of a water molecule. Cellulose is hydrolyzed at temperature above 200°C and hemicellulose is hydrolyzed at around 180°C. Alkaline condition gives higher hydrolysis reaction rate compared to acidic and neutral condition. Due to the higher amount of ether bonds, lignin is hydrolyzed at around 200°C (Funke & Ziegler, 2010). Unlike with pure model compounds, in real biomass interaction between various biomass components are unavoidable but very little is known about it.

Dehydration is removal of water molecules from biomass components which result in components with lower H/C and O/C ratios. It is generally explained by elimination of the hydroxyl group. Dehydration of glucose happens by formation of HMF or 1,6-anhydroglucose (Kabyemela, Adschiri, Malaluan, & Arai, 1999). In lignin, there is dehydroxylation of catechol and dehydration during cleavage of alcohol and phenolic groups above 150°C and 200°C, respectively (Murray & Evans, 1972; Ross, Loo, Doris, & Hirschon, 1991). Dehydration of cellulose can be expressed as follows (Funke & Ziegler, 2010):

$$4\left(C_6H_{10}O_5\right)_n \ \rightarrow \ 2\left(C_{12}H_{10}O_5\right)_n + 10H_2O$$

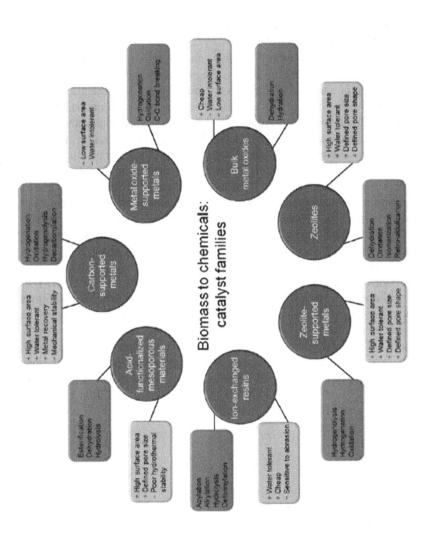

FIGURE 4.2 Variety of catalysts used in biomass conversion technologies, favored reactions in dark-gray box and advantages and disadvantages in light-gray box. (Reproduced from Dapsens, Mondelli, & Pérez-Ramírez, 2012.)

Under hydrothermal conditions above 150°C, carboxyl and carbonyl group degrade rapidly to form CO_2 and CO, respectively (Murray & Evans, 1972). One of the sources of CO_2 could be decomposition of formic acid which is formed by degrading cellulose under hydrothermal conditions (McCollom, Ritter, & Simoneit, 1999; Yu & Savage, 1998). It is also assumed that at elevated temperature above 300°C, water acts like an oxidizing agent. These oxidized products undergo thermal destruction and form CO_2 (Michels et al., 1996). Another possible explanation of CO_2 formation could be condensation reactions (Schafer, 1972) and cleavage of intramolecular bonds (Siskin & Katritzky, 1991).

Polymerization that forms solid products is an unwanted side reaction in other hydrothermal processes. The highly reactive fragments that polymerize are formed by the elimination of carboxyl and hydroxyl group (Funke & Ziegler, 2010). The formation of hydrochar through HTC is characterized by condensation polymerization, specifically aldol condensation (Kabyemela, Adschiri, Malaluan, & Arai, 1999; Nelson, Molton, Russell, & Hallen, 1984). In the case of lignin, polymerization happens in minutes at temperatures above 300°C while at room temperature it takes months (Aronovsky & Gortner, 1930). Fragments of hemicellulose stabilize fragments of lignin and thus slow down condensation reaction significantly (Bobleter, 1994). Condensation reactions of monosaccharides are slow as cross-linked polymerization competes with recondensation of oligosaccharides (Kuster, 1990). It has also been found that condensing molecules within the biomass blocks access of water to other macro-molecules and causes pyrolysis like reactions (Hashaikeh, Fang, Butler, Hawari, & Kozinski, 2007).

The rise in aromaticity of carbonaceous compounds with rise in severity of hydrothermal condition is shown using C-NMR measurements. Alkaline conditions favor aromatization and the reaction is significantly dependent on temperature. The aromatic compounds formed as a result of HTC are highly stable and are the building blocks of hydrochar. Lignin has a significant amount of aromatic bonds and thus hydrothermal treatment has less effect on its carbon content. Many other reactions like transformation, demethylation, pyrolytic reactions, and Fischer–Tropsch reactions are also said to have some role in HTC (Funke & Ziegler, 2010). Figure 4.3 represents the reaction pathway for hydrothermal carbonization of lignocellulosic biomass.

4.4 PROCESS DESIGN

Based on all the process parameters discussed and HTC chemistry, an outline can be drawn for designing the HTC process. First of all, the biomass to water ratio fed to the reactor should be as high as possible without compromising the yield and desired product properties. High biomass feed rate helps in process economics by reducing pumping energy cost and equipment size. The residence time should be high enough to have complete carbonization and minimum dissolved organic compounds in aqueous phase. Higher temperatures generally accelerate hydrothermal carbonization and might even allow for a higher achievable carbon content. Higher temperature causes higher energy losses and higher-pressure generation; this affects process economics. Thus, careful selection of temperature should be done considering the application of

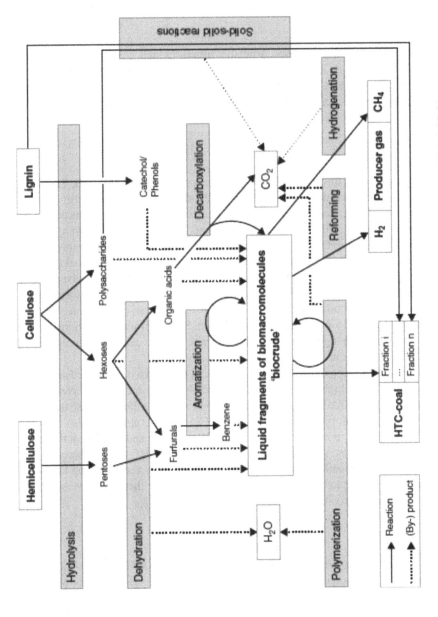

FIGURE 4.3 Reaction pathway for hydrothermal carbonization. (Reproduced from Kruse, Funke, & Titirici, 2013.)

end product. Biomass hydrolysis appears to be a diffusion-controlled rate determining step and thus, size of the feed could be an important factor in HTC. Grinding of biomass adds to energy demand and capital cost with no clear advantage in HTC performance.

4.5   HTC APPLICATION

Hydrochar consists of condensed aromatic structures and oxygenated functional groups (OFGs) (Liu & Zhang, 2009; Liu, Zhang, & Wu, 2010; Sevilla & Fuertes, 2009b). The presence of OFGs make hydrochar more hydrophilic and thus suitable for its application as catalyst, adsorbent, and precursor for activated carbon (Baccile et al., 2009; Titirici, Antonietti, & Baccile, 2008; Xue et al., 2012).

4.5.1   BIOFUELS

Hydrochar produced through HTC is a high energy density solid product. The heating value of hydrochar makes it competitive with other carbonaceous energy sources like lignite (Li, Flora, Caicedo, & Berge, 2015). Also, the hydrochar produced can be tuned in terms of ash, sulfur, and nitrogen content (He, Wang, Yang, & Wang, 2014). The combustion properties of hydrochar has been studied using Thermogravimetric Analyzer (TGA) in the lab. Some researchers have also blended it with coal to adjust fuel properties like ignition, emission, etc. Special attention should be given to ash behavior during the combustion of hydrochar. Ash could lead to operational issues including slagging, fouling, and ash melting. Other than direct combustion of hydrochar, its gasification to syngas is also investigated. Syngas produced from hydrochar can produce gas with higher content of H_2, CO, and in some cases CH_4, compared to untreated biomass (Román et al., 2018).

Energy densification of the HTC char produced from mixed wood feedstock was found to increase at higher process temperatures for experiments conducted in the range of 215–295°C. At 255°C, compared to raw feedstock, a 39% increase in energy density was achieved; this rose to 45% at 295°C. Higher temperature also requires higher pressure. Severe conditions need specialized process and equipment that adds to complexity and cost. At operating temperatures near 255°C and pressures near 5 MPa, the resulting char has coal-like properties and is expected to exhibit favorable behavior with respect to combustion, gasification, and other thermal conversion processes (Hoekman, Broch, & Robbins, 2011).

4.5.2   ADSORBENT

Liu et al. found that hydrochar derived from pinewood had OFGs about 340% higher compared to the pyrolytic char, and when employed as an adsorbent showed 62% higher uptake of copper ions. Even though pyrolysis char is more porous than hydrochar, OFGs made hydrochar a better adsorbent. Compared to pine wood derived hydrochar, rice husk derived hydrochar had lower OFGs content, depicting the importance of biomass precursor. Thus, pine wood hydrochar resulted in a 77% higher removal rate of lead (Liu et al., 2010).

The surface characteristics of hydrochar and efficient activation using various techniques are conducive to higher surface area and uptake capacities of hydrochar. Sevilla et al. produced high-surface area carbons (up to 2700 m²/g) by activation using KOH from furfural, glucose, starch, cellulose, and eucalyptus sawdust. This activated hydrochar had high storage capacity of H_2 up to 16.4 μmol/m² (Sevilla, Fuertes, & Mokaya, 2011). It was also found that hydrochar derived from rye straw had Brunauer–Emmett–Teller (BET) surface area of about 2200 m²/g and thus it could uptake 20 mmol CO_2/g at 25°C and 40 bar (Falco et al., 2013). Romero-Anaya et al. investigated the effect of various chemical and physical activating agents like KOH, NaOH, H_3PO_4, and CO_2 on porosity creation with glucose and sucrose derived hydrochar as precursors. The total oxygen content in the activated carbon with different activation methods varied in the following order: CO_2< H_3PO_4< KOH<NaOH (Romero-Anaya, Ouzzine, Lillo-Ródenas, & Linares-Solano, 2014).

Other than heavy metal removal, hydrochar is also investigated for its role in the removal of pathogens. Raw and KOH activated maize hydrochar removed up to 72% and 93% *E. coli*, respectively. The higher rate of the modified hydrochar is due to KOH improving the porous surface structure and decreasing negative surface charges on the hydrochar, as *E. coli* has a negative surface charges (Chung, Foppen, Izquierdo, & Lens, 2014).

4.5.3 ENERGY STORAGE

Hydrochar has been researched for its application as an electrode material i.e., for electrochemical capacitors or supercapacitors. Many carbon materials like carbon nanotubes, activated carbon, and composite materials are being investigated for this application, but the economics and environmental impact has made researchers explore a more sustainable alternative (Wang et al., 2017). The porosity of the hydrochar produced from biomass through HTC is generally not good enough for its direct application in energy storage devices. To improve the pore size distribution and surface area characteristics of hydrochar which is desired for its use as an electrode, hydrochar needs further treatment which takes away some of its advantage as a green solution. One notable advantage of hydrochar as electrode material is the homogeneous distribution of oxygen and nitrogen which enhances pseudocapacitance via additional Faradaic reactions and intrinsic catalytic activity in electrochemical devices (Román et al., 2018).

4.5.4 CATALYST

High stability and tunable surface properties of hydrochar make it an efficient catalyst or catalyst support in many processes (Román et al., 2018). Hydrochar has been widely investigated as a catalyst for production of H_2. Safari et. al. found that hydrochar as catalyst in hydrothermal gasification of macroalgae not only enhanced H_2 formation but also promoted phenol formation and inhibited acid production in aqueous phase (Safari, Norouzi, & Tavasoli, 2016). Hydrochar has also been successfully used as catalyst to enhance degradation of organic matter. Due to rich and tunable OFGs, hydrochar is used in many carbon catalyzed reactions where active sites are required (Román et al., 2018).

4.6 CONCLUSION AND WAY FORWARD

Hydrothermal carbonization of biomass has great potential to become an environmentally friendly process for the production of a wide variety of products. HTC can produce highly homogenized coal like product from moist biomass. Stopping side product formation like H_2O, CO_2, and organic compounds dissolved in the liquid phase is practically impossible. Hydrolysis, dehydration, decarboxylation, aromatization, and condensation polymerization have been identified as the major reactions during HTC. Process optimization and integration is required to make the process technically feasible and economically viable. The composition of biomass used as precursor has significant impact on the properties and yield of hydrochar produced through HTC. A wide range of biomass has been used for HTC with varying process parameters. Comparative study of product properties, process conditions, and biomass composition will help identify suitable feedstock and process conditions for desired properties in hydrochar. It is important to understand the influence of each biomass component on the structure of hydrochar as it affects its properties and thus potential application. The heteroatoms, or the inorganic components in the feedstock, can serve as sacrificial templates to produce well-defined porous structures. There are gaps in knowledge when it comes to detailed reaction network. The effects of various parameters on HTC has been investigated in isolation but for further technical development and implementation, reaction modeling is a necessity. The optimum ratio of biomass to water depends on both chemistry and process economics; there must be a trade-off between these two factors. Hydrothermal carbonization appears to be feasible only for waste streams with bad fuel properties (low heating value, high moisture content, heterogeneity) (Funke & Ziegler, 2010). Optimum carbonization should be wisely decided based on the benefits and the extra efforts required.

Hydrochar in unmodified form might be less effective in some applications. A number of technologies have been developed to engineer hydrochar to enhance its physiochemical properties. The concept of the engineered hydrochar should also be promoted in the research community to stimulate and foster the practical applications of the hydrochar technology in the real world. Whether HTC will prove to be an alternative solution to the problems like depletion of resources, global warming, and water and energy crises, will be closely observed in coming time. Overall, HTC seems to have great potential and many benefits over other hydrothermal processes and thus a detailed investigation is required to drive it toward commercialization.

REFERENCES

Abdullah, H., & Wu, H. W. (2009). Biochar as a fuel: 1. Properties and grindability of biochars produced from the pyrolysis of mallee wood under slow-heating conditions. *Energy & Fuels, 23*(8), 4174–4181.

Akhtar, J., & Amin, N. A. S. (2011). A review on process conditions for optimum bio-oil yield in hydrothermal liquefaction of biomass. *Renewable and Sustainable Energy Reviews, 15*(3), 1615–1624. doi:10.1016/j.rser.2010.11.054

Antal, M. J., Jr, Mok, W. S., & Richards, G. N. (1990). Mechanism of formation of 5-(hydroxymethyl)-2-furaldehyde from D-fructose and sucrose. *Carbohydrate Research, 199*(1), 91–109.

Aronovsky, S. I., & Gortner, R. A. (1930). The cooking process I—role of water in the cooking of wood1. *Industrial & Engineering Chemistry, 22*(3), 264–274. doi:10.1021/ie50243a017

Baccile, N., Laurent, G., Babonneau, F., Fayon, F., Titirici, M.-M., & Antonietti, M. (2009). Structural characterization of hydrothermal carbon spheres by advanced solid-state MAS 13C NMR investigations. *The Journal of Physical Chemistry C, 113*(22), 9644–9654. doi:10.1021/jp901582x

Bart, H. J. (2006). Buchbesprechung: Green separation processes. Von C. A. M. Afonso, J. G. Crespo (Eds.). *Chemie Ingenieur Technik, 78*(3), 320–320. doi:10.1002/cite.200690017

Bergius, F., & Specht, H. (1913). *Die Anwendung hoher Drucke bei chemischen Vorgängen und eine Nachbildung des Entstehungsprozesses der Steinkohle.*Verlag Wilhelm Knapp: Halle an der Saale, 58.

Blazsó, M., Jakab, E., Vargha, A., Székely, T., Zoebel, H., Klare, H., & Keil, G. (1986). The effect of hydrothermal treatment on a Merseburg lignite. *Fuel, 65*(3), 337–341.

Bobleter, O. (1994). Hydrothermal degradation of polymers derived from plants. *Progress in Polymer Science, 19*(5), 797–841. doi:10.1016/0079-6700(94)90033-7

Chung, J. W., Foppen, J. W., Izquierdo, M., & Lens, P. N. L. (2014). Removal of Escherichia coli from saturated sand columns supplemented with hydrochar produced from maize. *Journal of Environmental Quality, 43*(6), 2096–2103. doi:10.2134/jeq2014.05.0199

Chuntanapum, A., & Matsumura, Y. (2010). Char formation mechanism in supercritical water gasification process: A study of model compounds. *Industrial & Engineering Chemistry Research, 49*(9), 4055–4062. doi:10.1021/ie901346h

Dapsens, P. Y., Mondelli, C., & Pérez-Ramírez, J. (2012). Biobased chemicals from conception toward industrial reality: Lessons learned and to be learned. *ACS Catalysis, 2*(7), 1487–1499. doi:10.1021/cs300124m

Falco, C., Marco-Lozar, J. P., Salinas-Torres, D., Morallón, E., Cazorla-Amorós, D., Titirici, M. M., & Lozano-Castelló, D. (2013). Tailoring the porosity of chemically activated hydrothermal carbons: Influence of the precursor and hydrothermal carbonization temperature. *Carbon, 62*, 346–355. doi:10.1016/j.carbon.2013.06.017

Funke, A., & Ziegler, F. (2010). Hydrothermal carbonization of biomass: A summary and discussion of chemical mechanisms for process engineering. *Biofuels, Bioproducts and Biorefining, 4*(2), 160–177.

Hamid, S. B. A., Teh, S. J., & Lim, Y. S. (2015). Catalytic hydrothermal upgrading of α-cellulose using iron salts as a Lewis acid. *BioResources, 10*(3), 5974–5986.

Hashaikeh, R., Fang, Z., Butler, I. S., Hawari, J., & Kozinski, J. A. (2007). Hydrothermal dissolution of willow in hot compressed water as a model for biomass conversion. *Fuel, 86*(10–11), 1614–1622. doi:10.1016/j.fuel.2006.11.005

He, C., Wang, K., Yang, Y., & Wang, J.-Y. (2014). Utilization of sewage-sludge-derived hydrochars toward efficient cocombustion with different-rank coals: Effects of subcritical water conversion and blending scenarios. *Energy & Fuels, 28*(9), 6140–6150. doi:10.1021/ef501386g

Hoekman, S. K., Broch, A., & Robbins, C. (2011). Hydrothermal carbonization (HTC) of lignocellulosic biomass. *Energy & Fuels, 25*(4), 1802–1810.

Hu, B., Yu, S.-H., Wang, K., Liu, L., & Xu, X.-W. (2008). Functional carboneceous materials from hydrothermal carbonization of biomass: An effective chemical process. *Dalton Transactions.*

Hwang, I.-H., Aoyama, H., Matsuto, T., Nakagishi, T., & Matsuo, T. (2012). Recovery of solid fuel from municipal solid waste by hydrothermal treatment using subcritical water. *Waste Management, 32*(3), 410–416. doi:10.1016/j.wasman.2011.10.006

Inoue, S., Hanaoka, T., & Minowa, T. (2002). Hot compressed water treatment for production of charcoal from wood. *Journal of Chemical Engineering of Japan, 35*(10), 1020–1023.

Inoue, S., Uno, S., & Minowa, T. (2008). Carbonization of cellulose using the hydrothermal method. *Journal of Chemical Engineering of Japan, 41*(3), 210–215.

Kabyemela, B. M., Adschiri, T., Malaluan, R. M., & Arai, K. (1999). Glucose and fructose decomposition in subcritical and supercritical water: Detailed reaction pathway, mechanisms, and kinetics. *Industrial & Engineering Chemistry Research, 38*(8), 2888–2895. doi:10.1021/ie9806390

Karagöz, S., Bhaskar, T., Muto, A., & Sakata, Y. (2005a). Catalytic hydrothermal treatment of pine wood biomass: Effect of RbOH and CsOH on product distribution. *Journal of Chemical Technology & Biotechnology, 80*(10), 1097–1102. doi:10.1002/jctb.1287

Karagöz, S., Bhaskar, T., Muto, A., & Sakata, Y. (2005b). Comparative studies of oil compositions produced from sawdust, rice husk, lignin and cellulose by hydrothermal treatment. *Fuel, 84*(7–8), 875–884.

Karagöz, S., Bhaskar, T., Muto, A., & Sakata, Y. (2006). Hydrothermal upgrading of biomass: Effect of K2CO3 concentration and biomass/water ratio on products distribution. *Bioresource Technology, 97*(1), 90–98. doi:10.1016/j.biortech.2005.02.051

Khemchandani, G. V., Ray, T. B., & Sarkar, S. (1974). Studies on artificial coal. 1. Caking power and chloroform extracts. *Fuel, 53*(3), 163–167.

Kim, D., Lee, K., & Park, K. Y. (2014). Hydrothermal carbonization of anaerobically digested sludge for solid fuel production and energy recovery. *Fuel, 130*, 120–125. doi:10.1016/j.fuel.2014.04.030

Kishida, H., Jin, F., & Zhou, Z. (2005). Conversion of glycerin into lactic acid by alkaline hydrothermal reaction. *Chemistry Letters, 34*(11), 1560–1561.

Kong, L. Z., Li, G. M., & Wang, H. (2008). Hydrothermal catalytic conversion of biomass for lactic acid production. *Journal of Chemical Technology & Biotechnology, 83*, 383–388.

Kruse, A., Funke, A., & Titirici, M.-M. (2013). Hydrothermal conversion of biomass to fuels and energetic materials. *Current Opinion in Chemical Biology, 17*(3), 515–521. doi:https://doi.org/10.1016/j.cbpa.2013.05.004

Kumar, S., & Gupta, R. B. (2009). Biocrude production from switchgrass using subcritical water. *Energy & Fuels, 23*(10), 5151–5159.

Kuster, B. F. M. (1990). 5-Hydroxymethylfurfural (HMF). A review focussing on its manufacture. *Starch – Stärke, 42*(8), 314–321. doi:10.1002/star.19900420808

Lee, I., Kim, M., & Ihm, S. (2002). Gasification of glucose in supercritical water. *Industrial & Engineering Chemistry Research, 41*, 1182–1188.

Li, L., Flora, J. R. V., Caicedo, J. M., & Berge, N. D. (2015). Investigating the role of feedstock properties and process conditions on products formed during the hydrothermal carbonization of organics using regression techniques. *Bioresource Technology, 187*, 263–274. doi:10.1016/j.biortech.2015.03.054

Liu, Z., & Zhang, F.-S. (2009). Removal of lead from water using biochars prepared from hydrothermal liquefaction of biomass. *Journal of Hazardous Materials, 167*(1), 933–939. doi:10.1016/j.jhazmat.2009.01.085

Liu, Z., Zhang, F.-S., & Wu, J. (2010). Characterization and application of chars produced from pinewood pyrolysis and hydrothermal treatment. *Fuel, 89*(2), 510–514. doi:10.1016/j.fuel.2009.08.042

Lourdes, C., & David, V. (2002). Formation of organic acids during the hydrolysis and oxidation of several wastes in sub- and supercritical water. *Industrial & Engineering Chemistry Research, 41*, 6503–6509.

Marta, S., & Antonio, B. F. (2009). The production of carbon materials by hydrothermal carbonization of cellulose. *Carbon, 47*, 2281–2289.

Masaru, W., Takafumi, S., & Hiroshi, I. (2004). Chemical reactions of C1 compounds in near-critical and supercritical water. *Chemical Reviews, 104*, 5803–5821.

Matsumura, Y., Sasaki, M., Okuda, K., Takami, S., Ohara, S., Umetsu, M., & Adschiri, T. (2006). Supercritical water treatment of biomass for energy and material recovery. *Combustion Science and Technology, 178*, 509–536.

McCollom, T. M., Ritter, G., & Simoneit, B. R. (1999). Lipid synthesis under hydrothermal conditions by Fischer-Tropsch-type reactions. *Origins of Life and Evolution of the Biosphere, 29*(2), 153–166.

Michels, R., Langlois, E., Ruau, O., Mansuy, L., Elie, M., & Landais, P. (1996). Evolution of asphaltenes during artificial maturation: A record of the chemical processes. *Energy & Fuels, 10*(1), 39–48.

Mok, W. S. L., & Antal, M. J., Jr (1992). Uncatalyzed solvolysis of whole biomass hemicellulose by hot compressed liquid water. *Industrial & Engineering Chemistry Research, 31*(4), 1157–1161.

Möller, M., Nilges, P., Harnisch, F., & Schröder, U. (2011). Subcritical water as reaction environment: Fundamentals of hydrothermal biomass transformation. *ChemSusChem, 4*(5), 566–579. doi:10.1002/cssc.201000341

Motonobu, G., Ryusaku, O., & Tsutomu, H. (2004). Hydrothermal conversion of municipal organic waste into resources. *Bioresource Technology, 93*, 279–284.

Mumme, J., Titirici, M.-M., Pfeiffer, A., Lüder, U., Reza, M. T., & Mašek, O. e. (2015). Hydrothermal carbonization of digestate in the presence of zeolite: Process efficiency and composite properties. *ACS Sustainable Chemistry & Engineering, 3*(11), 2967–2974.

Murray, J. a., & Evans, D. (1972). The brown-coal/water system: Part 3. Thermal dewatering of brown coal. *Fuel, 51*(4), 290–296.

Nelson, D. A., Molton, P. M., Russell, J. A., & Hallen, R. T. (1984). Application of direct thermal liquefaction for the conversion of cellulosic biomass. *Industrial & Engineering Chemistry Product Research and Development, 23*(3), 471–475. doi:10.1021/i300015a029

Nizamuddin, S., Baloch, H. A., Griffin, G., Mubarak, N., Bhutto, A. W., Abro, R., … & Ali, B. S. (2017). An overview of effect of process parameters on hydrothermal carbonization of biomass. *Renewable and Sustainable Energy Reviews, 73*, 1289–1299.

Orem, W. H., Neuzil, S. G., Lerch, H. E., & Cecil, C. B. (1996). Experimental early-stage coalification of a peat sample and a peatified wood sample from Indonesia. *Organic Geochemistry, 24*(2), 111–125.

Parshetti, G. K., Kent Hoekman, S., & Balasubramanian, R. (2013). Chemical, structural and combustion characteristics of carbonaceous products obtained by hydrothermal carbonization of palm empty fruit bunches. *Bioresource Technology, 135*, 683–689. doi:10.1016/j.biortech.2012.09.042

Peterson, A. A., Vogel, F., Lachance, R. P., Fröling, M., Antal, M. J., Jr, & Tester, J. W. (2008). Thermochemical biofuel production in hydrothermal media: A review of sub- and supercritical water technologies. *Energy & Environmental Science, 1*(1), 32–65.

Román, S., Libra, J., Berge, N., Sabio, E., Ro, K., Li, L., … & Bae, S. (2018). Hydrothermal carbonization: Modeling, final properties design and applications: A review. *Energies, 11*(1), 216.

Romero-Anaya, A. J., Ouzzine, M., Lillo-Ródenas, M. A., & Linares-Solano, A. (2014). Spherical carbons: Synthesis, characterization and activation processes. *Carbon, 68*, 296–307. doi:10.1016/j.carbon.2013.11.006

Ross, D. S., Loo, B. H., Doris, S. T., & Hirschon, A. S. (1991). Hydrothermal treatment and the oxygen functionalities in Wyodak coal. *Fuel, 70*(3), 289–295.

Safari, F., Norouzi, O., & Tavasoli, A. (2016). Hydrothermal gasification of cladophora glomerata macroalgae over its hydrochar as a catalyst for hydrogen-rich gas production. *Bioresource Technology, 222*, 232–241. doi:10.1016/j.biortech.2016.09.082

Savage, P. E. (1999). Organic chemical reactions in supercritical water. *Chemical Reviews*, *99*, 603–621.

Savage, P. E. (2009). A perspective on catalysis in sub- and supercritical water. *The Journal of Supercritical Fluids*, *47*(3), 407–414. doi:10.1016/j.supflu.2008.09.007

Savovaa, D., Apakb, E., Ekinci, E., Yardimb, F., Petrov, N., Budinovaa, T., … & Minkovaa, V. (2001). Biomass conversion to carbon adsorbents and gas. *Biomass and Bioenergy*, *21*(2), 133–142.

Schafer, H. (1972). Factors affecting the equilibrium moisture contents of low-rank coals. *Fuel*, *51*(1), 4–9.

Schuhmacher, J., Huntjens, F., & Vankrevelen, D. (1960). Chemical structure and properties of coal. 26. Studies on artificial coalification. *Fuel*, *39*(3), 223–234.

Selhan, K., Thallada, B., & Akinori, M. (2005). Low-temperature catalytic hydrothermal treatment of wood biomass: Analysis of liquid products. *Chemical Engineering Journal*, *108*, 127–113.

Sevilla, M., & Fuertes, A. B. (2009a). Chemical and structural properties of carbonaceous products obtained by hydrothermal carbonization of saccharides. *Chemistry – A European Journal*, *15*(16), 4195–4203. doi:10.1002/chem.200802097

Sevilla, M., & Fuertes, A. B. (2009b). The production of carbon materials by hydrothermal carbonization of cellulose. *Carbon*, *47*(9), 2281–2289. doi:10.1016/j.carbon.2009.04.026

Sevilla, M., Fuertes, A. B., & Mokaya, R. (2011). High density hydrogen storage in superactivated carbons from hydrothermally carbonized renewable organic materials. *Energy & Environmental Science*, *4*(4), 1400–1410. doi:10.1039/C0EE00347F

Siskin, M., & Katritzky, A. R. (1991). Reactivity of organic compounds in hot water: Geochemical and technological implications. *Science*, *254*(5029), 231. doi:10.1126/science.254.5029.231

Takeda, N., Sato, S., & Machihara, T. (1990). Study of petroleum generation by compaction pyrolysis—I. Construction of a novel pyrolysis system with compaction and expulsion of pyrolyzate from source rock. *Organic Geochemistry*, *16*(1–3), 143–153.

Titirici, M.-M., Antonietti, M., & Baccile, N. (2008). Hydrothermal carbon from biomass: A comparison of the local structure from poly- to monosaccharides and pentoses/hexoses. *Green Chemistry*, *10*(11), 1204–1212. doi:10.1039/B807009A

Titirici, M.-M., Thomas, A., & Antonietti, M. (2007). Back in the black: Hydrothermal carbonization of plant material as an efficient chemical process to treat the CO_2 problem? *New Journal of Chemistry*, *31*(6), 787–789.

Tropsch, H., & Phillipovich, A. (1922). Über die künstliche Inkohlung von Cellulose und Lignin in Gegenwart von Wasser. *Gesammelte Abhandlungen zur Kenntnis der Kohle*, *7*, 84–102.

Varhegyi, G., Szabo, P., Till, F., Zelei, B., Antal, M. J. J., & Dai, X. (1998). TG, TG-MS, and FTIR characterization of high-yield biomass charcoals. *Energy & Fuels*, *12*, 969–974.

Wang, J., Nie, P., Ding, B., Dong, S., Hao, X., Dou, H., & Zhang, X. (2017). Biomass derived carbon for energy storage devices. *Journal of Materials Chemistry A*, *5*(6), 2411–2428. doi:10.1039/C6TA08742F

Wang, T., Zhai, Y., Zhu, Y., Li, C., & Zeng, G. (2018). A review of the hydrothermal carbonization of biomass waste for hydrochar formation: Process conditions, fundamentals, and physicochemical properties. *Renewable and Sustainable Energy Reviews*, *90*, 223–247. doi:10.1016/j.rser.2018.03.071

Xue, Y., Gao, B., Yao, Y., Inyang, M., Zhang, M., Zimmerman, A. R., & Ro, K. S. (2012). Hydrogen peroxide modification enhances the ability of biochar (hydrochar) produced from hydrothermal carbonization of peanut hull to remove aqueous heavy metals: Batch and column tests. *Chemical Engineering Journal*, *200–202*, 673–680. doi:10.1016/j.cej.2012.06.116

Yu, J., & Savage, P. E. (1998). Decomposition of formic acid under hydrothermal conditions. *Industrial & Engineering Chemistry Research*, *37*(1), 2–10.

Zhang, B., von Keitz, M., & Valentas, K. (2009). Thermochemical liquefaction of high-diversity grassland perennials. *Journal of Analytical and Applied Pyrolysis*, *84*(1), 18–24. doi:10.1016/j.jaap.2008.09.005

Zhao, P., Shen, Y., Ge, S., Chen, Z., & Yoshikawa, K. (2014). Clean solid biofuel production from high moisture content waste biomass employing hydrothermal treatment. *Applied Energy*, *131*, 345–367.

5 Supercritical Water Gasification of Biomass: Technology and Challenges

Anuj Thakkar and Sandeep Kumar

CONTENTS

5.1 INTRODUCTION

Water heated above 374°C and pressurized beyond 22.1 MPa possesses unique physical and chemical properties and is known as supercritical water. At supercritical conditions, water behaves like a nonpolar solvent and thus solvates organic molecules. Supercritical water has some of the desired properties of its liquid phase as well as the gaseous phase. Above the critical point, there is a drop in water properties like dielectric constant, viscosity, and density which enhance reaction kinetics along with the rise in diffusivity and free radical reactions. Supercritical water was first applied in hydrothermal conversion of organic compounds to gases in the 1970s (Amin, Reid, & Modell, 1975). The properties of water at various conditions are

TABLE 5.1

Properties of Water at Various Conditions

	Normal Water	Subcritical Water		Supercritical Water	
Temperature (°C)	25	250	350	400	400
Pressure (MPa)	0.1	5	25	25	50
Density, ρ (g/cm³)	1	0.8	0.6	0.17	0.58
Dielectric constant, ε (F/m)	78.5	27.1	14.07	5.9	10.5
Ionic product, pKw	14	11.2	12	19.4	11.9
Heat capacity, Cp (KJ/ Kg/K)	4.22	4.86	10.1	13	6.8
Dynamic viscosity, η (mPa s)	0.89	0.11	0.064	0.03	0.07

Source: Krammer & Vogel (2000), Kruse & Dinjus (2007).

shown in Table 5.1. These changes in the property of water above critical point move the reaction toward gas formation rather than solid by-products as described latter in this chapter. Supercritical water gasification (SCWG) is one of the emerging green technologies for conversion of biomass to gaseous biofuels. Since water is used as reactant and reaction media in the SCWG process, wet biomasses or organic wastes are the most suitable feedstock. It helps in avoiding cost intensive drying of biomass which is desired in conventional thermochemical gasification of biomass. It mainly produces H_2, CO_2, CH_4, and C_2H_6 at optimum conditions. Although the technology is promising, lower product yield is still a challenge. Higher product yield and lower operating cost can be achieved through reaction condition optimization and identifying an appropriate catalyst.

Supercritical water is being researched for its application in biomass oxidation as well as gasification. Supercritical water oxidation (SCWO) is generally used for neutralizing toxic organic matter. At supercritical conditions, organic matter and oxygen become fully miscible in water. Organic matter is fully oxidized within one minute at optimum reaction conditions. Supercritical water gasification (SCWG) is carried out in the absence of excess oxygen and the reactions occur in a reductive environment. Though SCWO and SCWG are similar, due to the difference in reaction conditions, reactor design considerations are very different in both the cases. SCWO is an exothermic process while SCWG is endothermic, thus heat recovery is vital in SCWG for process economics. SCWO technology is used for organic waste treatment and SCWG technology is used for fuel generation from organic matter (Kruse, 2009; Marrone & Hong, 2009; Pinkard et al., 2019).

The gases produced through SCWG can be used as fuel by direct combustion or can be upgraded to get higher H_2 through water gas shift reaction. The resultant mixture of gases can also be converted to liquid fuels using the Fischer–Tropsch reaction. As the reaction is carried out at higher pressure, the gaseous products obtained are highly pressurized and thus can be stored in smaller volumes without external pressurization (Byrd, Kumar, Kong, Ramsurn, & Gupta, 2011). SCWG/Hydrothermal gasification can be subdivided into three categories as shown in Table 5.2.

TABLE 5.2

Hydrothermal Gasification Subdivided into Three Categories

	Aqueous Phase Reforming	Near Critical Water Gasification	Supercritical Water Gasification
Temperature range	215–265°C	350–400°C	> 374°C
Major products	H_2 and CO_2	CH_4	H_2 and CO_2
Comments	Noble metal catalysts are necessary for the hydrogen formation	Catalysts help to enhance methane formation from CO hydrogenation	Varying amount of CH_4 and a small amount of short-chain hydrocarbons and CO are produced

Source: Rodriguez, Correa, & Kruse, 2018.

Biomass can be gasified using thermochemical processes like thermal gasification, pyrolysis, and SCWG. The advantage of SCWG over the other processes is that it eliminates the need to dry biomass before conversion and the process is conducted at lower temperature than the conventional gasification. Generally, thermochemical processes are more efficient in this conversion compared to biochemical processes like fermentation or anaerobic digestion (Ni, Leung, Leung, & Sumathy, 2006).

SCWG can convert more than 99% of biomass and produce H_2 up to 50% of product gas (Calzavara, Joussot-Dubien, Boissonnet, & Sarrade, 2005). For biomass with higher moisture content, SCWG has the highest energy conversion efficiency compared to other technologies (Lachos Perez, Prado, Mayanga, Forster-Carneiro, & Meireles, 2015). H_2 is the cleanest fuel and it can be produced by reforming hydrogen containing products like biomass (Ibrahim & Akilli, 2019) and natural gas.

5.2 CATALYST

SCWG without catalyst has been studied extensively. As water-gas shift reaction is promoted by catalysts, SCWG without catalyst produces a higher amount of unwanted CO. Also, for achieving the desired degree of conversion in the absence of catalyst, higher temperatures are required (Kruse, 2008). Thus, a techno-economic analysis for productivity and conversion in the presence and the absence of catalyst is required on a case-by-case basis. As catalysts available in the market significantly affect the gas yield, very few researchers have focused on catalyst development specifically for application in SCWG (Azadi & Farnood, 2011). Both homogenous and heterogenous catalyst have been widely tested for SCWG of biomass. Identifying an appropriate catalyst is necessary to drive the reaction toward desired product formation and mitigate severe reaction parameters. It is important to mitigate these parameters which have a direct implication on operating and capital cost. Supercritical water also acts as a catalyst and promotes acid-base catalytic reactions (Ikushima, Hatakeda, Sato, Yokoyama, & Arai, 2000, 2001; Watanabe, Osada, Inomata, Arai,

& Kruse, 2003). As stated earlier, the supercritical water has a dual role of reaction media as well as reactant (Azadi & Farnood, 2011).

5.2.1 HOMOGENOUS CATALYST

Homogenous catalysts widely tested for SCWG are NaOH, K_2CO_3, Na_2CO_3, $KHCO_3$, etc. These alkaline catalysts help in breaking the hydrocarbon carbon chain and enhance water-gas shift reaction (Y. Guo et al., 2010; Kruse, 2008; D. Xu et al., 2009). Alkali catalyst also promotes H_2 formation but may cause operational problems like corrosion, plugging, and fouling of reactor (Sınag, Kruse, & Rathert, 2003). Other homogenous catalysts like trona, red mud, borax, and dolomite also have a positive impact on gas yield from SCWG of lignocellulosic biomass (Okolie, Rana, Nanda, Dalai, & Kozinski, 2019).

5.2.2 HETEROGENOUS CATALYST

Heterogenous catalysts have a few benefits over homogenous catalysts such as selectivity, being environment friendly, and recyclability (Y. Guo et al., 2010). Heterogenous catalysts used in SCWG are mostly metals and metal oxides like Ni, Ru, Pt, Pd, Al_2O_3, and TiO_2. However, it has disadvantages like deactivation and poisoning due to the presence of sulfur, nitrogen, coke, and other heteroatoms (A. Okolie et al., 2019). Metal-based catalysts are mostly used with support material like alumina, activated carbon, titanium dioxide, etc. The nonmetallic heterogenous catalyst, activated carbon, was found to have enhanced water gas shift reaction and methanation (Xu, Matsumura, Stenberg, & Antal, 1996). Activated carbon catalyst itself gets decomposed in supercritical water condition in two to four hours and is not a great choice for catalyst if not used as a support for metal-based catalyst (Guo et al., 2010; . Lee & Ihm, 2009; . Xu et al., 1996). Most of the reactor material used for SCWG are alloys, desired or not, metal on the inner wall of the reactor plays a role in the reaction (Guo et al., 2010). Michael et al. published that Ni alloy reactors like Hastelloy catalyze gasification and steam reformation reaction (Antal, Allen, Schulman, Xu, & Divilio, 2000). The addition of Ni as catalyst could be beneficial but it tends to sinter and deactivate (Behnia, Yuan, Charpentier, & Xu, 2016; DiLeo & Savage, 2006; Guo et al., 2010; Lee, 2011; Lee, Chen, & Hwang, 2008). The poisoning of Ni catalyst could be due to deposition of carbon or salt on the catalyst surface (Guo et al., 2010). Ruthenium is found to be the most effective and stable catalyst in gasification of glucose, microalgae, and glycerol (Behnia et al., 2016; Byrd, Pant, & Gupta, 2007; Chakinala, Brilman, van Swaaij, & Kersten, 2010; May, Salvadó, Torras, & Montané, 2010; Osada, Sato, Watanabe, Adschiri, & Arai, 2004; Yamamura, Mori, Park, Fujii, & Tomiyasu, 2009). The high cost of ruthenium and its poisoning potential due to sulfur has prevented its widespread use (Chakinala et al., 2010; Guo et al., 2010).

5.3 REACTION PARAMETERS

The three important reaction parameters that govern the technology are temperature, pressure, and residence time of biomass. Unoptimized reaction parameters hamper H_2 yield and produce unwanted tars and chars that leads to operational challenges.

5.3.1 Temperature

The concentration of ionic products H^+ and OH^- increases as the temperature in SCWG rises to 300°C and decreases thereafter (Savage, 2009). At higher temperature, free radical reactions are favored causing dissolution of complex organic compounds and their degradation products to gaseous products (Chuntanapum & Matsumura, 2010). Acid and base catalyzed reactions, i.e., ionic product mechanism is favored in subcritical condition and free radical is favored in supercritical conditions (Okolie et al., 2019). SCWG can be subdivided into two, high-temperature supercritical water (550–700°C) and low-temperature supercritical water (374–550°C). With higher temperatures of critical water, complete gasification can be achieved without the use of a catalyst or by using an activated carbon or an alkali catalyst to inhibit tar formation (Fang, Minowa, Smith, Ogi, & Koziński, 2004; Osada, Sato, Watanabe, Shirai, & Arai, 2006; Schmieder et al., 2000). Complete gasification of cellulose and lignin has been reported at low-temperature critical water temperature but catalyst deactivation is a problem (Elliott, Sealock, & Baker, 1993). The ionic and free radical mechanism for lignocellulosic biomass during SCWG is shown in Figure 5.1.

It has been proven that biomass can be completely gasified in supercritical water conditions, but the product distribution profile is dependent on the temperature. Formation of CH_4 is preferred at a lower temperature while the reaction moves toward H_2 formation at a higher temperature. As per Le Chatelier's principle, the

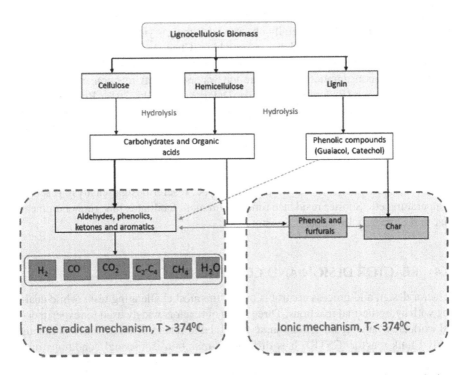

FIGURE 5.1 The ionic and free radical mechanism for lignocellulosic biomass during SCWG. (Reproduced from Okolie, Rana, Nanda, Dalai, & Kozinski, 2019.)

reaction can proceed toward the formation of H_2 even in lower temperature supercritical water if biomass loading to water ratio is kept low (Rodriguez, Correa, & Kruse, 2018). In the absence of a catalyst or less effective catalyst, the temperature has a significant effect on product distribution and efficiency. High heating rate favors SCWG while slower heating rate leads to the formation of char and tar. The possible explanation for this is that when the heating rate is low, the reaction mixture spends more time in a subcritical condition which leads to coke and char formation (Perez et al., 2015; Lu et al., 2006; Sınag et al., 2003).

5.3.2 PRESSURE

With the increase in pressure, it is observed that the free radical reaction rate decreases and the ionic reaction rate increases. At higher pressure, ionic and free radical reactions compete with each other due to ionic stabilization in high density of water which is caused by higher pressure (Anikeev & Fan, 2014). The formation of CH_4 increases while H_2 decreases in accordance with the Le Chatelier principle which says pressure favors the side with the least sum of molecules (Basu & Mettanant, 2009; Kruse, 2008).

5.3.3 RESIDENCE TIME

Residence time is the time for which the biomass stays in the reactor at the desired reaction temperature. It is generally observed that the gas yield is higher with higher residence time due to enhanced thermal cracking (Chen, Andries, Luo, & Spliethoff, 2003). With higher residence time, the gas generated has more of CH_4, CO_2, and H_2 due to the consumption of CO through water-gas shift reaction, methanation, and hydrogenation (Reddy, Ding, Nanda, Dalai, & Kozinski, 2014; Reddy, Nanda, Dalai, & Kozinski, 2014). It was observed that H_2 yield saturates at an optimum residence time beyond which increasing residence time has a negligible effect (D'Jesús, Boukis, Kraushaar-Czarnetzki, & Dinjus, 2006; Lu et al., 2006). Generally, it is observed that biomass mass with a longer hydrocarbon chain have longer optimum residence times for gasification to be complete (Anikeev & Fan, 2014). Optimum residence time is also a function of biomass type, reactor design, catalyst, reaction temperature, etc. Shorter residence time requirement facilitates lowering equipment size and ultimately lower capital cost.

5.4   REACTOR DESIGN AND CONFIGURATION

Reactor design and process control is one of the most challenging tasks while dealing with hydrothermal reactions. Three types of reactors widely used for experimental work are a batch reactor, continuous flow tubular reactor (CFTR,) and continuous stirred tank reactor (CSTR). It is difficult to maintain isothermal conditions in a batch reactor. Batch reactors, being the simplest form of the reactor, are widely used for parameter optimization. Minimizing the heating up and cooling down time of reaction mixture is also critical in all kinds of reactors (Okolie et al., 2019). Although

CFTR is a better design than the batch reactor for the continuous process with a shorter residence time of a few seconds, char formation, plugging, and nonuniform reaction mixture are some of the major issues associated with it (Reddy, Nanda, et al., 2014). CSTR combines the principle of both batch reactor and CFTR and thus provides better mass and heat transfer. Complexity in design and high cost of stirring are the disadvantages (Kruse & Faquir, 2007).

The VERENA pilot plant designed by Karlsruche Institute of Technology has a 100 kg hr^{-1} wet biomass throughput. The design temperature and pressure of the reactor are 700°C and 35 MPa, respectively. One of the unique features of VERENA is the removal of brines and solids from the lowest part of the reactor. This feature helps in reducing clogging and fouling of heat exchangers. The dry solid loading in VERENA is less than 10% which is not energetically optimum (Kruse, 2008). The process flow diagram of VERENA is shown in Figure 5.2. The University of Hiroshima has built a 50 kg hr^{-1} wet biomass throughput plant where catalyst-activated carbon moves with the liquid flow in the plant. The Dutch firm Gensos in Delft has developed a 50 kg hr^{-1} wet biomass throughput SCWG fluidized bed reactor (Okolie et al., 2019).

Ni alloy reactors were not found to be suitable for SCWG as Ni from Hastelloy leached into activated carbon bed in the reactor due to severe corrosion. (Antal et al., 2000). The material of construction for the reactor should be chosen such that it is chemically stable and does not corrode at SCWG reaction conditions. The conditions of high pressure and high-temperature supercritical water in the presence of alkali salts and heteroatoms is corrosive to a wide range of process equipment materials (Thomason & Modell, 1984). Corrosion can be kept in check by selecting corrosion-resistant material, tuning reaction parameters, applying corrosion-resistant coating, and avoiding contact of the corrosive mixture with sensitive components

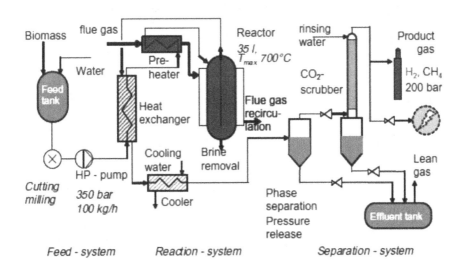

FIGURE 5.2 Process flow diagram of VERENA. (Reproduced from Möbius, Boukis, Galla, & Dinjus, 2012.)

(Marrone & Hong, 2009). If the catalytic effect of reactor material is undesired, reactors made of quartz or reactors with ceramic lining or seasoned reactors have been used (Kruse, 2008).

Pump selection is very critical in SCWG for high-pressure generation, precise flow rate, and capability to pump viscous fluid. For laboratory scale experiments, highly precise HPLC pumps are widely used. Although less precise, diaphragm pumps can be used to pump viscous slurries and highly concentrated biomass (Pinkard et al., 2019). The maximum dry matter or minimum moisture content of the feed that can be used also depends on the capability of the pump. Usually, it is 20% or less dry matter but it also depends on the type of biomass or pretreatment. The minimum concentration of dry matter should be such that the energy output is more than the input (Kruse, 2008). Liquefaction of biomass or solubilizing biomass through biochemical process can be used as biomass pretreatment if pumping is an issue (Kruse, 2009). Scale-up of the lab scale reactor to larger scales is associated with many uncertainties as there is very little or no effect of hydrodynamics, heat transfer, and mass transfer on small lab scale reactors (Basu & Mettanant, 2009).

5.5 SUPERCRITICAL WATER GASIFICATION CHEMISTRY

Overall, SCWG is an endothermal process and thus requires net positive energy for the reaction to be thermodynamically feasible. Water, the most abundant constituent of the reaction mixture has considerably high specific heat. Catalyst plays an important role in bringing down the temperature requirement for gasification (Azadi & Farnood, 2011). Higher biomass concentration in feed promotes higher CH_4 yield and lower H_2 yield as H_2 formation requires more water than CH_4 formation as shown in the equation that follows (Kruse, 2008). Like commercial production of H_2 from CH_4 through catalytic steam reforming reaction, ideally, biomass through SCWG should only form H_2, CO_2. In reality, there is formation of CH_4, CO, tar, char, and some longer hydrocarbons as biomass does not directly react with steam (Antal et al., 2000). Typical reaction pathways in SCWG of biomass are shown in Figure 5.3. The overall SCWG of biomass reactions can be represented in the simplified form as follows where x and y are elemental molar ratios of H/C and O/C, respectively (Guo et al., 2007).

1. Endothermic decomposition of biomass

    After this reaction, three intermediate competitive reactions may occur as shown below.

$$CH_xO_y + (2-y)H_2O \longrightarrow CO_2 + \left(2-y+\frac{x}{2}\right)H_2$$

2. Steam reformation

    Along with reaction 3, steam reformation is one of the major reactions.

$$CH_xO_y + (1-y)H_2O \rightleftharpoons CO + \left(1-y+\frac{x}{2}\right)H_2$$

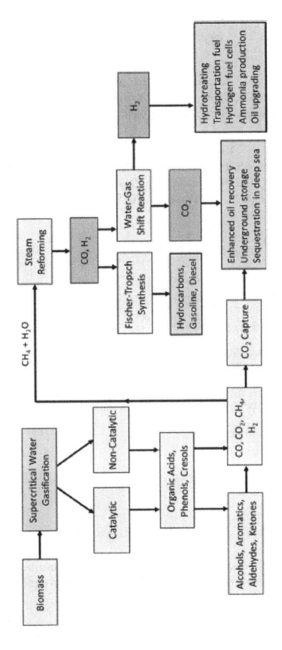

FIGURE 5.3 Typical reaction pathways in SCWG of biomass. (Reproduced from Okolie et al., 2019.)

3. Steam shift reaction

This reaction is favored at high temperature and low feed concentration (Nanda, Rana, Zheng, Kozinski, & Dalai, 2017; Susanti, Veriansyah, Kim, Kim, & Lee, 2010).

$$CO + H_2O \rightleftharpoons CO_2 + H_2$$

4. Methanation

Methanation is a secondary reaction and it requires longer residence time (Nanda, Reddy, Hunter, Dalai, & Kozinski, 2015; Reddy, Nanda, et al., 2014). It is undesired as it reduces H_2 yield.

$$CO + 3H_2 \rightleftharpoons CH_4 + H_2O$$

SCWG efficiency and selectivity can be calculated using the equations mentioned in what follows (lv et al., 2004; Nanda, Isen, Dalai, & Kozinski, 2016; Nanda et al., 2015; Seif, Tavakoli, Fatemi, & Bahmanyar, 2015).

1. Total gas yield $\left(\dfrac{\text{mmol}}{\text{gram feedstock}} \right)$

$$= \frac{\text{Total moles of gas produced}}{\text{Total feedstock in grams}}$$

2. Carbon gasification efficiency $(\%)$

$$= \frac{\text{Total number of carbon moles in CO, } CO_2 \text{ and } CH_4}{\text{Number of carbon moles in feed stock}}$$

3. Hydrogen selectivity $(\%)$

$$= \frac{\text{Moles of } H_2}{\text{Sum of moles of CO, } CO_2 \text{ and } CH_4}$$

4. Lower heating value $\left(\dfrac{\text{kJ}}{\text{Nm}^3} \right)$

$$= \left[(30 \times CO) + (25.7 \times H_2) + (85.4 \times CH_4) \right] \times 4.2$$

In the case of lignocellulosic biomass, gasification starts with hydrolysis of cellulose into glucose, hemicellulose into xylose, and glucuronic acid and lignin into phenolics (Okolie et al., 2019).

5.6   SUPERCRITICAL WATER GASIFICATION OF BIOMASS

Biomass is a sustainable source of energy that can help in regional economic development. Biomass can not only be converted to biofuels but also into high-value chemicals. Greenhouse gas emission from the use of fossil fuels can be mitigated by replacing it with biofuels (Demİrbas & Demİrbas, 2007). Biomass is seen as power banks of solar energy and can provide energy security to nations as its distribution is uniform across the globe, unlike fossil fuels. It is the fourth largest source of energy for the world and is projected to contribute 10–20 % of total energy by the end of the 21st century (Goldemberg, 2008). Energy security and environment concerns have made scientists around the world consider biomass as an alternative source of energy.

As there is good mass transfer in SCWG, ideally biomass feedstock size should not have a significant impact on the efficiency. But it was reported that smaller size biomass favored higher H_2 yield and conversion efficiency (Lu et al., 2006). For feedstock and water slurry to be pumpable, biomass should be crushed and milled to smaller particle size. Dilute starch gel also has been used to make sawdust pumpable for gasification experiments (Matsumura et al., 2005). In general, biomass feed for SCWG can be divided into lignocellulosic biomass and food waste, algae, and sewage sludge.

5.6.1 LIGNOCELLULOSIC BIOMASS AND FOOD WASTE

Lignocellulosic biomass, which makes up almost 90% of total land biomass available, mostly comes from agricultural by-products, forests, or dedicated energy crops (Rodriguez, Correa, & Kruse, 2018). Dedicated energy crops like switchgrass, poplar, etc., are also being grown to meet energy demands.

Lignocellulosic biomass is mainly composed of cellulose, hemicellulose, lignin, ash/salts, and protein in some cases (Kumar, Kothari, Kong, Lee, & Gupta, 2011; Salihu & Bala, 2011). Cellulose and hemicellulose are carbohydrates, and lignin includes aromatic rings. Thus, for investigation, carbohydrates like glucose and aromatic compounds like phenols are used as model compounds for cellulose and lignin, respectively (Kruse, 2008). At temperatures more than 190°C, hemicellulose and lignin readily undergo solvolysis. Hydrothermolysis of remaining lignocellulosic biomass (mainly cellulose) happens at relatively higher temperature. At temperatures below 600°C, these hydrolyzed products undergo isomerization, dehydration, fragmentation, and condensation to form gas and tar. At supercritical water condition with temperatures above 600°C, the primary products are H_2, CO_2, CH_4, CO, and some tar (Matsumura et al., 2005). At 400°C, cellulose can be hydrolyzed to its monomer glucose in time as short as three seconds. At higher temperatures, glucose is converted to intermediate compounds which are further gasified (Resende, Neff, & Savage, 2007; Sasaki, Adschiri, & Arai, 2004; Sasaki et al., 1998). Lower ash content in the biomass leads to a higher yield of CO and this can be decreased with the use of alkali salt catalyst (Kruse, 2008). Studies using model biomass as feed suggest that protein negatively impacts the gas yield (Kruse, Krupka, Schwarzkopf, Gamard, & Henningsen, 2005). A mixture of amino acid alanine was used with glucose. SCWG of this model biomass produced less gas and a large amount of dissolved organic compound as compared to feed without alanine. It was suggested that the presence of proteins led to the formation of amines which ultimately react with degradation products of carbohydrates and form N-heterocyclic compounds. N-heterocyclic compounds form stable free radical cations which act as a free radical scavenger. As discussed before, the presence of free radicals is vital in SCWG reactions (Kruse, 2008; Kruse, Maniam, & Spieler, 2007). The chemistry behind the decreased gas yield was not clear but experiments with model biomass containing lignin produced lesser H_2 and CH_4 than experiments with pure cellulose and xylan. The degradation products of cellulose and xylan reacted with ligin, resulting in lower H_2 yield (T. Yoshida & Matsumura, 2001).

Yoshida et al. conducted SCWG experiments using sawdust, rice straw, and model compounds for cellulose and lignin at 400°C and 25 MPa. It was concluded that carbon gasification efficiency in the real biomass was lower than the model compounds because of interactions between compounds in real biomass (Yoshida & Matsumura, 2001). Water hyacinth is a plant rich in hemicellulose, it is made up of about 25% cellulose, 33% hemicellulose, and 10% lignin (Istirokhatun et al., 2015). SCWG of water hyacinth at 600°C and 34.5 MPa in a packed bed reactor resulted in complete gasification with low level of CO (Antal, Matsumura, Xu, Stenberg, & Lipnik, 1995). Adam et al. evaluated SCWG of liquefied switchgrass using nickel, cobalt, and ruthenium catalysts supported on TiO_2, ZrO_2, and $MgAl_2O_4$ at 600°C and 250 bar. Catalysts supported on $MgAl_2O_4$ were found to be inappropriate as it charred immediately. Catalysts supported on TiO_2 were most resistant to charring. In terms of hydrogen gas yield, among the catalysts investigated, Ni/ZrO_2 gave the maximum yield whereas Ru/ZrO_2 gave the minimum (Byrd et al., 2011).

Food waste is one contributor to greenhouse gases (GHG) emission, about 20 million tons of CO_2 equivalent GHG is emitted from food waste every year (Chapagain and Keith, 2011). According to the Food and Agricultural Organization (FAO), about $750 billion worth of food waste is generated globally every year (FAO, 2013). One-third of total food produced for humans which is approximately 1.6 billion tons per year is wasted. 54% of this food wastage happens during production, post-harvest handling, and storage and about 46% happens when it enters into the industry (FAO, 2013, 2014).

Incineration and the use of landfills of food waste has been practiced widely and have created environmental concerns (Scialabba et al., 2013). Incineration causes the release of pollutants like dioxins, furans, and particulates while landfills lead to issues like odor, pests, and GHG emission (Deng et al., 2012). Considering thermodynamics, incinerating food waste with considerable water content is not the best option for disposal. Various food wastes generated like fruits, vegetables, sugarcane bagasse, food processing industry by-products, etc., can be used as feed for SCWG. Fruit waste is rich in sugars like glucose, fructose, and sucrose with some carbohydrates like cellulose and hemicellulose. Brazil, India, and China are the three major producers of sugarcane. Sugarcane bagasse is a by-product generated after crushing cane for sugar extraction. Sugarcane bagasse is a lignocellulosic biomass containing 32–48% cellulose, 19–24% hemicellulose, and 23–32% lignin.

Nanda et al. used SCWG for processing various types of food waste like aloe vera rind, banana pill, coconut shell, orange peel, pineapple peel, and sugarcane bagasse. The temperature in the range of 400–600°C, biomass to water ratio of 1:5 and 1:10, reaction time 15–45 minutes and pressure range of 22–25 MPa was used for these experiments. The catalysts NaOH and K_2CO_3 were investigated for their effect on hydrogen yields and selectivity. Using 2% K_2CO_3 as catalyst, SCWG of coconut shell at 600°C, 1:10 biomass to water ratio for 45 minutes reaction time produced superior H_2 yield (4.8 mmol/g), H_2 selectivity (45.8%), total gas yield (15 mmol/g) and lower heating value (1595 kJ/Nm³). The study concluded that SCWG technology can be an excellent process for food waste management and energy production (Nanda et al., 2016).

5.6.2 MICROALGAE

Microalgae can be grown in marine and freshwater in the presence of CO_2 and sunlight through photosynthesis. The biomass productivity of microalgae is much higher than other photosynthetic terrestrial plants. The productivity of microalgae can be 50 times higher than that of dedicated energy crops like switchgrass (Demirbaş, 2006). Algal cells are made up of proteins, carbohydrates, lipids, and ash in different proportions specific to the species and growth conditions.

Lower sulfur content in algae makes it a better feedstock than lignocellulosic biomass as the presence of sulfur causes corrosive reactions in the reactor. Generally, algae has a high protein content (Becker, 2007) and thus has higher nitrogen in the biomass. Nitrogen in the biomass ends up in the gas phase during gasification as NO_x, NH_3, and HCN. The presence of nitrogenous impurities makes the resultant gaseous product of lower quality. These impurities also cause catalyst deactivation during the production of liquid fuels through the Fischer–Tropsch process. Uniform transportation of highly concentrated algae solution into and out of the reactor is easier owing to its density and other physical properties compared to lignocellulosic biomass (Kumar, 2013).

Sherif et al. carried out SCWG of algae in a continuous reactor and were able to cultivate algae using recycled nutrients from the aqueous phase. Organic compounds like alcohols, aldehydes, and phenols formed during gasification in aqueous phase could be toxic for algae and should be removed. Activated carbon filtration and ultraviolet degradation were found to be effective in the removal of these toxic compounds. Algae cultivated in these recycled nutrients showed growth comparable to growth in fresh nutrients (Elsayed et al., 2016).

Algae can be converted into bio-crude through hydrothermal liquefaction (HTL) which also generates gaseous phase, solid residue, and carbon-rich aqueous phase. The fraction of biomass in the aqueous phase is generally in the range of 30–50% and can be as high as 68% (Biller et al., 2012). High N/C ratio in the aqueous phase makes it difficult for anaerobic digestion, opening the door for SCWG (Fricke, Santen, Wallmann, Hüttner, & Dichtl, 2007). Through hydrothermal gasification, chemical oxygen demand (COD) of the aqueous phase can be reduced by 98.8–99.8% (Elliott et al., 2013). The H_2 generated during SCWG can be used as fuel and for upgrading bio-crude (hydrotreating) produced from algae through HTL. The quantity of H_2 produced was found to be good enough to upgrade bio-crude produced in the same batch (Elliott et al., 2013; Frank, Elgowainy, Han, & Wang, 2013).

5.6.3 SEWAGE SLUDGE

Organic biomass like sewage sludge is the unavoidable product of water treatment plants and can be converted into gaseous fuel thus helping in its disposal. The United

States Office of Technology Assessment defines municipal solid waste (MSW) as the waste generated from residences, commercial establishments, and institutions but does not include construction/demolition debris or medical/pathological waste. Wastewater sludge is the largest contributor to MSW (Chandler et al., 1997). Wastewater sludge is a by-product of the domestic or commercial water treatment facility. Municipal sewage sludge (MSS) is made of carbohydrates, proteins, lipids, lignin, and ash. It also contains heavy metals and microbiological load. Treatment of MSS contributes toward significant cost in sewage treatment plants (Qian et al., 2016). The cost associated with the SCWG of bio-based organic waste is high, but the process can be made more economical by the recovery and recycling of phosphorus in solid residue. More than 95% of phosphorus can be recovered from the solid residue by leaching with acids (Acelas, López, Brilman, Kersten, & Kootstra, 2014). Biomass that has high disposal cost associated with it is an attractive feedstock choice for SCWG (Rodriguez, Correa, & Kruse, 2018).

The effect of temperature on the gasification of wastewater sludge was similar to other sources of biomass tested. It was observed that higher SCWG temperature not only helped improve gasification yields but also produced cleaner water (Ibrahim & Akilli, 2019).

Yunan et al. carried out SCWG of sewage sludge in a continuous flow fluidized bed reactor using alkali catalysts. It was concluded that higher temperatures and lower solid loading favored gasification and alkali catalyst enhanced H_2 formation in the product stream. The activities of alkali catalyst were found to be in the following order: KOH> K_2CO_3> NaOH> Na_2CO_3. Alkali catalyst promoted a water-gas shift reaction rather than steam reforming (Chen et al., 2013).

However, low hydrogen productivity prevents the commercial application of SCWG for energy recovery from sewage sludge (Gong, Zhu, Xu, Zhang, & Yang, 2014; Xu, Zhu, & Li, 2012). Activated carbon catalyst effectively improved H_2 yield at the cost of high operating cost (temperature >700°C) (Fan et al., 2016). A metal-based catalyst like Ni promotes steam reforming reaction and thus higher H_2 yields (Furusawa et al., 2007; Reddy, Nanda, et al., 2014). Inorganic salts from sewage sludge and char were found to deactivate catalyst by deposition and reduction in active sites. Other than catalyst deactivation, challenges like separation of heavy metals (metal and alkali metal catalyst) from the final reaction mixture were also reported. Fan et al. tested the effect of adding formic acid to the reaction mixture and found that H_2 yield increased from 0.16 moles/kg of organic matter to 10.07 moles/kg of organic matter with the use of 6 wt. % formic acid. Formic acid was also effective in reducing phenols and char formation. Two pathways by which formic acid was found to be beneficial in SCWG are as follows: 1. creating acidic conditions that facilitate rapid hydrolysis of organic matter and 2. providing free hydrogen radicals for H_2 production and suppression of polymerization or carbonization reactions (Fan et al., 2016). Gasafi et al. carried out an economic analysis of sewage sludge gasification using supercritical water. It was found that carrying charges have the largest contribution in revenue requirement. SCWG can be competitive due to revenues generated from the disposal of sewage sludge. Fuel cost has a small share as compared to carrying cost in terms of revenue requirement (Gasafi, Reinecke, Kruse, & Schebek, 2008).

5.7 CHALLENGES

Most of the research carried out for the production of gaseous fuels from biomass through SCWG is done using model compounds as the lab setup are better suited for it than actual biomass suspension. There is a lack of understanding about the interaction of various compounds in actual biomass during SCWG.

Plugging of reactor and deposition of salts in the reactor due to its lower solubility in supercritical water needs to be dealt with for smooth operation (Rodriguez, Correa, & Kruse, 2018). Clogging due to char formation or deposition of insoluble compounds is dominant in packed catalyst bed reactors as the flow path is not free (Kruse, 2008). Alkali metal salts melt at a temperature above 300°C causing catalyst deactivation and corrosion (Pinkard et al., 2019). To make the process more economical and environmentally friendly, catalysts need to be efficiently recycled and regenerated.

Identification of reaction parameters and catalysts for minimization of formation of unwanted products like tar and char can drive the research toward commercialization of the technology. Reactor design considering the corrosive nature of the reaction due to the presence of sulfur in some biomass and severe conditions (temperature and pressure) is one of the challenging aspects of research in SCWG (Okolie et al., 2019; Rodriguez, Correa, & Kruse, 2018). Not all biomass sources are suitable for SCWG, hence identification of biomass based upon its composition, flowability, pretreatment requirement, and quality of gas produce is critical.

The sizing of commercial-scale plant is very important, the higher the operating temperature, the larger the plant should be as losses increase with temperature (Kruse, 2009). The cost of transportation with higher water content is energy-consuming and expensive. High solid loading helps process economics but negatively impacts H_2 yields. To overcome this issue, a two-stage process involving hydrolysis in the first step and reforming in the second step is proposed (Kruse, 2009).

5.8 CONCLUSION AND WAY FORWARD

In terms of cost, other biomass conversion technologies are more attractive than SCWG but this can change with the optimization of heat exchangers and reduction in capital cost (Yoshida et al., 2003). The high production cost of standalone H_2 production through SCWG compared to fossil fuel based H_2 makes integrated biorefinery a more economically viable alternative. To bring SCWG of biomass close to commercialization, energy recovery and optimization is very important. SCWG and SWGO integrated technology also has been studied where heat generated from exothermic SCGO can be used to preheat feed to SCWG. Technologies like burning gas produced through SCWG to heat feed or generate electric power also have been patented (Casademont, García-Jarana, Sánchez-Oneto, Portela, & de la Ossa, 2017). The use of concentrated solar energy for biomass gasification through SCWG also has been successfully evaluated (Ibrahim & Akilli, 2019). Two main reasons for SCWG not being commercialized so far could be that SCWO is more efficient than SCWG in neutralizing waste and other techniques for producing H_2 are more economical than SCWG. A step-by-step solution to the technical challenges discussed in this chapter can make SCWG an economically viable technology like SCWO. Biomass feeding to the reactor which might be overlooked in batch experiments could be one of the challenges during scaling up of a continuous feeding system. A techno-economic analysis to determine tradeoff between solid loading and acceptable H_2 yield will give direction to the ongoing research. Specialized design of process equipment for SCWG considering the technical challenges is the way forward. Similarly, application-specific design of the catalyst can help overcome issues like deactivation, charring, sintering, clogging, etc. Overall, if the process economics is not considered, it is observed that most of the biomass tested as feed for SCWG can be efficiently converted into valuable fuel.

REFERENCES

Acelas, N. Y., López, D. P., Brilman, D. W. F., Kersten, S. R. A., & Kootstra, A. M. J. (2014). Supercritical water gasification of sewage sludge: Gas production and phosphorus recovery. *Bioresource Technology, 174*, 167–175. doi:10.1016/j.biortech.2014.10.003

Amin, S., Reid, R. C., & Modell, M. (1975). *Reforming and Decomposition of Glucose in an Aqueous Phase*. ASME PAPER 75-ENAS-21, Document ID: 19750056811.

Anikeev, V. I., & Fan, M. (2014). *Supercritical Fluid Technology for Energy and Environmental Applications*. Elsevier.

Antal, M. J., Allen, S. G., Schulman, D., Xu, X., & Divilio, R. J. (2000). Biomass gasification in supercritical water. *Industrial & Engineering Chemistry Research, 39*(11), 4040–4053. doi:10.1021/ie0003436

Antal, M. J., Matsumura, Y., Xu, X., Stenberg, J., & Lipnik, P. (1995). Catalytic gasification of wet biomass in supercritical water. *Preprints Of Papers-American Chemical Society Division Fuel Chemistry, 40*, 304–304.

Azadi, P., & Farnood, R. (2011). Review of heterogeneous catalysts for sub- and supercritical water gasification of biomass and wastes. *International Journal of Hydrogen Energy, 36*(16), 9529–9541. doi:https://doi.org/10.1016/j.ijhydene.2011.05.081

Basu, P., & Mettanant, V. (2009). *Biomass Gasification in Supercritical Water – A Review* (Vol. 7).

Becker, E. W. (2007). Micro-algae as a source of protein. *Biotechnology Advances, 25*(2), 207–210. doi:10.1016/j.biotechadv.2006.11.002

Behnia, I., Yuan, Z., Charpentier, P., & Xu, C. (2016). Production of methane and hydrogen via supercritical water gasification of renewable glucose at a relatively low temperature: Effects of metal catalysts and supports. *Fuel Processing Technology, 143*, 27–34. doi:10.1016/j.fuproc.2015.11.006

Biller, P., Ross, A. B., Skill, S. C., Lea-Langton, A., Balasundaram, B., Hall, C., … & Llewellyn, C. A. (2012). Nutrient recycling of aqueous phase for microalgae cultivation from the hydrothermal liquefaction process. *Algal Research, 1*(1), 70–76. doi:10.1016/j.algal.2012.02.002

Byrd, A. J., Kumar, S., Kong, L., Ramsurn, H., & Gupta, R. B. (2011). Hydrogen production from catalytic gasification of switchgrass biocrude in supercritical water. *International Journal of Hydrogen Energy, 36*(5), 3426–3433. doi:10.1016/j.ijhydene.2010.12.026

Byrd, A. J., Pant, K. K., & Gupta, R. B. (2007). Hydrogen production from glucose using Ru/Al2O3 catalyst in supercritical water. *Industrial & Engineering Chemistry Research, 46*(11), 3574–3579. doi:10.1021/ie070241g

Calzavara, Y., Joussot-Dubien, C., Boissonnet, G., & Sarrade, S. (2005). Evaluation of biomass gasification in supercritical water process for hydrogen production. *Energy Conversion and Management, 46*(4), 615–631. doi:10.1016/j.enconman.2004.04.003

Casademont, P., García-Jarana, M. B., Sánchez-Oneto, J., Portela, J. R., & de la Ossa, E. J. M. (2017). Supercritical water gasification: A patents review. *Reviews in Chemical Engineering*, *33*(3), 237–261.

Chakinala, A. G., Brilman, D. W. F., van Swaaij, W. P. M., & Kersten, S. R. A. (2010). Catalytic and non-catalytic supercritical water gasification of microalgae and glycerol. *Industrial & Engineering Chemistry Research*, *49*(3), 1113–1122. doi:10.1021/ie9008293

Chandler, A. J., Eighmy, T. T., Hjelmar, O., Kosson, D., Sawell, S., Vehlow, J., … & Hartlén, J. (1997). *Municipal Solid Waste Incinerator Residues* (Vol. 67): Elsevier.

Chapagain, A., & Keith, J. (2011). *The Water and Carbon Footprint of Household Food and Drink Waste in the UK: Waste & Resources Action Programme*. Retrieved from http://assets.wwf.org.uk/downloads/water_carbon_footprint_final_230311.pdf

Chen, G., Andries, J., Luo, Z., & Spliethoff, H. (2003). Biomass pyrolysis/gasification for product gas production: The overall investigation of parametric effects. *Energy Conversion and Management*, *44*(11), 1875–1884. doi:10.1016/S0196-8904(02)00188-7

Chen, Y., Guo, L., Cao, W., Jin, H., Guo, S., & Zhang, X. (2013). Hydrogen production by sewage sludge gasification in supercritical water with a fluidized bed reactor. *International Journal of Hydrogen Energy*, *38*(29), 12991–12999. doi:10.1016/j.ijhydene.2013.03.165

Chuntanapum, A., & Matsumura, Y. (2010). Char formation mechanism in supercritical water gasification process: A study of model compounds. *Industrial & Engineering Chemistry Research*, *49*(9), 4055–4062. doi:10.1021/ie901346h

Demirbaş, A. (2006). *Oily Products from Mosses and Algae via Pyrolysis* (Vol. Part A).

Demĭrbas, A. H., & Demĭrbas, I. (2007). Importance of rural bioenergy for developing countries. *Energy Conversion and Management*, *48*(8), 2386–2398. doi:10.1016/j.enconman.2007.03.005

Deng, G.-F., Shen, C., Xu, X.-R., Kuang, R.-D., Guo, Y.-J., Zeng, L.-S., … & Xia, E.-Q. (2012). Potential of fruit wastes as natural resources of bioactive compounds. *International Journal of Molecular Sciences*, *13*(7), 8308–8323.

DiLeo, G. J., & Savage, P. E. (2006). Catalysis during methanol gasification in supercritical water. *The Journal of Supercritical Fluids*, *39*(2), 228–232. doi:10.1016/j.supflu.2006.01.004

D'Jesús, P., Boukis, N., Kraushaar-Czarnetzki, B., & Dinjus, E. (2006). Gasification of corn and clover grass in supercritical water. *Fuel*, *85*(7), 1032–1038. doi:10.1016/j.fuel.2005.10.022

Elliott, D. C., Hart, T. R., Schmidt, A. J., Neuenschwander, G. G., Rotness, L. J., Olarte, M. V., … & Holladay, J. E. (2013). Process development for hydrothermal liquefaction of algae feedstocks in a continuous-flow reactor. *Algal Research*, *2*(4), 445–454. doi:10.1016/j.algal.2013.08.005

Elliott, D. C., Sealock, L. J., & Baker, E. G. (1993). Chemical processing in high-pressure aqueous environments. 2. Development of catalysts for gasification. *Industrial & Engineering Chemistry Research*, *32*(8), 1542–1548. doi:10.1021/ie00020a002

Elsayed, S., Boukis, N., Patzelt, D., Hindersin, S., Kerner, M., & Sauer, J. (2016). Gasification of microalgae using supercritical water and the potential of effluent recycling. *Chemical Engineering & Technology*, *39*(2), 335–342. doi:10.1002/ceat.201500146

Fan, Y. J., Zhu, W., Gong, M., Su, Y., Zhang, H. W., & Zeng, J. N. (2016). Catalytic gasification of dewatered sewage sludge in supercritical water: Influences of formic acid on hydrogen production. *International Journal of Hydrogen Energy*, *41*(7), 4366–4373. doi:10.1016/j.ijhydene.2015.11.071

Fang, Z., Minowa, T., Smith, R. L., Ogi, T., & Koziński, J. A. (2004). Liquefaction and gasification of cellulose with Na2CO3 and Ni in subcritical water at 350 °C. *Industrial & Engineering Chemistry Research*, *43*(10), 2454–2463. doi:10.1021/ie034146t

FAO. (2013). *Food Wastage Footprint: Impacts on Natural Resources*: FAO.

FAO. (2014). *Save Foods: Global Initiative on Food Loss and Waste Reduction*: FAO.

Frank, E. D., Elgowainy, A., Han, J., & Wang, Z. (2013). Life cycle comparison of hydrothermal liquefaction and lipid extraction pathways to renewable diesel from algae. *Mitigation and Adaptation Strategies for Global Change, 18*(1), 137–158. doi:10.1007/s11027-012-9395-1

Fricke, K., Santen, H., Wallmann, R., Hüttner, A., & Dichtl, N. (2007). Operating problems in anaerobic digestion plants resulting from nitrogen in MSW. *Waste Management, 27*(1), 30–43. doi:10.1016/j.wasman.2006.03.003

Furusawa, T., Sato, T., Sugito, H., Miura, Y., Ishiyama, Y., Sato, M., ... & Suzuki, N. (2007). Hydrogen production from the gasification of lignin with nickel catalysts in supercritical water. *International Journal of Hydrogen Energy, 32*(6), 699–704. doi:10.1016/j.ijhydene.2006.08.001

Gasafi, E., Reinecke, M.-Y., Kruse, A., & Schebek, L. (2008). Economic analysis of sewage sludge gasification in supercritical water for hydrogen production. *Biomass and Bioenergy, 32*(12), 1085–1096. doi:10.1016/j.biombioe.2008.02.021

Goldemberg, J. (2008). *Biomass and Energy* (Vol. 32).

Gong, M., Zhu, W., Xu, Z. R., Zhang, H. W., & Yang, H. P. (2014). Influence of sludge properties on the direct gasification of dewatered sewage sludge in supercritical water. *Renewable Energy, 66*, 605–611. doi:10.1016/j.renene.2014.01.006

Guo, L. J., Lu, Y. J., Zhang, X. M., Ji, C. M., Guan, Y., & Pei, A. X. (2007). *Hydrogen Production by Biomass Gasification in Supercritical Water: A Systematic Experimental and Analytical Study* (Vol. 129).

Guo, Y., Wang, S. Z., Xu, D. H., Gong, Y. M., Ma, H. H., & Tang, X. Y. (2010). Review of catalytic supercritical water gasification for hydrogen production from biomass. *Renewable and Sustainable Energy Reviews, 14*(1), 334–343. doi:10.1016/j.rser.2009.08.012

Ibrahim, A. B. A., & Akilli, H. (2019). Supercritical water gasification of wastewater sludge for hydrogen production. *International Journal of Hydrogen Energy, 44*(21), 10328–10349. doi:10.1016/j.ijhydene.2019.02.184

Ikushima, Y., Hatakeda, K., Sato, O., Yokoyama, T., & Arai, M. (2000). Acceleration of synthetic organic reactions using supercritical water: Noncatalytic Beckmann and Pinacol rearrangements. *Journal of the American Chemical Society, 122*(9), 1908–1918. doi:10.1021/ja9925251

Ikushima, Y., Hatakeda, K., Sato, O., Yokoyama, T., & Arai, M. (2001). Structure and base catalysis of supercritical water in the noncatalytic benzaldehyde disproportionation using water at high temperatures and pressures. *Angewandte Chemie International Edition, 40*(1), 210–213. doi:10.1002/1521-3773(20010105)40:1<210::AID-ANIE210>3.0.CO;2-7

Istirokhatun, T., Rokhati, N., Rachmawaty, R., Meriyani, M., Priyanto, S., & Susanto, H. (2015). Cellulose isolation from tropical water hyacinth for membrane preparation. *Procedia Environmental Sciences, 23*, 274–281. doi:10.1016/j.proenv.2015.01.041

Krammer, P., & Vogel, H. (2000). Hydrolysis of esters in subcritical and supercritical water. *The Journal of Supercritical Fluids, 16*(3), 189–206. doi: https://doi.org/10.1016/S0896-8446(99)00032-7

Kruse, A. (2008). Supercritical water gasification. *Biofuels, Bioproducts and Biorefining, 2*(5), 415–437. doi:10.1002/bbb.93

Kruse, A. (2009). Hydrothermal biomass gasification. *The Journal of Supercritical Fluids, 47*(3), 391–399. doi:10.1016/j.supflu.2008.10.009

Kruse, A., & Dinjus, E. (2007). *Hot compressed water as reaction medium and reactant. Properties and synthesis reactions* (Vol. 39).

Kruse, A., & Faquir, M. (2007). *Hydrothermal Biomass Gasification – Effects of Salts, Backmixing and Their Interaction* (Vol. 30).

Kruse, A., Krupka, A., Schwarzkopf, V., Gamard, C., & Henningsen, T. (2005). Influence of proteins on the hydrothermal gasification and liquefaction of biomass. 1. Comparison of different feedstocks. *Industrial & Engineering Chemistry Research, 44*(9), 3013–3020. doi:10.1021/ie049129y

Kruse, A., Maniam, P., & Spieler, F. (2007). Influence of proteins on the hydrothermal gasification and liquefaction of biomass. 2. Model compounds. *Industrial & Engineering Chemistry Research, 46*(1), 87–96. doi:10.1021/ie061047h

Kumar, S. (2013). Sub- and supercritical water technology for biofuels. In J. W. Lee (Ed.), *Advanced Biofuels and Bioproducts* (pp. 147–183). New York, NY: Springer New York.

Kumar, S., Kothari, U., Kong, L., Lee, Y. Y., & Gupta, R. B. (2011). Hydrothermal pretreatment of switchgrass and corn stover for production of ethanol and carbon microspheres. *Biomass and Bioenergy, 35*(2), 956–968. doi:10.1016/j.biombioe.2010.11.023

Lachos Perez, D., Prado, J., Torres Mayanga, P., Forster-Carneiro, T., & Meireles, M. A. (2015). *Supercritical Water Gasification of Biomass for Hydrogen Production: Variable of the Process* (Vol. 6).

Lee, I.-G. (2011). Effect of metal addition to Ni/activated charcoal catalyst on gasification of glucose in supercritical water. *International Journal of Hydrogen Energy, 36*(15), 8869–8877. doi:10.1016/j.ijhydene.2011.05.008

Lee, I.-G., & Ihm, S.-K. (2009). Catalytic gasification of glucose over Ni/activated charcoal in supercritical water. *Industrial & Engineering Chemistry Research, 48*(3), 1435–1442. doi:10.1021/ie8012456

Lee, W.-S., Chen, T.-H., & Hwang, H.-H. (2008). Impact response and microstructural evolution of biomedical titanium alloy under various temperatures. *Metallurgical and Materials Transactions A, 39*(6), 1435–1448. doi:10.1007/s11661-008-9514-5

Lu, Y. J., Guo, L. J., Ji, C. M., Zhang, X. M., Hao, X. H., & Yan, Q. H. (2006). Hydrogen production by biomass gasification in supercritical water: A parametric study. *International Journal of Hydrogen Energy, 31*(7), 822–831. doi:10.1016/j.ijhydene.2005.08.011

Lv, P., Xiong, Z., Chang, J., Wu, C. Z., Chen, Y., & Zhu, J. (2004). *An Experimental Study on Biomass Air-Steam Gasification in a Fluidized Bed* (Vol. 95).

Marrone, P. A., & Hong, G. T. (2009). Corrosion control methods in supercritical water oxidation and gasification processes. *The Journal of Supercritical Fluids, 51*(2), 83–103. doi:10.1016/j.supflu.2009.08.001

Matsumura, Y., Minowa, T., Potic, B., Kersten, S. R. A., Prins, W., van Swaaij, W. P. M., … & Antal Jr, M. J. (2005). Biomass gasification in near- and super-critical water: Status and prospects. *Biomass and Bioenergy, 29*(4), 269–292. doi:10.1016/j.biombioe.2005.04.006

May, A., Salvadó, J., Torras, C., & Montané, D. (2010). Catalytic gasification of glycerol in supercritical water. *Chemical Engineering Journal, 160*(2), 751–759. doi:10.1016/j.cej.2010.04.005

Möbius, A., Boukis, N., Galla, U., & Dinjus, E. (2012). Gasification of pyroligneous acid in supercritical water. *Fuel, 94*, 395–400. doi: https://doi.org/10.1016/j.fuel.2011.11.023

Nanda, S., Isen, J., Dalai, A. K., & Kozinski, J. A. (2016). Gasification of fruit wastes and agro-food residues in supercritical water. *Energy Conversion and Management, 110*, 296–306. doi:10.1016/j.enconman.2015.11.060

Nanda, S., Rana, R., Zheng, Y., Kozinski, J. A., & Dalai, A. (2017). *Insights on the Pathways for Hydrogen Generation from Ethanol* (Vol. 1).

Nanda, S., Reddy, S. N., Hunter, H. N., Dalai, A. K., & Kozinski, J. A. (2015). Supercritical water gasification of fructose as a model compound for waste fruits and vegetables. *The Journal of Supercritical Fluids, 104*, 112–121. doi:10.1016/j.supflu.2015.05.009

Ni, M., Leung, D. Y. C., Leung, M. K. H., & Sumathy, K. (2006). An overview of hydrogen production from biomass. *Fuel Processing Technology, 87*(5), 461–472. doi:10.1016/j.fuproc.2005.11.003

Okolie, A. J., Rana, R., Nanda, S., Dalai, A., & Kozinski, J. A. (2019). *Supercritical Water Gasification of Biomass: A State-of-the-Art Review of Process Parameters, Reaction Mechanisms and Catalysis* (Vol. 3).

Osada, M., Sato, T., Watanabe, M., Adschiri, T., & Arai, K. (2004). Low-temperature catalytic gasification of lignin and cellulose with a ruthenium catalyst in supercritical water. *Energy & Fuels, 18*(2), 327–333. doi:10.1021/ef034026y

Osada, M., Sato, T., Watanabe, M., Shirai, M., & Arai, K. (2006). Catalytic gasification of wood biomass in subcritical and supercritical water. *Combustion Science and Technology, 178*(1–3), 537–552. doi:10.1080/00102200500290807

Pinkard, B. R., Gorman, D. J., Tiwari, K., Rasmussen, E. G., Kramlich, J. C., Reinhall, P. G., & Novosselov, I. V. (2019). Supercritical water gasification: Practical design strategies and operational challenges for lab-scale, continuous flow reactors. *Heliyon, 5*(2), e01269. doi:10.1016/j.heliyon.2019.e01269

Qian, L., Wang, S., Xu, D., Guo, Y., Tang, X., & Wang, L. (2016). Treatment of municipal sewage sludge in supercritical water: A review. *Water Research, 89*, 118–131. doi:10.1016/j.watres.2015.11.047

Reddy, S. N., Ding, N., Nanda, S., Dalai, A. K., & Kozinski, J. A. (2014). Supercritical water gasification of biomass in diamond anvil cells and fluidized beds. *Biofuels, Bioproducts and Biorefining, 8*(5), 728–737. doi:10.1002/bbb.1514

Reddy, S. N., Nanda, S., Dalai, A. K., & Kozinski, J. A. (2014). Supercritical water gasification of biomass for hydrogen production. *International Journal of Hydrogen Energy, 39*(13), 6912–6926. doi:10.1016/j.ijhydene.2014.02.125

Resende, F. L. P., Neff, M. E., & Savage, P. E. (2007). Noncatalytic gasification of cellulose in supercritical water. *Energy & Fuels, 21*(6), 3637–3643. doi:10.1021/ef7002206

Rodriguez Correa, C., & Kruse, A. (2018). Supercritical water gasification of biomass for hydrogen production – Review. *The Journal of Supercritical Fluids, 133*, 573–590. doi:10.1016/j.supflu.2017.09.019

Salihu, A., & Bala, M. (2011). *Brewer's Spent Grain: A Review of Its Potentials and Applications* (Vol. 10).

Sasaki, M., Adschiri, T., & Arai, K. (2004). Kinetics of cellulose conversion at 25 MPa in sub- and supercritical water. *AIChE Journal, 50*(1), 192–202. doi:10.1002/aic.10018

Sasaki, M., Kabyemela, B., Malaluan, R., Hirose, S., Takeda, N., Adschiri, T., & Arai, K. (1998). Cellulose hydrolysis in subcritical and supercritical water. *The Journal of Supercritical Fluids, 13*(1), 261–268. doi:10.1016/S0896-8446(98)00060-6

Savage, P. E. (2009). A perspective on catalysis in sub- and supercritical water. *The Journal of Supercritical Fluids, 47*(3), 407–414. doi:10.1016/j.supflu.2008.09.007

Schmieder, H., Abeln, J., Boukis, N., Dinjus, E., Kruse, A., Kluth, M., … & Schacht, M. (2000). Hydrothermal gasification of biomass and organic wastes. *The Journal of Supercritical Fluids, 17*(2), 145–153. doi:10.1016/S0896-8446(99)00051-0

Scialabba, N., Jan, O., Tostivint, C., Turbé, A., O'Connor, C., Lavelle, P., … & Batello, C. (2013). *Food Wastage Footprint: Impacts on Natural Resources.* Summary Report.

Seif, S., Tavakoli, O., Fatemi, S., & Bahmanyar, H. (2015). Subcritical water gasification of beet-based distillery wastewater for hydrogen production. *The Journal of Supercritical Fluids, 104*, 212–220. doi:10.1016/j.supflu.2015.06.014

Sınag, A., Kruse, A., & Rathert, J. (2003). *Influence of the Heating Rate and the Type of Catalyst on the Formation of Key Intermediates and on the Generation of Gases During Hydropyrolysis of Glucose in Supercritical Water in a Batch Reactor* (Vol. 43).

Susanti, R. F., Veriansyah, B., Kim, J.-D., Kim, J., & Lee, Y.-W. (2010). Continuous supercritical water gasification of isooctane: A promising reactor design. *International Journal of Hydrogen Energy, 35*(5), 1957–1970. doi:10.1016/j.ijhydene.2009.12.157

Thomason, T. B., & Modell, M. (1984). Supercritical water destruction of aqueous wastes. *Hazardous Waste, 1*(4), 453–467. doi:10.1089/hzw.1984.1.453

Watanabe, M., Osada, M., Inomata, H., Arai, K., & Kruse, A. (2003). *Acidity and Basicity of Metal Oxide Catalysts for Formaldehyde Reaction in Supercritical Water at 673 K* (Vol. 245).

Xu, D., Wang, S., Hu, X., Chen, C., Zhang, Q., & Gong, Y. (2009). Catalytic gasification of glycine and glycerol in supercritical water. *International Journal of Hydrogen Energy, 34*(13), 5357–5364. doi:10.1016/j.ijhydene.2008.08.055

Xu, X., Matsumura, Y., Stenberg, J., & Antal, M. J. (1996). Carbon-catalyzed gasification of organic feedstocks in supercritical water. *Industrial & Engineering Chemistry Research, 35*(8), 2522–2530. doi:10.1021/ie950672b

Xu, Z. R., Zhu, W., & Li, M. (2012). Influence of moisture content on the direct gasification of dewatered sludge via supercritical water. *International Journal of Hydrogen Energy, 37*(8), 6527–6535. doi:10.1016/j.ijhydene.2012.01.086

Yamamura, T., Mori, T., Park, K. C., Fujii, Y., & Tomiyasu, H. (2009). Ruthenium(IV) dioxide-catalyzed reductive gasification of intractable biomass including cellulose, heterocyclic compounds, and sludge in supercritical water. *The Journal of Supercritical Fluids, 51*(1), 43–49. doi:10.1016/j.supflu.2009.07.007

Yoshida, T., & Matsumura, Y. (2001). Gasification of cellulose, xylan, and lignin mixtures in supercritical water. *Industrial & Engineering Chemistry Research, 40*(23), 5469–5474. doi:10.1021/ie0101590

Yoshida, Y., Dowaki, K., Matsumura, Y., Matsuhashi, R., Li, D., Ishitani, H., & Komiyama, H. (2003). Comprehensive comparison of efficiency and CO2 emissions between biomass energy conversion technologies—Position of supercritical water gasification in biomass technologies. *Biomass and Bioenergy, 25*(3), 257–272. doi:10.1016/S0961-9534(03)00016-3

6 Supercritical Water Oxidation of Hazardous Waste

Chen Li and Sandeep Kumar

CONTENTS

6.1 SCWO GENERAL BACKGROUND

Supercritical water oxidation (SCWO) technology was invented by Dr. Michael Modell in the 1980s, followed by a substantial increase in the amount of research on SCWO in areas including reaction kinetics [1–3], salt nucleation and growth [4, 5], materials compatibility and corrosion [6, 7], physical property measurements [1, 2], and reaction/system modeling [10, 11]. The first experiences in commercial application were operated mostly by the companies MODAR and MODEC. The early applications were for PCBs and hazardous wastes [3].

SCWO relies on the unique properties of water in supercritical conditions. It operates typically in the range of 500–650°C and 250–300 bar with reactor residence times under one minute for complete destruction into environmentally friendly small molecular compounds such as carbon dioxide, nitrogen, and water [4]. In general, the targeted materials to be treated are toxic or noxious ones, such as dioxins, pesticides, contaminated soils, wastewater, and biowastes. The oxidation reaction occurs in a homogeneous phase without much problems with mass transfer and interface crossing. Another feature of consequence is the heat generated in the exothermic oxidation process, which if properly handled is able to sustain all the energy requirements of the process and even leave some energy for external use [5].

The SCWO process consists of four main steps: feed preparation, reaction, salt separation, and heat recovery, as shown in Figure 6.1 [6]. The feed stream contains rich

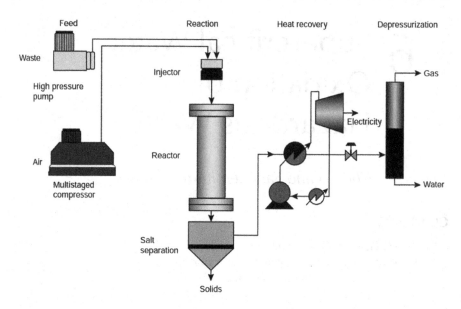

FIGURE 6.1   Schematics of an SCWO plant. (Adapted from Ref. [6].)

organic substances with typically air, oxygen, or hydrogen peroxide as oxidants. The material to be oxidized, as an aqueous slurry, with the oxidant are fed through a pressurizing pump into the reactor, where the exergonic oxidation happens, and the heat produced being sufficient to keep the reaction mixture at the required temperature. The effluent from the reactor is fed to a cyclone that separates solids (salts from the original feed as well as those formed in the SCWO reaction) from liquid effluent. The liquid effluent of the solid separator is a mixture of water, nitrogen gas, and carbon dioxide.

6.2   SCWO KINETIC MECHANISM AND INDUSTRIAL APPLICATION

6.2.1   STUDY ON SCWO KINETIC MECHANISM

SCWO kinetics of model compounds and real waste waters were studied by numerous authors [7]. The model compounds and waste water mainly include acetonitrile [56], acetic acid [17-23],butyric acid [8, 9], caprylic acid [10], formic acid [20, 22, 27], glyoxalic acid [11], lactic acid [12], methylphosphonic acid [30, 31], oleic acid [10], oxalic acid [11], propionic acid [18, 22], 3-hydroxypropionic acid [13], ammonia [32-34], ammonium [33, 35-37], aniline [14], benzene [14], biphenyl and PCBs [39-43], dyes [15], phenol derivatives [43, 45, 46–50], dichloroethane [16], diethanolamine [17], dimethyl methylphosphonate [18], EDTA [19], ethanol [33, 53-55], phenol [23, 45, 48, 56–73], glucose [20], methane [34, 74–76], methanol [2, 23, 55, 77–81], methylamine [21], nitro-alkanes [22], nitrobenzene [23], o-dichlorobenzene [24], pyridine [86–88], polyethylene glycol [25], propene [26], ion exchange resins [27], sodium 3,5,6-trichloropyridin-2-ol [28], sulfides [29], thiodiglycol [30], xylene [60, 86, 95], 2-chlorophenol [96–98], and 2,4,6-trinitrotoluene (TNT) [31].

Real wastewaters are a mixture of compounds and the kinetic study was very important to the design and operation of the SCWO process. Li et al. [32] proposed a generalized kinetic model to represent wastewaters containing nitrogen and chlorinated compounds, ammonia [33, 34] and methyl chloride being the corresponding representative intermediate compounds, respectively. Portela et al. [35] proposed a model based on carbon monoxide formation as a refractory intermediate compound, which was satisfactorily used in the case of a cutting oil used in the metalworking industry.

Organic matter destruction would take place in two steps that correspond to a pseudo-first-order kinetic, a fast step followed by a slow step. A two-step model was used by Shende and Mahajani [11] to fit the results of hydrothermal oxidation of glyoxalic acid, with and without catalyst. Shende and Levec [36] later used this two-step model to fit the hydrothermal oxidation of maleic and fumaric acids. Lei et al. [37] used it to model the hydrothermal oxidation of textile industrial wastewaters by the same method. In the case of cutting oils this model was the most adequate to represent the hydrothermal oxidation [38] with oxidant excess.

6.2.2 INDUSTRIAL WASTEWATERS STUDIED

Many industrial wastewaters with a high amount of organic compounds were also studied by numerous researchers. Pilot and full-scale SCWO process plants were satisfactorily applied with removal efficiency of up to 99.9% and residence times of order of seconds. Table 6.1 shows a compilation of representative industrial wastewaters studied for the last few years. Contrary to the facilities of mature subcritical oxidation, there are fewer facilities in supercritical water oxidation [39]. However, numerous pilot plant facilities have been built as can be seen in Table 6.2.

Since the first commercial SCWO plant was built by Eco Waste Technologies for the company Hunstman Chemical Corporation in Austin (Texas), numerous commercial SCWO plants have been built up. Table 6.3 shows SCWO industrial facilities commercialized to date, highlighting those that are inactive.

During the conventional SCWO process, there may problems such as corrosion and salt precipitation (Table 6.4); these challenges are being avoided by more and more companies who are involved in commercializing SCWO technology, and are mentioned in Table 6.5. Today, they remain very active in this business field. A summary of SCWO technology and recent activity at each of these companies is provided below [4].

General Atomics (GA), whose headquarters is in San Diego, CA, was built up in 1991, and has the longest tenure in SCWO of all active SCWO companies. In 1996, the first SCWO company (MODAR) and its knowledge assets in 1996 was purchased by GA. A vessel type reactor design is typically GA utilized. GA has had extensive experience with several different methods for controlling corrosion and salt precipitation/accumulation, such as the use of liners, coatings, feed additives, and mechanical scrapers [124, 125]. Therefore, the SCWO systems of GA could handle a wide variety of aggressive feeds such as chemical agent hydrolysates. Most of GA's work has been for government/military entities, but they have also worked with industrial customers. GA has unveiled a simpler, more cost effective version of SCWO for commercial applications referred to as industrial SCWO [59].

Hanwha Chemical Corp. whose headquarters is in Seoul, South Korea, has been working with these supercritical water-based technologies since 1994. Most of their

TABLE 6.1
Industrial Wastewaters Treated by SCWO

Waste	Reactor	Operation Mode	Capacity	Refs
Sewage sludge	Tubular	Batch		[40]
Alcohol distillery wastewater	Tubular	Continuous	0.06L/h	[40]
Olive mill wastewater	Tubular	Continuous	0.084kg/h	[41]
	Tubular	Continuous	3.56L/h	[42]
	Tubular	Continuous	2.8kg/h	[43]
Sulfonated lignin waste	Tubular	Semicontinuous	0.03kg/h	[44]
Sewage and industrial sludge	Tubular	Batch		[18]
Automobile manufacturer painting effluent	Tubular	Continuous	20kg/h	[45]
Wastewater from LCD manufacturing	Tubular	Continuous	0.63 kg/L	[18]
Wastewater from acrylonitrile manufacturing plant	Tubular	Continuous	1.15 L/h	[46]
Oily sludge	Stirred tank	Batch		[47]
Cutting oil wastes	Tubular	Continuous	2 L/h	[38]
PCB-contaminated mineral transformer oil	Tubular	Continuous	0.138L/h	[48]
	Tubular	Continuous	30 kg/h	[49]
Ion exchange resins	Double shell stirred	Continuous	0.5 kg/h	[27]
Strength coking wastewater	Tubular	Continuous	1.2 L/h	[50]
Coal	Tubular	Continuous	2 L/h	[51]
Landfill leachate	Transpiring wall	Continuous	2 L/h	[52]

customers for SCWO plants appear to be chemical manufacturing companies, as they have targeted industrial waste treatment applications. Both vessel and tubular reactor types were utilized in Hanwha systems. Hanwha has several bench and pilot scale systems on which they have performed research on different types of synthesis reactions and testing of various feeds by hydrothermal treatment [60]. It is one of the most versatile companies involved with hydrothermal technologies, with its SCWO Business Group having interest in SCWO and various hydrothermal synthesis processes such as growth of metal oxides and carbon nanotubes [61].

Paris-based Innoveox, whose headquarters is in France, having started in 2008, is the youngest of the active SCWO companies, but has built and improved on the developments and progress made by an earlier similar company, HOO (no official connection between the two) [62]. Innoveox has exclusive global rights to the version of SCWO developed and patented by Cansell's group at CNRS. The utilization of multiple injection points for oxidant along the length of its tubular reactor, instead of all at once at the beginning is the unique aspect of Innoveox's SCWO design.

This design could control temperature for the greatest reaction efficiency and avoid thermal runaway, and result in a temperature rise down the length of the reactor. Corrosion and salt precipitation are managed by limiting the concentration of key species for these phenomena in the feed. Innoveox's focus and business model is

TABLE 6.2
SCWO Pilot Plants, (P ≈25Mpa, T≈550°C)

Waste	Process	Reactor	Capacity	Refs
Wastewater and sewage sludge	Komatsu & Kurita Group (Japan)	Tank double shell	400 kg/h	[7]
PCB, dioxins, sewage sludge and radioactive waste	Organo Corporation (Japan)	Tank double tank	100 kg/day, 2 t/day	[7]
Wastewater from incineration plants	EBARA (Japan)	Tank with flame generation	60 kg/h	[7]
Propellant	U.S. Navy (USA)	Tank double tank	250 kg/h	[7]
Dangerous wastewaters from military	U.S. Navy (USA)	Tank double tank	250 kg/h	[7]
Toxics wastewaters	U.S. Navy (USA)	Tubular with porous wall	250 kg/h	[53]
Radioactive	U.S. Department of Defense (USA)	Coated tank	0.4 kg/h	[7]
Industrial wastewaters	Aquacritox (Sweden, Ireland)	Tubular	250 kg/h	[7]
Toxic wastewaters	PIOS HOO (France)	Tubular and tank combined	100 kg/h	[7]
Cutting oil, polyethylene terephthalate (PET), production wastewater, industrial wastewaters	Universidad de Valladolid(Spain)	Transpiring wall	40 kg/h	[54]
Industrial wastewater	Universidad de Valladolid (Spain)	Transpiring wall and cool wall	200 kg/h	[7]
Industrial wastewater	Universidad de Cadiz (Spain)	Tubular	25 kg/h	[55]
Sewage sludge	School of Energy and Power Engineering(China)	Transpiring wall combined with reverse flow tank	125 kg/h	[56]
Ammonia sulfate solution	University of British Columbia (Canada)	Tubular	120 kg/h	[57]
Explosive manufacture waste	Boreskov Institute of Catalysis (Russia)	Tubular	40–60 kg/h	[58]
Sewage sludge	Super Water solutions (USA)	Tubular	5 dried matter t/day	[7]

based on providing waste treatment as a service at a customer's site rather than just designing and selling SCWO equipment [63].

SRI International is a research and development organization, whose headquarters is in France; it developed the AHO version of SCWO. This technology is being utilized in the full-scale facility built for JESCO and currently in operation in Tokyo, Japan for destruction of PCBs. While AHO technology is still available, there does not appear to be any recent activity or developments in AHO and no further plans for additional full-scale plants at the present time.

TABLE 6.3
Commercially Designed Full-Scale SCWO Plants

Water	Customer	Company for Project	Reactor	Capacity	Dates
Hazardous industrial waste	Private company	Innoveox	Tubular	100 kg/h	2011
PCBs	Japan Environmental Safety Corp.	Mitsubishi Heavy Industries	Fluidized bed tank and tubular reactor	306 t/day	2005
Pink water	U.S. Army (Mcalester)	General Atomics	Tank	6.5 t/day	2001
Chemical weapons	U.S. Army (Richmond)	General Atomics	Tank	36/day	2011
Hydrolysate of conventional explosive devices	Toele Army (Toele)	General Atomics	Tank	18 t/day	2008
Explosives and propellants	Bluegrass Army Depot	General Atomics	Tank	50 t/day	—
Industrial wastewater	Pacific Environmental Corporation	General Atomics	Tank	10 t/day	—
Spent catalyst (recovery of tubular precious metals)	Johnson Matthey, Brimsdown (UK)	Chematur Engineering	Tubular	80 t/day	2002–2004
Wastewater from DNT production	Namhae Chemical Corp. (Korea)	Hanwha Chemical	Tank	53 t/day	2000–2005
Waste water from TPA production	Samah Petrochemical Corp. (Korea)	Hanwha Chemical	Tubular	145 t/day	2006–2007
Laboratory wastewater	University of Tokyo (Japan)	Organo (MODAR)	Tank	—	2002–2010
Municipal wastewater sludge	Shinko Pantec (Japan)	Eco Waste Technologies	Tubular	—	2000–2004
Alcohols glycols and amines	Hunstman chemical (USA)	Eco Waste Technologies	Tubular	29 t/day	1994–1999
Obsolete weapons	U.S. Army, Pinebluff Arsenal	Foster Wheeler Development Corp.	Tank	3.8 t/day	1998–2002

(Continued)

TABLE 6.3   (Continued)
Commercially Designed Full-Scale SCWO Plants

Water	Customer	Company for Project	Reactor	Capacity	Dates
Semiconductor industry waste	Nittetsu Semiconductor (Japan)	Organo (MODAR)	Tank	2 t/day	1998–2002
Sewage sludge	Harlingen wastewater treatment plant (USA)	Hydroprocessing	Tubular	150 t/day	2001–2002
Food industry wastewater	SYMPESA (France)	Hydrothermal Oxidation Option	Tubular	2.7 t/day	2004–2008

Source: Based on references [4].

TABLE 6.4
Conventional SCWO Process and Reactions Leading to Problems in the Particular Parts of the Plant

	Waste preheater	Oxidant Preheater	Reactor	Heat exchanger	Depressurization
P(bar), T(°C)	250bar, 25°C ⟶ 380°C	250bar,	250bar, 600°C	250bar, 600°C ↓ 25°C	250-1bar, 25°C
Reactions	Pyrolysis may occur		Oxidation	Formed acids dissociate	
Aggressive species	Salts, (HCl) no O_2	O_2	HCl; O_2	H^+;Cl^-;O_2	
Main Problem	-Slight corrosion -High energy demand in the start-up -Salt precipitation		- Salt precipitation -Corrosion -Hot spots corrosion	-Severe corrosion -Salt precipitation	-Limitation of equipment at industrial scale -Specialty valves for wastewaters with suspended solids

Source: From [16, 136].

TABLE 6.5

SCWO Corrosion Improvement Methods

Category	Approach	Purpose	Examples
Prevent corrosive species from reaching a solid surface.	Transpiring wall/ film-cooled wall reactor.	Use of thin film water barrier to protect reactor wall surface.	Foster Wheeler [142-144], FZK [110, 145], ETH [126, 146, 147], CEA [148, 149], University of Valladolid [150–152].
	Adsorption/reaction on fluidized solid phase (assisted hydrothermal oxidation).	Corrosive species formed during reaction retained and neutralized on solid particles.	SRI International [75], Mitsubishi Heavy Industries [76].
	Vortex/circulating flow reactor (conceptual).	Use of rotating fluid motion to keep hot (less dense) reacting fluid away from reactor wall.	Barber [77].
Form a corrosion -resistant barrier.	Use of high corrosion resistance materials (long-term applications).	Choice of primary construction material with acceptable strength and good overall corrosion resistance, typically nickel-based alloys.	Base case for corrosion control often used in conjunction with other methods; Kane [43] and Kritzer [12] provide guidelines for material choice.
	Liners (corrosion-resistant material).	Corrosion-resistant solid barrier installed to protect underlying structural or more permanent component; often used in pressure-balanced environment so only needs to withstand temperature and process fluid chemistry.	Base case for corrosion control often used in conjunction with other methods, Kane [78] and Kritzer [79] provide guidelines for material choice.
	Coatings.	Protective layer applied or formed on surface; usually thin and intimately bonded to surface.	Modar [158, 159], General Atomics.
Manage/ minimize Corrosion.	Liners (sacrificial material).	Inexpensive and disposable solid barrier that corrodes at predictable and acceptable rate while protecting underlying structural or more permanent component; usually pressure-balanced.	General Atomics [80].

(Continued)

TABLE 6.5   (Continued)
SCWO Corrosion Improvement Methods

Category	Approach	Purpose	Examples
	Use of adequate corrosion resistance materials (short-term applications).	Choice of primary construction material with acceptable strength and adequate corrosion resistance for expected duration of use; may be able to use less expensive materials such as stainless steel alloys.	Hayward et al. [81], Kane [78] and Kritzer [79] provide guidelines for material choice.
Adjust process conditions to avoid or minimize corrosion.	Pre-neutralization.	Typically involves addition of a base to the feed or entrance of reactor to neutralize corrosive acidic species as they form in the reactor. Alternatively, acid may be added to alkaline feeds.	Modar, General Atomic, Foster Wheeler [82].
	Cold (ambient temperature) feed injection.	Introduction of waste feed to the reactor without preheating; used to protect upstream feed lines.	Modar, General Atomics Foster Wheeler [80], Chematur [83], Organo
	Feed dilution with non-corrosive wastes.	Mixing of highly corrosive waste feeds with less corrosive waste feeds to reduce the overall concentration of aggressive species without increasing process time.	General Atomics.
	Effluent dilution/ cooling (quench water addition).	Addition of cooling water and/or base solution at reactor exit to reduce concentrations of aggressive species and rapidly drop temperature below range where corrosion is most severe; used to protect downstream effluent and heat exchanger lines.	Modar [84], General Atomics, Foster Wheel, Chematur.

(Continued)

TABLE 6.5   (Continued)
SCWO Corrosion Improvement Methods

Category	Approach	Purpose	Examples
	Optimization of process operating conditions.	Use of equilibrium and process-gathered data to adjust feed composition and/or operating parameters to region where corrosion is minimized.	Advocated by Latanision and coworkers [163, 164].
	Avoidance of corrosive feeds.	Constraint of system operation to non-aggressive feeds (i.e., CHON compounds) with minimal or no salts.	MODEC, HydroProcessing [85], EcoWaste Technologies, General Atomics
	Pretreatment to remove corrosive species.	For certain feeds, corrosive constituents are amenable to separation prior to the reactor.	Modar.

Source: From [70].

Super Critical Fluids International (SCFI), which is based in Cork, Ireland, is a relatively new company with a long history. Their SCWO technology began with Eco Waste Technologies (EWT) of Austin, TX, one of the original SCWO commercial companies. The Swedish firm Chematur AB first bought a license for the EWT SCWO process in Europe in 1995 and then bought the worldwide rights to EWT SCWO in 1999. With further development work, Chematur marketed their version of SCWO under the name Aquacritox®. They also developed and named different customized versions of the Aquacritox® process in collaboration with various clients. In 2007, Chematur sold their supercritical fluids division and equipment to SCFI. SCFI has continued to improve on the Chematur SCWO design, though they have consolidated Chematur's many versions of SCWO under the single Aquacritox® brand name for ease of marketing. SCFI utilizes a tubular reactor design and has chosen to focus primarily on sewage sludge and digestate feed applications [64]. While they have a sacrificial mixing pipe configuration that can be used at the entrance and exit to the reactor for dealing with corrosive feeds [65], SCFI prefers to limit applications to feeds that are relatively low in corrosion and salt formation potential. As such, they typically restrict salt levels in the feed to a few percent and do not process feeds with chlorinated materials. SCFI has partnered with Parsons to provide internal engineering support and marketing in North America [66], and with Rockwell Automation to provide control systems and construction support. SCFI has designed four different models of the Aquacritox® process based on nominal feed rate: 600, 2500, 10,000, and 20,000 kg/hr [67]. They are currently building their first commercial system (2500 kg/hr) for the waste treatment and recycling firm Eras Eco in Youghal, Ireland [48]. This system will include the option of power generation from the process effluent heat via a waste heat boiler and turbine.

SuperWater Solutions LLC, based in Wellington, FL, is the latest company that was cofounded by Dr. Michael Modell, whose experiments at MIT in the 1970s formed the basis of SCWO technology and who subsequently founded MODAR. Super Water Solutions was started in 2006 and the main focus has been on processing non-corrosive wastewater sludge. The SuperWater Solutions SCWO design is similar to that of Modell's previous company, MODEC. It features a tubular reactor system, and utilizes a high velocity flow and mechanical brushes for control/removal of salts/solids accumulation [68].

Commercialization of SCWO technology's potential was realized for the destruction of aqueous organic wastes and has been in progress for over three decades. As of early 2012, there are six commercial firms around the world that are working in SCWO. While this is less than the number of SCWO companies in the past, new companies continue to enter the field as other ones leave. However, each currently active company has at least one plant either in operation or in the design or construction stages at the present time. Each SCWO company has one or more unique features to their system design (for operation and control of corrosion and salt buildup) and/or business plan, and each one has targeted a specific feed niche. While not without its challenges, SCWO technology commercialization remains an area of great interest and activity.

6.3 SCWO CHALLENGE AND PROGRESS

Supercritical water oxidation (SCWO) is a promising technology, and a wide variety of industrial wastewaters could be treated by SCWO. Compared with other treatment methods, such as landfill, the main advantage of SCWO is that it is a destruction method. Based on oxidation of organic matter, activated carbon treatment, biological treatment, incineration, wet air oxidation, and supercritical water oxidation all account for destruction method. Depending on the organic content of wastewater content, the best method is chosen. Biological and activated carbon treatments are suitable for organic content of up to 1%. On the other hand, incineration is suitable to highly concentrated wastewaters, but in the range of 1–20% organic matter, SCWO is a better option, due to the toxic gases produced and the high cost of incineration [69].

The reason why SCWO has not yet become an important waste treatment technology could be classified into three kinds of aspects or shortcomings: technical drawbacks, high investment required, and high operational costs [7]. The following paragraphs describe each of the disadvantages in more detail. The technical drawbacks include reactor corrosion, salt precipitation, and plugging.

6.3.1 CORROSION PROBLEMS AND CONTROL

The greatest obstacle to successful implementation of SCWO is corrosion. Under highly oxidizing conditions during the oxidation of hydrocarbons, the formation of acids induces the chemically aggressive environments. Depending upon the particular feeds and materials of construction, the corrosion rate can reach several mils/h [70]. Moreover, the SCWO reactor utilizes a high partial pressure of oxygen in order to achieve rapid oxidation of the organic waste, giving rise to conditions that are highly oxidizing. There are several dominant types of corrosion found on SCWO equipment: general corrosion, pitting corrosion, stress corrosion cracking, deposits, intergranular corrosion, faceted grains, oxide dissolution and spallation, hydriding (associated mainly with titanium), as well as changes in color, weight, composition, and morphology are frequently observed in materials used in SCWO systems [71, 72]. The most severe corrosion appears at the regions where reactants/products are at the states just below the critical point and both high temperature and stable ionic species are present, for example, the beginning and ending parts of the reactor, the components and piping used for preheating, cooling, and heat exchange, and some unexpectedly formed subcritical areas in the system [70].

However, it is extremely unlikely that a single reactor material can withstand every conceivable acidic solution [73]. On the other hand, it was shown that some materials possess a satisfactory corrosion resistance against some acids. Unfortunately, these materials fail in the presence of others [73]. For example, titanium is almost not attacked by oxidizing HCl solutions at any temperature but shows poor resistance against H_2SO_4 or H_3PO_4 at temperatures above 400°C [74].

6.3.2   Salt Deposition and Plugging

At room temperature, water is an excellent solvent for most salts, typically several 100 g/l. However, on the other hand, the solubility of most salts is low in low-density supercritical water (typically 1–100 ppm) [4, 166, 167]. A fine-crystalline, slimy "shock precipitate" is formed when a subcritical salt-containing solution is rapidly heated to supercritical temperatures [86]. As a result, the precipitating salts could result in several problems as described below.

(1) Reactor Plugging

Since the solubility of salt is reduced markedly to lower than 100 mg/L in SCW, agglomerates and deposits are formed on internal surface of reactor during the precipitation process, which causes the reactor plugging, especially for smaller reactors with large or sticky salt crystals, such as calcium sulfate and aluminum phosphate and so on [8787]. Moreover, salt deposition also results in the heat transfer efficiency of reactor wall decreasing and pressure drop of reactor increasing.

(2) Reactor Corrosion

As previously discussed, salt precipitated in SCW mainly causes chemical corrosion through oxidation reaction. Salt in subcritical water mostly increases power-chemical corrosion and may start intergranular corrosion from the edge of metal grain. Depositing on the internal surface of reactor will severely erode reactor, especially when reactor material is sensitive to these corrosive species. Reactor corrosion should be prevented as much as possible, because it will lead to a short reactor life and a bad fluid treatment result. Due to the very low solubility of salt, two contradictions between preventing salt deposition and minimizing reactor corrosion rate are displayed as follows. First, preventing salt deposition needs high density

SCW because it will exhibit a relatively improved solvent property for precipitated salts [157, 170]. However, minimizing corrosion rate requires low SCW density [157, 171, 172] for decreasing the content of salt in the form of ion. Second, adding alkali compounds independently or in feedstock before reactor is helpful for inhibiting reactor corrosion, but the possibility of reactor plugging increases because of salt deposition [88]. That is why some alkali compounds are delivered into reaction systems from reactor outlets. Therefore, it is necessary to prevent corrosive reaction fluid from contacting reactor's internal surface, no matter whether it is under supercritical or subcritical condition.

(3) Catalyst Deactivation

The function of catalyst in SCWO is decreasing the reaction temperature and pressure, improving reactant conversion efficiency, and accelerating the reaction rate. The types of catalysts includes zeolite, CuO, MnO_2 , V_2O_5, nickel, platinum, and so on. These catalysts may be poisoned and/or polluted quickly by precipitated salts [174–176]. Also, it is difficult to replace catalyst in traditional reactor configuration. Thereby, it is important to separate these precipitated salts before they contact heterogeneous catalyst.

Thus, efforts are put into two aspects for preventing the salt deposition problem; the first is salt separation technology. Theoretically, salt separation can be performed before, during, and after reaction through centrifuge reactor, hydrocyclone, or a reactor with rotational spin. However, these apparatuses still may be plugged due to the deposition of sticky salt [89]. Additionally, salt separation was also conducted in the preheating step using two typical biorefinery residues before catalytic reaction. Overall, for ideal and efficient results, salt separation technology should be combined with other technics such as filtrating. The second is to improve the reactor configuration for preventing salt deposition. For example, the uses of a cool wall reactor [90] and a transpiring wall reactor [126, 179] can help to solve salt deposition problems. These two kinds of reactors are divided into pressure-bearing wall and non-load-bearing wall. With the use of cool wall reactor, precipitated salts are redissolved in a subcritical water film on the internal surface of the non-load-bearing wall. A transpiring wall reactor can not only prevent salt deposition but also decrease reactor corrosion rate through a porous transpiring wall element. Clean water flows across the porous wall to form a protective film to continuously dilute corrosive species, redissolve precipitated salt particles, and/or sweep them away from the internal surface of the reactor.

6.3.3 HIGH INVESTMENT AND OPERATION COST

High investment costs are necessary based on the nature and characteristics of the SCWO process. On one side, in order to have equipment able to work at high pressure and temperatures and on the other side, the application of high corrosion resistance alloys to build reactors and heat exchangers (preferably alloys with high nickel content) due to the corrosion problem. Besides, the maintenance of high-pressure operational conditions of the material and repair of the equipment working under extreme conditions makes the cost very high. Therefore, initial investment is a lot and the best way to make the SCWO process feasible is to reach the autothermal regime during the reaction. During the design of a SCWO plant, reactor cost is one of the main costs, so the objective is to design it with a small volume [7].

6.4   OUTLOOK AND SUGGESTIONS

In order to make SCWO a waste treatment technology, there are still many questions that need to be solved. For example, the research about measurements on the solubility of most oxides is rare and still at the initial stage. As a result, SCWO in some cases is still an unpredictable process and thus may be "dangerous" for technical applications, especially when evaluating the lifetimes of reactors.

It is not possible for SCWO to become a "general" treatment technology for all kinds of waste streams. Consequently, the wastes suitable for SCWO need to be found and selected carefully. Considering the characteristic of each waste, an "ideal" reactor material is needed for building the reactor, which may be resistant over long useful lifetimes.

Considering an industrial application, it doesn't matter if the destruction rate of a certain waste is 99.99 or 99.999%. However, the long-time applicability of the industrial process absolutely is required to be proven. Thus, these unfinished tests must neither be performed with safe model compounds nor over reaction times of only several hours. Additionally, they have to be performed with exactly the real wastes selected and need to be treated before.

Above all, when all these requirements are fulfilled successfully, SCWO will surely become a waste treatment process of industrial interest.

REFERENCES

1. Watson, J.T.R., R.S. Basu, and J.V. Sengers, An improved representative equation for the dynamic viscosity of water substance. *Journal of Physical and Chemical Reference Data*, 1980. **9**(4): p. 1255–1290.
2. Lamb, W.J., G.A. Hoffman, and J. Jonas, Self-diffusion in compressed supercritical water. *The Journal of Chemical Physics*, 1981. **74**(12): p. 6875–6880.
3. Hodes, M. et al., Salt precipitation and scale control in supercritical water oxidation—Part A: Fundamentals and research. *The Journal of Supercritical Fluids*, 2004. **29**(3): p. 265–288.
4. Marrone, P.A., Supercritical water oxidation—Current status of full-scale commercial activity for waste destruction. *The Journal of Supercritical Fluids*, 2013. **79**: p. 283–288.
5. Vogel, G.H., Supercritical water. A green solvent: Properties and uses. Edited by Yizhak Marcus. *Angewandte Chemie International Edition*, 2013. **52**(8): p. 2158–2158.
6. Maria Dolores Bermejo, D.R.V.V.M.J.C., Supercritical water oxidation: Fundamentals and reactor modeling. *Chemical Industry and Chemical Engineering Quarterly*, 2007. **13**(2): p. 79–87.
7. Vadillo, V. et al., Problems in supercritical water oxidation process and proposed solutions. *Industrial & Engineering Chemistry Research*, 2013. **52**(23): p. 7617–7629.
8. Williams, P.E.L. et al., Wet air oxidation of low molecular weight organic acids. *Water Quality Research Journal*, 1973. **8**(1): p. 224–236.

9. Sanchez-Oneto, J. et al., Kinetics and mechanism of wet air oxidation of butyric acid. *Industrial & Engineering Chemistry Research*, 2006. **45**(12): p. 4117–4122.

10. Sánchez-Oneto, J. et al., Wet air oxidation of long-chain carboxylic acids. *Chemical Engineering Journal*, 2004. **100**(1): p. 43–50.

11. Shende, R. and V. Mahajani, Kinetics of wet air oxidation of glyoxalic acid and oxalic acid. *Industrial & Engineering Chemistry Research*, 1994. **33**(12): p. 3125–3130.

12. Li, L.X. et al., Oxidation and hydrolysis of lactic acid in near-critical water. *Industrial & Engineering Chemistry Research*, 1999. **38**(7): p. 2599–2606.

13. Shende, R.V. and J. Levec, Wet oxidation kinetics of refractory low molecular mass carboxylic acids. *Industrial & Engineering Chemistry Research*, 1999. **38**(10): p. 3830–3837.

14. Dinaro, J.L. et al., Analysis of an elementary reaction mechanism for benzene oxidation in supercritical water. *Proceedings of the Combustion Institute*, 2000. **28**(2): p. 1529–1536.

15. Chen, G., L. Lei, and P. Yue, Wet oxidation of high-concentration reactive dyes. *Industrial & Engineering Chemistry Research*, 1999. **38**(5): p. 1837–1843.

16. Baillod, C.R., B.M. Faith, and O. Masi, Fate of specific pollutants during wet oxidation and ozonation. Where do the pollutants wind up in oxidational purification processes? Here is a well documented experimental study. *Environmental Progress*, 1982. **1**(3): p. 217–227.

17. Mishra, V.S., V.V. Mahajani, and J.B. Joshi, Wet air oxidation. *Industrial & Engineering Chemistry Research*, 1995. **34**(1): p. 2–48.

18. Veriansyah, B., J.D. Kim, and Y.W. Lee, Decomposition kinetics of dimethyl methylphospate(chemical agent simulant) by supercritical water oxidation. *Journal of Environmental Sciences (China)*, 2006. **18**(1): p. 13–16.

19. Lee, H.-C. et al., Decomposition of ethylenediaminetetraacetic acid by supercritical water oxidation. *Industrial & Engineering Chemistry Research*, 2004. **43**(13): p. 3223–3227.

20. Shishido, M., K. Okubo, and M. Saisu, Effect of oxygen concentration on supercritical water oxidation of glucose. *Kagaku Kogaku Ronbunshu*, 2001. **27**(6): p. 806–811.

21. Benjamin, K. and P. Savage, Detailed chemical kinetic modeling of methylamine in supercritical water. *Industrial & Engineering Chemistry Research*, 2005. **44**(26): p. 9785–9793.

22. Anikeev, V., A. Yermakova, and M. Goto, Decomposition and oxidation of aliphatic nitro compounds in supercritical water. *Industrial & Engineering Chemistry Research*, 2004. **43**(26): p. 8141–8147.

23. Zhang, G. and I. Hua, Supercritical water oxidation of nitrobenzene. *Industrial & Engineering Chemistry Research*, 2003. **42**(2): p. 285–289.

24. Svishchev, I.M. and A. Plugatyr, Supercritical water oxidation of o-dichlorobenzene: Degradation studies and simulation insights. *The Journal of Supercritical Fluids*, 2006. **37**(1): p. 94–101.

25. Otal, E. et al., Integrated wet air oxidation and biological treatment of polyethylene glycol-containing wastewaters. *Journal of Chemical Technology & Biotechnology*, 1997. **70**(2): p. 147–156.

26. Broll, D., A. Kramer, and H. Vogel, Partial oxidation of propene in subcritical and supercritical water. *Chemie Ingenieur Technik*, 2002. **74**(1–2): p. 81–85.

27. Leybros, A. et al., Ion exchange resins destruction in a stirred supercritical water oxidation reactor. *The Journal of Supercritical Fluids*, 2010. **51**(3): p. 369–375.

28. Liu, N., H.-Y. Cui, and D. Yao, Decomposition and oxidation of sodium 3,5,6-trichloropyridin-2-ol in sub- and supercritical water. *Process Safety and Environmental Protection*, 2009. **87**(6): p. 387–394.

29. Wang, T. et al., Supercritical water oxidation of sulfide. *Environmental Science & Technology*, 2003. **37**(9): p. 1955–1961.
30. Lachance, R. et al., Thiodiglycol hydrolysis and oxidation in sub- and supercritical water. *The Journal of Supercritical Fluids*, 1999. **16**(2): p. 133–147.
31. Chang, S.-J. and Y.-C. Liu, Degradation mechanism of 2,4,6-trinitrotoluene in supercritical water oxidation. *Journal of Environmental Sciences*, 2007. **19**(12): p. 1430–1435.
32. Li, L., P. Chen, and E.F. Gloyna, Generalized kinetic model for wet oxidation of organic compounds. *AIChE Journal*, 1991. **37**(11): p. 1687–1697.
33. Webley, P.A., J.W. Tester, and H.R. Holgate, Oxidation kinetics of ammonia and ammonia-methanol mixtures in supercritical water in the temperature range 530–700. degree.C at 246 bar. *Industrial & Engineering Chemistry Research*, 1991. **30**(8): p. 1745–1754.
34. Li, L., Kinetic model for wet oxidation of organic compounds in subcritical and supercritical water. *ACS Symposium Series*, 1993. **514**: p. 305–313.
35. Portela, J.R., E. Nebot, and E. Martínez de la Ossa, Generalized kinetic models for supercritical water oxidation of cutting oil wastes. *The Journal of Supercritical Fluids*, 2001. **21**(2): p. 135–145.
36. Shende, R.V. and J. Levec, Subcritical aqueous-phase oxidation kinetics of acrylic, maleic, fumaric, and muconic acids. *Industrial & Engineering Chemistry Research*, 2000. **39**(1): p. 40–47.
37. Lei, L. et al., Wet air oxidation of desizing wastewater from the textile industry. *Industrial & Engineering Chemistry Research*, 2000. **39**(8): p. 2896–2901.
38. Sánchez-Oneto, J. et al., Hydrothermal oxidation: Application to the treatment of different cutting fluid wastes. *Journal of Hazardous Materials*, 2007. **144**(3): p. 639–644.
39. Aymonier, C., *Traitement hydrothermal de déchets industriels spéciaux. Données pour le dimensionnement d'installations industrielles et concepts innovants de réacteurs sonochimique et électrochimique.* 2000, Université Sciences et Technologies: Bordeaux I.
40. Goto, M. et al., Kinetic analysis for ammonia decomposition in supercritical water oxidation of sewage sludge. *Industrial & Engineering Chemistry Research*, 1999. **38**(11): p. 4500–4503.
41. Erkonak, H., O.Ö. Söğüt, and M. Akgün, Treatment of olive mill wastewater by supercritical water oxidation. *The Journal of Supercritical Fluids*, 2008. **46**(2): p. 142–148.
42. Portela, J. and F. Beltran, Supercritical water oxidation of olive oil mill wastewater. *Industrial & Engineering Chemistry Research*, 2001. **40**(16): p. 3670–3674.
43. Chkoundali, S. et al., Hydrothermal oxidation of olive oil mill wastewater with multi-injection of oxygen: Simulation and experimental data. *Environmental Engineering Science*, 2008. **25**(2): p. 173–173.
44. Drews, M., M. Barr, and M. Williams, A kinetic study of the SCWO of a sulfonated lignin waste stream. *Industrial & Engineering Chemistry Research*, 2000. **39**(12): p. 4784–4793.
45. Abeln, J. et al., Supercritical water oxidation (SCWO) using a transpiring wall reactor: CFD simulations and experimental results of ethanol oxidation. *Environmental Engineering Science*, 2004. **21**(1): p. 93–99.
46. Shin, Y.H. et al., Supercritical water oxidation of wastewater from acrylonitrile manufacturing plant. *Journal of Hazardous Materials*, 2009. **163**(2): p. 1142–1147.
47. Cui, B. et al., Oxidation of oily sludge in supercritical water. *Journal of Hazardous Materials*, 2009. **165**(1): p. 511–517.
48. Marulanda, V. and G. Bolaños, Supercritical water oxidation of a heavily PCB-contaminated mineral transformer oil: Laboratory-scale data and economic assessment. *The Journal of Supercritical Fluids*, 2010. **54**(2): p. 258–265.

49. Kim, K. et al., Environmental effects of supercritical water oxidation (SCWO) process for treating transformer oil contaminated with polychlorinated biphenyls (PCBs). *Chemical Engineering Journal*, 2010. **165**(1): p. 170–174.

50. Du, X. et al., Treatment of high strength coking wastewater by supercritical water oxidation. *Fuel*, 2013. **104**: p. 77–82.

51. Wang, S. et al., Supercritical water oxidation of coal: Investigation of operating parameters' effects, reaction kinetics and mechanism. *Fuel Processing Technology*, 2011. **92**(3): p. 291–297.

52. Gong, W. and X. Duan, Degradation of landfill leachate using transpiring-wall supercritical water oxidation (SCWO) reactor. *Waste Management*, 2010. **30**(11): p. 2103–2107.

53. Crooker, P. et al., Operating results from supercritical water oxidation plants. *Industrial & Engineering Chemistry Research*, 2000. **39**(12): p. 4865–4870.

54. Cocero, M. et al., Supercritical water oxidation in a pilot plant of nitrogenous compounds: 2-Propanol mixtures in the temperature range 500–750 degrees C. *Industrial & Engineering Chemistry Research*, 2000. **39**(10): p. 3707–3716.

55. García Jarana, B. et al., Simulation of supercritical water oxidation with air at pilot plant scale. *International Journal of Chemical Reactor Engineering*, 2010. **8**(1): A58.

56. Xu, D. et al., Design of the first pilot scale plant of China for supercritical water oxidation of sewage sludge. *Transactions of the Institution of Chemical Engineers Part A: Chemical Engineering Research and Design*, 2012. **90**(2): p. 288–297.

57. Asselin, E., A. Alfantazi, and S. Rogak, Thermodynamics of the corrosion of alloy 625 supercritical water oxidation reactor tubing in ammoniacal sulfate solution. *Corrosion*, 2008. **64**(4): p. 301–314.

58. Anikeev, V. and A. Yermakova, Technique for complete oxidation of organic compounds in supercritical water. *Russian Journal of Applied Chemistry*, 2011. **84**(1): p. 88–94.

59. Wellig, B., K. Lieball, and P. Rudolf von Rohr, Operating characteristics of a transpiring-wall SCWO reactor with a hydrothermal flame as internal heat source. *The Journal of Supercritical Fluids*, 2005. **34**(1): p. 35–50.

60. Serikawa, R.M. et al., Hydrothermal flames in supercritical water oxidation: Investigation in a pilot scale continuous reactor. *Fuel*, 2002. **81**(9): p. 1147–1159.

61. Steeper, R.R. et al., Methane and methanol diffusion flames in supercritical water. *The Journal of Supercritical Fluids*, 1992. **5**(4): p. 262–268.

62. Pohsner, G.M. and E.U. Franck, Spectra and temperatures of diffusion flames at high pressures to 1000 bar. *Berichte der Bunsengesellschaft für Physikalische Chemie*, 1994. **98**(8): p. 1082–1090.

63. Steinle, J.U. and E.U. Franck, High pressure combustion – Ignition temperatures to 1000 bar. *Berichte der Bunsengesellschaft für Physikalische Chemie*, 1995. **99**(1): p. 66–73.

64. Narayanan, C. et al., Numerical modelling of a supercritical water oxidation reactor containing a hydrothermal flame. *The Journal of Supercritical Fluids*, 2008. **46**(2): p. 149–155.

65. Bermejo, M.D. et al., Experimental study of hydrothermal flames initiation using different static mixer configurations. *The Journal of Supercritical Fluids*, 2009. **50**(3): p. 240–249.

66. Bermejo, M.D. et al., Analysis of the scale up of a transpiring wall reactor with a hydrothermal flame as a heat source for the supercritical water oxidation. *The Journal of Supercritical Fluids*, 2011. **56**(1): p. 21–32.

67. Cabeza, P. et al., Experimental study of the supercritical water oxidation of recalcitrant compounds under hydrothermal flames using tubular reactors. *Water Research*, 2011. **45**(8): p. 2485–2495.

68. Ding, Z.Y. et al., Catalytic oxidation in supercritical water. *Industrial & Engineering Chemistry Research*, 1996. **35**(10): p. 3257–3279.

69. Kritzer, P. and E. Dinjus, An assessment of supercritical water oxidation (SCWO): Existing problems, possible solutions and new reactor concepts. *Chemical Engineering Journal*, 2001. **83**(3): p. 207–214.

70. Marrone, P.A. and G.T. Hong, Corrosion control methods in supercritical water oxidation and gasification processes. *The Journal of Supercritical Fluids*, 2009. **51**(2): p. 83–103.

71. Dong, J., F.R. Cook, and W. Zhu, Equine infectious anemia virus in Japan: Viral isolates V70 and V26 are of North American not Japanese origin. *Veterinary Microbiology*, 2014. **174**(1–2): p. 276–278.

72. Kritzer, P., Separators for nickel metal hydride and nickel cadmium batteries designed to reduce self-discharge rates. *Journal of Power Sources*, 2004. **137**(2): p. 317–321.

73. Kritzer, P., N. Boukis, and E. Dinjus, Factors controlling corrosion in high-temperature aqueous solutions: A contribution to the dissociation and solubility data influencing corrosion processes. 1999. p. 205–227.

74. Boukis, N., C. Friedrich, and E. Dinjus, Titanium as reactor material for SCWO applications – First experimental results, in *Corrosion 98*. 1998, NACE International: San Diego, CA. p. 7.

75. Ross, D.S., I. Jayaweera, and D. Bomberger, On-site disposal of hazardous waste via assisted hydrothermal oxidation. 高圧力の科学と技術, 1998. **7**: p. 1386–1388.

76. Mitton, D.B. et al., Corrosion mitigation in SCWO systems for hazardous waste disposal. *Conference: Corrosion '98*, San Diego, CA (United States), 22–27 Mar 1998; Other Information: PBD: 1998; Related Information: Is Part of Corrosion '98: p. 53. Annual Conference and Exposition, Proceedings; PB: [6600] p. 1998: NACE International, Houston, TX (United States). Medium: X; Size: p. 15, Paper 414; Quantity: 1 CD-ROM; OS: Windows 3.1; Windows95; Windows98; Macintosh; UNIX; Compatibility: 486 processor, 8MB RAM, 2X CD-ROM drive; Macintosh, 68020, 4MB RAM, 2X CD-ROM drive; Sun SPARCstati.

77. Martin, A., M. Dolores Bermejo, and M. Jose Cocero, Recent developments of supercritical water oxidation: A patents review. *Recent Patents on Chemical Engineering*, 2011. **4**(3): p. 219–230.

78. Kane, R.D., Pick the right materials for wet oxidation. *Chemical Engineering Progress*, 1999. **95**(3): p. 51–58.

79. Kritzer, P., Corrosion in high-temperature and supercritical water and aqueous solutions: A review. *The Journal of Supercritical Fluids*, 2004. **29**(1): p. 1–29.

80. Marrone, P.A., S.D. Cantwell, and D.W. Dalton, SCWO system designs for waste treatment: Application to chemical weapons destruction. *Industrial & Engineering Chemistry Research*, 2005. **44**(24): p. 9030–9039.

81. Hayward, T.M., I.M. Svishchev, and R.C. Makhija, Stainless steel flow reactor for supercritical water oxidation: Corrosion tests. *The Journal of Supercritical Fluids*, 2003. **27**(3): p. 275–281.

82. Crooker, P.J. et al., Operating results from supercritical water oxidation plants. *Industrial & Engineering Chemistry Research*, 2000. **39**(12): p. 4865–4870.

83. Lavric, E.D. et al., Delocalized organic pollutant destruction through a self-sustaining supercritical water oxidation process. *Energy Conversion and Management*, 2005. **46**(9): p. 1345–1364.

84. Garcia, K.M., *Supercritical Water Oxidation Data Acquisition Testing*. 1996, Idaho National Engineering and Environmental Laboratory, Idaho Falls, ID (US). p. Medium: P; Size: p. 45.

85. Griffith, J.W. and D.H. Raymond, The first commercial supercritical water oxidation sludge processing plant. *Waste Management*, 2002. **22**(4): p. 453–459.

86. Armellini, F.J., J.W. Tester, and G.T. Hong, Precipitation of sodium chloride and sodium sulfate in water from sub- to supercritical conditions: 150 to 550 °C, 100 to 300 bar. *The Journal of Supercritical Fluids*, 1994. **7**(3): p. 147–158.

87. Brunner, G., Near and supercritical water. Part II: Oxidative processes. *The Journal of Supercritical Fluids*, 2009. **47**(3): p. 382–390.

88. Peterson, A.A. et al., In situ visualization of the performance of a supercritical-water salt separator using neutron radiography. *The Journal of Supercritical Fluids*, 2008. **43**(3): p. 490–499.

89. Elliott, D.C., Catalytic hydrothermal gasification of biomass. *Biofuels, Bioproducts and Biorefining*, 2008. **2**(3): p. 254–265.

90. Cocero, M.J. and J.L. Martínez, Cool wall reactor for supercritical water oxidation: Modelling and operation results. *The Journal of Supercritical Fluids*, 2004. **31**(1): p. 41–55.

7 Hydrothermal Mineralization

Elena Barbera and Sandeep Kumar

CONTENTS

7.1 INTRODUCTION: BACKGROUND AND PRINCIPLES

The term "hydrothermal mineralization" (HTM), also known as "hydrothermal precipitation," "hydrothermal crystal growth," or more recently as "hydrothermal synthesis," has its origins in the geological field. It was introduced in the 18th century to describe the action of high-temperature and high-pressure water on the formation of various rocks and minerals forming the Earth's crust, including ore deposits [1,2]. In fact, the first HTM experiments were mainly of geological interest and were carried out with the aim to mimic natural conditions existing under the Earth's crust in order to understand the complex processes of minerals formation [3,4]. In recent years, this technique has become popular among scientists and engineers of different fields for the precipitation of a wide variety of inorganic compounds with and without natural analogs, (from native elements to complex oxides, hydroxides, silicates, phosphates, etc.), and lately for the synthesis of advanced ceramic materials as well as novel nanoparticles.

The term "hydrothermal mineralization" refers to heterogeneous reactions (crystallization) carried out in an aqueous solvent, sometimes with the presence of a mineralizer, under high temperature and pressure conditions (above 100°C and 1 bar) [5–8]. The process can be carried out either under sub- or supercritical water conditions, however the tendency is, when possible, to keep milder operating conditions. Under hydrothermal conditions, the physical properties of the solvent (i.e., water) change, so that high solvation power, high compressibility and mass transport allow the occurrence of various different reactions, e.g., synthesis/stabilization of new phases, production of crystals with desired shape and size (including nanocrystals), and growth of several inorganic compounds that are characterized by low solubility at ambient conditions. For example, it is the only method that can be employed for the precipitation of large crystals of α-quartz, or of compounds with elements in oxidation states otherwise difficult to obtain (e.g., transitional metal compounds such as chromium (VI) oxide) [2]. Moreover, HTM offers several advantages compared to conventional crystallization carried out under ambient or near-ambient conditions in terms of purity, homogeneity, reproducibility, and symmetry of the obtained crystals.

7.1.1 THERMODYNAMICS AND KINETICS OF CRYSTALLIZATION UNDER HYDROTHERMAL CONDITIONS

When considering hydrothermal crystal growth, both thermodynamic and kinetic aspects need to be taken into account, in order to understand whether the chemical reaction is controlled by the former or the latter.

The most important thermodynamic parameter in precipitation processes is the solubility, i.e., the maximum amount of a solute that can be dissolved in a solvent. Its value mainly depends on the physical and chemical properties of the solute and of the solvent, as well as on temperature, pressure, and pH of the solution. The solubility of a compound can either increase (positive) or decrease (negative) with increasing temperature. Some compounds exhibit a positive solubility up to a certain temperature and then show a negative trend, or vice versa. Usually, a sharp change in solubility with changing temperature or pressure is desired [2,9]. Another important aspect to be considered is the interaction between solvent and solute particles (i.e., the solvation). Clearly, a hydrothermal solution is far from ideal solutions behavior. In particular, close to the critical point water properties such as density (ρ), dielectric constant (ε), and ion product vary drastically with little changes in T and P. This strongly influences the equilibrium, the solubility, and the phase behavior of the chemical species involved. Because of the decrease in ε with increasing temperature, electrolytes that are completely dissociated at ambient conditions may become highly associated in the supercritical region [9]. Finally, the knowledge of equilibrium phase diagrams is important to identify the growth conditions necessary to obtain the formation of desired single crystals. According to the Gibbs rule, at equilibrium the number of degrees of freedom F is given by:

$$F = NC + NP + 2 \tag{10.1}$$

with *NC* and *NP* indicating the number of components and the number of phases, respectively. This relation is valid when temperature, pressure, and composition are the only variables. Phase diagrams have been determined and are now available for several systems [10–12]. They can be expressed in terms of composition (i.e., with reagent vs solvent concentration) or as temperature vs concentration diagrams. The latter highlight the effect of temperature on the formation of different phases as well as various polymorphic modifications of the compounds.

Concerning kinetics of hydrothermal crystallization, it is known that temperature, pressure, concentration, and type of precursor reagent have a fundamental role. Pressure itself does not appear to have a direct effect on the rate of crystal growth, but rather it shows an influence possibly through an indirect effect on other parameters such as solubility and mass transfer. Thanks to the marked decrease in the viscosity (η) of the medium under hydrothermal conditions, the molecular mobility increases: in fact, diffusion is known to be inversely proportional to the viscosity. Because of this, hydrothermal crystallization is expected not to be diffusion-limited, with the primary role being played by surface interactions occurring directly at the crystal-solution interface. This allows generally higher growth rates, and the formation of single crystals with a well-developed morphology and reduces the risk of dendritic growths [9].

Thermodynamic and kinetic information under hydrothermal conditions are not trivial to obtain due to the difficulty in observation and sampling of such a system. However, in the latest years extensive investigations have been carried out, which led to the development of thermodynamic and kinetic models and commercial softwares that are now successfully used for the design and optimization of synthesis conditions to obtain phase-pure materials with controlled shape and size.

7.1.2 HYDROTHERMAL MINERALIZATION APPARATUS

The growth of crystals under hydrothermal conditions requires a vessel that is able to sustain high temperature and pressure conditions as well as to resist possiblly corrosive media.

Several types of such autoclaves have been designed and built according to their specific applications. The most commonly used materials for HTM pressure vessels are high-strength metal alloys, such as austenitic stainless steel, nickel, cobalt, titanium or molybdenum alloys, and Hastelloy [2]. For highly corrosive systems, a liner made of Teflon or other fluorocarbon resins is also necessary.

The majority of the autoclaves used worldwide for hydrothermal mineralization are test tube-type pressure vessels, i.e., *batch reactors*. These systems are versatile, simple to operate, and are used to carry out phase equilibria and solubility studies, as well as nanocrystal synthesis, etc. For example, the Tuttle vessels, developed in the late 1950's, (see Figure 7.1A) can hold temperatures of around 900°C and pressures of 1000 bar, while the so-called TZM vessels can reach up to 1150°C and 4000 bar [2,13]. Today, batch autoclaves are also equipped with stirring systems and rapid cooling systems, which allow better control and understanding of the HTM process.

A slightly different system is an *in situ separation-type* apparatus (see Figure 7.1B) [14,15], in which the reacting suspension is led by the autogenous pressure

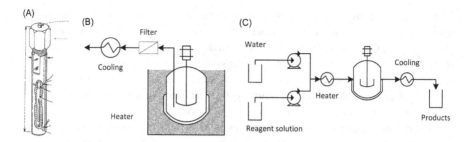

FIGURE 7.1   Schematic of different HTM apparatus: Tuttle autoclave (A) [13], *in situ* separation system (B) and continuous-flow reactor (C).

to a sampling tube, where the precipitates are directly separated from the suspension through a metal filter, and immediately cooled. This system allows for higher recovery yields in minerals that exhibit negative solubility behavior, by eliminating redissolution phenomena that would occur during the cooling period.

In the last two decades, also *continuous-flow* hydrothermal mineralization reactors have developed and become very popular for the synthesis of small micro/nanocrystals with controlled shape and size [2,13,16]. In these systems (see Figure 7.1C) a stream containing the reagents solution and another one containing water are pumped into a pre-heater and then into the reactor. After the reactor, the products are filtered and rapidly cooled. The pressure inside the reactor is controlled by a back-pressure regulator. One of the main advantages of such a system is the rapid and more efficient heating and cooling, which eliminates the effect of the respective periods on the HTM precipitation process.

7.2   HYDROTHERMAL SYNTHESIS OF BULK CRYSTALS

The first applications of hydrothermal mineralization, during the past century, were mainly directed toward the synthesis of large-size bulk crystals of compounds of interest, carried out at very high temperatures and pressures in order to favor crystal growth over nucleation.

One of the main industrial applications was the synthesis of quartz crystals, which is one of the most relevant technological materials. Quartz exists in over 22 polymorphic forms, however only α-quartz has piezoelectric characteristics that are of interest for the production of electronic materials. Due to its properties, only hydrothermal crystallization is suitable for the growth of this polymorph. Typically, SiO_2 precursors such as silica glass, silica gel, silica sand, or small-size α-quartz crystals are dissolved in alkali water solutions (using NaOH, KOH, or Na_2CO_3) [2,5,17]. Hosaka and Taki [18] showed that satisfactory growths and fairly large crystals could also be obtained using pure water solvent. Depending on the seed, growth can be oriented toward specific directions. Operating temperature and pressure range between 400°C–500°C and 1–1.7 kbar.

Currently, the largest producers of synthetic quartz crystals are in Japan (which alone makes 50% of the world production), the United States, and China, for applications such as improved resonator devices, oscillators, or high precision polishing devices.

7.3 HYDROTHERMAL SYNTHESIS OF ADVANCED MATERIALS

With the 21st century, the interest has shifted from the production of large-size bulk crystal to the development of fine powders and nanoparticles. These materials, thanks to the presence of significant surface effects, possess unique optical, mechanical, thermal, and electrical properties that cannot be found in corresponding bulk samples. In this regard, the desired properties greatly depend on the shape and size of the particles, so that precise control of these two variables is extremely important. At the same time, the increasing awareness of sustainability issues has driven the search for green, environmentally benign, and less energy demanding processes for the production of these materials. Hydrothermal mineralization represents a great solution, combining the possibility of controlling the process variables in order to obtain the desired products with a greener technology. In fact, thanks to a deeper knowledge of the physical chemistry and PVT relationship under hydrothermal conditions, the temperature and pressure conditions have been drastically reduced in the latest years, so that HTM now consumes less energy, produces little to no waste, and has a high selectivity toward the compounds of interest, without involving any hazardous material [19].

Among the advantages of using hydrothermal methods over other techniques to synthetize advanced fine and nano- materials, are the high control of process conditions so that products with high purity and quality, controlled morphology, narrow size distributions, high crystallinity, and less defects can be obtained in a very reproducible way. Moreover, by choosing appropriate precursors and surface modifiers agents, the growth can be oriented to obtain desired morphologies, such as whiskers, rods, tubes, needles, plates, or spheres, depending on the application.

A wide variety of fine powders and nanomaterials are nowadays synthetized with the hydrothermal mineralization methods, including native elements, advanced ceramic materials (metal oxides, perovskites), and biomaterials such as hydroxyapatite (HAp). As it would be impossible to discuss all the materials synthetized by HTM during the past decades, the following sections will provide an overview of a few selected materials of interest, mainly belonging to the advanced ceramic family.

7.3.1 CERIA AND ZIRCONIA OXIDES (ZrO_2-CeO_2)

Mixed ceria and zirconia oxides nanoparticles represent a widely studied advanced ceramic material, thanks to their exceptional optical, electronic, and catalytic properties which make them interesting for electrochemistry, optics, UV absorption, catalysis, and biology applications [20–22].

The properties of these material are clearly related to their geometric features, such as shape and size. In order to obtain the desired morphology, phase diagrams have to be known. These have been investigated by several authors [23,24]. However, the knowledge of the metastable regions, which generally is not included in conventional phase diagram representations, is also essential to understand the structural changes of zirconia-ceria crystals. For example, Yashima et al. [25] have characterized the transition between cubic c and a metastable tetragonal phase t', which occurs in a diffusionless way, and is different from the stable tetragonal phase t that is formed diffusionally (see Figure 7.2) [19].

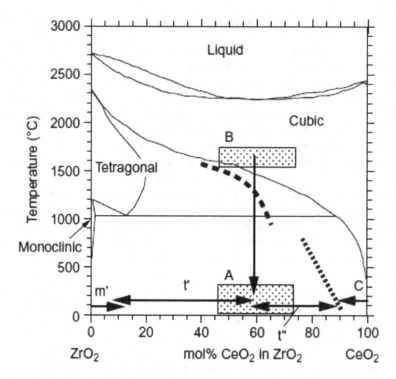

FIGURE 7.2   Metastable phase diagram for the ZrO_2-CeO_2 system [19].

Mixed ceria-zirconia oxides synthesis by hydrothermal mineralization can be conducted either under sub- or supercritical water conditions. The formation of metal oxides under hydrothermal conditions occurs in two steps [20]:

1. Hydrolysis: $M(NO_3)_{x\,(aq)} + xH_2O \rightarrow M(OH)_{x\,(s)} + xHNO_3$ (10.2)

2. Dehydration: $M(OH)_{x\,(s)} \rightarrow MO_{x/2} + \dfrac{x}{2}H_2O$ (10.3)

Subcritical water is generally preferred as it allows softer conditions. Ahniyaz et al. [26] were able to obtain 5 ± 1 nm size $Zr_{0.5}Ce_{0.5}O_2$ crystals of t'' tetragonal form by hydrothermal treatment of 2M $Ce(NO_3)_3$ and $Zr(NO_3)_2$ at 120°C and 6hours, in a batch autoclave. However, it has been seen that under subcritical water conditions the reaction time strongly affects the particle size, even at high conversion values (>95%), which has repercussions on the properties of the obtained material. This is likely due to redissolution of precipitated particles and subsequent re-precipitation as larger crystals. On the other hand, under supercritical water conditions this phenomenon is not observed, and smaller particles of controlled size and narrow distribution are obtained. Indeed, supercritical conditions decrease the solubility and increase the degree of supersaturation, so that nucleation is strongly enhanced rather than crystal growth [19] (see Figure 7.3). Kim et al. [20] synthetized CeO_2, ZrO_2, and

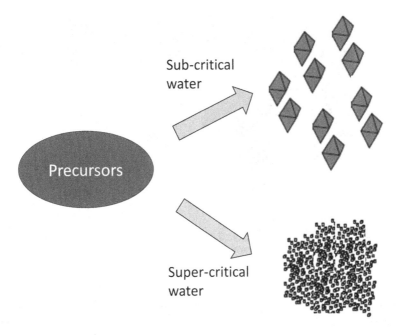

FIGURE 7.3 Schematics of Zr-Ce crystals obtained by sub- or supercritical water mineralization.

various mixed Ce-Zr oxides with high thermal stability and oxygen storage capacity under supercritical water conditions (400°C, 6 seconds) in a continuous-flow apparatus, which had superior catalytic performances compared to the same compounds produced by the coprecipitation method.

7.3.2 TITANIA (TiO$_2$) NANOPARTICLES AND NANOTUBES

Titanium dioxide nanoparticles have gained increasing interest after the discovery of photocatalytic water splitting on TiO$_2$ electrodes in 1972 [27]. Indeed, this material has several applications in the field of photocatalysis, in dye-sensitized solar cells, and even in the cosmetic sector.

TiO$_2$ nanoparticles (rutile or anatase phases) can be easily obtained by hydrothermal mineralization, which is more environmentally benign and allows a better manipulation of process parameters compared to other fabrication methods, such as sol-gel, microwave, microemulsions, etc. The synthesis of titanium oxide nanocrystals is usually carried out in general use batch autoclaves, under temperatures ranging between 120–180°C, sometimes up to 250°C, or around 400°C in the case of supercritical hydrothermal conditions, and autogenous saturation pressures. The treatment duration can range between 4 hours and 24 hours depending on the precursor and conditions used [19]. Typical precursors are stabilized TiCl$_4$ solutions, or alternatively Ti(SO$_4$)$_2$ has also been used [28]. The solvents employed are usually acidic (HNO$_3$, HCl, HCOOH) or alkaline (KOH, NaOH). Often, *in situ* surface modification is performed through the addition of organic compounds, such

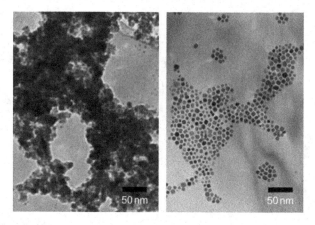

FIGURE 7.4 Hydrothermal synthesis of TiO2 nanoparticles at 400°C with (right) and without (left) surface modifiers addition [19].

as hexaldehyde, which helps prevent aggregation phenomena between particles that normally occur, allowing the production of nicely dispersed particles with narrow size distributions (see Figure 7.4) [19]. The addition of organic compounds is facilitated under hydrothermal conditions thanks to the reduced dielectric constant of water when increasing temperature.

Sometimes, organic solvents such as alcohols or glycols are employed instead of water (solvothermal method), allowing for significantly lower temperatures with respect to conventional hydrothermal crystallization.

Despite the outstanding properties of TiO_2 nanocrystals in the photocatalytic field, their large-scale application is limited by the difficulty in recovering and recycling these small particles. To overcome this issue, recently new titania nanostructures have been developed, such as nanotubes. TiO_2 nanotubes possess many advantages compared to nanopowders, including a higher surface area [29,30]. They are also fabricated by hydrothermal methods under mild temperatures, starting from TiO_2 powders under strongly alkaline conditions. The properties of the resulting nanotubes are strongly influenced by the operating parameters, such as type of precursor and alkali used, their concentrations, and by temperature and duration of the treatment.

Different TiO_2 powders can be used as precursors, namely rutile, anatase, or brookite crystallites, amorphous titanium dioxide, commercial P25 nanopowders, or even titanates [29]. Initially, it was believed that only crystalline precursor could yield nanotubes formation, however, by extending the treatment duration to three days it was found that also amorphous TiO_2 could lead to the production of nanotubes. Smaller precursor particles generally result in higher crystallinity nanotubes. Recently, a study on the dissolution kinetics of anatase and rutile TiO_2 precursors revealed that the former is dissolved according to zero-order kinetics, while the latter is significantly slower, following a second-order kinetics [30]. Concerning the alkaline solution, only NaOH resulted in the formation of nanotubes, while this did not occur when using KOH (which lead to wire-like products) or LiOH. Similarly, the concentration of NaOH in the solution has to be in the range between 5–20 M (better between 10–15 M).

Treatment time and temperature exert a combined effect on the yield of titania nanotubes produced. Several authors found that the appropriate temperature range varies between 100°C and 180°C: at lower temperatures only nanosheets are obtained, while at T > 180°C (sometimes already at 160–170°C) the nanotubes are converted to nanorods. Even though higher temperatures increase the product yield, many authors agree that optimum temperatures are in the range between 130–150°C [29,30]. Similarly, increasing the treatment time (typically 24h–48 h) has a positive effect on the amount of nanotubes produced. The required reaction time decreases when the temperature increases.

7.3.3 PEROVSKITE-TYPE CERAMICS

The general formula of perovskite minerals is ABO_3, where the A cation is a low-valence, relatively large alkali-earth metal (Ba^{2+}, Ca^{2+}, Pb^{2+}, Sr^{2+}) and the B cation is a relatively small, high-valence transition metal (Ti^{4+}, Zr^{4+}, Mn^{3+}). Among perovskites, high-purity titanates are widely employed thanks to their ferroelectric properties.

These materials can be produced by hydrothermal method at moderate T and P conditions (>100°C and 1 bar). In concentrated solutions, the predominant reaction is the following:

$$Me^{2+} + TiO_2 + 2OH^- \rightarrow MeTiO_3 + H_2O \qquad (10.4)$$

where Me = Ca, Ba, Pb, Sr. Accordingly, 2 moles of OH⁻ are consumed to form one mole of $MeTiO_3$. If the metal hydroxide is used as precursor at high concentration, the required amount of OH⁻ ions is directly provided by the precursor, and no additional mineralizer is required. On the other hand, if low metal hydroxide concentrations or other types of precursors (e.g., nitrates, chlorides, etc.) are used, the addition of a suitable mineralizer is necessary, which could result in impurities in the mineral lattice [19].

Among the various perovskite titanates, $BaTiO_3$ has received much attention for electronic applications. In particular, its high dielectric permittivity and ferroelectric response make it an attractive material for multilayer ceramic capacitors (MLCCs). In this regard, high-quality particles with uniform size of the order of 100 nm are desired. Conventional methods, such as solid-state reaction between $BaCO_3$ and TiO_2, carried out at T > 900°C, are not suitable to achieve this goal, as they lead to generally larger particles with uncontrolled morphologies [31]. In this regard, HTM has been found to be a very promising method for the synthesis of ultrafine $BaTiO_3$ particles at lower temperatures (100–240°C), showing good performances in terms of purity, reproducibility, particle size distribution, and stoichiometry control.

In order to obtain the desired products and performances, it is necessary to know the thermodynamics of the system and the mineral formation mechanism. Factors such as temperature, pH, type and concentration of precursors, and environmental conditions affect the performances.

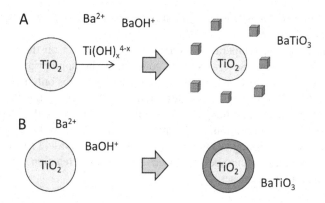

FIGURE 7.5 BaTiO3 precipitation mechanisms under hydrothermal conditions: dissolution-precipitation (A) and *in situ* reaction (B).

Two possible pathways have been proposed in the literature for the formation of $BaTiO_3$ particles: (i) a dissolution-precipitation mechanism and (ii) an *in situ* reaction pathway (see Figure 7.5 A and B) [19,31–33]. According to the first model, TiO_2 particles dissolve into soluble $Ti(OH)_x^{4-x}$, which then reacts with barium ions (Ba^{2+} or $BaOH^+$) to form $BaTiO_3$ precipitates. The nucleation stage can either occur in the bulk solution (homogeneous nucleation) or at the TiO_2 substrate surface (heterogeneous nucleation) [19]. The *in situ* transformation model assumes that the TiO_2 particles directly react with Ba to form an inwardly growing shell of $BaTiO_3$, which can be either a porous or a dense layer. Ba must therefore diffuse through the product layer to continue the reaction until the TiO_2 is exhausted [32,33]. In the latter case, the produced $BaTiO_3$ particles preserve the shape of the starting TiO_2 precursor. Eckert et al. [33] have extensively reviewed these two precipitation models and developed a kinetic study of $BaTiO_3$ formation under hydrothermal conditions. They found out that the formation mechanism changed from dissolution-precipitation at the early stages toward *in situ* transformation happening at longer reaction times [31].

The knowledge of stability diagrams is fundamental in order to know the optimum conditions at which the desired products are thermodynamically stable. Riman and coworkers carried out an extensive work in this regard. Figure 7.6 displays the stability diagram of the Ba-Ti system at 90°C, either without (A) or with the presence of CO_2 (B) in the reaction environment [34]. The presence of carbon dioxide often causes the presence of $BaCO_3$ impurities [31].

Normally, below the Curie temperature of 130°C $BaTiO_3$ perovskite structure undergoes a cubic-to-tetragonal phase transition, with the tetragonal phase being the thermodynamically stable phase at room temperature. However, different authors [31,35] have observed that $BaTiO_3$ particles obtained by HTM preserved their cubic structure. Zhu et al. [31] observed that their $BaTiO_3$ nanocrystalline particles synthetized by hydrothermal method at 100°C for five hours, with $Ba(OH)_2 \cdot 8H_2O$ and $Ti(OH)_4$ starting materials, exhibited a spherical morphology and a metastable cubic structure at room temperature. They attributed this phenomenon to lattice defects such as Ba vacancies. Kržmanc et al. [36] investigated the formation of tetragonal $BaTiO_3$ nanoparticles by HTM in the temperature range 100–240°C for 12 hours,

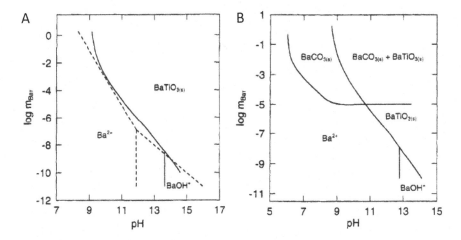

FIGURE 7.6 Stability of the Ba-Ti system at 90°C without (A) and with (B) the presence of CO2 in the air [34].

starting from $Ba(CH_3COO)_2$ and $K_2Ti_6O_{13}$ nanowires precursors. They reported that at low temperature (100°C), the obtained particles had a mixed cubic (22%) and tetragonal (78%) phase, while an increased temperature led to the formation of a dominating tetragonal phase. The authors also conclude that the $BaTiO_3$ nanoparticles formed via a dissolution-precipitation mechanism, as the nanowire shape of the $K_2Ti_6O_{13}$ precursor could not be preserved, whereas star-like, dendritic, and seaweed-like particles were obtained, depending on the reaction temperature.

7.3.4 Hydroxyapatite (HAp) Bioceramic Materials

Hydroxyapatite ($Ca_5(PO_4)_3(OH)$, HAp), is the most stable calcium phosphate mineral under neutral to alkaline conditions. HAp is one of the most employed bioceramic materials, thanks to its similarity to the mineral constituents of hard tissue (teeth and bones) [37–39]. Owing to its excellent biocompatibility, it is widely employed in a large number of biomedical applications. Due to the poor mechanical reliability, HAp is often employed in the form of thin film coating on bioinert supports, or in the form of whiskers. The latter are long filamentary crystals (aspect ratio ~ 5–20), characterized by a low dislocation density, which results in high tensile properties with respect to dense HAp bioceramics, and makes them attractive for bone fracture toughness improvement [19,40].

A number of methods exist for the synthesis of hydroxyapatite minerals, such as solid-state reaction, electrodeposition, sol-gel methods, and so on [37]. Among these, hydrothermal precipitation from aqueous solutions offers several advantages in terms of morphology controllability, crystallinity, particle size, product composition, and stoichiometry.

Similarly to the hydrothermal synthesis of other compounds, the knowledge of phase diagrams is important to identify and design proper operation conditions. Riman et al. [39] evaluated the phase diagram of the system $CaO-P_2O_5-NH_4NO_3-H_2O$

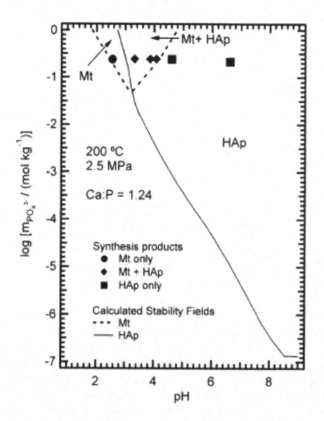

FIGURE 7.7 Phase equilibrium of the CaO-P$_2$O$_5$-NH$_4$NO$_3$-H$_2$O system at 200°C as a function of precursor concentration and pH [39]. HAp = hydroxyapatite, Mt = monetite.

in the temperature range 25–200°C (see Figure 7.7), where it is seen that alkaline pH is necessary to obtain pure HAp, otherwise monetite (CaHPO$_4$, Mt) is also obtained.

Hydrothermal precipitation of HAp whiskers can be carried out according to two main routes, namely homogeneous precipitation with urea or via decomposition of chelating agents (e.g., EDTA, lactic acid, or citric acid) [19]. The latter method allows obtaining single whiskers crystals without the presence of large amounts of CO$_3^{2-}$ in the lattice. Suchanek and Yoshimura [41] obtained HAp crystals of controlled morphology by HTM carried out at 200°C and 20 bar for five hours, starting from H$_3$PO$_4$, Ca(OH)$_2$, and lactic acid. They found that lactic acid/Ca and Ca/P ratio in the starting solution influenced the aspect ratio of the crystals (5–20) as well as the Ca/P in the product crystals. Often, values of Ca/P = 1.59–1.62 are reported, which are lower than the stoichiometric ratio of HAp (= 1.67), probably due to the formation of other phosphate phases or of amorphous compounds [19]. Liu et al. [40] investigated the influence of temperature (60–140°C) and pH (6–14) on the homogenous precipitation of HAp whiskers under hydrothermal conditions, starting from Ca(OH)$_2$ and CaHPO$_4$·2H$_2$O and using urea to control the pH. They found that if pH is not controlled, the resulting HAp product was not pure but contained also some monetite.

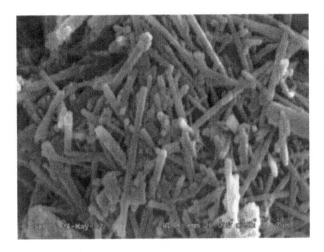

FIGURE 7.8   SEM micrographs of HAp crystals synthetized at pH = 9 and T = 140°C [40].

At the same time, higher temperatures resulted in higher crystallinity. Overall, they were able to obtain well-elongated crystals (aspect ratio > 20) under pH = 9 and T = 120–140°C (see Figure 7.8).

Other authors [37,38] report the synthesis of hydroxyapatite crystals with different morphologies (e.g., microspheres, nanorods, hexagonal prisms, or flower-like particles) by homogeneous hydrothermal precipitation with urea, highlighting how they can be obtained according to different reaction conditions (namely temperature, precursor types and ratios and time). The addition of urea for pH adjustment inevitably leads to carbonate substitutions in the lattice.

7.4 REMOVAL AND RECOVERY OF HAZARDOUS/ RARE ELEMENTS FROM WASTEWATER

From an environmental point of view, hydrothermal mineralization has been proved to be a very promising technique in wastewater treatment, being able to reduce the concentration of several toxic pollutants from industrial wastewaters [14,15,42–45]. HTM looks particularly interesting as, besides successfully removing various oxyanions that are very difficult to treat by conventional methods because of their high stability, it enables recovery of many kinds of rare elements in the form of mineral resources that can be recycled for industrial or agricultural applications. High purity and effective recovery of precious elements from wastes for efficient reuse is of tremendous importance in building an environmentally sustainable society, and it should be encouraged in opposition to landfill disposal of solids generated by other wastewater treatment methods (e.g., coagulation or adsorption).

Extensive work on the assessment of removal and recovery of some critical elements from waste sources by hydrothermal mineralization has been carried out by the Japanese group of Itakura et al. The main results are discussed in the following paragraphs, and are summarized in Table 7.1.

TABLE 7.1

Summary of Results of HTM for Removal/Recovery of Pollutants from Waste Streams

Element	Ion	Discharge Limit (ppm)	Conc. Reached (ppm)	HTM Conditions	Mineral Recovered
As	$As^VO_4^{3-}$ $As^{III}O_3^{3-}$	0.1	0.016	150°C, 2h	$Ca_5(AsO_4)_3(OH)$
Sb	$Sb^VO_4^{3-}$	0.02	0.06	230°C, 12h	$Ca_2(Sb_2O_7)$
Cr	$Cr^{VI}O_4^{2-}$	0.5	0.48	230°C, 12h	Ca_2CrO_4
P	$P^VO_4^{3-}$	16	n.a.	280°C, 1.5 h	$Ca_5(PO_4)_3(OH)$
	$P^{III}O_3^{3-}$		0.2	200°C, 12 h	$Ca(HP^{III}O_3)(H_2O)$
	$P^IO_2^{3-}$		0.2	200°C, 12 h	$Ca(HP^{III}O_3)(H_2O)$
B	$B(OH)_4^-$	10	4	150°C, 1 h	$Ca_2B_2O_5 \cdot H_2O$
F	F^-	8	n.a.	200°C, 4 h	CaF_2
	BF_4^-		4	200°C, 30 h	$Ca_2B_2O_5 \cdot H_2O$

n.a. = not available.

7.4.1 ARSENIC

Arsenic is an important element in advanced electronic materials production. The electronic industry, as well as petroleum refining and power plants produce As containing wastewaters, which are extremely toxic for both human health and the environment. For this reason, the World Health Organization (WHO) has set a discharge limit for effluent wastewaters of 0.1 ppm As. Arsenic is present in the water as pentavalent ($As^VO_4^{3-}$) or trivalent ($As^{III}O_3^{3-}$) anions. Conventional removal methods, such as adsorption with Fe, are not efficient for As^{III} removal.

Using HTM at 100–200°C with the addition of $Ca(OH)_2$ as mineralizer easily removes AsO_4^{3-} ions, reaching concentrations of 0.02 ppm of As within 15 hours [14,15,43]. The mineral recovered is $Ca_5(AsO_4)_3(OH)$, i.e., Johnbaumite: this compound is also known as arsenic apatite, because its structure is very similar to that of hydroxyapatite. The removal of AsO_3^{3-} is slightly less trivial. The addition of $Ca(OH)_2$ causes in fact the precipitation of $Ca_5(AsO_3)_3 \cdot 4H_2O$, leaving a residual As concentration in the water of 4 ppm at ambient conditions, which decreases to a minimum of 0.3 ppm after HTM treatment (100°C, 12 hours) [15,43]. Nonetheless, by adding a proper amount of oxidizer (e.g., H_2O_2), only arsenic apatite is formed, as AsO_3^{3-} is oxidized to AsO_4^{3-} ions. Water itself has been shown to partially act as oxidizer but is not sufficient to remove all the arsenite from the wastewater. Using an *in situ-sampling* apparatus, As concentrations as low as 0.016 ppm can be reached within two-hour treatment time at 150°C.

7.4.2 ANTIMONY

Antimony is industrially used as a fire retardant, or in the fabrication of ceramics and other advanced materials. As an element it has chemical properties similar to those of arsenic. For this reason, there are no effective conventional techniques available to remove antimonic anions (such as SbO_4^{3-}) from industrial aqueous effluents. On the other hand, given its toxicity and adverse effects on human health, the discharge limit for antimonic species in water is set to 0.02 ppm [15].

Antimonic ions tend to be quite stable, so that addition of $Ca(OH)_2$ at ambient conditions does not result in any precipitation. However, the concentration of Sb in water can be drastically reduced under hydrothermal conditions, reaching values down to 0.06 ppm after 12-hour treatment at 230°C [15]. Most importantly, the mineral recovered was identified as $Ca_2Sb_2O_7$, which has crystal structure very similar to natural monimolite (($Pb,Ca)_2Sb_2O_7$). It was also shown that $Ca_2Sb_2O_7$ precipitates as fine crystals on the surface of larger hexagonal $Ca(OH)_2$, which hence acts both as a mineralizer and as seeding material.

7.4.3 CHROMIUM

Chromium is widely used as a reagent in electroplating and metal finishing, as well as in tannery industries, so that Cr containing wastewaters are generated from these activities. Cr can be found in these waters in trivalent (Cr^{III}) or hexavalent (Cr^{VI}) forms, the latter being highly toxic to living beings. Conventional treatment methods (membrane separation, ion exchange, adsorption on activated carbon, among others), generally require the reduction of Cr^{VI} to Cr^{III}.

Using HTM, at a temperature between 200–230°C and by addition of $CaCl_2$, it is possible to reduce the concentration of CrO_4^{2-} in the water effluent, while recovering chromium in the form of $CaCrO_4$, which is valuable to be recycled in electroplating processes [14,45]. In fact, in the treatment of this element, a conventional batch type apparatus has shown not to be sufficient to reach the required concentration in the effluent (i.e., 0.5 ppm as in the Japanese regulation), attaining Cr concentration equal to 400 ppm, or 200 ppm when increasing the cooling rate, even with 12 hours of treatment. On the other hand, by using the *in situ-sampling* reactor, an outlet concentration of 0.48 ppm could be reached by the authors.

7.4.4 PHOSPHORUS

Phosphorus represents a particularly critical element. It is in fact a major constituent of agricultural fertilizers, being one of the essential nutrients required for intensive crop cultivation. Serious concerns are related to the sustainability of fertilizers production, since these are mainly derived from phosphate rock mines, which are expected to be depleted in the near future [46]. On the other hand, when large amounts of P are released in the waters, for example from detergents or from printing industries, besides fertilizers, the high concentrations of this nutrient are known to cause severe environmental damages by eutrophication [47,48]. For these reasons the discharge limit of total P in water is 16 ppm in Japan [45], and even 2 ppm in Europe [49]. At the same time, the recovery of valuable mineral phosphorus compounds is crucial for the sustainability of our society.

P can be present in wastewaters in different oxidation states, namely orthophosphates $P^V O_4^{3-}$, phosphites $P^{III}O_3^{3-}$, or phosphinates $P^I O_2^{3-}$ anions [45,50]. Regarding phosphates, also atmospheric precipitation methods have been demonstrated to efficiently remove P from the water in the form of Magnesium Ammonium Phosphate (MAP) salts, known as struvite, which has potential applications as a slow-release fertilizer [51,52]. This compound has also been efficiently precipitated from microalgae-derived hydrolysate by addition of $Mg(OH)_2$ or $MgCl_2$, showing P removal efficiencies around 65 wt% [53]. Through HTM, however, as detailed in Section 7.3.4, high-value phosphate compounds such as hydroxyapatite can be recovered from these waste streams. Itakura et al. [50] obtained high removal (final concentration 0.1 ppm of P) and recovery of $Ca_5(PO_4)_3(OH)$ from water by HTM batch treatment at 150°C and addition of $Ca(OH)_2$ mineralizer. They also observed that a treatment temperature of at least 130°C was necessary to convert all the calcium phosphate $(CaHPO_4)$ into hydroxyapatite [42]. Teymouri et al. [53,54] demonstrated that more than 97 wt% of P could be removed also from a complex organic substrate such as algal hydrolysate through HTM at 280°C and 90 mins of reaction time using a rapidly heated and cooled batch apparatus and $Ca(OH)_2$ as mineralizer. The authors identified the precipitates as carbonate-substituted hydroxyapatite, with some whitlockite present as secondary phase (i.e., $Ca_9(Mg,Fe^{2+})(PO_4)_6(PO_3OH)$).

The removal of phosphites and phosphinates is less trivial compared to orthophosphate. By addition of $Ca(OH)_2$, precipitates of $Ca(HP^{III}O_3)(H_2O)$ are formed also at ambient conditions, for both trivalent and monovalent P ions [45,50]. However, the measured residual concentration in the effluent remains higher than 100 ppm, hence not meeting the discharge requirements. HTM treatment, on the other hand, is found to be more effective in further reducing the P concentration in the water, as a result of the negative solubility behavior of the precipitated compound, which is reduced under hydrothermal conditions. Even in this case, the addition of a suitable oxidizer improves the removal and recovery yields in PO_2^{3-}-containing waters, by converting it into PO_3^{3-} ions. It was found that using an *in situ*-sampling apparatus, the final P concentration of wastewaters containing 2000 ppm of $PO_2^{3-} + PO_3^{3-}$ could be reduced to 0.2 ppm after 12 hours at 200°C, compared to 18 ppm obtained with a conventional batch system.

7.4.5 BORON AND FLUORINE

Boron is an important micronutrient for both plants and humans. However, an excess intake of this element can cause damage to the living beings. Its content in industrial wastewaters discharged into superficial waters is therefore subject to regulations. For example, in Japan the limit of B concentration in waters generated from manufacturing plants such as semiconductor, ceramic, plating, and glass industries is set to 10 ppm [42]. Similarly, these wastewaters often contain other pollutants such as fluorine (F^- ions), which is also subject to discharge limits (0.8–8 ppm) [44]. Several boron removal and defluorination techniques have been proposed. The first ones comprise coprecipitation, ion exchange, or coagulation-sedimentation methods, while the latter includes lime precipitation, electrochemical methods, and ion exchange, among others. However, these methods appear either not sufficiently efficient or highly expensive.

Boric acid and borate ions ($B(OH)_4^-$) can be precipitated under hydrothermal conditions by the addition of $Ca(OH)_2$ mineralizer, and recovered in the form of

$Ca_2B_2O_5 \cdot H_2O$, i.e., parasibirskite, which is a valuable precursor for borax production [14,16,42,44]. Evidence has shown that the presence of calcium hydroxide as a mineralizer is essential for the effective recovery of this mineral, which is suggested to be formed as needle crystals by heterogeneous nucleation onto the flat-plate surface of undissolved $Ca(OH)_2$. A conventional batch apparatus resulted even in the case ineffective in meeting the discharge limits of B in the water, reaching values of about 80 ppm after 14 hours of treatment at 150°C [42]. However, by using an *in situ* separation-type system, concentration values below 10 ppm could be reached within one hour at the same temperature [14]. An attempt at developing a more practical industrial process, it was verified that hydrothermal mineralization of B can be carried out also in continuous-flow system, achieving a stable boron removal under optimal pH of 12, 150°C, and $CaCl_2$ as a mineralizer [16].

Concerning fluoride ions, the addition of $Ca(OH)_2$ allows recovering fluorite (CaF_2) crystals even at ambient conditions. However, the residual fluorine concentration is too high to meet the required specifications. On the other hand, by carrying out HTM treatment at 200°C for four hours, fluorine concentration drops below the detection limit [44].

When fluorine and boron coexist in water, they tend to form highly stable ions, such as fluoroborate BF_4^-. The removal of these ions is not straightforward. In fact, the simple addition of calcium hydroxide is not sufficient to induce precipitation. Nevertheless, by carrying out precipitation under hydrothermal conditions (200°C) for 30 hours, precipitates of both parasibirskite and fluorite were identified [44]. More precisely, the precipitates were observed to be arranged in a three-layer structure, with $Ca(OH)_2$ large crystals as the bottom layer, $Ca_2B_2O_5 \cdot H_2O$ constituting the middle layer, and fine particles of CaF_2 as the upper layer. The latter showed a slow sedimentation rate, due to their very small size. Overall, 99.9% of F and 98% of B could be removed by wastewaters containing fluoroboric acid, regardless of the initial concentration [44].

7.4.6 Neodymium

Another rare-earth element whose recovery and recycling is desirable is Neodymium: this metal is widely used for the production of sintered permanent magnets of Nd-Fe-B applied in electromagnetic and electronic devices [55]. The recovery of this element has hence to be carried out not from aqueous wastes, but from solid ones. However, by dissolving the magnets in a proper aqueous solution, Nd can be recovered by HTM with a purity that is sufficient for its subsequent recycling. In fact, the issue related to other Nd recovery methods, such as extraction with molten Mg metal, is that impurities such as Ni or other elements compromise the properties of the recycled magnets.

Itakura et al. [55] showed that a commercially available Ni-coated $Nd_2Fe_{14}B$ magnet could be dissolved in a mixed solution of hydrochloric acid and oxalic acid. The acidic environment is necessary for the magnet dissolution, while oxalic acids serves as a Nd precipitating agent. Alternative compounds to be employed could be NaCl, or ethanol. The optimal concentration of HCl was found to be 3 mol/L, as higher values yield lower amounts of precipitates due to increased solubility, while lower concentration cause the coprecipitation of iron oxalates, which occurs when the pH is higher. Another important operating parameter for the process is the temperature. The authors reported that more than 40 hours are required at a temperature of 100°C. On the other hand, only six hours were sufficient at temperatures over 110°C. Temperatures higher than 150°C did not produce any precipitate, likely due to the decomposition of oxalic acid. The authors concluded that HTM carried out at 110°C for six hours could be an efficient method for the recovery of Nd from permanent magnets in the form of $Nd_2(C_2O_4)_3 \cdot H_2O$, easy to convert to pure Nd_2O_3 at lower temperature for recycling in the production of further magnets.

7.4.7 SUMMARY

According to the specific cases discussed, hydrothermal mineralization shows a tremendous potential in environmental applications aiming at the simultaneous removal of several toxic pollutants from waste sources and their recovery into valuable minerals, for a sustainable management of resources. The overall results are summarized in Table 7.1.

7.5 INDUSTRIAL APPLICATIONS OF HTM

Despite being originally studied mainly by scientists and geologists who wanted to understand naturally occurring hydrothermal precipitation phenomena, in the second half of the 19th century HTM started to be acknowledged as an important technology for materials synthesis at industrial level. Initially, HTM was used predominantly for the synthesis of single crystals, such as α-quartz for frequency control and optical applications. Sawyer Research Products, Inc. (currently Sawyer technical materials, LLC) was the first commercial company to produce single α-quartz crystals, in 1956 [56,57]. Currently, due to the high energy required by the severe conditions that are necessary to synthetize single crystals, only a few commercial realities exist for these applications, limited to, e.g., ZnO for UV- and blue light-emitting devices (Tokyo Denpa Co. Ltd), or $KTiOPO_4$ for nonlinear optical applications (Northrop Grumman Synoptics) [56]. However, at present the main commercial HTM applications regard inorganic ceramic powder production. A famous example is the Bayer process for alumina purification, which applies hydrothermal conditions to dissolve bauxite and precipitate aluminum hydroxide, later treated at high temperature to crystallize as α-alumina, with production of around 40 million tons/year [57]. Another example is provided by Sawyer Technical Materials, LLC, who recently developed hydrothermal production of high-purity α-Al_2O_3 powders. In particular, they have developed a product line called DiamoCor, which includes α-Al_2O_3 powders of different crystal sizes, ranging from 100 nm to 40 μm.

By means of hydrothermal precipitation they are able to obtain ultra-pure crystals (> 99.9%) with highly controlled morphology and narrow particle size distributions, with a very high reproducibility.

Another promising commercial area is the production of perovskite-based and zirconia-based structural dielectric ceramics, with companies such as Cabot Corporation, Sakai Chemical Company, Murata Industries, Ferro Corporation, among others, which have established commercial hydrothermal production processes for making dielectric ceramic powders for capacitors.

REFERENCES

1. K. Byrappa, M. Yoshimura, Hydrothermal technology – Principles and applications, in: *Handbook of Hydrothermal Technology*, 2013: pp. 1–49. doi:10.1016/B978-081551445-9.50002-7.

2. K. Byrappa, N. Keerthiraj, S.M. Byrappa, *Hydrothermal Growth of Crystals-Design and Processing*, Second Edition, Elsevier B.V., 2014. doi:10.1016/B978-0-444-63303-3.00014-6.

3. R.W. Goranson, Solubility of water in granite magmas, *Eos Trans. Am. Geophys. Union* 12 (1931) 183. doi:10.1029/TR012i001p00183-1.

4. R.M. Barrer, Syntheses and reactions of mordenite, *J. Chem. Soc.* (1948) 2158–2163. doi:10.1039/JR9480002158.

5. R.A. Laudise, *The Growth of Single Crystals*, Prentice-Hall, 1970. doi:10.1016/0022-0248(83)90150-1.

6. A. Rabenau, The role of hydrothermal synthesis in preparative chemistry, *Angew. Chem.* (English Edition) 24 (1985) 1026–1040.

7. A.N. Lobachev, *Crystallization Processes under Hydrothermal Conditions*, Springer, 1973.

8. M. Yoshimura, H. Suda, No titlehydrothermal processing of hydroxyapatite: Past, present, and future, in: *Hydroxyapatite and Related Materials*, CRC Press, Boca Raton, FL, 1993: pp. 45–72.

9. K. Byrappa, M. Yoshimura, Physical chemistry of hydrothermal growth of crystals, in: *Handbook of Hydrothermal Technology*, 2013: pp. 139–175. doi:10.1016/B978-0-12-375090-7.00004-9.

10. O.V. Dimitrova, *Investigations of the Phase Formations in the System Na2O-RE2O3-SiO2-H2O Under Hydrothermal Conditions*, Moscow State University, Russia, 1985.

11. K. Byrappa, J.R. Paramesha, Crystal growth and characterisation of rare earth phosphates, *Mater. Sci. Forum* 315–317 (1999) 514–518.

12. E.D. Kolb, R.A. Laudise, The phase diagram, LiOH-Ta2O5-H2O and the hydrothermal synthesis of LiTaO3 and LiNbO3, *J. Cryst. Growth* 33 (1976) 145–149. doi:10.1016/0022-0248(76)90089-0.

13. K. Byrappa, M. Yoshimura, Apparatus, in: *Handbook of Hydrothermal Technology*, 2013: pp. 75–137. doi:10.1016/B978-0-12-375090-7.00003-7.

14. H. Itoh, R. Sasai, T. Itakura, Consolidation recovery of rare/hazardous elements from polluted water by the hydrothermal mineralization process, *Waste Manag. Environ. V* 140 (2010) 369–378. doi:10.2495/WM100331.

15. T. Itakura, R. Sasai, H. Itoh, Detoxification of wastewater containing As and Sb by hydrothermal mineralization, in: *Proceedings of International Symposium on EcoTopia Science*, 2007: pp. 902–907.

16. R. Sasai, Y. Matsumoto, T. Itakura, Continuous-flow detoxification treatment of boron-containing wastewater under hydrothermal conditions, *J. Ceram. Soc. Jpn.* 119 (2011) 277–281. doi:10.2109/jcersj2.119.277.

17. G.W. Morey, P. Niggli, The hydrothermal formation of silicates: A review, *J. Am. Chem. Soc.* 35 (1913) 1086–1130. doi:10.1021/ja02198a600.

18. M. Hosaka, S. Taki, Hydrothermal growth of quartz crystals in pure water, *J. Cryst. Growth* 51 (1981) 640–642.

19. K. Byrappa, M. Yoshimura, Hydrothermal technology for nanotechnology—A technology for processing of advanced materials, in: *Handbook of Hydrothermal Technology*, 2013: pp. 615–762. doi:10.1016/B978-0-12-375090-7.00010-4.

20. J.-R. Kim, K.-Y. Lee, M.-J. Suh, S.-K. Ihm, Ceria–zirconia mixed oxide prepared by continuous hydrothermal synthesis in supercritical water as catalyst support, *Catal. Today* 185 (2012) 25–34. doi:10.1016/J.CATTOD.2011.08.018.

21. D. Devaiah, L.H. Reddy, S.-E. Park, B.M. Reddy, Ceria–zirconia mixed oxides: Synthetic methods and applications, *Catal. Rev.* 60 (2018) 177–277.

22. I. Kosacki, T. Suzuki, V. Petrovsky, H.U. Anderson, Electrical conductivity of nanocrystalline ceria and zirconia thin films, *Solid State Ionics* 136–137 (2000) 1225–1233. doi:10.1016/S0167-2738(00)00591-9.

23. P. Duwez, F. Odell, Phase relationships in the system Zirconia-Ceria, *J. Am. Ceram. Soc.* 33 (1950) 274–283.

24. F. Zhang, C.-H. Chen, J.C. Hanson, R.D. Robinson, I.P. Herman, S.-W. Chan, Phases in ceria-zirconia binary oxide (1-x)CeO2–xZrO2 nanoparticles: The effect of particle size, *J. Am. Ceram. Soc.* 89 (2006) 1028–1036. doi:10.1111/j.1551-2916.2005.00788.x.

25. M. Yashima, K. Ohtake, M. Kakihana, H. Arashi, M. Yoshimura, Determination of tetragonal-cubic phase boundary of Zr(1-x)R(x)O(2-x/2) (R=Nd,Sm,Y,Er and Yb) by Raman scattering, *J. Phys. Chem. Solids* 57 (1996) 17–24. doi:10.1016/0022-3697(95)00085-2.

26. A. Ahniyaz, T. Watanabe, M. Yoshimura, Tetragonal nanocrystals from the Zr0.5 Ce0.5 O2 solid solution by hydrothermal method, *J. Phys. Chem. B* 109 (2005) 6136–6139. doi:10.1021/jp050047c.

27. A. Fujishima, K. Honda, Electrochemical photolysis of water at a semiconductor electrode, *Nature* 238 (1972) 37–38.

28. T. Mousavand, J. Zhang, S. Ohara, M. Umetsu, T. Naka, T. Adschiri, Organic-ligand-assisted supercritical hydrothermal synthesis of titanium oxide nanocrystals leading to perfectly dispersed titanium oxide nanoparticle in organic phase, *J. Nanopart. Res.* 9 (2007) 1067–1071.

29. C.L. Wong, Y.N. Tan, A.R. Mohamed, A review on the formation of titania nanotube photocatalysts by hydrothermal treatment, *J. Environ. Manage.* 92 (2011) 1669–1680. doi:10.1016/j.jenvman.2011.03.006.

30. N. Liu, X. Chen, J. Zhang, J.W. Schwank, A review on TiO2-based nanotubes synthesized via hydrothermal method: Formation mechanism, structure modification, and photocatalytic applications, *Catal. Today* 225 (2014) 34–51. doi:10.1016/j.cattod.2013.10.090.

31. X. Zhu, J. Zhu, S. Zhou, Z. Liu, N. Ming, Hydrothermal synthesis of nanocrystalline BaTiO3 particles and structural characterization by high-resolution transmission electron microscopy, *J. Cryst. Growth* 310 (2008) 434–441. doi:10.1016/j.jcrysgro.2007.10.076.

32. W. Hertl, Kinetics of barium titanate synthesis, *J. Am. Ceram. Soc.* 71 (1988) 879–883. doi:10.1111/j.1151-2916.1988.tb07540.x.

33. J.O. Eckert, B.L. Hung-Houston, C.C. Gersten, M.M. Lencka, R.E. Riman, Kinetics and mechanisms of hydrothermal synthesis of barium titanate, *J. Am. Ceram. Soc.* 79 (1996) 2929–2939. doi:10.1111/j.1151-2916.1996.tb08728.x.

34. M.M. Lencka, R.E. Riman, Thermodynamic modeling of hydrothermal synthesis of ceramic powders, *Chem. Mater.* 5 (1993) 61–70. doi:10.1021/cm00025a014.

35. I.A. Aksay, C.M. Chun, T. Lee, Mechanism of BaTiO3 formation by hydrothermal reactions, in: *Proceedings of the Second International Conference on Solvothermal Reactions*, 1996: pp. 76–79.

36. M. Maček Kržmanc, D. Klement, B. Jančar, D. Suvorov, Hydrothermal conditions for the formation of tetragonal BaTiO3 particles from potassium titanate and barium salt, *Ceram. Int.* 41 (2015) 15128–15137. doi:10.1016/j.ceramint.2015.08.085.

37. Y. Yang, Q. Wu, M. Wang, J. Long, Z. Mao, X. Chen, Hydrothermal synthesis of hydroxyapatite with different morphologies: Influence of supersaturation of the reaction system, *Cryst. Growth Des* (2014) 14, 4864-4871.

38. K. Lin, J. Chang, Y. Zhu, W. Wu, G. Cheng, Y. Zeng, et al., A facile one-step surfactant-free and low-temperature hydrothermal method to prepare uniform 3D structured carbonated apatite flowers, *Cryst. Growth Des.* (2009) 9, 177–181.

39. R.E. Riman, W.L. Suchanek, K. Byrappa, C.-W. Chen, P. Shuk, C.S. Oakes, Solution synthesis of hydroxyapatite designer particulates, *Solid State Ionics* 151 (2002) 393–402. doi:10.1016/S0167-2738(02)00545-3.

40. J. Liu, X. Ye, H. Wang, M. Zhu, B. Wang, H. Yan, The influence of pH and temperature on the morphology of hydroxyapatite synthesized by hydrothermal method, *Ceram. Int.* 29 (2003) 629–633. doi:10.1016/S0272-8842(02)00210-9.

41. W.L. Suchanek, M. Yoshimura, Preparation of fibrous, porous hydroxyapatite ceramics from hydroxyapatite whiskers, *J. Am. Ceram. Soc.* 81 (1998) 765–767.

42. T. Itakura, R. Sasai, H. Itoh, Precipitation recovery of boron from wastewater by hydrothermal mineralization, *Water Res.* 39 (2005) 2543–2548. doi:10.1016/j.watres.2005.04.035.

43. T. Itakura, R. Sasai, H. Itoh, Arsenic recovery from water containing arsenite and arsenate ions by hydrothermal mineralization, *J. Hazard. Mater.* 146 (2007) 328–333. doi:10.1016/j.jhazmat.2006.12.025.

44. T. Itakura, R. Sasai, H. Itoh, A novel recovery method for treating wastewater containing fluoride and fluoroboric acid, *Bull. Chem. Soc. Jpn.* 79 (2006) 1303–1307. doi:10.1246/bcsj.79.1303.

45. T. Itakura, H. Imaizumi, R. Sasai, H. Itoh, Chromium and phosphorous recovery from polluted water by hydrothermal mineralization, *Waste Manage. Environ. IV* 109 (2008) 781–788. doi:10.2495/WM080791.

46. D. Cordell, J.O. Drangert, S. White, The story of phosphorus: Global food security and food for thought, *Glob. Environ. Chang.* 19 (2009) 292–305. doi:10.1016/j.gloenvcha.2008.10.009.

47. R.E. Hecky, P. Kilham, Nutrient limitation of phytoplankton in freshwater and marine environments: A review of recent evidence on the effects of enrichment. Nutritional requirements of phytoplankton. Algal cells require elements in relatively, *Limnol. Oceanogr.* 33 (1988) 796–822.

48. M.S. Massey, J.G. Davis, J.a. Ippolito, R.E. Sheffield, Effectiveness of recovered magnesium phosphates as fertilizers in neutral and slightly alkaline soils, *Agron. J.* 101 (2009) 323–329. doi:10.2134/agronj2008.0144.

49. Commission Directive 98/15/EC, amending Council Directive 91/271/EEC with respect to certain requirements established in Annex I thereof *Off. J. Eur. Communities* 4 (1998) 29–30.

50. T. Itakura, H. Imaizumi, R. Sasai, H. Itoh, Phosphorus mineralization for resource recovery from wastewater using hydrothermal treatment, *J. Ceram. Soc. Jpn.* 117 (2009) 316–319. doi:10.2109/jcersj2.117.316.

51. N. Marti, L. Pastor, A. Bouzas, J. Ferrer, A. Seco, Phosphorus recovery by struvite crystallization in WWTPs: Influence of the sludge treatment line operation, *Water Res.* 44 (2010) 2371–2379. doi:10.1016/j.watres.2009.12.043.

52. P.J. Talboys, J. Heppell, T. Roose, J.R. Healey, D.L. Jones, P.J.a Withers, Struvite: A slow-release fertiliser for sustainable phosphorus management?, *Plant Soil* 401 (2016) 109–123. doi:10.1007/s11104-015-2747-3.

53. A. Teymouri, B.J. Stuart, S. Kumar, Hydroxyapatite and dittmarite precipitation from algae hydrolysate, *Algal Res.* 29 (2018) 202–211. doi:10.1016/j.algal.2017.11.030.

54. A. Teymouri, B.J. Stuart, S. Kumar, Effect of reaction time on phosphate mineralization from microalgae hydrolysate, *ACS Sustain. Chem. Eng.* 6 (2018) 618–625. doi:10.1021/acssuschemeng.7b02951.

55. T. Itakura, R. Sasai, H. Itoh, Resource recovery from Nd-Fe-B sintered magnet by hydrothermal treatment, *J. Alloys Compd.* 408–412 (2006) 1382–1385. doi:10.1016/j.jallcom.2005.04.088.

56. W.L. Suchanek, R.E. Riman, Hydrothermal routes to advanced ceramic powders and materials. In: *Ceramics Science and Technology, Volume 3: Synthesis and Processing,* Wiley, 2011.

57. W.L. Suchanek, R.E. Riman, Hydrothermal synthesis of advanced ceramic powders, *Adv. Sci. Technol.* 45 (2006) 184–193. doi:10.1017/cbo9780511565014.009.

8 Supercritical Water Processing of Coal to Liquid Fuels

Maoqi Feng

CONTENTS

8.1 INTRODUCTION

Coal, especially low-rank coals are a major domestic resource and account for about 48% of the demonstrated U.S. coal reserves. Characterized by their high moisture content, high oxygen content, and lower calorific value than bituminous coals, low-rank coals are more difficult to convert to liquid fuels through conventional gasification than bituminous coals. Although gasification can use numerous feedstocks, including low-rank coals, biomass, and municipal waste. The quantity of CO$_2$ emissions from coal gasification plants is one of the biggest hurdles for the coal-to-liquids (CTL) processes.

Different fossil fuels have different hydrogen/carbon ratios, as shown in Figure 8.1 (Vasireddy et al., 2011). Coal has a carbon-to-hydrogen (C/H) weight ratio ranging

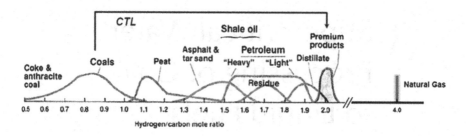

FIGURE 8.1   H/C ratios of various hydrocarbon sources. (From Vasireddy et al., 2011.)

from 12 for lignite to 20 for bituminous. Either by addition of hydrogen or by rejection of carbon, the C/H ratio can be lowered to 10 to produce the syncrudes. Syncrudes are an intermediate stage of liquid fuel production wherein unconventional source materials are converted to liquids suitable for refining into the finished fuels of interest. Hydrogen to support the reactions is produced by reacting steam with char (see Equation 8.1):

$$C + H_2O \rightarrow H_2 + CO \qquad (8.1)$$

Generally, clean liquid fuels can be produced from coal via five processes:

- Donor solvent processing (direct catalytic hydrogenation)
- Solvent extraction (noncatalytic liquid phase dissolution)
- Pyrolysis
- Liquid hydrocarbon catalytic synthesis (Fischer–Tropsch, indirect coal liquefaction)
- SCW Coal Processing

SCW coal processing is a direct coal conversion pathway. Lumpkin reviewed the direct coal conversion work prior to 1988 (Lumpkin, 1988). Direct coal processes include (Vasireddy et al., 2011):

1) Single-stage processes, which were mainly developed from the mid-1960's to the early 1980's. These processes produce liquid fuels through one primary reactor or reactors in series operating under the same conditions. Examples are:
   - Kohleoel—Ruhrkohle, Germany
   - NEDOL—NEDO, Japan, Pilot Plant operation late 1990's
   - H-Coal—HRI, USA
   - Exxon Donor Solvent—Exxon, USA
   - Solvent Refined Coal—SRC-I and SRC-II, Gulf Oil, USA

and (2) Two-stage processes, which produce products via two reactor stages in series that operate under different conditions. The primary function of the first stage is coal dissolution and is operated either with a manufactured catalyst or with a low activity, iron-based disposable catalyst. The heavy coal liquids produced in the first stage are hydrocracked and hydrotreated in the second stage reactor in the presence of high activity, albeit more expensive catalyst. Examples are:

- Brown Coal Liquefaction—NEDO, Japan
- Catalytic Two-Stage Liquefaction—DOE, HTI, USA
- Integrated Two-Stage Liquefaction—Lummus, USA
- Liquid Solvent Extraction—British Coal Corporation, UK
- Supercritical Gas Extraction—British Coal Corporation, UK

Among the above processing methods, pyrolysis, solvent refining coal (SRC) processing, and SCW coal processing accept low-rank coals. However, the pyrolysis processes produce significant quantities of by-product gas and char, which must be disposed of economically—a challenging requirement. Both hydrogasification and the Fischer–Tropsch processes have high capital and energy costs. With the crude oil price under \$60/barrel, the Fischer–Tropsch process is not competitive (DOE, 2007). Coal liquefaction by SRC process may harm the environment by unavoidable solvent emissions. In contrast to these currently known techniques, water is an environmentally friendly solvent with great power at elevated pressure and temperature. Green chemistry and sustainability have drawn interest to the application of water in coal liquefaction. Temperature and pressure have a great effect on the properties of water without any environmental detriment. In the supercritical region, water (SCW) becomes completely miscible with gases and most hydrocarbons. For these reasons, SCW has much technological potential as a reactant and reaction medium.

8.2 EARLY WORK

The hydrothermal processing of coal developed from different lines of investigation. In 1970, Barton and Fenske showed that aromatic hydrocarbons were much more soluble in water near its critical point than at lower temperatures (Barton and Fenske, 1970), which indicates that water should be a good solvent for coal extraction because coal is highly aromatic. In 1973, Steward and Dyer patented a process for hydrogenating coal in supercritical water which anticipated total pressure as high as 10,000 psi (Steward, 1974). In 1978, Modell, et al. reacted high sulfur coal with supercritical water obtaining significant yields of gases and liquids (Modell et al., 1978). Ross et al. investigated some aspects of the coal hydrolysis and hydrogenation mechanisms in an aqueous environment (Ross et al., 1984a, b). KOH and Na_2MoO_4 catalyzed the reaction, KOH favoring production of H_2. The Na_2MoO_4 was a more effective catalyst for liquefaction when CO was used as the reducing agent rather than H_2.

Another line of research was the "hot water" or "steam drying" technique for improving the quality of low-rank coals. Termed "hydrothermal coal processing," it has been known since the 1920's to reduce the inherent moisture of low-rank coals. In Fleissner's semi-batch drying process, patented in 1927 and 1928 (Fleissner, 1927, 1928), the low-rank coal was exposed to steam at 400 psig for about an hour. Moisture content was typically reduced from about 35% to 15%, and the process was used commercially in eastern Europe (Little, 1954). Research done since that time shows potential for removing other impurities including sulfur. Murray et al. heated Australian brown coals in water and noted that the sodium content decreased continually as the water temperature was increased to 300°C (Murray and Evans, 1972). The chlorine content dropped sharply until the temperature reached about 150°C, then remained constant until 300°C, where most of the remaining chlorine was removed. Some calcium was also removed at the highest temperature. Paulson et al. compared analyses of 15 coals before and after treatment at 330°C and sulfur reductions ranged from 0 to 50% (Paulson et al., 1985). Takacs et al., working with Hungarian lignites achieved 40–50% reduction in organic sulfur at temperatures up to 300°C, and they also observed some reduction of inorganic sulfur at 340°C (Takacs, 1985).

Stenberg et al. liquified low-rank coals in supercritical water and found that the presence of both CO (syngas) and H_2S were beneficial (Stenberg et al., 1982, 1984). Information pertaining to the mechanism was reviewed by Sondreal et al. (1982), who pointed out that the results were promising for the development of an aqueous process for liquefying low-rank coals.

Later, Brockrath and Davis obtained a quite high conversion of the bituminous Illinois No. 6 coal using supercritical water, hydrogen, and a catalyst, at 0.09 g/mL water density (Brockrath and Davis, 1987). In contrast, Berkowitz and Calderon extracted a Canadian coal with water and CO to provide H_2 via the shift reaction (Berkowitz and Calderon, 1987), using a "disposable" $FeCl_3$ catalyst, and reported lower conversions, but noted that the reaction was complete within ten minutes.

Barton addressed the advantages of an aqueous based process, indicating that it would have reduced hydrogen cost, fewer solid-liquid separation difficulties, and a wider range of effective hydrogenation catalysts, and reported on the liquefaction of Illinois coal, using $SnCl_2$ or MoS_3 as catalysts (Barton, 1983). Deshpande et al. investigated some of the important process variables and found that water densities near the supercritical density or higher were preferred (Deshpande et al., 1984, 1985). Slomka et al. observed a limiting conversion for coal liquefaction in supercritical water which was temperature-dependent, but insensitive to density changes above 0.20 g/mL (Slomka et al., 1985). Swanson working with low- and medium-rank coals found limiting conversions for supercritical extractions approximately equal to the coal volatile matter (Swanson et al., 1986).

8.3   RECENT DEVELOPMENT

The studies done previously with coal used static batch systems, or high-pressure extraction apparatus. They did not use oxygen or hydrogen in the relatively small quantities needed for process heat in a deep well reactor, and they generally did not use a gradual heating rate representative of what would occur in such a reactor. The types of laboratory reactors they used would be too expensive if scaled up to commercial size. Their results are encouraging for deep well reactor application, but it needs to be demonstrated in a representative reactor to determine what can be achieved.

Some experimental work in supercritical water has been done in batch reactors. Small (10 mL) capacity tubing reactors, or ampules, were used for isothermal work, and a 1.6-liter batch reactor was used to obtain samples in the 500–1000 mL size range. While these provide useful information about the process, they can functionally represent only a small part of it at any one time. A continuous flow, laboratory-scale reactor would better represent the entire process, and it could be used to produce gallon-size samples.

Continuous-flow supercritical water processing requires handling both aqueous and gas phase reactants. In addition, all the components must be designed for handling slurries, such as mixtures of water with particles of wood, plastics, coal, oil shale, or soil, so that slurry feeding and receiving capabilities can be added in the future and slurries can be processed without modifying the basic processing system. The reactor is to be capable of operation at temperatures up to 427°C and total pressures up to 5000 psi, either isothermally, or with a graded temperature increase as material flows through it. In isothermal operation, it would represent a surface reactor; in graded temperature operation, it would represent a deep well reactor. In each case, the temperature, pressure, and composition would be controlled to represent conditions achievable in industrially sized reactors.

SCW coal processing can be used to produce synthetic petroleum or syncrude from coal using synthesis gas (syngas) in an aqueous reaction matrix. In the literature, this approach has not been developed as much as either direct liquefaction of coal using hydrogen in an oil matrix, or indirect liquefaction by making paraffins from the reaction of CO and hydrogen over a suitable catalyst. Yet aqueous processing has some attractive features including the ability to use carbon monoxide directly in the liquefaction reactor, and the fact that water provides an inexpensive reaction medium. CO_2 produced in the liquefaction reactor could be recycled to the gasifier to provide additional carbon-monoxide-rich feed gas to use in SCW coal processing. Literature reports provided encouragement that such an aqueous approach could work and considering previous experience with reactions in supercritical and near-critical water, the present work centered on this approach. Laboratory data from the literature (Feng et al., 2010) was used with a process economic model to determine ability to meet the required cost and environmental criteria, and the use of syngas in an aqueous matrix appeared most likely to achieve those goals.

The information lacking concerned primarily the rates of reaction in the coal liquefaction reactor and the product quality in terms of its value as feedstock for conventional fuel and chemical processes, particularly to produce specification jet fuel. The reaction rates reported in the literature varied widely, but generally indicated good liquefaction yields with reaction times equal to or less than one hour of residence time in the reactor. In some cases, experiments may have continued for one hour because it was inconvenient to achieve the reaction conditions for short periods of time. Good yields in short reaction times are critically important for meeting the production cost requirements of commercialization.

In terms of quality, it was hoped to produce synthetic petroleum or syncrude that had sufficiently low concentrations of oxygen, sulfur, and nitrogen (heteroatoms) that it would separate from water readily and would be miscible with conventional light hydrocarbons. Since the conventional refining work needed for meeting the specified properties of JP-8 fuel work only with an oil matrix, any hypothetical product that would be liquid, but water soluble rather than oil soluble, would require an additional processing step with reducing gases to make oil. Processing that would go from coal directly to oil in a single step would be far more economically viable that any more complex combination of processing steps to reach the same end.

A novel method that liquefies coal directly to syncrude by reduction with syngas under supercritical or near supercritical water conditions was developed recently (Moulton and Erwin, 2014, Feng, 2013), see Figure 8.2 for the block flow diagram of the process. In this method, coal is directly liquefied to liquid fuels by reduction with synthesis gas (syngas, H_2 + CO) in 85% yield in near/supercritical water conditions. The syngas used in the coal liquefaction step was produced from plasma gasification of low-rank coals in high carbon conversion (>99%) yield. All the CO_2 emitted from the gasification process and from the direct liquefaction process can be used in the plasma reactor to be converted to CO by reacting with coal through the reverse Boudouard reaction. High quality coal-derived liquids produced this way can be further upgraded to jet fuel by hydrotreating and hydrocracking for upgrading to jet fuel and associated products.

Sodium hydroxide could function as a "once through" or disposable catalyst because of its low cost. It is also a strong base and the mechanism of aqueous hydrogenation involves reaction between the hydroxide ion in solution and carbon monoxide to produce a formate ion. The formate ion either donates hydrogen to the coal thus liberating carbon dioxide, or the formate ion could interact with its cation then

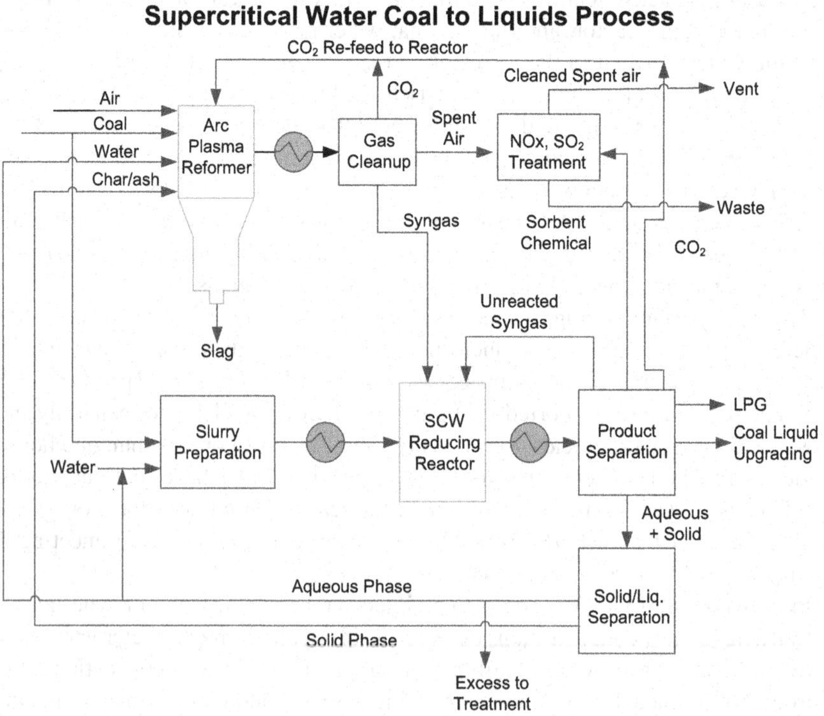

Supercritical Water Coal to Liquids Process

FIGURE 8.2   Process flow diagram for the new method of SCW coal liquefaction.

decompose to form H_2 and CO_2. The hydrogen might then attack coal molecules to potentially produce liquid products.

It was expected to find some oil, along with water in the separator vessel. Heavier oil, catalyst, ash, and char would remain in the reactor. The heavier oil would be separated from the other reactor contents, and with the oil from the separator would be measured as a yield of oil-soluble material as the product.

Formic acid, a hydrogen donor generated *in situ*, facilitates the coal conversion. The syngas used in the above conversion step is made from coal gasification in a plasma arc reactor. The CO_2 generated from the gasification process in the plasma reactor and from the direct liquefaction reactor was utilized for conversion to CO. Previous work reported in the literature shows that the additional CO in the syngas feed to the direct liquefaction reactor increased the yield of oil relative to low CO feed using both coal and organic wastes (Young, 2006). Because coals generally contain some water, and chemical water consumption will be low, the coal conversion process itself should be a net producer of water, and may exceed the water requirement for the gasifier, depending upon the coal properties.

Supercritical water (SCW, $Tc = 647$ K and $Pc = 22.1$ MPa) has a liquid-like density, which gives a large capacity for solvation, and it has a high molecular diffusivity and low viscosity, which makes it an ideal medium for efficient mass transfer. Further, its high compressibility gives large density variations with very small pressure changes, yielding extraordinary selectivity characteristics, which are most important in the removal of nitrogen from coal or coal liquids.

SCW is miscible with light gases such as H_2, CO, and O_2. As water temperature approaches the critical level, it becomes miscible with oils and aromatics since the dielectric constant increases from 2 to above 20, like that of polar organic solvents at room temperature. The miscibility of water with H_2, CO, aromatics, and oils provides a unique, homogeneous reaction medium for coal liquefaction.

The supercritical water phase, as well as liquid water at temperatures slightly below the critical point, provide a strong solvating action dissolving both the coal particles and the reducing gases. This brings the reactants together in a condition favoring rapid reaction, and reaction times of less than five minutes might be achievable. By comparison, nonaqueous, hydrogen donor solvents for coal generally require more than 30 minutes for effective liquefaction.

Hydrogenation of coals through the water–gas shift reaction (WGS reaction, $CO + H_2O \rightarrow CO_2 + H_2$) was studied in the literature (Penninger, 1989). For the reaction of Wyoming coal with a CO–SCW and a N_2–SCW mixture at 450°C and 7.1–10.9 MPa, the coal conversion rate reached 53.9 wt.% for dry ash-free (DAF) coal at 7.1 MPa, and the yields of aromatics and polar compounds were much higher in CO–SCW than in N_2–SCW. The results were attributed to formic acid formation in the reaction since formic acid is an active hydrogen donor.

Hydrogenation of coal in the presence of CO under SCW condition was also reported as a very attractive method for coal liquefaction under these conditions (Penninger, 1989):

Temperature range: $1:12 > T/Tc > 1:04$
Pressure range: $0:99 > P/Pc > 0:32$

Here SCW does not seem to work as an extraction solvent, since the density of SCW at these conditions is low (0.02–0.12 g/cm^3). If the reaction could be conducted in denser SCW, a higher coal conversion could probably be obtained, because in general the solubility increases with increasing solvent density, and hydrolysis is promoted in high dielectric constant medium, high-density SCW. In another report, Adschiri et al. (2000) used HCOOH–SCW (380°C, 35 MPa) for the extraction of coal with HCOOH–SCW, with the coal conversion 80 wt% DAF. This demonstrated that coal could be converted to the lighter oils probably through hydrolysis and hydrogenation in HCOOH–SCW. This work also showed a high conversion in only the time required to heat the mixture to 380°C.

8.4   COAL SLURRY PREPARATION

In the SCW coal process, usually coal slurry is flowing into the system. Here is an example of coal slurry preparation. A lignite coal sample was purchased from a commercial power plant in Texas and ground to 74% passing 200 mesh, a typical level of grinding for use in pulverized coal power plants by a commercial company in Aurora, Indiana. A sample analyzed per ASTM Method D-5373-02 provided the composition given in Table 8.1.

The organic portion of the coal, or the portion excluding both the ash and moisture is 68.75% of the weight of the coal, as received. This portion is referenced as "DAF Coal" in other parts of this report, representing the phrase "dry, ash-free coal."

The coal was always used as a slurry composed of the coal, water, a trivially small amount of a surfactant (Neutrad), and a thickener to prevent settling and separation of the coal particles. The thickener may not be required in an industrial setting where the slurry would be mixed immediately prior to use. But for lab experiments, storing the slurry as a weak gel prevented any significant particulate settling for several days. Kelgum, a common food thickener, was used as the slurry thickener with either of two procedures for hydration. In the first procedure, Kelgum hydration resulted from boiling it in water for about 15 minutes, then the other components were added along with sufficient additional water to obtain the required composition. In the second procedure, a Kelgum-water paste was prepared, and separately the rest of the

TABLE 8.1
Lignite Coal Analysis Per ASTM D-5373-02

Component	As Received, Wt.%	Dry Basis, Wt. %
Carbon	50.87	64.48
Hydrogen[a]	3.93	4.98
Nitrogen	0.92	1.16
Oxygen (by difference)[a]	12.09	15.32
Ash	10.14	12.85
Sulfur	0.95	1.21
Moisture	21.11	0

[a] Values do not include hydrogen and oxygen from the moisture.

slurry was mixed and warmed, and finally the paste and the rest of the slurry were mixed in a blender for several minutes. Both procedures gave acceptable results. No standard Kelgum concentration was established because higher coal concentration slurries required higher Kelgum concentrations for several days of storage stability, the blending procedure seemed to require slightly higher concentrations than the boiling method, and the catalyst concentration affected the Kelgum concentration required for acceptable gel strength.

8.5   SCW COAL PROCESSING EXAMPLE

Batch experiments for SCW coal processing with the above coal slurry as feed succeeded in making a water-insoluble but oil-soluble, product using zinc chloride as the catalyst (Feng et al., 2010). This result came as a surprise since zinc chloride is acidic in solution and was not expected to provide the hydroxyl ions for making the kind of hydrogen donor species usually associated with coal liquefaction. $ZnCl_2$ is a well-known coal hydrogenation catalyst in systems containing concentrated catalyst and hydrogen gas. In contrast, the present results showed that with $ZnCl_2$ in the reactor, carbon monoxide was consumed and coal hydrogenation took place.

During the batch experiments, design of the de-asher reactor and the required ancillary equipment proceeded in a parallel effort. With the success of the zinc chloride in the batch reactor, a modified plan to include zinc chloride in the de-asher continuous flow experiments was developed. Even with a more promising catalyst for the liquefaction reaction, there were some differences in the physical chemistry involving a concentrated zinc chloride matrix compared with a more dilute sodium hydroxide-in-water matrix that had to be dealt with. With the concentrated zinc chloride, the reaction matrix is still liquid, water is still present, and it performs a pivotal role in the chemistry, but the water vapor pressure is much lower. At 330°C, an 85 wt% $ZnCl_2$ solution has a water equilibrium vapor pressure of only about 300 psi, whereas pure water has about 1867 psi vapor pressure. The concentrated $ZnCl_2$ has a liquid density more than 50% higher than the dilute sodium hydroxide, it is more viscous, and as discovered later, it tends to form stable foams.

8.6   PROBLEMS ASSOCIATED WITH SCW COAL PROCESSING

Some SCW reactors have encountered significant operational problems. These mainly concern the deposition of salts at the point in the reactor or feed delivery system where the feed goes above the critical temperature and inorganic salts become insoluble. Salt buildup has plugged reactors and forced unplanned shutdowns, and every coal-water mixture will contain inorganic salts.

An example of the SCW reactor that permits feed delivery and salt withdrawal is shown in Figure 8.3. This is a long vertical, downflow reactor, subcritical at the top where the feed enters and at the bottom where the salt (including ash and char) leaves, and with a side arm leading from near the bottom of the supercritical section out some distance horizontally to the product recovery port. Pulverized coal is fed to the top of the SCW reactor. Ash is removed through a lock hopper system. We can get feed in and salt out if we can maintain the conditions at the ends not only

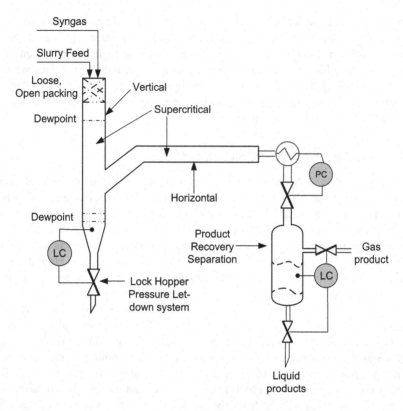

FIGURE 8.3 An example setup of coal liquefaction in supercritical water.

subcritical, but below the water dewpoint, thereby providing liquid for carrying in
and out. Loose, open packing in the subcritical feed section is needed to prevent
slugs of feed from dropping like a rock in an uncontrolled way.

When salt deposition or another issue caused specifically by SCW operation
becomes a major obstacle, an option solution is to decrease the temperatures to near,
but just below, the critical temperature. While the reaction rate may be slower, or less
complete below the critical point, eliminating the SCW problems may make it the
better choice and enable the project to meet the criteria for success. In previous work
at SwRI using both subcritical water and SCW, the differences have been small when
the temperature remained only slightly below the critical temperature (374°C). At
temperatures much higher than the critical temperature, thermodynamic equilibrium
favors hydrogen removal from the molecules and hydrogen addition requires higher
hydrogen pressure (Yui and Sanford, 1985).

One obvious problem in the continuous processing that we did not experience in
the batch runs was the matter of liquid level in the reactor with the apparently associated problem of temperature stability. A definitive diagnosis has not been found, but
the behavior observed during the single distillation experiment made for oil recovery may have given an important clue. The slurry sample, coming from the reactor
where it had been exposed to gas at high pressure, contained a lot of foam. Sitting in

the lab overnight did not disperse the foam at all. During the first attempt at distillation in one-liter, three-neck flasks, the foam expanded as the initial boiling point approached. When the material in the flask began boiling, the foam pushed up the walls of the flask and the heat was cut off just before it entered the fractionating column. Raschig rings were then added to the flask to a height greater than the top of the foam, and a slow flow of nitrogen was added to the bottom of the flask to assist the distillation. When the heat was applied again, the flask began boiling without foam production and the distillation proceeded smoothly to the conclusion. The inside diameter of the reactor was 1.5 inches wide and was not expected to cause plug flow. Even so the behavior of the foam in the distillation flask suggested otherwise. Slugs, or blocks of foam might have been rising through the reactor, separated from each other by a large bubble of syngas, then collapsing as they encountered a section with a different temperature. If that did happen, and the material then dropped quickly back down the reactor, it could explain the temperature excursions often experienced during the runs. It would also have produced a less efficient contact between the gas and the slurry liquid than anticipated.

8.7 DE-ASHING DESIGN

Two wet de-asher systems are currently in worldwide use (Burke et al., 2001). Flushing chamber de-ashers and drag-chain wet de-ashers also known as submerged scraper conveyors (SSC). The drag-chain wet de-ashers (submerged scraper conveyors) are the most frequently used systems, flushing chamber de-ashers are now rarely planned in new power plants. Drag-chain wet de-ashers (SSC) are preferred because of their easier maintenance.

8.8 SCW COAL PROCESSING IN PILOT PLANT SCALE

After achieving success in the batch reactor, larger quantities of product are preferred to be produced in a continuous-flow reactor. With the pumps, heat exchangers, valves, and product separators needed for safe and realistic operation, the result would be a small coal liquefaction pilot plant. Much of the equipment for setting up the continuous-flow reactor and its process control came from an existing experimental facility for supercritical processing.

The continuous flow reactor, or coal liquefaction pilot plant, was developed mostly in parallel with the batch reactor work. Its purpose was to provide a processing environment embodying more of the functions of a larger-scale processing unit than could be modeled by making small batch experiments. Data from the continuous plant and the samples that it would produce would have provided a more reliable basis for determining process costs than would the batch data, and product yields and quality would be more realistic. A continuous process would also provide product in sufficient quantity for processing in a hydrotreater facility to more accurately determine that portion of the overall cost of producing drop-in fuels that come from upgrading the syncrude product.

A continuous flow reactor system was built for operation at about 4500 psi. The reactor is a ten-foot tubular reactor designed for 7000 psi. A diaphragm reciprocating pump for slurry use works at pressures up to 7500 psi. These two units, the tubular reactor, and the slurry pump became the foundation units of the pilot plant and ancillaries to work with them were designed. The initial reactor concept was to make it function somewhat like a reactive distillation column at high pressure. The purpose was to introduce both the slurry and gas into the reactor above an unstirred section of the reactor tube where it was hoped the ash would collect. The reactor tube was named the "De-asher". Above the unstirred section, the reactor was heated, and the liquefaction reactions would take place there. At the top, a cooled section would provide some reflux of high boiling products while permitting the lower boiling products to go out with the product gas overhead and into a condenser, still at pressure. The higher boiling point products could then be further reacted to products in the desired boiling point range.

The reactor was operated continuously for periods up to a week to get all the systems operating properly and to produce samples of significant size. Safe and successful operation required functional systems for preparing and pumping the slurry, compressing the syngas, heating and cooling the reactant and product streams, removal of both gaseous and slurry products from high-temperature-high-pressure environments, separation and collection of products, acquiring product samples, and doing it all without releasing syngas to the laboratory because of its toxicity and flammability. A process control interface cabinet was used to control most aspects of the process. The capacity of the main items is shown in Table 8.2.

The main process flow can be followed by starting with the syngas, purchased in standard, high-pressure gas cylinders and stored outside the building. The compressor C-1 took the syngas from the cylinder pressure and compressed it to 5800 psi in the indoor syngas storage cylinder. From there, the pressure was regulated to approximately 200 psi above the experimental design pressure and through a flow rate controlling regulator FCV-31 to the mixing tee near the bottom of the reactor, PV-8.

TABLE 8.2
Main Processing Equipment in Continuous Flow Reactor System

Item	Capacity	Comment
Syngas compressor, C-1	17 SCFH	1000 to 6000 Psi
Syngas storage vessel, PV-1	6000 psi	1.6 actual cubic feet
Slurry preparation tank, T-6	10 Gal.	Temp. controlled, stirred
Slurry feed vessel, PV-2	7.5 Gal	Pressure vessel
Slurry feed pump, P-1	6 L/hr	10:1 turndown
"De-asher" reactor, PV-8	3.48 L	7000 Psi
Bottoms pressure let-down, AV-1, AV-2	0.8 L/hr	4000 Psi to atmospheric pressure
Product condenser, HE-4	2 Ft²	Counterflow, 20,000 Psi, process side
Condensate pressure let-down, AV-3, AV-4	2.4L/hr	4000 Psi to Atmospheric Pressure
System pressure regulator, PCV-5	To 6000 Psi	—
Overhead product tank, T-3	2 Gal.	Equipped with dip tube for transfer
Dual alternate ash traps, T-10, T-12	1 Gal.	—
Ash product stripper, T-9	0.5 Gal.	—
Bottoms product tank, T-11	2 Gal.	Equipped with dip tube for transfer

The coal was weighed and mixed into the slurry in the slurry preparation tank T-6, then pumped into the slurry feed tank, temporarily under vacuum to degas the slurry as it entered the tank. This helped control foam that can interfere with pumping. The pump P-1 sends the slurry to the mixing tee at a controlled rate. From the mixing tee, both the gas and the slurry enter the reactor together a few inches above the bottom and still within the cool zone.

Spent syngas and overhead products leave the reactor at the top and go to the condenser HE-4, then to PV-5, the high-pressure separator. From there, the gas goes to the system back pressure regulator, PCV-5, through the cold trap, HE-6, the low-pressure regulator PCV-6, then through the gas dryer, T-6, and either of two flow rate measurement devices that cover different ranges of flow rate before going to the vent. The liquids in the overhead products leave the high-pressure separator at the bottom, flow out through the pressure let down valves AV-3 and AV-4 to the low-pressure separator, PV-6, which also functions as a stripper for removing residual syngas. From the low-pressure separator, the stripping gas joins the vent line and the liquids go to the overhead product tank, T-3.

8.8.1 SLURRY CHARGE PUMP CHALLENGE FOR SCW COAL PROCESSING

A slurry pump is used for feeding coal to a supercritical water reactor. Abundant experiences had been gained from the operation of the H-Coal pilot plant project, though not a SCW process (DOE, 1984), these are still relevant. Acceptable on-stream times (30 days) for the reactor high-pressure slurry charge pumps had been achieved. The slurry pumps use inlet/outlet check valves around a reciprocating plunger arrangement to raise the feed slurry pressure from nominally 50–90 psi to 3200 psi. As expected, some problems with these pumps are found due to wearing of wetted parts, i.e., plunger, plunger packing, and inlet/outlet valves, caused by the abrasive slurry particles in the pumped fluids.

8.8.2 AN EXAMPLE OF MASS BALANCE FOR LARGE-SCALE SCW COAL PROCESSING

An example of mass balance large-scale SCW coal processing is described below. The operating conditions for the SCW and subcritical water reactor are listed in Table 8.3.

A mass balance for the coal reaction in the SCW reactor was estimated for each of three coal compositions: a midwestern, high sulfur, bituminous coal, a western

TABLE 8.3
SCW Conditions

Parameters	SCW	Subcritical Water
Temperature	380–450°C	325–375°C
Pressure	3500–4500 psig	2000–4000 psig
Residence time	5 Minutes	5–30 Minutes

TABLE 8.4

Estimated Mass Balances for SCW Reactor Making 100,000 bbl/Day

Feed	Midwestern High Sulfur Bituminous	Western Subbituminous	Lignite
Raw Coal, ton/day	30,477	27,976	48,223
CO, ton/day	9,946	8,070	12,101
H_2, ton/day	581	930	697
H_2O consumed, ton/day	4,565	519	2,376
Total feed, ton/day	45,529	37,495	63,397
Products			
Coal Liquid, ton/day	17,079	17,076	17,078
CO_2, ton/day	15,629	12,681	19,017
H_2S, ton/day	1,100	142	292
N_2, ton/day	250	177	180
H_2O, included, ton/day	3,779	1,788	17,360
Ash, ton/day	4,328	1,970	5,690
Unreacted DAF coal, ton/day	3,352	3,683	3,776
Total Products	45,517	37,517	63,393
Net water balance, ton/day	−786	+1,269	+14,984
Notes:			
DAF Coal Fed, ton/day	22,348	24,219	25,172
DAF Coal Reacted, ton/day	18,996	20,586	21,397

subbituminous coal, and a lignite. The coal liquid hydrocarbon was assumed to have about 12% hydrogen content and 0.90 specific gravity (Sefer and Erwin, 1989), and we assume 85% carbon conversion. The results as shown in Table 8.4 indicate an efficient process with small net water consumption for the bituminous coal and net water production for the others. The CO_2 production is low, and a large quantity can be recycled to the gasifier to maintain the desired high concentration of CO in the syngas feed.

In the SCW reactor, the syngas feed had the carbon monoxide to hydrogen ratios given in Table 8.2, and the amount of CO was set to match the oxygen in the DAF coal on a molar basis. This was an arbitrary assumption that took care of the problem of deciding which syngas component reacted with the oxygen. The reactor conversion was taken as 85% of the DAF coal and the coal liquids were modeled as C_9H_{15}, an arbitrary formula to represent an aromatic hydrocarbon liquid with a 0.90 specific gravity. This is a conservative density for the product, and a lower density may likely be obtained. The "raw coal" feed includes both the ash and the included water in addition to the DAF coal.

8.8.3 WATER CONSUMPTION

The total coal and water consumption for a 100,000 bbl/d coal liquids plant are listed in Table 8.5. Three different types of coals are used as the feed. For the three cases,

TABLE 8.5
Coal and Water Consumption for a 100,000 bbl/d Coal Liquids Plant

Feed	Midwestern High Sulfur Bituminous	Western Sub-Bituminous	Southern Lignite
Coal consumption in gasifier, ton/day	10,880	9,106	11,721
Coal consumption in SCW reactor, ton/day	30,477	27,976	48,223
Water consumption in gasifier, with CO_2 re-feed, ton/day	5,395	5,718	2,012
Water consumption in SCW reactor (assume 3% water loss), ton/day	923	0	0
Total coal consumption, ton/day	41,357	37,082	59,944
Total water consumption, ton/day	6,318	5,718	2,012
Water consumption per bbl, kg/bbl	57.3	51.9	18.2

Note: Ton = short ton (S/T).

the water consumption (kilograms) per bbl of coal liquid is 57.3 kg/bbl, 51.9 kg/bbl, and 18.2 kg/bbl, respectively.

Because of the included moisture in coal, the reducing reactor will be a net producer of water. In the unlikely event that this water production should exceed any process losses and the water requirement for the gasifier, some coal drying might be done if that proves to be the least cost method of managing the water balance, but it should not be a requirement for any process related reason. The main water requirement in the coal reducing step will be for cooling water makeup. This novel combination of processing steps has the potential of dramatic cost savings over conventional processing.

8.8.4 CO_2 EMISSION

Zero CO_2 emission can be achieved by feeding CO_2 into the gasifier reactor to react with coal for CO production. CO_2 emitted from the gasifier and the SCW reactor will be captured and fed to the gasifier. Other pollutants such as mercury, sulfoxides, and oxides of nitrogen (NOx) are to be removed in acid gas scrubbers.

8.8.5   COMPARISON WITH THE FISCHER–TROPSCH PROCESS

Indirect coal liquefaction methods, such as the Fischer–Tropsch process, are common practice for liquid fuels production. Coal liquefaction by supercritical treatment can tolerate high pollutants in the syngas, while the Fischer–Tropsch process is very sensitive to catalyst poison in the syngas feed (e.g., sulfur < 1 ppm). This provides more cost saving for the syngas-cleaning step.

DOE reported (DOE-NETL-2007/1260) coal consumption for a 50,000 bbl/day F-T plant to be at 24,533 ton/day, or 49,066 ton/day for a 100,000 bbl/day capacity. Total coal estimated for this process, using a comparable coal, is only 41,357 ton/day. Relative to other processes, much of the CO_2 can be recycled, and coal drying is not required. The solvent water molecules are much smaller than the hydrogen donor solvents used in some processes and may present as less of a diffusion barrier to rapid reaction rates with related cost savings for reactor size. The process should be adaptable to a wide variety of feedstocks, even including high cellulosic wastes, such as sugar cane fiber, particularly as gasifier feed, but possibly as SCW reactor feed also. Novel reactor designs for gas-slurry combinations could also reduce costs. For years, isolated experiments have indicated a potential for useful chemistry in a supercritical or subcritical reactor, but the lack of data on the process fundamentals means that it remains mostly unexplored territory. Similar work has not been done before, and this project may uncover totally unexpected results that could result in process improvements or cost savings.

8.9   DEEP WELL REACTOR FOR SCW COAL PROCESSING

Supercritical water processing studies had been expanded from batch reactor unit to continuous flow systems, capable of representing conditions in large-scale reactors. The technology is developing rapidly, and many potential applications will occur with requirements for feasibility studies and design data. The continuous flow processing systems allow the processing of a wide variety of reactant materials.

A new reactor concept facilitates the processing of large volumes, as well as solid-liquid mixtures or slurries. Previously, reactors for supercritical water applications have been large and expensive to fabricate to safely contain the pressure of steam in that temperature range. A process using supercritical water may easily require design for 4,000 psi at 426°C with all the bulk and safety measures those conditions imply. The new reactor concept is called a deep well reactor because it would be constructed within a well several thousand feet deep. The high temperature portion of the reactor is to be confined to the lowest part, where the hydrostatic head would provide much of the required pressure. Safety is greatly enhanced because the surrounding earth provides emergency containment. A large volume reactor could be made using thinner metal than in a reactor on the surface because the diameter of the deep well reactor can be smaller with the volume being obtained by the increased reactor length, or depth, into the earth. Since most of the pressure is provided hydrostatically, pumping to high pressures is not required, eliminating one of the major stumbling blocks to any process requiring high pressure.

A feasible method for processing large volumes of slurries is made even more exciting by the low heat requirements. Many substances are difficult to separate from either water or soil, and the heat requirement for drying, or for heat that must be discarded with soil, can be enormous in conventional processing or waste destruction technology. The deep well reactor concept requires no drying. Soils would be intentionally slurried with water for processing, and the heat conservative properties of the reactor would allow it to operate on very low heat input.

Some applications for supercritical water processing do not involve large volumes. Small volumes do not justify the use of a large reactor and it would be more economical to process them in surface reactors designed for the purpose. A greater understanding of the basic principles of supercritical water processing and developments in reactor design are widening the range of applications for surface reactors.

The deep well reactor concept allows safe, high-volume processing at critical, or near critical, conditions and at low cost. In addition to coal processing, there are several potential applications in waste destruction, alternative fuels production, chemical processing, and resource recovery.

The main impediment to designing reactors for supercritical water processing is the requirement for pressure generation and containment. At 374°C, the vapor pressure of water is 3200 psi and in that temperature range the density is strongly pressure dependent. At lower densities, the fluid behaves like a gas with little dissolving power for substances which are not gases. To maintain the dissolving power at higher temperatures, higher pressures must be used to give the fluid a higher density, so it behaves more like a liquid. The pressures of any gaseous reactants such as hydrogen or oxygen must be added to the water vapor pressure to obtain the total pressure, which could be about 4000 psi in many applications. In surface reactors all the reactants must be pumped or compressed to the total pressure. In the 315–426°C temperature range, the tensile strength of metals is lower than at ambient temperature, so conventional surface reactors are heavy, expensive to construct and operate, and require stringent safety procedures.

One potential disadvantage of subcritical operation concerns reactor design. By introducing the gas into the reactor, the rising bubbles make the reactor a stirred, rather than a plug flow reactor. A stirred reactor always has a lower conversion than a plug flow reactor because some feed can reach the exit after only a short time in the reactor. To achieve high conversions, we can run several reactors in a series or make a separation and recycle unreacted feed.

An alternative, novel reactor design may allow plug flow operation with significant cost and operating cost savings. We include a description only as an example of potential benefits as the design has had only one commercial application and cannot be considered a proven technology. The design has worked commercially to destroy solid organic wastes fed as slurry to the reactor where they were oxidized by oxygen in a subcritical water matrix (see Providential website, www.providentia-environment-solutions.nl).

The reactor concept, pioneered by chemist James Burleson, is called the "deep well reactor" (Burleson, 1986); see the diagram in Figure 8.4. In this concept, the reactor consists of a long pipe positioned vertically within a cased, cemented, well bore, possibly as deep as 8000 feet. The weight of water above the reaction zone provides the required pressure, so the pumping cost would be little more than that required for pumping water through about three miles of pipe, but still producing about 3000 psi or more, in the reaction zone at the bottom. Concentric partitions within the reactor would direct the flow of feed to the bottom of the reactor before it turns back upward for the return trip to the surface and the reactor exit. All the products would exit in a mixed stream above the ground. This reactor concept has the following main advantages over a conventional reactor of comparable volume:

- Fabrication from mass-produced pipe.
- Gravity provides most of the pressure required for the reaction zone at the bottom of the reactor, reducing the slurry feed pumping requirements and costs.
- Efficient heat exchange between unreacted feed and product.
- Safety shielding provide by the well bore.

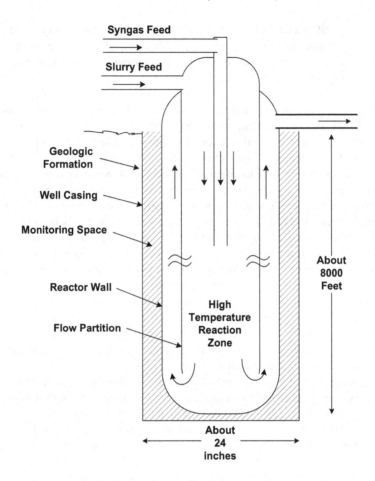

FIGURE 8.4   Diagram of a deep well reactor.

These advantages could reduce both the cost of construction and the operating costs. The deep well reactor concept may present an economic advantage large enough to make the method commercially viable or provide a greater return than processing in conventional reactors.

The deep well reactor concept avoids some problems associated with conventional surface reactors. This type of reactor is suspended in a well several thousand feet deep, and most of the pressure in the lower part of the reactor results from the hydrostatic head of water, for example, an 8000 foot well would provide about 3500 psi. Only the lower part is heated, and the flow in and out is through concentric tubes which function as a long, counter-flow heat exchanger, retaining the heat in the lower part. The outer wall of the reactor contains the entire pressure, pumped plus hydrostatic, and because it has a relatively small diameter, it can be made from thinner metal than would be practical for a surface reactor with a comparable volume. The well casing and the surrounding cement and rock provide emergency containment in the event of a leak or rupture. A small annular space between the reactor's outer wall and the well casing provides for process and environmental monitoring, and it can be pressurized with nitrogen to reduce the net pressure on the wall. The greatest tensile stress on the wall occurs at the top, where the temperature is lowest.

Heat is produced in the lower part of the reactor by the chemical reaction, such as oxidation or hydrogenation. The hot, up-flow effluent exchanges heat with the cold, down-flow feed. The net pressure difference across the internal tubular separating the flows is small and is due mostly to the friction in the flow. As a result, the tubular can be made of relatively thin material and heat exchange is efficient. The only heat losses from the reactor are due to the difference in heat content between the efflu- ent and feed streams, plus the loss across the reactor wall, through the well casing, and into the earth. In steady-state operation, heat loss to the earth is small, and with efficient transfer across the internal tubular, loss with the effluent can also be small. Modeling indicates that this type of reactor can be operated with a feed containing as low as 30 Btu/lb higher heating value (Kodra, 1992).

The pumping and compressing requirements are low. The weight of the efflu- ent is approximately the same as the feed because the densities differ only slightly due to compositional or small net temperature differences. Hypothetically, pumping is only required to overcome the friction due to flow through the reactor. In prac- tice, additional pumping will be used to maintain flow control at the outlet and any downstream processes, and because pumping to a low pressure is less expensive than achieving it hydrostatically with additional drilling. A gaseous reactant must be compressed to the full pressure if it is introduced at the bottom of the reactor. However, in many cases, the volumetric flow rate of the gas is small enough that it can be entrained into the liquid feed part-way down the reactor at less than the pres- sure at the bottom of the reactor.

8.10   COAL CLEANING AND UPGRADING

Research has shown the effectiveness of "hot water" or "hydrothermal" drying of low quality coals, but it has not been demonstrated in a type of reactor scalable to a commercial size. Lignite coal is high in sodium and alkali metals, carboxyl groups i.e., oxygen, and moisture. The moisture is "bound" or incorporated into the coal matrix and is difficult to remove by drying. During hydrothermal treatment, the oxygen leaves the coal as CO_2 (Paulson et al., 1985) and most of the alkali met- als and chlorine are leached out (Murray and Evans, 1972), leaving the coal less polar and less hygroscopic. Some of the sulfur can also be removed (DOE, 1989). When the water is removed, the improvement in heating value is dramatic, and the lower sodium content allows reduced corrosion during subsequent combustion. The technique is not applied commercially because of the difficulties with conventional surface reactors mentioned above.

If lignite coal could be partially oxidized to provide the heat required, the reaction would be suited for a deep well reactor. To get the oxidation reaction to occur, it may be necessary to use higher temperatures than were used in most of the experimental work upgrading low-rank coals; in that event it may be possible to remove other ash components and impurities. Hydrogenation reactions are also exothermic, and more coal upgrading, or even catalytic liquefaction using hydrogen are possibilities.

8.11 GEOLOGICAL EXAMPLES

Supercritical geothermal resources are a frontier for the next generation of geothermal electrical power plant due to the high heat capacity of supercritical fluids (SCF). Shallow crust (0–1 km) high-enthalpy (>150°C) geothermal reservoirs have been exploited as renewable sources of energy for over a century. Steam entrapped within the rock matrix is used for electricity production and heating. In the past decade, the exploitation of fluids close to the supercritical conditions (i.e., 374°C and 22.1 MPa, for H_2O) has increased, because it can lead to a ten-fold increase in energy extraction. Two examples are the Icelandic project IDDP-1, which is located at a shallow depth (>900°C at 2104 m depth) and achieved SCF flow-rates up to 50 kg/s, and the Lardello geothermal field (Agostinetti et al., 2017).

8.12 RECOMMENDATIONS FOR FURTHER RESEARCH ON COAL LIQUEFACTION IN WATER

The following recommendations address needs in understanding the basic chemistry as well as modifications to the overall process design. Further work with the zinc chloride catalyst is justified because of its ability to produce an oil product in a single vessel using syngas of any composition. Further work with sodium aluminate is also justified because of its high yield.

Concentrated zinc chloride does not appear to require a gum or stabilizer for slurry stability. The coal particles are probably less dense than the solution. Slurry compositions should be tried to determine what concentrations can be pumped and how long they could remain stable.

More batch runs need to be done (preferably with equipment that supports faster operation such as the tube reactors) to explore the relationship between the zinc chloride-to-water ratio, other process variables, and the global reaction rates and product yields and quality.

The attempt to begin product separation within the reactor vessel did not produce any good result, and it probably contributed to the temperature instabilities that plagued the first two runs. The process should be modified to make the entire reactor an approximately isothermal vessel with both gaseous and nongaseous products exiting through a pressure let down valve into a flash unit to begin the product separation.

To provide better gas and liquid mixing, the entire reactor should be packed with Raschig rings.

Measurements of slurry density and viscosity during reaction conditions coupled with modeling work could help understand the prospects for operation in a deep well or gravity well reactor. Superficially, the density of zinc chloride opens the possibility of such a reactor that would not need to be as deep as one for a slurry with a lower density.

Batch runs using zinc chloride and syngas should be made to explore applicability to other materials such as higher rank coals, heavy oils, biomass, cellulose, and cellulosic waste.

The product made in the sodium aluminate runs had a lot of high boiling material in it, a higher percentage than many heavy crude oils. Research with this catalyst should include catalyst or process modifications to increase cracking and hydrogenation reactions. This should include operation at higher temperatures.

Research with sodium aluminate should also include a search for an acidic cocatalyst that would not react with the sodium aluminate but would contribute to cracking or hydrocracking.

Another avenue of research with sodium aluminate concerns its own preparation. It might be possible to decrease the sodium content or substitute another metal for sodium that would make the catalyst more active.

The discussions on coal can be extended to oil-shale. A major problem in retorting oil-shale to obtain the oil is the recovery of heat from the solid, spent shale. By heating pulverized shale in a water slurry, and using a deep well reactor, the oil would be separated with little heat loss. It may be possible to use either hydrogen or oxygen reactions to provide the heat. In a preliminary experiment, two small samples of oil-shale were slurried in water and heated to 398°C in separate reactors. One reactor contained hydrogen and the other contained nitrogen (inert). Oil was extracted in both cases (Feng, 2015). However, the odor of hydrogen sulfide from the reactor which contained hydrogen feed indicated that the hydrogen did react with the oil-shale.

REFERENCES

Adschiri, T., Sato, T., Shibuichi, H., Fang, Z., Okazaki, S., and Arai, K., "Extraction of Taiheiyo Coal with Supercritical Water–HCOOH Mixture," *Fuel*, Vol. 79, p.243–248 (2000).

Agostinetti, N.P., Licciardi, A., Piccinini, D., Mazzarini, F., Musumeci, G., Saccorotti, G., and Chiarabba, C., "Discovering Geothermal Supercritical Fluids: A New Frontier for Seismic Exploration," *Scientific Reports*, Vol. 7, p.14592 (2017).

Barton, P., "Supercritical Separation in Aqueous Coal Liquefaction with Impregnated Catalyst," *Industrial Engineering and Chemistry, Process Design and Development*, Vol. 22, p.589 (1983).

Barton, P., and Fenske, M.R., "Hydrogenation Extraction of Saline Water," *Industrial and Engineering Chemistry, Process Design and Development*, Vol. 9, No. 1, p.18 (1970).

Berkowitz, N., and Calderon, J., "On Partial Coal Conversion by Extraction with Supercritical H_2O," *Fuel Processing Technology, American Chemical Society, Division of Fuel Chemistry Preprints*, Vol. 16, p.245 (1987).

Brockrath, B.C., and Davis, H.M., "Catalysis of Coal Conversion Using Water or Alcohols," *American Chemical Society, Division of Fuel Chemistry Preprints*, Vol. 32, No. 1, p.598 (1987).

Burke, F.P., Brandes, S.D., McCoy, D.C., and Winschel, R.A., "Modular Rud Systems for Submerged Scraper Conveyors (SSC)," *Summary Report of the DOE Direct Liquefaction Process Development Campaign of the Late Twentieth Century: Topical Report*, DOE Contract DE-AC22-94PC93054, Chapter 4, p.27, H-COAL PROCESS, July 2001.

Burleson, J.C., "Method and Apparatus for Disposal of a Broad Spectrum of Waste Featuring Oxidation of Waste," U.S. Patent 4,564,458 (Jan. 14, 1986).

Deshpande, G.V., et al., "Effect of Solvent Density on Coal Liquefaction Under Supercritical Conditions," *American Chemical Society, Division of Fuel Chemistry Preprints*, Vol. 30, No. 3, p.112 (1985).

Deshpande, G.V., et al., "Extraction of Coal Using Supercritical Water," *Fuel*, Vol. 63, p.956 (1984).

DOE Report, *Department of Energy, Applied Research Development, Demonstration, Testing and Evaluation Plan, (Draft)*, Department of Energy, Washington, DC, p.104, Nov. 1989.

DOE Report DOE-NETL-2007/1260, 2007.

DOE Report, "H-Coal Pilot Plant," *Final Report, Volume III.4.0 Equipment Performance, Report prepared by Ashland Synthetic Fuels, Inc., Ashland, KY, DOE Contract No. DEAC05-76ET10143*, 1984.

Feng, M., "Coal to Liquid Fuels by Supercritical Water Processing," *BIT's 4th Annual International Symposium of Clean Coal Technology −2015, Xi 'An, China*, Sept. 24–26, 2015.

Feng, M., "Enhancements of Syngas Production in Coal Gasification with CO_2 Conversion Under Plasma Conditions," Patent 8,435,478 (2013).

Feng, M., Moulton, D., Timmerman, T., Huang, F., and Erwin, J., "Coal to Liquids: A Hybrid Approach of Direct Liquefaction and Indirect Liquefaction for Jet Fuel Production," *Final Report for DARPA Project BAA 08-58*, p.254, Sept. 14, 2010.

Fleissner, H., U.S. Patent 1,632,829 (1927).

Fleissner, H., U.S. Patent 1,679,078 (1928).

Kodra, D., "Vemuri Balakotaiah, Modeling of Supercritical Oxidation of Aqueous Wastes in a Deep-Well Reactor," *AIChE Journal*, Vol. 38, No. 7, p.988 (1992).

Little, A.D. Inc., "Technology of Lignite," *Report to the Resources Research Committee*, Report No. C-58882-1, Cambridge, MA, 1954.

Lumpkin, R.E., "Recent Progress in the Direct Liquefaction of Coal," *Science*, Vol. 239, p.873–877 (1988).

Modell, M., et al., U.S. Patent 4,113,446 (1978).

Moulton, D., and Erwin, J., "Synthetic Hydrocarbon Production by Direct Reduction of Carbonaceous Materials with Synthesis Gas," US Patent 8,679,368 (2014).

Murray, J.B., and Evans, D.G., "The Brown-Coal Water System: Part 3. Thermal Dewatering of Brown Coal," *Fuel*, Vol. 51, p.290 (1972).

Paulson, L.E., Sears, R.E., Baker, G.G., Maas, D.J., and Potas, T.A., "Preparation of Alternate Low-Rank Coal Water Fuels," *Presented at Second Annual Pittsburgh Coal Conference, Pittsburgh, PA*, Sept. 1985.

Penninger, J.M.L., "Selective Effects in Aqueous Supercritical Fluid Extraction of Subbituminous Coal," *Fuel*, Vol. 68, p.983 (1989).

Providentia Environment Solutions, website: www.providentia-environment-solutions.nl

Ross, D.S., et al., "Conversion of Bituminous Coal in CO/H_2O Systems, 1. Soluble Metal Catalysts," *Fuel*, Vol. 63, p.1206 (1984a).

Ross, D.S., et al., "Conversion of Bituminous Coal in CO/H_2O Systems, 2. pH Dependence," *Fuel*, Vol. 63, p.1201 (1984b).

Sefer, N.R., and Erwin, J., "Hydroprocessing of Direct Coal Liquefaction Product for Diesel Engine Fuel," Society of Automotive Engineers, *1989 International Fuels and Lubricants Meeting and Exposition*, SAE Paper No. 892131, Baltimore, MD, Sept. 27, 1989.

Slomka, B., et al., "Aqueous Liquefaction of Illinois No. 6 Coal," *American Chemical Society, Division of Chemistry Preprints*, Vol. 30, No. 2, p.368 (1985).

Sondreal, E., et al., "Mechanisms Leading to Process Improvements in Lignite Liquefaction using CO and H_2S," *Fuel*, Vol. 61, p.925 (1982).

Stenberg, V.I., et al., "Hydrogen Sulfide Catalysis of Low-Rank Coal Liquefaction," *American Chemical Society, Division of Fuel Chemistry Preprints*, Vol. 27, Nos. 3–4, p.22 (1982).

Stenberg, V.I., et al., "Novel Liquefaction Solvent," *American Chemical Society, Division of Chemical Preprints*, Vol. 29, No. 5, p.63 (1984).

Steward, T., Jr., and Dyer, G.H., U.S. Patent 3,850,738 (1974).

Swanson, M.L., et al., "Extraction of Low-Rank Coals with Supercritical Water," *American Chemistry Society, Division of Fuel Chemistry Preprints*, Vol. 31, No. 4, p.43 (1986).

Takacs, P., Wolf, G., and Bognar, T., "Lowering the Sulfur Content of Brown Coals and Lignites by Applying Hydro-Thermal Treatment," *Central Institute for Mining Development*, P.O. Box 83, H-1525, Budapest, Hungary (1985).

Vasireddy, S., Morreale, B., Cugini, A., Songc, C., and Spivey, J.J., "Clean Liquid Fuels from Direct Coal Liquefaction: Chemistry, Catalysis, Technological Status and Challenges," *Energy & Environmental Science*, Vol. 4, p.311–345 (2011). doi:10.1039/c0ee00097c

Young, G.C., "Zapping MSW with Plasma Arc," *Pollution Engineering*, Vol. 38, No. 11, p.26 (2006).

Yui, S.M., and Sanford, E.C., "Kinetics of Aromatics Hydrogenation and Prediction of Cetane Number of Synthetic Distillates," *Proceedings of API Refining Department*, May 1985.

9 Hydrothermal Processing of Heavy Oil and Bitumen

Eleazer P. Resurreccion and Sandeep Kumar

CONTENTS

9.1 INTRODUCTION

Petroleum (conventional petroleum, crude oil) is the most commonly used source of energy, usually in liquid fuel form. As a thick black liquid, it is found beneath the earth's surface, within the microscopic pores of sedimentary rocks. Petroleum consists of naturally occurring hydrocarbons of various molecular weights and is formed from fossilized remains of organisms subjected to extreme heat and pressure (i.e., fossil fuel). It is extracted from oilfields via drilling provided that such oilfields: (1) have enough crude oil volume, (2) are in tight formation where hydraulic fracturing can be applied, and (3) are within reasonable depth such that economic benefits outweigh drilling costs. In contrast, heavy oil is more viscous than conventional petroleum with high levels of sulfur (IEA, 2005; Ancheyta and Speight, 2007). As such, heavy oil is expensive to drill and refine. It contains low amounts of volatile compounds of low molecular weight but with high amounts of non-volatile/low-volatile compounds of high molecular weight (e.g. paraffins, asphaltenes). Heavy oil has low levels of straight-chain alkanes. It has low mobility and high specific gravity (low API gravity). Bitumen is a type of heavy oil which is a mixture of clay, sand, water, and tar (extremely viscous form of petroleum) found in large quantities in Canada (Bauquis, 2006; Alberta's Oil Sands, 2008). It is colloquially known as asphalt used for road surfacing and roofing. This chapter details hydrothermal processing of heavy oil and bitumen, with emphasis on supercritical water (SCW) and subcritical water as solvent, reactors, catalysts, and in hydrogen production.

9.2   PETROLEUM

Petroleum (crude oil) are naturally occurring hydrocarbon (hydrogen, carbon) materials containing variable amounts of oxygen, nitrogen, sulfur, and other elements (ASTM D4175; Speight, 2012; US EIA, 2014). Petroleum is generally in liquid form. Viscous crude oils may also contain impurities such as nickel and vanadium with concentrations of up to a thousand parts per million, resulting in problems during processing (Speight, 1984; Speight, 2014). Most of these impurities are removed during refining. Petroleum has varying constituents and proportions and its color varies from colorless to black. They have a wide range of boiling points and carbon numbers. Boiling point boundaries and carbon numbers of the different constituents are arbitrarily defined. Crude oil can have high or low proportions of lower boiling components (heavy oil, bitumen) and higher boiling components (asphaltic components, residuum). Petroleum products not only include gaseous and liquid fuels but also high-value products such as lubricants and asphalt (a residue of the refining process used for highway surfaces, roofing materials, and miscellaneous waterproofing uses). Paraffin is the simplest hydrocarbon. Increasing the number of hydrocarbons per chain produces other petroleum compounds such as methane (natural gas), gasoline, and waxes. Naphthene is a series of ring-shaped hydrocarbons found in petroleum, which includes the volatile liquid naphtha and high molecular-weight asphalt. Aromatics, such as benzene, is also a series of ring-shaped hydrocarbons found in petroleum.

Petroleum is deposited in underground reservoirs at varying depths. It also exists in gaseous form as a result of extreme pressure underneath the Earth's surface. Petroleum originated from the remains of plants and animals that lived millions of years ago, the remains of which were deposited as sedimentary rocks. Over time, the organic matters were converted into petroleum. Petroleum accumulations are referred to as *reservoirs*. A collection of separate reservoirs located near each other is called an *oilfield*; such oilfields are positioned in a single geologic sedimentary basin.

Extraction of petroleum is achieved through primary recovery (drilling) using modern rotary equipment. Oil reservoirs (wells) can be as deep as 30,000 feet (9,000 m). Recovery is facilitated by the pressure created by natural gas or water within the reservoir. Water or stream is injected to the reservoir to artificially raise the pressure, thus, bringing the crude oil to the surface. More recently, pressure elevation is achieved via the injection of carbon dioxide, polymers, and solvents which reduce crude oil viscosity making extractions easier. Another method used to enhance heavy crude oil production is thermal recovery. Resistance to heavy crude oil extraction is associated with viscous resistance to flow at reservoir temperatures. There are three methods of crude oil recovery: primary, secondary, and tertiary. Primary recovery forces the petroleum in the reservoir trap to the surface by the natural pressure contained in the trap. At some point, pressure decreases, and petroleum extraction dwindles. Pressure reduction is caused by less force (reservoir energy) which drives the oil toward the well. Pressure reduction also minimizes rock surface permeability due to the movement of gas into the emptied pore spaces resulting in difficult crude oil flow to the production well. Pressure drop and the loss of dissolved gases increases the surface tension and viscosity of the oil.

Secondary recovery is another method used to recover crude oil. Natural gas is pumped into the reservoir above the oil which forces the oil downward. Water is then injected below the oil to force it upward. The gases utilized in the secondary recovery are produced from the primary recovery. Such use avoids the heavy cost of transporting the gases as market products. Carbon dioxide or nitrogen are examples of such gases. Additional methods such as hydraulic fracturing can be employed to extract residual crude oil (Speight, 2015). The process involves pumping large quantities of fluids at high pressure down a wellbore and into the target rock formation. Fracturing fluid consists of water, proppant, and chemical additives that open and enlarge fractures within the rock formation. These fractures can extend several hundred feet away from the wellbore. Sand, ceramic pellets, and other small incompressible particles (proppants) hold open the newly created fractures.

The last approach is tertiary recovery which involves injecting steam, detergents, solvents, bacteria, or bacterial nutrient solutions into the remaining oil. High-pressure steam heats the oil, thereby decreasing its density and viscosity and increasing the rate of flow. The goal of the injected material is to reduce the ability of the oil to stick to the rock surface for easy flushability. Microbial enhanced oil recovery (MEOR) is a tertiary technique that fills the water-filled pores of the reservoir rock by polymeric substances produced by some bacteria to effectively force the oil out. Some bacteria generate CO_2 which increases rock pore pressure which facilitates more oil recovery.

Around 300 of the largest oil fields contain almost 75% of the available crude oil. Ninety nations in total produce petroleum. Major locations of petroleum production are the Persian Gulf, North and West Africa, the North Sea, and the Gulf of Mexico. About 100 countries produce crude oil. However, in 2018, five countries accounted for about half of the world's total crude oil production (US EIA, 2019a, b). The top five crude oil producers and their shares of world crude oil production in 2018 were the United States (13.2%), Russia (13%), Saudi Arabia, (12.6%), Iraq (5.6%), and Canada (5.2%).

9.3 HEAVY OIL

Heavy oil is a type of petroleum that is difficult to extract. It is a collective term for unconventional crude oil with a gravity smaller than 20 API and a viscosity greater than 100 cP (Santos et al., 2014; Shah et al., 2010). It includes heavy oil, high viscosity oil, bitumen, oil sand, oil shale, residue, etc. (Rana et al., 2007; Caniaz and Erkey, 2014). According to the statistical reports of the United States Geological Survey (USGS), in 2003, more than 70% of the remaining 10,000 billion barrels of petroleum in the world is from heavy oil resources (Santos et al., 2014; Kapadia et al., 2015). They are formed from biodegradation of organic deposits, contained within shallow reservoirs formed by unconsolidated sands. Although difficult to drill, heavy oil has high permeability. It has high specific gravity, low hydrogen-to-carbon ratios, high carbon residues, and high contents of asphaltenes, heavy metals, sulfur, and nitrogen relative to conventional crude oil. It contains a very low amount of hydrogen and has high carbon, sulfur, and heavy metals content which makes its consistency thick necessitating additional processing (upgrading) to become a suitable refinery feedstock. Heavy oil is too viscous at the surface to be transported through conventional pipelines, so it requires heated pipelines for transportation. It can be diluted with a light hydrocarbon (e.g., aromatic naphtha) to create a mixture appropriate for transportation and recovery/upgrading via thermal recovery technique. Some heavy oil is sufficiently liquid to be recovered by pumping operations and some is already being recovered using this method. Countries with large reserves of heavy oil are the United States, Canada, Russia, and Venezuela. Venezuela has 47 to 76 billion bbls of heavy oil reserves (Mares and Altamirano, 2007). Arcaya (2001) reports that Venezuela claims to have 1.2 trillion bbls of unconventional oil reserves located adjacent to the Orinoco River near Trinidad all the way to the eastern Andes mountains (Arcaya, 2001). Although unverified, only parts have been explored, containing approximately three to four billion bbls of heavy oil.

While conventional petroleum is recoverable via pumping operations as a free-flowing dark- to light-colored liquid, heavy oil is different in terms of the difficulty of recovery from subsurface reservoir. Heavy oil determination is based on API gravity or viscosity. Classification of fossil fuel products into petroleum, heavy oil, bitumen, and residue utilizes API gravity, sulfur content, or viscosity (Speight, 2007). Heavy oil falls between 10–15° API while tar sand bitumen (i.e., extra heavy oil) usually has an API gravity in the range of 5–10° (Athabasca bitumen = 8° API). Residue is typically determined based on the temperature at which distillation was terminated, with values ranging between 2 to 8° API (Speight, 2000; Speight, 2014; Speight and Ozum, 2002; Parkash, 2003; Hsu and Robinson, 2006; Gary et al., 2007). Figure 9.1

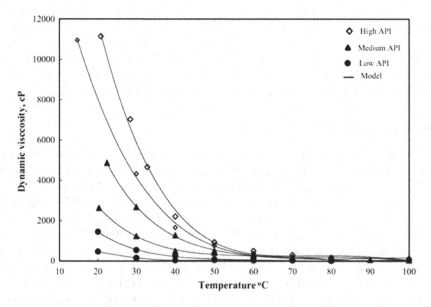

FIGURE 9.1 Evaluation of measured dynamic viscosity with temperature-dependent model of heavy fuel oils with different API values. (From Alomair et al., 2016.)

depicts the dynamic viscosity behavior of selected heavy oil at various API in response to temperatures (Alomair et al., 2016).

According to BP, proven global oil reserve has decreased by 0.5 billion bbls (−0.03%) in 2017 from previous year. However, reserves in Venezuela (~ 1.4 billion bbls) has been offset by decrease in Canadian reserve (−1.6 billion bbls) and other non-OPEC countries (BP, 2019).

9.4 BITUMEN

Bitumen, also called native asphalt, is a collective term for all naturally occurring reddish-brown to black materials of semisolid, viscous to brittle character that can contain up to 50 wt% mineral. It is also known as tar sand, oil sand (in Canada), bituminous sand, or sandstone/friable sand (quartz) impregnated with viscous bitumen (a hydrocarbonaceous material soluble in carbon disulfide). Clay may be present which typically fills pores and crevices of sandstone, limestone, or argillaceous sediments. If such is the case, bitumen is referred to as *rock asphalt* (Abraham, 1945; Hoiberg, 1964). The average boiling point of tar sand bitumen is 350°C (660°F). Bitumen makes up a significant portion of tar sand from which liquid fuels can be derived (Speight, 2013; Speight, 2014). Bitumen and tar sands are synonymous. However, bitumen is not like tar or pitch. Tar is black bituminous material and is not associated with natural materials. Tar is applied to the heavy product remaining after the destructive distillation of coal or other organic matter. Pitch, on the other hand, is the distillation residue of the various types of tar. Tar sand bitumen in the Athabasca deposit is water-wet (rather than bitumen-wet), that is, water surrounds the sand grain and bitumen fills the voids of the wet grains. Other types of tar sand bitumen have no water layer; the sand and bitumen are in direct contact with each other. Tar sand bitumen contains large amount of minerals and clay which makes upgrading difficult. There are two types of bitumen: (1) layer of viscous material without capping layer promoting fugitive hydrocarbon emissions; and (2) formations where liquid petroleum seeps into a near-surface reservoir in replacement of fugitive hydrocarbon emissions. The less-volatile components are modified by air, bacteria, and groundwater. Bitumen cannot be defined by a single property (API gravity). Unlike other forms of heavy oil, bitumen cannot be recovered in its natural state (without any conversion) by conventional oil well production methods including currently used enhanced recovery techniques. Tar sand deposits occur throughout the world, the largest of which occur in Alberta, Canada (the Athabasca, Wabasca, Cold Lake, and Peace River areas), and in Venezuela. There are some smaller deposits in the United States: Utah, California, New Mexico, and Kentucky. Table 9.1 shows the physical properties of different tar sand bitumen in North America (Speight, 2016).

Heavy oil deposits and bitumen in sand deposits are found in countries around the world, with Venezuela having the largest heavy oil deposit (Faja del Orinoco, Venezuela) and Canada having the largest tar sand deposit (Ft. McMurray, Northeast Alberta, Canada). There is an estimated 1.7 trillion bbls (1.7×10^{12} bbls or 270×10^9 m^3) of tar sand bitumen in Canada, around 97% of it is in Alberta or the Canadian Athabasca tar sand deposits (US EIA, 2019a, b). Venezuelan Orinoco tar sand deposits have 1.8 trillion bbls (1.8×10^{12} bbls or 280×10^9 m^3) in reserve (Speight, 2016).

TABLE 9.1

Specific Gravity, API Gravity, and Viscosity of Tar Sand Bitumen

Source	Specific Gravity	API Gravity	Viscosity, cP	F
Athabasca (Canada)				
Mildred-Ruth Lakes	1.025	6.5	35,000	100
Abasand	1.027	6.3	500,000	100
	1.034	5.4	570,000	100
Ells River	1.008	8.9	25,000	
Utah (United States)				
Asphalt Ridge			610,000	140
Tar Sand Triangle			760,000	140
Sunnyside			1,650,000	100
California				
Arroyo Grande	1.055	2.6	1,300,000	220

Source: Speight, (2016).

In perspective, the reserve estimate of the world's conventional crude oil as of 2017 is 1.69 trillion bbls (1.69×10^{12} bbls or 268×10^9 m^3), most of it in Saudi Arabia and other Middle Eastern countries (BP, 2019). Saudi Arabia's recoverable oil reserve is approximately 2.5×10^{11} bbls while Venezuela's oil in place is 1.2×10^{12} bbls and Canada's oil in-place is 2.2×10^{12} bbls (Dusseault, 2002). Oil in-place is not a recoverable reserve. Venezuela's heavy oil is high in sulfur, coke, and metals and needs more refining than conventional oil.

Bitumen in tar sands is a potential large supply of energy. However, this energy reserve is difficult to extract, and other refinery methods must be applied to convert these materials to low-sulfur liquid products. Fracturing procedures are required in bitumen recovery followed by subsequent thermal recovery methods. As of date, more is known about the Alberta, Canada tar sand reserves than any other reserves in the world and is therefore used as basis for discussion. Tar sand deposits are widely distributed throughout the world with a total of $>3.5 \times 10^{12}$ bbls of petroleum equivalent (Speight, 1990; Speight, 2014). In contrast, the Unites States has $>54 \times 10^6$ bbls of petroleum equivalent (Crysdale and Schenk, 1990). Canadian and Unites States' bitumen deposits are considerably different in terms of accessibility and recoverability; therefore, reserve amount is based on recoverable energy using current technology. Concentration of bitumen within tar sands and accessibility dictate the economic future of bitumen-based energy exploration. Mining plus some further *in-situ* processing or operation on the oil sands is the main recovery technique for bitumen. It is applicable to bitumen in shallow deposits, characterized by an overburden ratio (depth-to-thickness ratio of bitumen). Bitumen fills veins and fissures in fractured rocks or impregnates relatively shallow sand, sandstone, and limestone strata. Some deposits contain as much as 20 wt% bituminous material. The bitumen (tar, pitch) content of various inorganic material mixtures is determined using

ASTM D4 [ASTM D4]. The organic components of such rocks are usually refractory and are only slightly affected by most organic solvents.

Recovery and treatment processes depend on the properties and composition of bitumen. High concentrations of heteroatoms (nitrogen, oxygen, sulfur, and metals) increase viscosity, increase the bonding of bitumen with minerals, reduce yields, and make processing more difficult. Technologies such as selective catalytic reduction, flue gas desulfurization, and electrostatic precipitation are suitable for cleanup of the exhaust emissions (Mokhatab, et al. 2006).

9.5 SUPERCRITICAL FLUID (SCF) AND SUBCRITICAL FLUID

A fluid is considered supercritical or subcritical when its temperature and pressure go above (supercritical) or below (subcritical) some critical point where distinct liquid and gaseous phase are nonexistent. Supercritical fluids (SCFs) and subcritical fluids possess unique properties. They can effuse through solids like gases. They are also ideal solvents because they dissolve materials not normally soluble in either liquid or gaseous solvents. Because of these properties, SCFs and subcritical fluids promote gasification and liquefaction reaction (Xu and Etcheverry, 2008). Close to the critical point, minute changes in either temperature or pressure result in significant changes in density. Water and CO_2 are the most popular SCFs and subcritical fluids with a wide range of industrial and laboratory applications. Figure 9.2 (left) and Figure 9.2 (right) show the phase diagrams of both water and CO_2, respectively (Canıaz and Erkey, 2014; Laboureur et al., 2015). Water is the cheapest and most widely used SFC and subcritical fluid in hydrothermal processing. One of the challenges associated with using supercritical water (SCW) or subcritical water is the high cost of producing them, albeit lower temperature and environmentally friendly process. There have been research projects that explore other SCFs, but they are mainly used for liquefaction of biomass and are performed on lab- or pilot-scale. Glycerol is one that has been tested for delignification (Demirbas and Celik, 2005; Demirbas, 2008; Kűcűk, 2005), bio-oil separation (Li et al., 2009), and for the

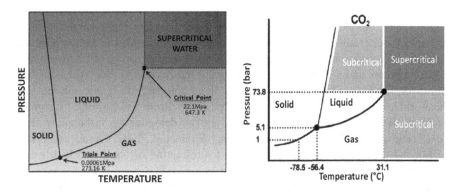

FIGURE 9.2 Phase diagram for water (left) and CO_2 (left) showing the supercritical regions. Critical temperatures: water = 647.3 K, CO_2 = 304.2 K; critical pressures: water = 22.1 MPa, CO_2 = 7.38 MPa. (From Canıaz and Erkey, 2014; Laboureur et al., 2015.)

improvement liquefaction conversion of biomass into bio-oil (Demirbas, 2009; Gan et al., 2010; Xiu et al., 2010). Other polar solvents such as methanol, ethanol, and fur-fural (Radlein et al., 1996; Diebold and Czenik, 1997; Boucher et al., 2000; Oasmaa et al., 2004) are used to lower biomass oil viscosity; however, because of their cost, they are applied to bio-oil on lab-scale and not to petroleum heavy oils and tar sand bitumens on an industrial-/large-scale.

9.6 HYDROTHERMAL PROCESSING IN SCW AND SUBCRITICAL WATER

SCW and subcritical water (T_c = 647 K and P_c = 22.1 MPa) are commonly used to upgrade heavy oil due to its distinct physical and chemical properties. First, water is a cheap source of hydrogen. It has a dielectric coefficient = 10 making it highly soluble to organics and gas. SCW and subcritical water conveniently remove hetero-atoms because of its acidic and catalytic activity (Kozhevnikov et al., 2010; Ma et al., 2003). However, SCW and subcritical water are still not widely used in industrial scale upgrading of heavy oil due to gaps in fundamental understanding on chemical mechanism and catalysis. This has led to active research on this area. Regardless, the use of SCW and subcritical water as media in heavy oil pyrolysis enhances elimina-tion of heteroatoms in the absence of H_2 or catalyst, suppresses coke formation, and enhances light product yields (Kozhevnikov et al., 2010; Han et al., 2011; Scott et al., 2001; Watanabe et al., 2010; Ximing, 2004; Rudyk and Spirov, 2014). Interestingly, SCW and subcritical water are applicable to different kinds of heavy oil without necessary feedstock pretreatment.

9.6.1 HEAVY OIL

Heavy oil is recently being considered as an alternative energy source due to the decrease of conventional crude oil reserves and the increase of crude oil prices (Meng et al., 2006). The goal for upgrading heavy oil is to produce light oil suitable for further utilization, achieved through cracking and removal of macromolecules, elimination of the heteroatoms, lowering of the viscosity, and increasing of the H/C ratio (Shah et al., 2010; Strauz et al., 1999). In the past few years, numerous tech-nologies for upgrading heavy oil had been developed, based on the carbon rejection and hydrogen addition route. Deasphalting is often used due to its effective removal of heteroatoms and asphaltene in the extraction processes (Ali and Abbas, 2006). However, the extraction process consumes large amounts of organic solvents and heavy oil losses is around 30% (Ali and Abbas, 2006; Brons and Yu, 1995). Common technologies in upgrading heavy oils are as follows: deasphalting, visbreaking, ther-mal cracking, coking, and catalytic cracking (Rana et al., 2007; Caniaz and Erkey, 2014). All these techniques use fixed bed hydrotreating or ebulliated bed hydro-cracking as a hydrogen addition mechanism.

The use of SCW or subcritical water may provide a way to upgrade heavy oil. They have ubiquitous solvent properties and the potential ability to donate hydrogen (Furimsky, 2013). Unlike deasphalting, upgrading in SCW and subcritical water is a green process, without any auxiliary chemicals required. SCW and subcritical water are excellent solvent, thanks to their tunable dielectric constant (2 to 30) with temperature and pressure (Reddy et al., 2014). SCW and subcritical water are miscible in most organic compounds and gases (Reddy et al., 2014; Muthukumaran and Gupta, 2000; Guo et al., 2010; Azadi and Farnood, 2011; Marrone, 2013), thereby enhancing heavy oil recovery. The dissolution of organic compounds is also improved in SCW and subcritical water because the hydrogen bonds and phase boundaries are decreased with increasing temperature (Reddy et al., 2014; Guo et al., 2010). SCW and subcritical water can act as an acid or base catalyst (Guo et al., 2010; Kruse and Dinjus, 2007). Research indicates that upgrading heavy oils in SCW and subcritical water produces cleaner light fractions and less coke than those via conventional pyrolysis.

Various lab-scale hydrothermal processing experiments using SCW and subcritical water revealed promising results. Canıaz Research on upgrading heavy oils in SCW from a process intensification standpoint was reviewed by Canıaz and Erkey (2014). The Timko research group summarized the results of upgrading and desulfurization of heavy oils with SCW (Timko et al., 2015). Patwardhan et al. studied the rate of desulfurization in various heavy oil samples and found out that the decomposition rate of aliphatic sulfides is 90 wt%, while that of aromatic sulfides was less than 3 wt% (Patwardhan et al., 2013). A high mass transfer rate and good solubility of feedstock in SCW was demonstrated by Ding et al. while achieving simultaneous decrease in residue viscosity, sulfur, and nickel content (Ding et al., 2010). In another study, heavy oils and polyolefins were co-pyrolyzed resulting in the radical formation of heavy oils (Tan et al., 2014; Bai et al., 2013). These radicals are easily saturated which: (1) promotes the decomposition of heavy oils; and (2) prevents the transformation of light oil products into coke precursors. Any condensation of resins or asphaltenes was considerably suppressed.

9.6.2 Bitumen

An effective method of upgrading bitumen is via hydrogenation through a water gas shift (WGS) reaction, a process that donates hydrogen. It is a relatively rapid chemical reaction that interacts with macromolecules and suppresses coke formation. The intermediate of the WGS reaction in SCW and subcritical water is formic acid (HCCOH). Sato et al. evaluated the effects of HCOOH on asphaltene decomposition and coke formation (Sato et al., 2010). It was determined that higher conversions of asphaltene and lower coke yields were achieved in systems of SCW+HCOOH than in pyrolytic systems with only SCW. This study also found that higher asphaltene conversion and lower coke yields were achieved at a higher water/oil ratio.

The partial oxidation of hydrocarbons results in the formation of CO and hydrogenation via WGS. In another study by Sato et al., the effect of temperature (from 653 to 723 K) on the partial oxidation of bitumen in SCW was examined at various water/oil ratios (from 0 to 3) and air pressures (up to 5.1 MPa) (Sato et al., 2012). Results revealed that higher temperatures favored asphaltene transformation and coke and gas formation, while lower temperatures contributed to the selective partial oxidation of hydrocarbons. The ratio of $CO/(CO+CO_2)$ achieved a maximal value at 653 K. The increase in water/oil ratio significantly affects $CO/(CO+CO_2)$, but not asphaltene transformation and coke formation.

Reverse WGS reaction can be an alternative method of hydrogen release during bitumen upgrading in SCW. The Sato team evaluated the effect of reverse WGS reaction on asphaltene decomposition and coke formation by performing batch experiments of bitumen upgrading in SCW and its mixtures with hydrogen and carbon dioxide (SCW+H_2+CO_2) (Sato et al., 2013, 2003). The results revealed that upgrading of bitumen in SCW+H_2+CO_2 resulted in a lower coke yield and a higher asphaltene yield than those in SCW, an observation attributed to the inhibitory effects of hydrogen formed from reverse WGS on the polymerization of active fused-ring units.

In terms of tar liquefaction in both SCW and subcritical water, Sasaki and Goto have demonstrated successful upgrading through the production of phenolic compounds: phenol (3.44 wt%), biphenyl (2.23 wt%), diphenylether (13.70 wt%), and diphenylmethane (1.30 wt%) (Sasaki and Goto, 2008). The conversion of tar in SCW was determined to promote the formation of light oils. Ma et al., obtained a 51.55 wt% yield without the formation coke (Ma et al., 2003). The team found out that the use of SCW cracks weak hydrocarbons in coal tar such as tert-butyl benzene, heptylbenzene, and heptyl-naphthalene at low and high temperatures, promoting high yields of light oil. While C-O, C-S, and C-N bonds are easily broken by SCW, the high-energy C-C bond is not. This fact was demonstrated by reacting biphenyl methyl naphthalene at 733 K for one hour resulting in no chemical change. A comparison between the use of SCW versus N_2 as medium in high temperature coal tar processing revealed that SCW-mediated reaction achieved higher asphaltene conversion, higher maltene yield, and higher H/C atom ratio in products than the N_2-mediated one (Han et al., 2011, Han and Zhang, 2008, 2009). Bitumen can be cracked into alkanes and aromatics (light and desulfurized products) without the production of sulfur compounds such as benzothiophene and dibenzothiophene (Kishita et al., 2003). The Canadian Athabasca tar sand bitumen was upgraded in a batch reactor via SCW-mediated pyrolysis and neat pyrolysis (Watanabe et al., 2010). Results revealed that conversion was higher and coke formation was lower in SCW than in neat pyrolysis. Finally, the products that are soluble in hexane obtained through the hydrothermal conversion of asphaltene was found to be chemically identical to diesel fractions and vacuum gas oil (Kozhevnikov et al., 2010).

Vilcáez et al. upgraded bitumen employing a continuous hydrothermal extraction method using near SCW in a column flow reactor. The results at 300°C, 3-6 MPa, and 3–10 g/min water flow showed a higher degree of upgrading than in the autoclave reactor. Significant conversion of asphaltene to maltene with lighter maltene composition was obtained, while coke formation was completely suppressed (Vilcáez et al., 2012). The study demonstrated that a continuous hydrothermal extraction method using a column flow reactor can solve the problem of incompatibility between enhancing the degree of upgrading heavy oil and suppressing the formation of coke.

9.7   CHEMICAL EFFECTS OF SCW AND SUBCRITICAL WATER ON THE UPGRADING PROCESS

The role of SCW and subcritical water on heavy oil upgrading is still unclear despite their use as solvent. SCW and subcritical water form a homogeneous phase with heavy oil. Currently, it is unclear whether SCW and subcritical water donate hydrogen in hydrothermal processing.

A number of researchers have investigated the chemical effects of SCW on heavy oil upgrading. Xu sought to verify the mechanism through which SCW donates hydrogen by picking model compounds p-benzoquinone, naphthalene, and azobenzene as probes (Xu et al., 2013). The hydrogenation products for p-benzoquinone and naphthalene were not detected. Results indicate that SCW cannot donate H· radicals; instead, generation of H· in the temperature range of SCW is attributed to the condensation of carbonaceous materials. H$^+$ seems to play a role at a higher extent of water ionization. Kida et al. obtained an opposite result. His group utilized hexyl sulfide as reactant with and without SCW. The products from each experiment were substantially different although both experiments were governed by free radical reactions (Kida et al., 2014). Only in the SCW experiment were pentane, CO, and CO$_2$ detected. In all experiments utilizing hexyl sulfides and other di-n-alkyl sulfides, aldehydes and H$_2$S were formed, an indication that water is a hydrogen donor in H$_2$S formation.

A comprehensive study performed by Morimoto et al. utilized SCW, supercritical toluene, and N$_2$ in upgrading Canadian oil sand bitumen mined by the steam assisted gravity drainage method (Morimoto et al., 2010). The conditions were: 420–450 °C and 20–30 MPa for up to 120 minutes. Analysis of the product revealed the following components: gas, middle distillate, distillation residue, and coke. Comparative characterization of products showed that the composition and properties of middle distillates were similar for SCW and N$_2$ but different for toluene. The distillation product obtained for SCW had a larger number of aromatic rings, shorter side chains, and lower molecular weight distribution compared to that for N$_2$. The results seemed to indicate that two simultaneous processes occurred in SCW-mediated oil sand bitumen upgrade: intramolecular dehydrogenation and suppression of heavy components formation as a result of the dispersion of heavy fractions in SCW.

These investigations highlight critical conclusions on the chemical effects of SCW and subcritical water on heavy oil upgrading. First, SCW does not provide H· or HO· radicals for hydrocarbon hydrogenation. Second, the partial oxidation of hydrocarbons and the water-gas shift reaction (WGSR) provide hydrogen for hydrocarbon hydrogenation. Finally, free radicals generated from hydrocarbon pyrolysis do not couple with each other in the SCW system.

9.8 PHYSICAL EFFECTS OF SCW AND SUBCRITICAL WATER ON THE UPGRADING PROCESS

There are two effects of SCW and subcritical water on heavy oil upgrading: solvent effects and dispersion effects (Cheng et al., 2009). In the conventional pyrolysis process, hydrogen transfer is suppressed, and the radicals are condensed. In the SCW-mediated upgrading of heavy oil, coke formation is restrained. In the SCW system, the reaction is homogeneous (solvent effects) and there exists mass transfer of radicals (dispersion effects) (Canıaz and Erkey, 2014). The coke precursors are diluted by the light fraction/unreacted oil-water mixture, effectively preventing coke formation of the heavy fraction. The heavy fraction particles are "caged" from reacting with each other to form coke but allow the passage of small radicals.

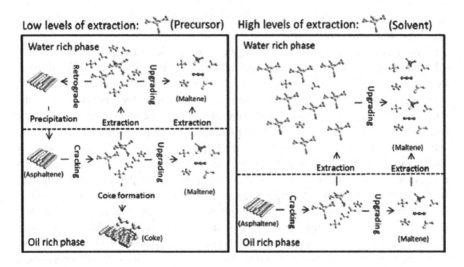

FIGURE 9.3 Schematic representation of the proposed mechanism of coke suppression. Low levels of extraction of asphaltene core (left); high levels of extraction of asphaltene core (right). (Vilcáez et al., 2012.)

In the Vilcáez study mentioned above, maltene, asphaltene, and coke components in raw bitumen, treated bitumen (573 K, 3–6 MPa, 3–10 g/min water flow), and hydrothermally extracted bitumen were investigated (Vilcáez et al., 2012). Results suggest that bitumen extracted from the column flow reactor at 613 K and 13 MPa has the lowest percentage of coke, which is an indication of simultaneous heavy oil conversion and coke suppression. The mechanism of coke suppression is detailed in Figure 9.3. In their view, the formation of coke was closely related to the extraction levels of asphaltene core in the oil-rich phase. Low-level extraction of asphaltene core resulted in the formation of coke in the oil-rich phase, whereas high-level extraction of asphaltene core would suppress coke formation. Liu et al. determined that coke formation is prevented in an upgrading process with a continuous water phase (Liu et al., 2013). In another bitumen conversion study, the bitumen (673 K and 30 MPa) was fed on top of a vertically aligned tubular reactor while the SCW was supplied at the bottom (Fedyaeva et al., 2014a). Because the temperature at the top portion of the reactor is much lower than that at the bottom portion, part of the bitumen flowing downward was converted into light hydrocarbons which was subsequently removed by the countercurrent SCW flow. Increasing the temperature of the reactor converts the deposited heavy bitumen components at the bottom into light hydrocarbons.

9.9 UPGRADING USING CATALYSTS

Despite SCW's ability to promote light oil formation and suppress coke generation during heavy oil upgrade, large amount of energy is needed to reach the critical point of water. At the supercritical or subcritical conditions, water is corrosive (Han et al., 2009). It is therefore necessary to add solvents that have low critical points and are noncorrosive. However, unlike SCW, these solvents have lower activity such that the addition of catalysts is necessary. Some catalysts may inadvertently be deactivated by the addition of certain types of solvents. In addition, sulfur heteroatoms in heavy oils tend to poison some metal catalysts (Adschiri et al., 1998, Adschiri et al., 2000; Savage, 2009; Yeh et al., 2013; Yuan et al., 2005). At the supercritical or subcritical conditions, Si and Al catalyst support are deteriorated (Elliott et al., 1994). The most commonly used catalysts in heterogenous catalysis in conjunction with SCW are activated carbon and transition metals and their oxides (Yeh et al., 2013; Azadi and Farnood, 2011). The oxides of Ce, Co, Fe, Mo, and Zn are a few examples.

The quality of the final products of heavy oil upgrade is enhanced using several catalysts/additives with SCW. Effects of such additives are detailed in Table 9.2. Effects on feedstock solubility, hydrogenation, desulfurization, and H transfer are achieved using different catalysts.

Upgrade of heavy oils is governed by free radical mechanism in SCW and subcritical water and SCW. In the case of heavy oil upgrade, hydrocarbon radicals are generated by the cleavage of the aliphatic side chains and the weak bonds (C–O, C–S, C–N) connecting aromatic groups. The reactions are as follows: C–C cleavage, ß-scission, isomerization, H-abstraction and addition to olefins, as shown in Figure 9.4 (Liu et al., 2013).

Catalysts action for the upgrading of heavy oil in SCW is best exemplified by the splitting of H_2O to produce active oxygen or active hydrogen. Dejhosseini et al. carried out experiments to classify the effect of CeO_2 nanoparticles on the cracking of Canadian oil sand bitumen in SCW at 723 K (Dejhosseini et al., 2013). Results demonstrated that the addition of CeO_2 nanoparticles could significantly increase the conversion of asphaltene and lower the coke yield. Such increase was intensified

---

TABLE 9.2

Effects of Different Catalysts Using SCFs

Additives	Solvents	Effects	Reference
Activated carbon	SC-n-hexane	Hydrogenation	Scott et al., 2001; Viet et al., 2012; Viet et al., 2013
	SC-n-dodecane	H-transfer	
CO	SCW	WGSR	Adschiri et al., 1998; Sato et al., 2013; Arai et al., 2000; Cheng et al., 2003; Sato et al., 2004
CO_2-H_2 HCOOH		Improves the solubility	
NaOH	SCW	Increases the solubility	Li et al., 2015
NaY zeolite	SC-gasoline SC-xylene	Hydrogenation	Gu et al., 2012
$Ni/Mo/Al_2O_3$	SCW	Desulfurization	Adschiri et al., 1998; Fedyaeva et al., 2014a
ZnO MoO_3 MoS	SCW	Desulfurization	Ates et al., 2014

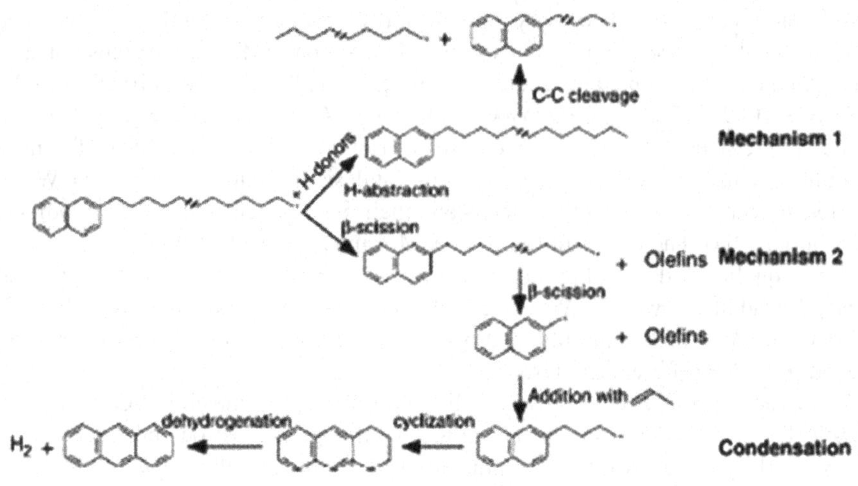

FIGURE 9.4 Reaction routes in the upgrade of heavy oil. (From Liu et al., 2013.)

by the increase of the CeO_2 loading amounts. It is apparent that coke suppression via CeO_2 nanoparticles addition was closely related to the active oxygen or active hydrogen formed from the redox reaction among the CeO_2, H_2O, and bitumen. The presence of the active oxygen enabled the enhanced absorption and release of oxygen via the Ce^{4+}/Ce^{3+} redox cycle. The oxygen on the surface of the CeO_2 catalyst was unstable and would crack the heavy oil via oxidation.

The catalytic effect of silica-supported hematite iron oxide nanoparticles on the cracking of heavy petroleum residue in SCW was studied by Hosseinpour et al. (2015). Results highlighted the fact that the addition of silica-supported hematite iron oxide nanoparticles effectively suppress coke formation. The oxygen species generated from H_2O over magnetite particles were found on the surface where the heavy oil decomposition occurred while the hydrogen species was transferred to the lighter molecules. SCW over iron oxide catalyst as a solvent: (1) has its physical (solvation and dispersion effects) and chemical effects; and (2) can donate hydrogen. These characteristics allow it to suppress coke formation. Figure 9.5 provides a schematic of the overall process, showing catalytic and pyrolytic cracking of heavy constituents.

In a study employing ZnO catalyst, 25% DBT decomposition was achieved (Ates et al., 2014). The study desulfurized heavy crude oil and two model compounds (hexyl sulfde in hexadecane and dibenzothiophene (DBT) in hexadecane) in the absence and presence of ZnO, MoO_3, and MoS_2. Using SCW alone, 6–7 wt% of the sulfur in crude oil was removed. A 12% sulfur removal increase was achieved when MoS_2 catalyst was added into the system. The research by Fedyaeva et al. using Zn and Al catalysts confirmed the conversion of asphaltite: 56.3% to 98.3 wt% increase in volatile and liquid products yield and 20.3% to 72.3% increase in desulfurization rate (Fedyaeva et al., 2014a). Gai et al. cracked bitumen under sub- and supercritical water environments in H_2 or N_2 atmosphere using activated carbon (AC) supported nickel catalyst. Water at supercritical and subcritical conditions provided a unique

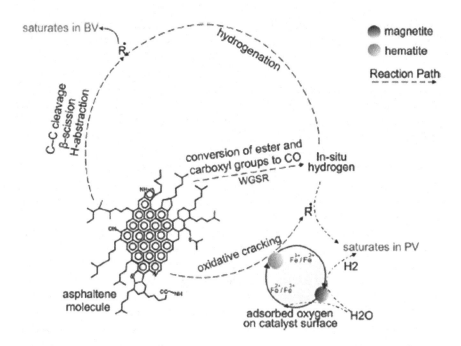

FIGURE 9.5 Schematic illustration of proposed pathway for the catalytic cracking of bitumen in SCW. (From Hosseinpour et al., 2015.)

homogeneous, acidic reaction system for the bitumen cracking reaction promoting the bitumen conversion and selectivity of the middle distillates (kerosene, gasoline), and restraining the formation of carbon deposition, as compared to the conventional pyrolysis reaction without water addition (Gai et al., 2016).

Metal catalyst deactivation by carbon deposition is the main challenge in SCW-mediated catalysis. Therefore, the use of activated carbon catalyst in heavy oil hydrothermal processing has been recently employed. An alkane-rich hydrocarbon-activated carbon system could supply hydrogen for the efficient hydrocracking of heavy oil. Activated carbon is employed in reactions at mild conditions because it has high surface area, variable pore structure, and surface functional groups (Scott and Radlein, 2001; Fukuyama and Terai, 2007; Xu et al., 2007). Scott and Radlein upgraded bitumen using activated carbon catalysts in supercritical solvents rich in hydrogen at 7 MPa and 673–723 K (Scott et al., 2001). The yield of distillable liquids achieved was 82–88 wt%, with only a 6–8 wt% coke yield. Rates of demetallization and desulfurization were almost 100% and 80%, respectively. Four kinds of activated carbon catalysts in supercritical m-xylene was investigated by Viet et al. to hydrocrack vacuum residue (Viet et al., 2013). Catalyst A was activated carbon based on coal tar pitch and catalyst B was catalyst A treated with sulfuric acid. Catalyst C was activated carbon based on petroleum pitch and catalyst D was catalyst B treated with sulfuric acid. Fe_2O_3, $NiSO_4$, and $LiC_2H_3O_2$ were added to modify the activated carbon catalysts. Results showed that Catalyst D impregnated with 10 wt% Fe had the best catalytic effect; the conversion and the light product (naphtha and middle distillate) yield were the highest. High conversion and coke suppression were determined to be caused by impediments in the mesopores of the activated carbon. The other study by Viet et al. investigated the hydrocracking of vacuum residue in supercritical aromatics (m-xylene, toluene) and normal alkanes (n-hexane, n-dodecane) by applying acid-treated activated carbon as catalyst at 673 K and 6.89 MPa, with a H_2 partial pressure 3.45 MPa (Viet et al., 2012). Meso-pores and macropores were all found to have significant effect on the hydrocracking process. The results confirmed high residue conversion (69.2 wt%), low coke formation (13.5 wt%), and high-quality light oils (13.0 wt% of naphtha, 34.9 wt% of middle distillate, 27.1 wt% of vacuum gas oil, and 11.2 wt% of residue) for the trial in supercritical m-xylene with the modified activated carbon catalyst. Upgrading process using activated carbon catalysts also ensures a lower hydrogen consumption and a higher pitch conversion relative to the process employing Co/Mo catalysts.

The use of SCW in hydrothermal processing of heavy oil is highly dependent on temperature. Conversion yield and rate of coke formation is insignificantly affected by the amount of hydrogen. SCW-mediated upgrading confirmed that both modified and unmodified activated carbon facilitates excellent catalysis. The mesopores and macropores of activated carbon are responsible for limiting coke formation. Free radicals generated in heavy oil upgrade (see Figure 9.3) can be diffused in the mesopores and can be absorbed on the active sites of the activated carbon. Additionally, to improve desulphurization and denitrification reactions in vacuum residues, the use of activated carbon catalysts with metal ligands is beneficial. Thus, in order to upgrade heavy oil to generate high-quality paraffinic and naphthenic hydrocarbons, activated carbon, hydrogen, and SCW are essential.

9.10   INDUSTRIAL PERSPECTIVE AND DIRECTIONS

Upgrading heavy oil in SCW and subcritical water may be a promising alternative technique to meet the constantly increasing demand for clean light oil. These solvents effectively convert hydrocarbon, suppress coke, and remove heteroatoms. Despite research progress within the last decade related to hydrothermal processing using SCW and subcritical water, no industrial scale application has been implemented to date. Extensive studies show that the product yield for heavy oil upgrading in SCW and subcritical water depends on the nature of heavy oil, reaction conditions, and operation mode (Canter et al., 2015). Currently, the commonly used operation modes for reactions in SCW and subcritical water include batch, semi-batch, or continuous reactors, all of which are lab- or pilot-scale. Compared with the batch and semi-batch reactors, continuous reactors offer efficient extraction of hydrocarbons. SCW and subcritical water prevents the recombination of radicals produced during hydrolysis, which in turn increase yield and suppresses coke (Morimoto et al., 2014; Fedyaeva et al., 2014a). Because SCW- and subcritical water-mediated heavy oil upgrade is a comprehensive reaction, optimization of processes in the lab is warranted prior to industrial scale-up.

Part of the challenge is the balance between additional benefits associated with light oil production through upgrade and the high operational cost on high-pressure equipment. Outlook and future direction in on scale-up, to which research emphasis on certain aspects must be met. These aspects include: (1) Development of novel catalysts that induce active hydrogen generation via water splitting. Hydrogen is a crucial component in heavy oils upgrading in SCW, driving both economics and environmental effects. Enough supply of hydrogen from SCW and subcritical water ensures effective conversion of heavy oils to light oil fractions, coke suppression, and heteroatoms removal. In most industrial applications, provision for *in situ* hydrogen dictates economic profitability. (2) Design and deployment of modern SCW and subcritical water devices. Advancement in equipment design ensures the collection of reliable data. It also facilitates automated collection and separation of the upgrading products while allowing for batch, semi-batch, or continuous operation mode. (3) Research on removal of Ni and V. These heteroatoms are ubiquitous in heavy oil, but they have hazardous effects once transferred in the light oil fraction.

REFERENCES

Abraham H. (1945) *Asphalts and Allied Substances*. Van Nostrand, New York, NY.

Adschiri T., Sato T., Shibuichi H., Fang Z., Okazaki S., Arai K. (2000) Extraction of Taiheiyo coal with supercritical water-HCOOH mixture. *Fuel* 79, (3–4), 243–248.

Adschiri T., Shibata R., Sato T., Watanabe M., Arai K. (1998) Catalytic hydrodesulfurization of dibenzothiophene through partial oxidation and a water-gas shift reaction in supercritical water. *Ind Eng Chem Res* 37, (7), 2634–2638.

Alberta's Oil Sands: Opportunity, Balance (2008). Government of Alberta, Alberta, Canada. ISBN 9780778573487. http://www.assembly.ab.ca/lao/library/egovdocs/2008/alen/165 630.pdf

Ali M. F., Abbas S. (2006) A review of methods for the demetallization of residual fuel oils. *Fuel Process Technol* 87, (7), 573–584.

Alomair O., Jumaa M., Alkoriem A., Hamed M. (2016) Heavy oil viscosity and density prediction at normal and elevated temperatures. *J Petrol Explor Prod Technol* 6, 253–263.

Ancheyta J., Speight J. G. (2007) *Hydroprocessing of Heavy Oils and Residua*. CRC Press, Taylor & Francis Group, Boca Raton, FL.

Arai K., Adschiri T., Watanabe M. (2000) Hydrogenation of hydrocarbons through partial oxidation in supercritical water. *Ind Eng Chem Res* 39, (12), 4697–4701.

Arcaya I. (2001) *Venezuela, the United States, and Global Energy Security. WTC Featured Speakers*. Address at the Windsor Court Hotel, New Orleans, LA.

ASTM D4 Standard test method for bitumen content. Annual Book of Standards. American Society for Testing and Materials, West Conshohocken, PA.

ASTM D4175 Standard terminology relating to petroleum, petroleum products, and lubricants. Annual Book of Standards. American Society for Testing and Materials, West Conshohocken, PA.

Ates A., Azimi G., Choi K.-H., Green W. H., Timko M. T. (2014) The role of catalyst in supercritical water desulfurization. *Appl Catal B* 147, 144–155.

Azadi P., Farnood R. (2011) Review of heterogeneous catalysts for sub- and supercritical water gasification of biomass and wastes. *Int J Hydrogen Energy* 36, (16), 9529–9541.

Bai F., Zhu C.-C., Liu Y., Yuan P.-Q., Cheng Z.-M., Yuan W.-K. (2013) Co-pyrolysis of residual oil and polyethylene in sub- and supercritical water. *Fuel Process Technol* 106, 267–274.

Bauquis P.-R. (2006) *What the Future for Extra Heavy Oil and Bitumen: The Orinoco Case*. World Energy Council, London, United Kingdom.

Boucher M. E., Chaala A., Roy C. (2000) Bio-oils obtained by vacuum pyrolysis of softwood bark as a liquid fuel for gas turbines. Part I: Properties of bio-oil and its blends with methanol and a pyrolytic aqueous phase. *Biomass Bioenergy* 19, (5), 337–350.

BP (2019) *Oil Reserves*. https://www.bp.com/en/global/corporate/energy-economics/statistic al-review-of-world-energy/oil.html#oil-reserves

Brons G., Yu J. M. (1995) Solvent deasphalting effects on whole cold lake bitumen. *Energy Fuels* 9, 641–647.

Canıaz R. O., Erkey C. (2014) Process intensification for heavy oil upgrading using supercritical water. *Chem Eng Res Des* 92, 1845–1863.

Canter D. A., Bermejo M. D., Cocero M. J. (2015) Reaction engineering for process intensification of supercritical water biomass refining. *J Supercrit Fluids* 96, 21–35.

Cheng J., Liu Y.-H., Luo Y.-H., Liu G.-X., Que G.-H. (2003) Hydrocracking of Gudao residual oil in suspended bed using supercritical water-syngas as hydrogen source. I. The effect of catalyst on hydrocracking. *J Fuel Chem Technol* 31, 574–578.

Cheng Z.-M., Ding Y., Zhao L.-Q., Yuan P.-Q., Yuan W.-K. (2009) Effects of supercritical water in vacuum residue upgrading. *Energy Fuels* 23, 3178–3183.

Crysdale B. L., Schenk C. J. (1990) *US Geological Survey Bulletin 1885: Heavy Oil Resources of the United States*. United States Geological Survey, United States Department of the Interior, Washington, DC. https://pubs.usgs.gov/bul/1885/report.pdf

Dejhosseini M., Aida T., Watanabe M., Takami S., Hojo D., Aoki N., Arita T., Kishita A., Adschiri T. (2013) Catalytic cracking reaction of heavy oil in the presence of cerium oxide nanoparticles in supercritical water. *Energy Fuels* 27, (8), 4624–4631.

Demirbas A. (2008) Liquefaction of biomass using glycerol. *Energy Sources Part A* 30, (12), 1120–1126.

Demirbas A. (2009) Biofuels from agricultural biomass. *Energy Sources Part A* 31, 1573–1582.

Demirbas A., Celik A. (2005) Degradation of poplar and spruce wood chips using alkaline glycerol. *Energy Sources Part A* 27, (11), 1073–1084.

Diebold J. P., Czernik S. (1997) Additives to lower and stabilize the viscosity of pyrolysis oils during storage. *Energy Fuels* 11, 1081–1091.

Ding Y.-H., Chen H., Wang D.-F., Ma W.-G., Wang J.-F., Xu D.-P., Wang Y.-G. (2010) Supercritical fluid extraction and fractionation of high-temperature coal tar. *J Fuel Chem Technol* 38, 140–143.

Dusseault M. B. (2002) *CHOPS: Cold Heavy Oil Production with Sand in the Canadian Heavy Oil Industry*. Prepared for the Alberta Department of Energy, Alberta, Canada. https://open.alberta.ca/publications/2815953

Elliott D. C., Phelps M., Sealock Jr. L. J., Baker E. G. (1994) Chemical processing in high-pressure aqueous environments. 4. Continuous-flow reactor process development experiments for organics destruction. *Ind Eng Chem Res* 33, (3), 566–574.

Fedyaeva O. N., Antipenko V. R., Vostrikov A. A. (2014a) Conversion of sulfur-rich asphaltene in supercritical water and effect of metal additives. *J Supercrit Fluids* 88, 105–116.

Fedyaeva O. N., Shatrova A. V., Vostrikov A. A. (2014b) Effect of temperature on bitumen conversion in a supercritical water flow. *J Supercrit Fluids* 95, 437–443.

Fukuyama H., Terai S. (2007) An active carbon catalyst prevents coke formation from asphaltenes during the hydrocracking of vacuum residue. *Pet Sci Technol* 25, (1–2), 231–240.

Furimsky E. (2013) Hydroprocessing in aqueous phase. *Ind Eng Chem Res* 52, 17695–17713.

Gai X.-K., Arano H., Lu P., Mao J.-W., Yoneyama Y., Lu C.-X., Yang R.-Q., Tsubaki N. (2016) Catalytic bitumen cracking in sub- and supercritical water. *Fuel Process Technol* 142, 315–318.

Gan J., Yuan W., Nelson N. O., Agudelo S. C. (2010) Hydrothermal conversion of corn cobs and crude glycerol *J Biol Eng* 2, (4), 197–210.

Gary J. G., Handwerk G. E., Kaiser M. J. (2007) *Petroleum Refining: Technology and Economics* (5th edition). CRC Press, Taylor & Francis Group, Boca Raton, FL.

Gu Z., Chang N., Hou X., Wang J., Liu Z. (2012) Experimental study on the coal tar hydro-cracking process in supercritical solvents. *Fuel* 91, (1), 33–39.

Guo Y., Wang S. Z., Xu D. H., Gong Y. M., Ma H. H., Tang X. Y. (2010) Review of catalytic supercritical water gasification for hydrogen production from biomass. *Renew Sustain Energy Rev* 14, (1), 334–343.

Han L., Zhang R., Bi J.-C. (2009) Experimental investigation of high-temperature coal tar upgrading in supercritical water. *Fuel Process Technol* 90, (2), 292–300.

Han L., Zhang R., Bi J.-C., Cheng L. (2011) Pyrolysis of coal-tar asphaltene in supercritical water. *J Anal Appl Pyrolysis* 91, (2), 281–287.

Han L.-N., Zhang R., Bi J.-C. (2008) Upgrading of coal-tar pitch in supercritical water. *J Fuel Chem Technol* 36, (1), 1–5.

Hoiberg A. J. (1964) *Bituminous Materials: Asphalts, Tars, and Pitches*. John Wiley & Sons Inc., Hoboken, NJ.

Hosseinpour M., Ahmadi S. J., Fatemi S. (2015) Successive co-operation of supercritical water and silica-supported iron oxide nanoparticles in upgrading of heavy petroleum residue: Suppression of coke deposition over catalyst. *J Supercrit Fluids* 100, 70–78.

Hsu C. S., Robinson P. R. (2006) *Practical Advances in Petroleum Processing* (Volumes 1 and 2). Springer, New York, NY.

IEA (2005) *Resources to Reserves: Oil & Gas Technologies for the Energy Markets of the Future*. International Energy Agency, Paris, France. https://www.iea.org/newsroom/news/2005/

Kapadia P. R., Kallos M. S., Gates I. D. (2015) A review of pyrolysis, aquathermolysis, and oxidation of Athabasca bitumen. *Fuel Process Technol* 131, 270–289.

Kida Y., Class C. A., Concepcion A. J., Timko M. T., Green W. H. (2014) Combining experiment and theory to elucidate the role of supercritical water in sulfide decomposition. *Phys Chem Chem Phys* 16, 9220–9228.

Kishita A., Takahashi S., Kamimura H., Miki M., Moriya T., Enomoto H. (2003) Upgrading for bitumen by hydrothermal visbreaking in supercritical water with alkali. *J Jpn Pet Inst* 46, 215–221.

Kozhevnikov I. V., Nuzhdin A. L., Martyanov O. N. (2010) Transformation of petroleum asphaltenes in supercritical water. *J Supercrit Fluids* 55, (1), 217–222.

Kruse A., Dinjus E. (2007) Hot compressed water as reaction medium and reactant— Properties and synthesis reactions. *J Supercrit Fluids* 39, 362–380.

Kűčűk M. M. (2005). Delignification of biomass using alkaline glycerol. *Energy Sources Part A* 27, (13), 1245–1255.

Laboureur L., Ollero M., Toubol D. (2015) Lipidomics by supercritical fluid chromatography. *Int J Mol Sci* 16, 13868–13884.

Li D., Zhao Y., Yao F., Guo Q. (2009) Green solvent for flash pyrolysis oil separation. *Energy Fuels* 23, 3337–3338.

Li N., Yan B., Zhang L., Quan S.-X., Hu C., Xiao X.-M. (2015) Effect of NaOH on asphaltene transformation in supercritical water. *J Supercrit Fluids* 97, 116–124.

Liu Y., Bai F., Zhu C.-C., Yuan P.-Q., Cheng Z.-M., Yuan W.-K. (2013) Upgrading of residual oil in sub- and supercritical water: An experimental study. *Fuel Process Technol* 106, 281–288.

Ma C.-X., Zhang R., Bi J.-C. (2003) Upgrading of coal tar in supercritical water. *J Fuel Chem Technol* 31, 103–110.

Mares D. R., Altamirano N. (2007) *Venezuela's PDVSA and World Energy Markets: Corporate Strategies and Political Factors Determining Its Behavior and Influence*. https://www.bakerinstitute.org/media/files/page/9c4eb216/noc_pdvsa_mares_altamirano.pdf

Marrone P. A. (2013) Supercritical water oxidation-current status of full-scale commercial activity for waste destruction. *J Supercrit Fluids* 79, 283–288.

Meng M., Hu H. Q., Zhang Q. M., Ding M. (2006) Extraction of Tumuji oil sand with sub- and supercritical water. *Energy Fuels* 20, (3), 1157–1160.

Mokhatab S., Poe W. A., Speight J. G. (2006) *Handbook of Natural Gas Transmission and Processing*. Elsevier, Amsterdam, Netherlands.

Morimoto M., Sugimoto Y., Saotome Y., Sato S., Takanohashi T. (2010) Effect of supercritical water on upgrading reaction of oil sand bitumen. *J Supercrit Fluids* 55, 223–231.

Morimoto M., Sugimoto Y., Sato S., Takanohashi T. (2014) Bitumen cracking in supercritical water upflow. *Energy Fuels* 28, (2), 858–861.

Muthukumaran P., Gupta R. B. (2000) Sodium-carbonate-assisted supercritical water oxidation of chlorinated waste. *Ind Eng Chem Res* 39, 4555–4563.

Oasmaa A., Kuoppala E., Selin J. F., Gust S., Solantausta Y. (2004) Fast pyrolysis of forestry residue and pine. 4. Improvement of the product quality by solvent addition. *Energy Fuels* 18, (5), 1578–1583.

Parkash S. (2003) *Refining Processes Handbook*. Gulf Professional Publishing, Elsevier, Amsterdam, Netherlands.

Patwardhan P. R., Timko M. T., Class C. A., Bonomi R. E., Kida Y., Hernandez H. H., Tester J. W., Green W. H. (2013) Supercritical water desulfurization of organic sulfides is consistent with free-radical kinetics. *Energy Fuels* 27, 6108–6117.

Radlein D. J., Piskorz J., Majerski P. (1996) *Method of Upgrading Biomass Pyrolysis Liquids for Use as Fuels and as Sources of Chemicals Reaction with Alcohol*, Patent: CA2165858.

Rana M. S., Sámano V., Ancheyta J., Diaz J. A. I. (2007) A review of recent advances on process technologies for upgrading of heavy oils and residua. *Fuel* 86, 1216–1231.

Reddy S. N., Nanda S., Dalai A. K., Kozinski J. A. (2014) Supercritical water gasification of biomass for hydrogen production. *Int J Hydrogen Energy* 39, 6912–6926.

Rudyk S., Spirov P. (2014) Upgrading and extraction of bitumen from Nigerian tar sand by supercritical carbon dioxide. *Appl Energy* 113, 1397–1404.

Santos R., Loh W., Bannwart A., Trevisan O. (2014) An overview of heavy oil properties and its recovery and transportation methods. *Braz J Chem Eng* 31, 571–590.

Sasaki M., Goto M. (2008) Kinetic study for liquefaction of tar in sub- and supercritical water. *Polym Degrad Stab* 93, 1194–1204.

Sato T., Adschiri T., Arai K., Rempel G. L., Ng F. T. T. (2003) Upgrading of asphalt with and without partial oxidation in supercritical water. *Fuel* 82, 1231–1239.

Sato T., Mori S., Watanabe M., Sasaki M., Itoh N. (2010) Upgrading of bitumen with formic acid in supercritical water. *J Supercrit Fluids* 55, 232–240.

Sato T., Tomita T., Trung P. H., Itoh N., Sato S., Takanohashi T. (2013) Upgrading of bitumen in the presence of hydrogen and carbon dioxide in supercritical water. *Energy Fuels* 27, 646–653.

Sato T., Trung P. H., Tomita T., Itoh N. (2012) Effect of water density and air pressure on partial oxidation of bitumen in supercritical water. *Fuel* 95, 347–351.

Sato T., Watanabe M., Smith R. L., Adschiri T., Arai K. (2004) Analysis of the density effect on partial oxidation of methane in supercritical water. *J Supercrit Fluids* 28, (1), 69–77.

Savage P. E. (2009) A perspective on catalysis in sub- and supercritical water. *J Supercrit Fluids* 47, (3), 407–414.

Scott D., Radlein D., Piskorz J., Majerski P., deBrujin Th. J. W. (2001) Upgrading of bitumen in supercritical fluids. *Fuel* 80, 1087–1099.

Shah A., Fishwick R., Wood J., Leeke G., Rigby S., Greaves M. (2010) A review of novel techniques for heavy oil and bitumen extraction and upgrading. *Energy Environ Sci* 3, 700–714.

Speight J. G. (1984) In: Kaliaguine S., Mahay A. (Eds.) *Characterization of Heavy Crude Oils and Petroleum Residues*. Elsevier, Amsterdam, Netherlands, p. 515.

Speight J. G. (2000) *The Desulfurization of Heavy Oils and Residua* (2nd edition). Marcel Dekker Inc., New York, NY.

Speight J. G. (2007) *Natural Gas: A Basic Handbook*. GPC Books, Gulf Publishing Company, Houston, TX.

Speight J. G. (2012) *Crude Oil Assay Database*. Knovel, Elsevier, New York, NY.

Speight J. G. (2013) *Heavy and Extra Heavy Oil Upgrading Technologies*. Gulf Professional Publishing, Elsevier, Oxford, United Kingdom.

Speight J. G. (2014) *The Chemistry and Technology of Petroleum* (5th edition). CRC Press, Taylor & Francis Group, Boca Raton, FL.

Speight J. G. (2015) *Handbook of Hydraulic Fracturing*. John Wiley & Sons Inc., Hoboken, NJ.

Speight J. G. (2016) *Introduction to Enhanced Recovery Methods for Heavy Oil and Tar Sands*. Elsevier, Amsterdam, Netherlands, p. 34.

Speight J. G. (Ed.) (1990) Chaps. 12–16. In: *Fuel Science and Technology Handbook*. Marcel Dekker Inc., New York, NY.

Speight J. G., Ozum B. (2002) *Petroleum Refining Processes*. Marcel Dekker Inc., New York, NY.

Strausz O. P., Mojelsky T. W., Payzant J. D., Olah G. A., Prakash G. K. S. (1999) Upgrading of Alberta's heavy oils by superacid-catalyzed hydrocracking. *Energy Fuels* 13, 558–569.

Tan X.-C., Zhu C.-C., Liu Q.-K., Ma T.-Y., Yuan P.-Q., Cheng Z.-M., Yuan W.-K. (2014) Co-pyrolysis of heavy oil and low-density polyethylene in the presence of supercritical water: The suppression of coke formation. *Fuel Process Technol* 118, 49–54.

Timko M. T., Ghoniem A. F., Green W. H. (2015) Upgrading and desulfurization of heavy oils by supercritical water. *J Supercrit Fluids* 96, 114–123.

US EIA (2014) *Crude Oils and Different Quality Characteristics*. Energy Information Administration, United States Department of Energy, Washington, DC. https://www.eia.gov/todayinenergy/detail.php?id=7110

US EIA (2019a) Canada. Energy Information Administration, United States Department of Energy, Washington, DC. https://www.eia.gov/beta/international/analysis.php?iso=CAN

US EIA (2019b) *Total Energy*. Energy Information Administration, United States Department of Energy, Washington, DC. https://www.eia.gov/totalenergy/data/browser/?tbl=T11.01B

Viet T. T., Lee J. H., Ma F., Kim G. R., Ahn I. S., Lee C. H. (2013) Hydrocracking of petroleum residue with activated carbon and metal additives in a supercritical m-xylene solvent. *Fuel* 103, 553–561.

Viet T. T., Lee J.-H., Ryu J. W., Ahn I.-S., Lee C.-H. (2012) Hydrocracking of vacuum residue with activated carbon in supercritical hydrocarbon solvents. *Fuel* 94, 556–562.

Vilcáez J., Watanabe M., Watanabe N., Kishita A., Adschiri T. (2012) Hydrothermal extractive upgrading of bitumen without coke formation. *Fuel* 102, 379–385.

Watanabe M., Kato S.-N., Ishizeki S., Inomata H., Smith Jr. R. L. (2010) Heavy oil upgrading in the presence of high density water: Basic study. *J Supercrit Fluids* 53, (1–3), 48–52.

Ximing C. (2004) Study on supercritical fluid extraction technology in tar processing. *Fuel Chem Process* 4, 024.

Xiu S., Shabazi A., Shirley V. B., Mims M. R., Wallace C. W. (2010) Effectiveness and mechanisms of crude glycerol on the biofuels production from swine manure through hydrothermal pyrolysis. *J Anal Appl Pyrolysis* 87, (2), 194–198.

Xu C., Etcheverry T. (2008) Hydro-liquefaction of woody biomass in sub- and super-critical ethanol with iron-based catalysts. *Fuel* 87, 335–345.

Xu C., Hamilton S., Mallik A., Ghosh M. (2007) Upgrading of Athabasca vacuum tower bottoms (VTB) in supercritical hydrocarbon solvents with activated carbon-supported metallic catalysts. *Energy Fuels* 21, (6), 3490–3498.

Xu T., Liu Q., Liu Z., Wu J. (2013) The role of supercritical water in pyrolysis of carbonaceous compounds. *Energy Fuels* 27, (6), 3148–3153.

Yeh T. M., Dickinson J. G., Franck A., Linic S., Thompson L. T., Savage P. E. (2013) Hydrothermal catalytic production of fuels and chemicals from aquatic biomass. *J Chem Technol Biotechnol* 88, (1), 13–24.

Yuan P.-Q., Cheng Z.-M., Jiang W.-L., Zhang R., Yuan W.-K. (2005) Catalytic desulfurization of residual oil through partial oxidation in supercritical water. *J Supercrit Fluids* 35, (1), 70–75.

10 Potential Use of Subcritical Water Technology in Polymer Fibers Recycling

Florin Barla and Sandeep Kumar

CONTENTS

10.1 INTRODUCTION

Reducing CO_2 emissions by developing alternative energy resources to fossil fuels is highly recommended. Biomass has been receiving much attention as a renewable energy source and studies on the conversion of biomass to high-performance and easy-to-use fuel or energy are being strongly supported. On the other hand, the recycling of some kinds of wastes is very necessary from the viewpoint of environmental conservation and effective uses of resources. Particularly, the recycling of waste plastics is important for the prevention of the exhaustion of fossil resources.

With the rapid development of petrochemical industry and manufacturing industry plastic products are broadly used and waste plastics have been known for triggering serious pollution issues. Plastic products are extensively used in nearly all fields of production such as agriculture, electronics, and households, since the first synthesis back in the early 1900s (Smit & Nasr, 1992; Li et al., 2015; Riber et al., 2009). These various plastic products have greatly improved our quality of lives because they are light, durable, corrosion-resistant, easy to process, low-cost, and have some other unique properties (Wong et al., 2015; Wang et al., 2014; Cao, 2016a). Quick growth of the demand on plastic products triggered environmental problems, due to the accumulation of plastic wastes. They are difficult to degrade and will occupy the landfill space for hundreds of years before their disposal. Plastic products account for 8–12% of the total plastic solid wastes that are discarded in landfills (Lettieri & Al-Salem, 2011).

Generally, the disposal methods for plastic wastes are divided into four categories, namely landfill, incineration, mechanical recycling, and chemical recovery (Ma et al., 2016). Among these choices, the chemical recovery, mainly including pyrolysis and hydrothermal treatment, is an economically feasible and environmentally friendly technology which can convert plastic wastes into fuels or chemical raw materials and has attracted considerable attention (Wu et al., 2014; Wu & Williams, 2010).

Feedstock recycling is a promising route for the management of waste plastic (Wang et al., 2017; Ziming et al., 2017), and the pyrolysis to convert waste plastics into chemical intermediates gets more attention nowadays. Direct pyrolysis in inert atmospheric conditions is a valuable method, but it has many disadvantages such as the technology allows only batch production, the yield is low, poor quality, long residence time, second pollution of dust, exhaust air, sewage, solid waste, and high carbonization to block tubing, thus, its industrial application is a huge challenge. Recently, it is found that pyrolysis of waste plastic in sub- and supercritical fluids can avoid the disadvantages of direct pyrolysis and shows good prospects for recycling waste polymer to produce chemicals (Goto, 2009. The polystyrene pyrolysis in sub- and supercritical water shows that products of pyrolysis can dissolve in water due to the decrease of polarity of water at high temperature, therefore, it can obtain a higher yield of low molecular-weight to avoid further condensation reaction and carbonization. In addition, after the reaction when the temperature decreases to the ambient temperature, some nonpolar organic products can auto segregate from water (Chen et al., 2006.

Polyethylene-terephthalate (PET), a saturated polyester of terephthalic acid (TPA) and ethylene glycol (EG), was widely used in daily life such as in drink bottles, so treatment of waste PET becomes very important. As a polymer of an ester bond link, it is found that PET can hydrolyze at ester bond to produce TPA and EG (Yoshiaki et al., 1994; de Carvalho et al., 2006). But up to now, the main method of hydrolysis of waste PET is acid or alkali catalysis, and thus, the pollution of acid or alkali sewage and corrosion lead to a very problematic industrial application. Now it is known that high temperature and pressure water presents some unique properties such as dissolution of nonpolar organic, low dielectric properties, high diffusibility, and high concentrations of H^+ and OH^- ions from dissociation of water, thus, it is a clean and effective medium for chemical reactions and a powerful catalysis based on acid or alkali, and even a reactant or product participated in chemical reaction (Akiya & Savage, 2002; Brunner, 2009; Cheng et al., 2008). Due to these unique properties and environmental cleanliness, the high-pressure and high-temperature water, have great potential applications for hydrolysis of PET to produce TPA and EG without the addition of acid and alkali.

There are well-established systems for the material recycling of glass, metals, and paper, but not for more complex materials such as textiles. Even though there are systems for textiles collection (Elander & Ljungkvit, 2016), a major barrier to accomplishing textile recycling is the large mix of materials, coatings, dyes, and non-textile objects (Wang, 2006). No commercial-scale process to chemically recycle fiber-to-fiber textiles exists today, however, small-scale projects are ongoing such as Worn Again, Evrnu, Eco Circle, Re:newcell, Ioncell (Elander & Ljungkvit, 2016), or Tyton Biosciences that developed a hydrothermal based process to recycle mixed complex fibers such as polycotton (PET and cotton in varying percentages).

An increase in the amount of dyestuffs and chemicals used in the dyeing industry has led to increase in the complexity of their wastewater. Today, synthetic dyes, as well as other substrates such as paper, leather, fur, hair, drugs, cosmetics, waxes, greases, and plastics, are used extensively in the textile dyeing process. Unfortunately, the exact amount of the synthetic dyes produced in the world is not clear (Forgacs et al., 2004), it is estimated to be more than 7×10^5 tons per year (Reife, 1993; Diorio et al., 2008). Also, it is estimated that a total of 10–30% of the world's dyes is lost during the dyeing process and is released into textile wastewaters and consequently into the environment (Wong et al., 2004; Secula et al., 2008). Dyes are compounds that are difficult to degrade in the environment (Forgacs et al., 2004), and are usually characterized by high toxicity (Perkowski et al., 2003] Besides, the presence of dyestuff in the wastewater causes reduced sunlight transmittance, which results in a negative influence on the assimilation process in water plants. A major concern with color is its aesthetic character at the point of wastewater discharges with respect to visibility in rivers (Barros et al., 2006). A wide range of methods has been developed to remove the synthetic dyes from wastewaters and to decrease their impact on the environment and to provide cost-effective solutions for the textile wastewaters. Unfortunately, most of the conventional methods including biological and chemical methods are either inefficient or intensive and time consuming (Kirby et al., 1995; Sheetal et al., 2008). Common methods were employed for treating textile wastewater including biological, physical, and chemical processes (Pazdzior et al., 2009; Van der Bruggen et al., 2004). Chemical decomposition of the wastewaters has extensively been promoted (Salem et al., 2009; Behnajady & Modirshahla, 2006; Rezaee et al., 2008) because of the simplicity of the handling; (catalytic) oxidation methods are most commonly used for the chemical decolorization and decomposition process. These methods include electrochemical methods (Zhao et al., 2010), ozonation, advanced oxidation using UV/H_2O_2, UV/TiO_2 (Selcuk, 2005; Georgiu et al., 2002; Liu et al., 2005) and Fenton's reagent (H_2O_2/Fe^{2+}) (Kurbus et al., 2003). Hydrogen peroxide can oxidize organic compounds; therefore, TOC and toxicity of the water will be reduced directly in this way. Because it does not add any other substance to the water like permanganate or hypochlorite, nor does it form other toxic substances like chlorinated organic compounds, hydrogen peroxide is usually chosen as oxidizer. Further, organic substances can be partly oxidized with hydrogen peroxide, resulting in reaction products that are easily biodegradable (Salem et al., 2009; Su et al., 2009). Besides the advantage with regards to decolorization and toxicity reduction, these methods also possess disadvantages. In most conventional methods, pH plays an important role in the decolorization process and is one of the main parameters. On the other hand, some of those are time consuming. Fenton oxidation process involves flocculation; thus, impurities are transferred from the wastewater to the sludge, which still needs ecologically questionable land-deposition. Waste treatment processing including hydrolysis and oxidation using subcritical water technology may offer an alternative to the conventional methods. Subcritical water has been increasingly used for the decomposition and degradation of a variety of waste materials. Subcritical water has properties that are different from water under normal conditions. At high temperature and pressure, the value of the ion product increases considerably while dynamic viscosity decreases (Assael et al., 1993; Oomori et al., 2004). Because of these properties, along with the low relative dielectric constant comparable with that of methanol or acetone under ambient conditions, water offers a good alternative medium for a variety of reactions.

10.2   CHEMICAL DEPOLYMERIZATION OF WASTE PET

PET is a semicrystalline polyester with excellent thermal and mechanical properties, very vulnerable to chemical degradation. The growing interest in recycling PET is due to its wide use in packaging for the food industry, mainly in soft-drink bottles although its main application is in the textile industry. PET is very vulnerable to chemical degradation; a schematic representation of four possible groups of reactions is shown in Figure 10.1.

10.2.1   METHANOLYSIS

Under methanolysis (methanol at high temperatures and high pressures) the PET is degraded to dimethyl terephthalate (DMT) and EG which are the raw materials necessary for the production of PET. Currently, methanolysis is successfully applied to PET scrap including bottles, fiber waste, used films, and plant waste (Scheirs, 1998). The advantage of this method is that a processing unit could be installed in the polymer production line since the DMT produced via methanolysis has a product quality identical to virgin DMT and the EG is easily recovered and recycled. The main disadvantage is related with the fact that most of the PET is produced from TPA and not DMT as raw material, and the conversion of DMT to TPA increases considerably the cost of the methanolysis process. Methanolysis is conducted at 2–4 MPa and temperatures of 180–280°C in the presence of a transesterification catalyst such as zinc acetate, magnesium acetate, cobalt acetate, and lead dioxide, the polymer degradation takes place with the release of EG (Paszun & Spychaj, 1997). A schematic diagram of PET production via methanolysis is shown in Figure 10.2. Mitsubishi Heavy Industries, Ltd., Japan, has developed a chemical recycling process using supercritical methanol for depolymerization postconsumer PET bottles at pilot scale (Goto, 2016).

hydrolysis

alcoholysis

acidolysis

aminolysis

FIGURE 10.1 Schematic representation of the PET degradation reactions.

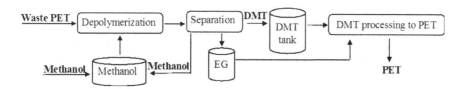

FIGURE 10.2 Flowchart of PET production via methanolysis.

10.2.2 GLYCOLISIS

The main PET chemical recycling processes that reached commercial maturity are glycolysis and methanolysis. Glycolysis involves the insertion of EG (or diethylene glycol and propylene glycol) in PET chains to form bis(hydroxyethyl)terephthalate (BHET) which is a precursor for PET synthesis and its oligomers. The process is conducted at 180–250°C and the reaction time 0.5–8 hours, with an addition of about

0.5% by weight catalyst (usually zinc acetate). The main advantage of glycolysis is that it can be easily integrated into conventional production of PET and the recovered BHET can be blended with fresh BHET. On the other hand, the main disadvantage is that the reaction products are not discrete chemicals but BHET along with conventional techniques (Karayannidis et al., 2002).

10.2.3 HYDROLYSIS

During hydrolysis, the PET waste is basically hydrolyzed to TPA and EG. The method attracted much interest due to development of PET synthesis directly from TPA and EG. In this way, the methanol is eliminated from the technological cycle. However, commercially hydrolysis process is not used to produce food grade recycled PET due to the cost associated with the purification needed for the recycled TPA. PET hydrolysis could be done as alkaline hydrolysis, acid, or neutral hydrolysis (Karayannidis et al., 2002).

10.2.3.1 Alkaline Hydrolysis

PET hydrolysis is usually performed using an aqueous alkaline solution of NaOH or KOH at a concentration of 4–20 wt% obtaining disodium terephthalate salt (TPA-Na$_2$) and EG. The EG is recovered via distillation at 340°C and pure TPA is precipitated during neutralization of the reaction with a mineral acid (e.g., H$_2$SO$_4$), then filtered off, rinsed, and dried. The process is conducted at 210–250°C and the reaction time three to five hours under pressure of 1.4–2MPa (Alter, 1986). Also, aqueous ammonia solution at 200°C was used and successfully hydrolyzed PET forming a solution of TPA diammonium salt which after filtration and acidification with sulfuric acid, TPA with a high purity (99 wt%) is obtained (Datye et al., 1984). Addition of ether (dioxane or tetrahydrofuran) as mixed solvent with methanol or ethanol significantly increased the reaction rate. An improved reaction rate could be achieved when water is mixed with methanol, ethanol, or any other alcohol, knowing that water co-solvents mix in general, alter the subcritical point of water, and for some applications better outcomes are achieved. The time to complete the reaction, PET yield as solid of more than 96%, with NaOH in methanol at 60°C was 40 minutes when dioxane was used as a cosolvent, and seven hours without dioxane addition (Hu et al., 1997).

During alkaline hydrolysis of PET, a significant amount of TPA impurities are generated; to limit this, an additional oxidation step could be performed to convert them into insoluble forms (Rollick, 1995).

Some experimental reaction conditions for PET hydrolysis in the aqueous NaOH solution are shown in Table 10.1. A significant increase in the TPA yield was observed when the reaction temperature was higher. The reaction yield was calculated based on the theoretical amount of TPA that should be produced after complete decomposition, about 86% (Karayannidis et al., 2002). In Table 10.2, the yield and unreacted PET are presented as a function of KOH concentration. The PET hydrolysis was carried out in methyl cellosolve, as the KOH concentration increases a significant increase in TPA yield and hydrolysis rate was observed (Karayannidis et al., 2002).

Fabrics woven made of PET coated with a layer of polyvinyl chloride (PVC) are commonly used as ceiling materials due to their high durability and low weight

TABLE 10.1

Aqueous Alkaline Hydrolysis of PET—Various Experimental Reaction Conditions

Temperature (t°C)	Preheating (min)	Reaction time (h)	TPA (g)	Reaction Yield (%)
200	42	1	42.31	97.9
150	33	1	14.95	34.6
150	34	3	25.74	59.6
150	33	5	32.96	76.3
150	33	7	36.29	84
120	23	3	7.22	16.7
120	23	4	8.75	20.2
120	23	5	10.79	25
120	23	7	14.26	33

Source: Adapted from Karayannidis et al. (2002).

TABLE 10.2

Methyl Cellosolve Alkaline Hydrolysis of PET—Various Experimental Reaction Conditions

KOH Concentrantion (mol/L)	TPA (g)	Reaction Yield (%)	Unreacted PET (g)
3	0.854	10.3	7.789
4	4.577	55.1	3.724
5	5.237	63.1	3.004
6	5.792	69.8	2.447
7	5.836	70.3	2.293
9	6.813	82.1	1.122

Source: Adapted from Karayannidis et al. (2002).

compared with standard building materials. The PET fibers were completely hydrolyzed under alkaline conditions. The reaction was conducted at 120–180°C in the presence of 1M NaOH solution, quantitatively yielding TPA. A dechlorination rate of 1% of PVC was observed at 120°C. The kinetic analysis of PET hydrolysis and PVC dechlorination showed that both processes progressed simultaneously and independently, with no interaction between the PET, PVC, and their degradation products, this fact simplifying the design of a recycling process (Kumagai et al., 2018).

A chemical separation of polycotton into PET monomers (TPA and EG) and to preserve cotton residue, is shown in Figure 10.3. PET was completely hydrolyzed within 40 minutes in 10% NaOH at 90°C in the presence of 52 mmol BTBAC/ kg (benzylltributylammonium chloride, a phase transfer catalyst) hydrolysis solution, into its monomers. Increasing the temperature, NaOH concentration, and the

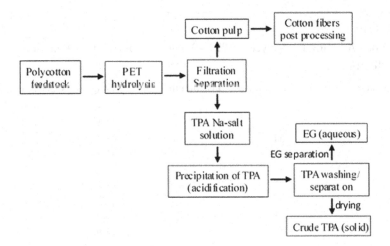

FIGURE 10.3 Basic flowchart of polycotton separation process. (Adapted from Palme et al., 2017.)

BTBAC increased the reaction rate, and a hydrolysis without BTBAC leads to the same streams but increased the reaction time and decreased the cotton yield (Palme et al., 2017).

10.2.3.2 Acid Hydrolysis

Concentrated sulfuric acid (>14.5M), nitric acid, or phosphoric acid are the most common acids used to perform the acid hydrolysis of PET. If concentrated acid is used, high temperatures and high pressures in the reaction vessel can be avoided. However, the acid should be recycled together with the EG, and this increases the cost. When a relatively diluted sulfuric acid was used, the reaction time needed was five hours and the temperature increased to 150°C, with the intent to recover the sulfuric acid via dialysis (Yoshioka et al., 1994). A schematic diagram of PET production via acid hydrolysis is shown in Figure 10.4. Polyester powder from waste bottles could be depolymerized using nitric acid 7-13M at 70–100°C for 72 hours. During this process the EG is oxidized to oxalic acid that has a better commercial value than TPA and EG (Yoshioka et al., 1998).

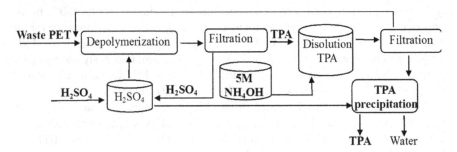

FIGURE 10.4 Flowchart of PET production via acid hydrolysis.

10.2.3.3 Neutral Hydrolysis

The neural hydrolysis is conducted in highly pressurized autoclaves at temperatures of 200–300°C and pressure 1–4MPa in excess of water or steam, obtaining high purity monomers TPA and EG (Scheirs, 1998). The pH at the end of the reaction is 3.5–4 due to formation of TPA monoglycol ester (Michalski, 1993). It is also confirmed that PET hydrolysis at higher temperature, more than 245°C, is advantageous since the hydrolysis process is considerably faster in the molten state than in solid. During PET hydrolysis in addition to TPA, a monoester of glycol is formed as a byproduct (about 2%) that dissolves well in water at 95–100°C, whereas the TPA is practically insoluble at this temperature making the separation easy (Michalski, 1993). An efficient continuous hydtolysis process that can produce PET oligomers (containing 2–3 repetitive units) was developed using a twin-screw extruder as a reactor (Kamal et al., 1994). The products with end carboxyl groups have higher melting temperatures as compared with virgin TPA, while those with one carboxyl group and one or two hydroxyl groups have lower melting points. Using cold or hot water in the process does not induce a reasonable degree of depolymerization (Kamal et al., 1994). Neutral hydrolysis has the main advantage, over alkaline and acid hydrolysis, that it avoids the formation of significant amounts of organic salts which are difficult to dispose of, therefore the process can be considered environmentally friendly. The main drawback of this process is that all the impurities from PET are left in TPA, thus the product can be considered of lower purity (Karayannidis et al., 2002).

10.2.4 Ammonolysis

TPA amide can be produced from PET when reacted with anhydrous ammonia in an ethylene glycol environment that can be converted to terephthalonitrile and further to other chemical substances such as p-xylylenediamine or 1,4-bis(aminoethyl) cyclohexane. The reaction is conducted at 120–180°C and pressure of about 2MPa for one to seven hours, usually using postconsumer PET bottles. The produced amide is filtered, rinsed with water, and then dried at a temperature of 80°C, the purity is not less than 99% and the yield higher than 90% (Blackmon et al., 1988).

10.2.5 Aminolysis

Aminolysis is the reaction of PET waste with different amines to yield the corresponding diamides TPA and EG. The reaction is usually conducted at a temperature range of 20–100°C, using primary amine aqueous solutions such as methylamine, ethylamine, and ethanolamine Paszun & Spychaj, 1997. If triethanolamines is used for degradation, a different ester is produced that may constitute potentially raw materials for the synthesis of rigid polyurethane foams (Spychaj & Paszun, 1998).

10.3 HYDROTHERMAL DEPOLYMERIZATION OF PET

Water under specific conditions such as high temperature and pressure, presents some unique properties such as dissolution of nonpolar organic, low dielectric properties, and high diffusibility. Due to these unique properties and environmentally friendly solvent, the high-pressure and high-temperature water, have great potential applications for hydrolysis of PET to produce TPA and EG without addition of acid and alkali. Some tests indicated that under hydrothermal conditions PET melts at 217°C, but at 260°C, at a ratio of water: PET of 10:1 and a reaction time of 15 minutes, the hydrolysis conditions were optimum with almost 100% PET degradation and 91.64% TPA recovery yield (Chen et al., 1991).

In a different experiment, PET decomposition was conducted in a pressurized Parr-reactor at temperatures above 100°C, equipped with digital temperature control system, an agitator, and a pressure gage indicator. The PET decomposition reaction was performed in water in the presence of 1.125M NaOH solution. The reaction was then quenched with cold water for about five minutes, neutralized with H_2SO_4 to pH 6.5, and filtered to remove any unreacted PET and the TPA was precipitated with H_2SO_4 to a pH 2.5. Finally, the mixture was filtered and washed with methanol and the solid TPA was dried at 80°C in an oven, and could be further repolymerized to PET (Karayannidis et al., 2002). Also, PET was successfully depolymerized to its monomers in near critical water or methanol, the yield was close to 100% and the purity greater than 97% under the conditions of 200°C, 40MPa, and 30 minutes residence time (Goto et al., 2006). Some secondary products were identified such as benzoic acid, diethylene glycol, 1,4-dioxane, acetaldehyde, and crotonic acid. Methanolysis in supercritical methanol produced both monomers DMT and EG with almost 100% yield in 30 minutes without catalyst (Goto et al., 2006).

Due to water's unique properties and cleanness for the environment, at high pressure and high temperature it has great potential application for hydrolysis of PET to produce TPA and EG without addition of acid and alkali or minimizing the chemicals loading, considerably speeding up the reaction time and decreasing the reaction temperature, with a use or not of a catalyst.

10.4   SUB- AND SUPERCRITICAL WATER DEPOLYMERIZATION OF NYLON 6

Nylon 6 is a polymer synthesized by open-ring polymerization of ε-caprolactam, and has been depolymerized by hydrolysis in subcritical and supercritical water, with ε-caprolactam and ε-aminocaproic acid being detected in the product liquid phase (Goto, 2016). The yield of the monomers was about 100%. The hydrolysis reaction was conducted at 300°C for 60 minutes and at 330°C for 30 minutes. The ε-aminocaproic acid yield rapidly decreased as the reaction time increased. It was found that Nylon 6 was decomposed via hydrolysis to ε-aminocaproic acid followed by cyclodehydration to ε-caprolactam or decomposition further to smaller molecules, this clearly indicates that cyclodehydration reaction occurs in water near its critical temperature (Goto et al., 2006).

10.5   SWT IN RECYCLING WASTE PLASTICS

Lu & Chiang (2017) did research to recover hydrogen energy and aluminum from non-recycled plastic material through gasification, and cold gas efficiency reached approximately 72.2%. Diaz, Silvarrey, & Phan (2016) conducted pyrolysis experiments on plastic waste, and presented a kinetic model which successfully predicted the rate of reaction and conversion at low and intermediate heating rates. In the chemical recovery process, supercritical fluids were considered an excellent chemical medium for their unique physicochemical properties and were widely used in various hydrothermal reactions such as hydrolysis, gasification, depolymerization, synthesis, and hydrogenation (Ma et al., 2016; Guan et al., 2017; Liu et al., 2017). For instance, supercritical water (SCW) that uses temperatures above 374°C and pressure above 22.1 MPa, has been extensively used in organic wastes disposal (Cao, 2017a; Shibasaki et al., 2004) due to its special properties (Cao, 2016b; Jin et al., 2017; Liu et al., 2016; Chen et al., 2017) like high solubility, high reactivity, high diffusivity, low dielectric constant, and low viscosity (Cao, 2017b; Jin et al., 2018; Xu et al., 2017; Yang et al., 2017). Onwudili & Williams (2009) conducted experiments to study the degradation of two brominated flame-retarded plastics, Br-acrylonitrile-butadiene-styrene and Br-high impact polystyrene, at 450°C and 31 MPa, in a batch Hastelloy-C reactor. The reaction products contained gas, oil, and solid residues, with up to 99wt% of the bromine atoms in the plastics removed into the aqueous phase, and hydrogen and hydrocarbon gases were detected in the gaseous products. Lilac & Lee (2001) conducted SCW partial oxidation experiments of polystyrene to investigate the kinetics and reaction mechanisms at 382°C and 24 MPa. Under such conditions, the polystyrene was successfully decomposed into its monomer, oligomer, and other useful hydrocarbons within a relatively short residence time, with the highest styrene selectivity of 71%. There are mainly two ways for the SCW to be used in the treatment of plastic wastes. One is to convert the organic compounds present in plastic into hydrogen, methane, carbon monoxide, and other hydrocarbon fuel gases to achieve the resource utilization of waste, and the other is to decompose plastic into corresponding polymeric monomers to achieve its recovery and recycling. According to the literature, most studies were focused on the later, and there is little research work on the SCW gasification of plastic for fuel gas. The supercritical water gasification (SCWG) of acrylonitrile-butadiene-styrene plastic at relevantly high temperatures for fuel gas production was conducted as the supplement in this field, and the effects of some of the important operating parameters on the gasification results were discussed. Besides, the subcritical water hydrolysis (SCWH) at relatively low temperatures for oil products was also studied, and optimal reaction condition for monomer recovery was obtained (Liu et al., 2018). The SCWG experiments were conducted at 450–700°C, 23MPa, 20–80 minutes, showing that increasing the reaction time, temperature, and material/water ratio significantly promoted the gasification reaction. The SCWH for oil products was conducted at 350–450°C, 21MPa, for 0.5–60 minutes, and the studies indicated that most of the monomers were converted into more stable substances at longer residence time (Liu et al., 2018).

10.6   SWT IN RECYCLING TEXTILE DYES

SWT may offer an alternative to the conventional methods in waste treatment processing including hydrolysis and oxidation. Subcritical water has been increasingly used for the decomposition and degradation of a variety of waste materials due to its properties that are different from water under normal conditions. At high temperature and pressure, the value of the ion product increases considerably while dynamic viscosity decreases, its dielectric constant is comparable with that of methanol or acetone under ambient conditions. Acid orange 7 (AO7) was used as a model for the textile wastewater and its decolorization and decomposition were conducted under subcritical water conditions in the presence of hydrogen peroxide at a temperature of 100–175°C and residence time of six minutes (Daneshvar et al., 2013). The decolorization was completed at 135°C, and the conclusion was that subcritical water treatment in the presence of hydrogen peroxide decolorized AO7 via breaking it into small molecules and further oxidase them into final gaseous products. The optimum reaction conditions for degrading AO7 were 135°C for about six minutes in the presence of 0.5% hydrogen peroxide. In addition, ten types of dyes were selected to be treated under the optimal conditions obtained from subcritical water treatment of AO7. Acid dyes (Brilliant Blue R; Acid Orange 52), direct dyes (Brilliant Red 120; Reactive Black 5), basic dyes (Methylene Blue Hydrate; Crystal Violet), and unclassified dyes (Variamine Blue B; Bromophenole Blue) were selected for this experiment that was conducted under subcritical conditions at 135°C for 15 minutes in the presence of 0.4% hydrogen peroxide. Maximum decolorization and decomposition were observed for most of the dyes except Acid Orange 52, under the mentioned reaction conditions. Under moderate subcritical water temperatures and short residence time in the presence of hydrogen peroxide, efficient decolorization and decomposition of textile dyes could be achieved (Daneshvar et al., 2013).

10.7   SUPER/SUBCRITICAL FLUIDS IN RECYCLING THERMOSETTING PLASTICS

Limited attention has been paid to the thermosetting plastics and composite plastics because they are not generally considered to be recyclable due to their low fluidity and poor moldability. Typically, the thermosetting plastics waste ends in the landfill and advanced recycling methods have been expected for a long time. Recently, new techniques using super/subcritical fluids were developed for recycling waste plastics including composite materials or thermosetting plastics such as carbon fiber reinforced plastic (CFRP), Laminate films composed of polyamide and polyethylene, or Silane-cross-linked polyethylene. CFRP has flame resistance and contains refractory thermosetting plastic such as epoxy resin and therefore is usually landfilled at high cost. Cyclohexane-1,2-dicarboxylic anhydride was used as a curing agent and a trimethylamine as an accelerator was used in subcritical water and high-pressure steam decomposition of thermosetting epoxy resin (Okajima & Sako, 2014). The best catalyst was potassium carbonate 2.5 wt% and the optimum conditions were 400°C, 20MPa for 45 minutes residence time with a maximum yield of phenolic monomer of 71%. Under subcritical methanol the optimum reaction conditions were 270°C, 10MPa for 60 minutes, without catalyst and the epoxy resin was decomposed and dissolved completely in methanol. Laminate films are used in the food packaging industry due to their special characteristics (waterproof; strength for liquid food, gas barrier). The laminate film consisting of Nylon 6 and polyethylene layers, glued using unsaturated fatty acid degeneration polyethylene adhesive, were used for the experiment. The decomposition of Nylon 6 to ε-caprolactam yielded 94% at 330°C and did not change by increasing the temperature. The yield of pure polyethylene was 95% in the region of 280–350°C and could be recovered pure since the Nylon 6 decomposed to Ɛ-caprolactam and dissolved in water. Cross-linked polyethylene (XLPE) is used as electric wires and cables insulator. Silane-XLPE was treated under subcritical methanol at 320–360°C, the minimum reaction time was 20 minutes. More than 90% of the decomposition efficiency of the cross-linking points occurred at 320°C, 10MPa (Okajima & Sako, 2014).

10.8   INDUSTRIAL APPLICATIONS

In 2017, Unilever committed to all of its plastic packaging being reusable, recyclable, or compostable by 2025. Europe-based consumer products maker Unilever has announced a partnership with Ioniqa Technologies and PET (polyethylene terephthalate) resins producer Indorama Ventures to pioneer a new technology designed to convert PET scrap back into virgin-grade material for use in food packaging.

DEMETO, a new European project on the chemical recycling of PET has officially launched with the aim of enabling chemical depolymerization of PET at industrial scale based on its microwave-based process intensification, focusing as a start on colored bottles waste.

DEMETO and The Coca-Cola Company share a common interest in closing the loop of the Plastics Circular Economy.

Ioniqa indicates it has developed a technology designed to convert any PET scrap, including colored packaging, back into what it calls transparent virgin-grade material. "The technology has successfully passed its pilot stage and is now moving toward testing at an industrial scale," the company states.

10.9   CONCLUSIONS

Sub- and supercritical fluids technologies are promising reactions media for green chemistry. Various polymers could be degraded easily in sub- or supercritical water, alcohol, ether, etc. Alkali decomposition of soft drink bottles (PET) is an efficient method for regeneration of PET monomers TPA and EG that could be repolymerized back to PET. Proton NMR spectrum reveled that TPA could be recovered in a high purity, of about 98% (Karayannidis et al., 2002). It is well known that water, ethanol, and even methanol are considered green solvents. Various applications use subcritical water, methanol, ethanol, or other alcohols, and their mixtures. Such an approach could be a feasible application when SWT is used in textile and dyes recycling. However, at industrial scale there are challenges related to solvents flammability and recovery. PET recycling become economically viable when applied to large volumes such as industrial polymer waste or in the case of manufacturing special products of low or medium tonnage (Paszun & Spychaj ,1997). Moderate subcritical water temperatures in the presence of hydrogen peroxide could successfully and efficiently decompose and decolorate textile dyes, compounds that are otherwise difficult to degrade, usually characterized by high toxicity, and are released into the textile wastewater and consequently into the environment. There are developments in chemical recycling of waste plastics via decomposition reactions in sub-and supercritical fluids. Condensation polymerization plastics are easily depolymerized to their monomers, cross-linked polymers could be recycled by selective decrosslinking, and fiber reinforced plastics can also be recycled by depolymerization of the resin component to yield recovered fiber and monomers (Goto, 2016).

REFERENCES

Akiya N. & Savage P.E. (2002) Roles of water for chemical reactions in high-temperature water. *Chem. Rev.* 102: 2725–2750.

Alter H. (1986) *Encyclopedia of Polymer Science and Engineering.* New York, NY: Wiley I. Vol. 5, p. 103.

Assael M.J., Polimatidou S.K. & Wakeham W.A. (1993) The viscosity of liquid water at pressures up to 32MPa. *Int. J. Thermophys.* 14: 795–803.

Barros A.L., Pizzolato T.M., Carissimi E. et al. (2006) Decolorizing dye wastewater from the agate industry with Fenton oxidation process. *Miner. Eng.* 19: 87–90.

Behnajady M.A. & Modirshahla N. (2006) Kinetic modeling on photooxidative degradation of C.I. Acid Orange 7 in a tubular continuous-flow photoreactor. *Chemosphere* 62: 1543–1548.

Blackmon K.P., Fox D.W. & Shafer S.J. (1988) Process for converting PET scrap to diamide monomers. Eur. Patent 365: 842.

Brunner G. (2009) Near subcritical water. Part I. Hydrolytic and hydrothermal Processes. *J. Supercrit. Fluids* 47: 373–381.

Cao C., He Y., Chen J. et al. (2017a) Evaluation of effect of evaporation on supercritical water gasification of black liquor by energy and energy analysis. *Int. J. Hydrogen Energy* 43: 13788–13797.

Cao C., Xu L., He Y. et al. (2017b) High-efficiency gasification of wheat straw black liquor in supercritical water at high temperatures for hydrogen production. *Energy Fuels* 31: 3970–3978.

Cao Q., Yuan G., Yin L. et al. (2016a) Morphological characteristics of polyvinyl chloride (PVC) dechlorination during pyrolysis process: Influence of PVC content and heating rate. *Waste Manag.* 58: 241–249.

Cao W., Cao C., Guo L. et al. (2016b) Hydrogen production from supercritical water gasification of chicken manure. *Int. J. Hydrogen Energy* 41: 22722–22731.

Chen G., Yang X., Chen S. et al. (2017) Transformation of heavy metals in lignite during supercritical water gasification. *Appl. Energy* 187: 272–280.

Chen J., Zhang H., Zheng H. et al. (2006) In situ visualization of transformation of organic matter in water at high pressures and temperatures. *J. Anal. Appl. Pyrol.* 76: 260–264.

Chen J.Y., Ou Y.C. & Lin C.C. (1991) Depolimerization of Poly(ethylene terephthalate) resin under pressure. *J. Appl. Polym. Sci.* 42: 1501.

Cheng H., Zhu X., Zhu C. et al. (2008) Hydrolysis technology of biomass waste to produce amino acids in sub-critical water. *Bioresour. Technol.* 99: 3337–3341.

Daneshvar S., Hidemi N., Salak F. et al. (2013) Degradation of textiles dyes under subcritical water conditions in the presence of hydrogen peroxide. *Can. J. Chem. Eng.* 92: 615–622.

Datye K.V., Raje H.M. & Sharma N.D. (1984) Poly(ethylene terephthalate) waste and its utilization: A review. *Resour. Conserv.* 11: 136.

de Carvalho G.M., Muniz E.C. & Rubira A.F. (2006) Hydrolysis of post-consume poly(ethylene terephthalate) with sulfuric acid and product characterization by WAXD, 13C NMR and DSC. *Polym. Degrad. Stabil.* 91: 1326–1332.

Diaz Silvarrey L.S. & Phan A.N. (2016) Kinetic study of municipal plastic waste. *Int. J. Hydrogen Energy* 41: 16352–16364.

Diorno L.A., Mercuri A.A., Nahabedian D.E. et al. (2008) Development of a bioreactor system for the decolorization of dyes by Coriolus versicolor f. antarticus. *Chemosphere* 72: 150–156.

Elander M. & Ljungkvist H. (2016) Critical aspects in design for fiber-to-fiber recycling of textiles. Mistra Future Fashion Report Number: 2016:1.

Forgacs E., Cserhati T. & Oros G. (2004) Removal of synthetic dyes from wastewaters: A review. *Environ. Int.* 30: 953–971.

Georgiou D., Melidis P., Aivasidis A. et al. (2002) Degradation of azo-reactive dyes by ultraviolet radiation in the presence of hydrogen peroxide. *Dyes Pigments* 52: 69–78.

Goto M. (2009) Chemical recycling of plastics using sub- and supercritical fluids. *J. Supercrit. Fluids* 47: 500–507.

Goto M. (2016) Subcritical and supercritical fluid technology for recycling waste plastic. *J. Jpn. Petrol. Inst.* 59: 254–258.

Goto M., Sasaki M. & Hirose T. (2006) Reactions of polymers in supercritical fluids for chemical recycling of waste plastics. *J. Mater. Sci.* 41: 1509–1515.

Guan Q., Chen S., Chen Y. et al. (2017) High performance noble-metal-free NiCo/AC bimetal for gasofication in supercritical water. *Int. J. Hydrogen Energy* 42: 6511–6518.

Hu L.C., Oku A., Yamada E. et al. (1997) Alkali-decomposition of poly (ethylene terephthalate) in mixed media of nonaqueous alcohol and ether. Study on recycling of poly (ethylene terephthalate). *Polym. J.* 29: 708–712.

Jin H., Chen B., Zhao X. et al. (2018) Molecular dynamic simulation of hydrogen production by catalytic gasification of key intermediates of biomass in supercritical water. *J. Energy Resour. Technol.* 140.

Jin H., Fan C., Guo L. et al. (2017) Experimental study on hydrogen production by lignite gasification in supercritical water fluidized bed reactor using external recycle of liquid residual. *Energy Convers. Manage.* 145: 214–219.

Kamal M.R., Lai-Fook R.A. & Yalcinyuva T. (1994) Reactive extrusion for the hydrolytic depolymerisation of polyethylene terephthalate. *ANTEC Conference*, pp. 2896–2900.

Karayannidis G.P., Chatziavgoustis A.P. & Achilias D.S. (2002) Poly(ethylene terephthalate) recycling and recovery of pure terephtalic acid bu alkaline hydrolysis. *Adv. Polym. Technol.* 21: 250–259.

Kirby N., Mcmullan G. & Marchant R. (1995) Decolorization of an artificial textile effluent by *Phanerochaete chrysosporium. Biotechnol. Lett.* 17: 761–764.

Kumagai S., Hirahashi S., Grause G. et al. (2018) Alkaline hydrolysis of PVC-coated PET fibers for simultaneous recycling of PET and PVC. *J. Mater. Cycles Waste Manage.* 20: 439–449.

Kurbus T., Le Marechal A.M. & Brodnjak V.D. (2003) Comparison of H_2O_2/UV, H_2O_2/O_3 and H_2O_2/Fe^{2+} processes for the decolorization of vinyl sulphone reactive dyes. *Dyes Pigments* 58: 245–252.

Lettieri P. & Al-Salem S.M. (2011) Thermochemical treatment of plastic solid waste. In *Waste: A Handbook of Waste Management and Recycling.* Amsterdam, Netherland: Elsevier, pp. 233–242.

Li J., Yang J. & Liu L. (2015) Development potential of e-waste recycling industry in China. *Waste Manage. Res.* 33: 533–542.

Lilac W.D. & Lee S. (2001) Kinetics and mechanisms of styrene monomer recovery from waste polystyrene by supercritical water partial oxidation. *Adv. Environ. Res.* 6: 9–16.

Liu S., Jin H., Wei W. et al. (2016) Gasification of indole in supercritical water: Nitrogen transformation mechanism and kinetics. *Int. J. Hydrogen Energy* 41: 15985–15997.

Liu S., Li L., Guo L. et al. (2017) Sulfur transformation characteristics and mechanisms during hydrogen production by coal gasification in supercritical water. *Energy Fuels* 31: 12046–12053.

Liu Y., Chen X., Li J. et al. (2005) Photocatalytic degradation of azo dyes by nitrogen-doped TiO$_2$ nanocatalysts. *Chemosphere* 61: 11–18.

Liu Y., Fan C., Zhang H. et al. (2018) The resource utilization of ABS plastic waste with subcritical and supercritical water treatment. *Int. J. Hydrogen Energy* 44: 15758–15675.

Lu C.H. & Chiang K.Y. (2017) Gasification of non-recycled plastic packaging material containing aluminum. Hydrogen energy production and aluminum recovery. *Int. J. Hydrogen Energy* 42: 27532–27542.

Ma C., Yu J., Wang B. et al. (2016) Chemical recycling of brominated flame retarded plastics from e-waste for clean fuels production: A review. *Renew. Sustain. Energy Rev.* 61: 433–450.

Michalski A. (1993) Poly(ethylene terephthalate) Waste utilization by hydrolysis in neutral environment. *Ekoplast.* 2: 52 (in Polish).

Okajima I. & Sako T. (2014) Chapter 13 – Energy conversion of biomass and recycling of waste plastic using supercritical fluid, subcritical fluid and high-pressure superheated steam. In *Supercritical Fluid Technology for Energy and Environmental Applications* doi:10.1016/B978-0-444-62696-7.00013-7.

Onwudili J.A. & Williams P.T. (2009) Degradation of brominated flame-retarded plastics (Br-ABS and Br-HIPS) in supercritical water. *J. Supercrit. Fluids* 49: 356–368.

Oomori T., Khajavi S.H., Kimura Y. et al. (2004) Hydrolysis of dissccharides containing glucose residue in subcritical water. *Biochem. Eng. J.* 18: 143–147.

Palme A., Peterson A., de la Motte H. et al. (2017) Development of an efficient route for combined recycling of PET and cotton from mixed fabrics. *Text. Cloth. Sustain.* 3: 4.

Paszun D. & Spychaj T. (1997) Chemical recycling of Poly(ethylene terephthalate). *Ind. Eng. Chem. Res.* 36: 1373–1383.

Pazdzior K., Klepacz-Smolka A., Ledakowicz S. et al. (2009) Integration of nanofiltration and biological degradation of textile wastewater containing azo dye. *Chemosphere* 75: 250–255.

Perkowski J., Lech K., Ledakowics S. et al. (2003) Decomposition of anthraquinone dye acid blue 62 by the decoloration of textile wastewater by advanced oxidation process. *Fibers Text. East. Eur.* 11: 88–95.

Reife A. (1993) *Encyclopedia of Chemical Technology.* New York, NY: Wiley, pp. 672–783.

Rezaee A., Ghenian M., Hashemian S.J. et al. (2008) Decolorization of reactive blue 19 dye from textile wastewater by the UV/H$_2$O$_2$ process. *J. Appl. Sci.* 8: 1108–1112.

Riber C., Petersen C. & Christensen T.H. (2009) Chemical composition of material fractions in Danish household waste. *Waste Manage.* 29: 1251–1257.

Rollick K.L. (1995) Process for recovering dicarboxylic acid with reduced impurities from polyester polymer. *WO Patent 95 10499, CA 123.*

Salem M.A., Abdel-Halim T., El-Sawy A.E.M. et al. (2009) Kinetics of degradation of allura red, ponceau 4R and carmosine dyes with potassium ferrioxalate complex in the presence of H$_2$O$_2$. *Chemosphere* 76: 1088–1093.

Scheirs J. (1998) Polymer recycling. In *Wiley Series in Polymer Science.* Sussex: Wiley. Ch. 4, pp. 119–182.

Secula M.S., Suditu G.D., Poulios C. et al. (2008) Response surface optimization of the photocatalytic decolorization of a simulated dyestuff effluent. *Chem. Eng. J.* 141: 18–26.

Selcuk H. (2005) Decolorization and detoxification of textile wastewater by ozonation and coagulation processes. *Dyes Pigments* 64: 217–222.

Sheetal U.J., Mital U.J. & Anuradha N.K. (2008) Decolorisation of Brilliant Blue G dye mediated by degradation of the microbial consortium of *Galactomyces geotrichum* and *Bacillus* sp. *J. Chin. Inst. Chem. Eng.* 39: 563–570.

Shibasaki Y., Kamimori T., Kadokawa J. et al. (2004) Decomposition reactions of plastic model compounds in sub- and supercritical water. *Polym. Degrad. Stabil.* 83: 481–485.

Smit J. & Nasr J. (1992) Urban agriculture for sustainable cities: Using wastes and idle land and water bodies as resources. *Environ. Urban.* 4: 141–152.

Spychaj T. & Paszun D. (1998) New trends in chemical recycling of Poly(ethylene terephthalate) – PET. *Macromol. Symp.* 135: 137–145.

Su R., Sun J., Sun Y.P. et al. (2009) Oxidative degradation of dye pollutants over a broad pH range using hydrogen peroxide catalyzed by FePz(dtnCl$_2$)$_4$. *Chemosphere* 77: 1146–1151.

Van der Bruggen B., Curcio E. & Drioli E. (2004) Process intensification in the textile industry: The role of membrane technology. J. Environ. Manage. 73: 267–274.

Wang S.J., Zhang H., Shao L.M. et al. (2014) Thermochemical reaction mechanism of lead oxide with poly(vinyl chloride) in waste thermal treatment. *Chemosphere* 117: 353–359.

Wang Y. (2006) *Recycling Textiles.* Cambridge and Boca Raton, FL: Woodhead Publishing and CRC Press.

Wang Y., Li X., Wang W. et al. (2017) Experimental and in-situ estimation on hydrogen and methane emission from spontaneous gasification in coal fire. *Int. J. Hydrogen Energy* 42: 18728–18733.

Wong S.L., Ngadi N., Abdullah T.A.T. et al. (2015) Current state and future prospects of plastic waste as source of fuel: A review. *Renew. Sustain. Energy Rev.* 50: 1167–1180.

Wong Y.C., Szeto Y.S. & Cheung W.H. (2004) Adsorption of acid dyes on chitosan – Equilibrium isotherm analyses. Process Biochem. 39: 693–702.

Wu C., Nahil M.A., Miskolczi N. et al. (2014) Processing real-world waste plastics by pyrolysis-reforming for hydrogen and high-value carbon nanotubes. *Environ. Sci. Technol.* 48: 819–826.

Wu C. & Williams P.T. (2010) Pyrolysis-gasification of post-consumer municipal solid plastic waste for hydrogen production. *Int. J. Hydrogen Energy* 35: 949–957.

Xu D., Ma Z., Guo S. et al. (2017) Corrosion characteristics of 316L as transpiring wall material in supercritical water oxidation of sewage sludge. *Int. J. Hydrogen Energy* 42: 19819–19828.

Yang J., Wang S., Xu D. et al. (2017) Effect of ammonium chloride on corrosion behavior of Ni-based alloys and stainless steel in supercritical water gasification process. *Int. J. Hydrogen Energy* 42: 19788–19797.

Yoshiaki T., Sato T. & Okuwaki A. (1994) Hydrolysis of waste PET by sulfuric acid at 150°C for a chemical recycling. *J. Appl. Polym. Sci.* 52: 1353–1355.

Yoshioka T., Okayama N. & Okuwaki A. (1998) Kinetics of hydrolysis of PET powder in nitric acid by a modified shrinking-core model. *Ind. Eng. Chem. Res.* 37: 336–340.

Zhao G.H., Gao J.X., Shi W. et al. (2010) Electrochemical incineration of high concentration azo dye wastewater on the in situ activated platinum electrode with sustained microwave radiation. *Chemosphere* 77: 188–193.

Ziming C., Fuqiang W., Yinmo X. et al. (2017) Investigation of optical properties and radiative transfer of sea water-based nanofluids for photocatalysis with different salt concentrations. *Int. J. Hydrogen Energy* 42: 26626–26638.

Index

Printed in the United States
by Baker & Taylor Publisher Services